Praise for *The Organic Grain Grower*

"Jack Lazor did not wait for a new movement to inspire him. Jack inspired the movement. Jack began reclaiming the small farm's grain heritage right from the start of his farm many years ago. That is why this book is such a delight. These are the words of someone who has talked to all the old-timers and done it all himself. It is like acquiring hundreds of years of knowledge in one book. And he presents everything in an appealing, storytelling manner that will have you sitting up late reading page after page."

—from the Foreword by ELIOT COLEMAN, author of *The Winter Harvest Handbook*

"Given our industrial agriculture, most of us assume that grain can be grown only in huge monocultures devoted to producing as much as possible, unmindful of the quality. But in *The Organic Grain Grower*, Jack Lazor provides us with a practical and attractive alternative. As a farmer he has demonstrated that one can provide an emerging market with a diversity of superior-quality grains, grown on a small scale, using heirloom varieties and modest investment."

—FREDERICK KIRSCHENMANN, author of *Cultivating an Ecological Consciousness*

"I believe I can safely say, without losing any money, that if you know of one fact truly necessary to growing grains organically in the United States that is not in this book, I'll pay you five bucks out of my own pocket. Plus there's a whole bunch of stuff about how to process and use grains in the barn or on the table that I have not found all in one place before."

—GENE LOGSDON, author of *Small-Scale Grain Growing*

"Jack writes from the top of a mountain—the mountain of his life. His long years of experience are longer than his very beard, and the wisdom and distillation of his farming life are written here with clarity and graceful articulation. As he says in the book, 'people are hungry for meaning as well as food.' In this classic book, Jack provides not only the meaning, but also the methods required to succeed as a small-scale grower of organic grains."

—JEFFREY HAMELMAN, director, King Arthur Flour Bakery, and author of *Bread: A Baker's Book of Techniques and Recipes*

"*The Organic Grain Grower* is quite possibly the most complete and extensive text ever written on grain production in the Northeast. Jack Lazor's deep passion and knowledge create an astounding story, and he shares his wisdom and experience generously. If you have ever wanted to grow grain, this is a book to own and cherish."

—DR. HEATHER DARBY,
University of Vermont Extension agronomist

THE ORGANIC GRAIN GROWER

Small-Scale, Holistic Grain Production for the Home and Market Producer

JACK LAZOR

Foreword by Eliot Coleman

Chelsea Green Publishing
White River Junction, Vermont

Copyright © 2013 by Jack Lazor.
All rights reserved.

Unless otherwise noted, all photographs by Jack Lazor.
Unless otherwise noted, all illustrations copyright © 2013 by Elayne Sears.

No part of this book may be transmitted or reproduced in any form by any means without permission in writing from the publisher.

Project Manager: Patricia Stone
Editor: Makenna Goodman
Copy Editor: Laura Jorstad
Proofreader: Helen Walden
Indexer: Margaret Holloway
Designer: Melissa Jacobson

Printed in the United States of America.
First printing July, 2013.
10 9 8 7 6 5 4 3 2 1   13 14 15 16 17

Our Commitment to Green Publishing
Chelsea Green sees publishing as a tool for cultural change and ecological stewardship. We strive to align our book manufacturing practices with our editorial mission and to reduce the impact of our business enterprise in the environment. We print our books and catalogs on chlorine-free recycled paper, using vegetable-based inks whenever possible. This book may cost slightly more because it was printed on paper that contains recycled fiber, and we hope you'll agree that it's worth it. Chelsea Green is a member of the Green Press Initiative (www.greenpressinitiative.org), a nonprofit coalition of publishers, manufacturers, and authors working to protect the world's endangered forests and conserve natural resources. *The Organic Grain Grower* was printed on FSC®-certified paper supplied by Maple Press that contains 30% postconsumer recycled fiber.

Library of Congress Cataloging-in-Publication Data
Lazor, Jack, 1951-
  The organic grain grower: small-scale, holistic grain production for the home and market producer / Jack Lazor; foreword by Eliot Coleman.
      p. cm.
  Small scale, holistic grain production for the home and market producer
  Includes bibliographical references and index.
  ISBN 978-1-60358-365-7 (hardcover)—ISBN 978-1-60358-366-4 (ebook)
1. Grain—Northeastern States. 2. Grain—Canada. 3. Organic farming—Northeastern States. 4. Organic farming—Canada.
5. Farms, Small—Northeastern States. 6. Farms, Small—Canada. I. Title. II. Title: Small scale, holistic grain production for the home and market producer.

SB189.L29 2013
633.1—dc23
                                2013011765

Chelsea Green Publishing
85 North Main Street, Suite 120
White River Junction, VT 05001
(802) 295-6300
www.chelseagreen.com

Chelsea Green Publishing is committed to preserving ancient forests and natural resources. We elected to print this title on paper containing at least 30% postconsumer recycled paper, processed chlorine-free. As a result, for this printing, we have saved:

42 Trees (40' tall and 6-8" diameter)
19,462 Gallons of Wastewater
1 million BTUs Total Energy
1,303 Pounds of Solid Waste
3,589 Pounds of Greenhouse Gases

Chelsea Green Publishing made this paper choice because we are a member of the Green Press Initiative, a nonprofit program dedicated to supporting authors, publishers, and suppliers in their efforts to reduce their use of fiber obtained from endangered forests. For more information, visit www.greenpressinitiative.org.

Environmental impact estimates were made using the Environmental Defense Paper Calculator. For more information visit: www.papercalculator.org.

Contents

FOREWORD IX
ACKNOWLEDGMENTS XI
INTRODUCTION XIII

CHAPTER ONE
A History of Grain Growing and Consumption in the Northeast — 1
Grain Production in Early America — 2
Grain Production in Canada — 9
Organic Grains in Vermont — 11
Growing Grains to Offset Feed Costs — 15
The (Re)Birth of the Local Food Movement — 16
A Growth Spurt in Research and Community — 17

CHAPTER TWO
Soil Fertility Considerations — 21
A Brief History of Chemical Fertilizers — 22
The Importance of Soil Fertility — 23
Managing Soil Fertility Organically — 24
The Soil Test — 25
Reading the Soil Test — 27
Soil Fertility as Part of a Holistic System — 27

CHAPTER THREE
Getting Started with Tillage — 29
Choosing Your Crops — 30
Preparing the Soil for Planting — 31
The Benefit of Cover Crops — 32
The Dangers of Tillage — 32
Tilling Your Grain Plot — 32
Secondary Tillage — 33
The Final Fitting — 33
Preparing the Seedbed — 34

CHAPTER FOUR
Sourcing and Planting Seeds — 35
The Value of Canadian Varieties — 35
Producing Your Own Seed — 36
Germination Potential — 37
The Mechanics of the Seed-Planting Process — 38
Sowing Seeds in the Home or Research Plot — 39
The Grain Drill — 39
Caring for Your Grain Drill — 43

CHAPTER FIVE
Early Growth and Weed Control — 45
The First Signs of Growth — 45
Dealing with Weeds — 46
The Tine Weeder — 48
How to Operate a Tine Weeder — 51
Understanding Your Crop — 53
When to Weed Again? — 53
Weeds as Soil Messengers — 54

CHAPTER SIX
The Growth Cycle and Harvest of Cereal Grains — 55
The Vegetative Growth Period — 55
The Reproductive Period — 56
Pollination — 56
Grain Filling — 56
The Harvest — 57
Binding and Reaping — 57
Stooking Grain — 58
Swathing Grain — 59
Harvesting in a Wet Season — 60
Threshing — 60
The Combine — 65

CHAPTER SEVEN
From Field to Storage — 77

Storage Container Options — 77
Grain Moisture — 80
Measuring Grain Density — 82
Grain Precleaning — 83
The Grain Auger — 84
Grain Storage and Elementary Drying — 85
The Screw-In Grain Aerator — 86
Drying Grain in Silos — 86
Pressure-Cure Drying — 90
Drying Small Amounts of Grain — 92
Stationary Grain Dryers — 94
Additional Drying Techniques — 95

CHAPTER EIGHT
Preparing Grain for Storage, Sale, or End Use — 99

The Butterworks Granary — 100
Preparation for Cleaning — 101
The Basics of Cleaning and Grading — 101
Air Screen Cleaners — 102
Additional Notes on Cleaning Infrastructure — 107
The Gravity Table Separator — 109
Other Types of Cleaning Equipment — 111

CHAPTER NINE
Corn — 113

Growing Corn in Colder Climates — 113
The Basic Biology of Corn — 114
Ensilage — 116
Breeding for Productivity — 116
The Growth of Corn Agriculture — 118
The Rise of Genetic Modification — 118
My Personal Corn Odyssey 1975–2011 — 119
Planting Corn — 128
Emergence, Early Growth, and Primary Weed Control — 136
Cultivating Corn — 143
Late Vegetative and Reproductive Stages of the Corn Plant — 149
A Note on the Full-Season Nature of Corn — 152
Dry-Down and Harvest — 153
Drying and Storage — 164
Processing Corn for Animal Consumption — 168
Processing Corn for Human Consumption — 169
Corn Products for Good Eating — 170
Corn Breeding and Seed Saving — 171

CHAPTER TEN
Wheat and Its Relatives — 175

Wheat: A Brief History — 175
Agronomic Considerations for Growing Wheat — 178
Planting Your Wheat — 183
Types of Wheat — 185
Wheat Diseases — 189
Modern and Heirloom Wheat Breed Varieties — 196
How to Harvest High-Quality Wheat — 201
Wheat Processing and Flour Production — 204
Wheat for Livestock Feed and Other Uses — 217

CHAPTER ELEVEN
Barley — 219

Barley: A Brief History — 219
Barley for Human Consumption — 220
Barley for Livestock Rations — 220
Types of Barley — 221
Agronomic Considerations for Barley — 224
Post-Harvest Considerations — 233
Processing Barley for Animal Feed — 234
Malting Barley — 237
Barley for Human Consumption — 240

CHAPTER TWELVE
Oats — 245

Oats: A Brief History — 245
The Basic Biology of Oats — 246
Finding and Choosing Seed — 250
The Culture of Oats — 252
Processing Oats for Animal Consumption — 260
Processing Oats for Human Consumption — 262
Advances in Oat Breeding — 272

CHAPTER THIRTEEN
The Winter Cereals: Rye, Spelt, and Triticale — 275
Rye — 276
Spelt — 284
Triticale — 293

CHAPTER FOURTEEN
Soybeans — 297
Soybean Origins — 297
My Search for a Short-Season Variety — 300
Soybean Maturity Zones — 301
Things to Consider Before Getting Started — 303
Choosing the Right Variety — 303
Planting Soybeans — 305
Early Growth and Cultivation of Soybeans — 312
Insects and Diseases in Soybeans — 314
Fruiting and Maturation of Soybeans — 316
Soybean Harvest — 317
Storage and Post-Harvest Treatment of Soybeans — 321
Organic Soybeans in an Ever-Increasing GMO Environment — 324

CHAPTER FIFTEEN
Dry Beans — 327
Dry Bean Production in the Late 1800s — 327
The Benefits of Growing Dry Beans — 328
Determinate and Indeterminate — 328
Fertility Considerations — 329
Planting Dry Beans — 330
Growth and Cultivation of Dry Beans — 333
Dry Bean Harvest — 338
Growing Beans at Butterworks Farm — 342
After the Harvest — 344

CHAPTER SIXTEEN
Oilseeds — 347
Lessons from Québec — 347
Sunflowers — 348
Flax — 359
Canola — 364

CHAPTER SEVENTEEN
The Minor Grains — 373
Buckwheat — 373
Grass Seed — 378
Legume Seed Production — 379

CHAPTER EIGHTEEN
Preparing Livestock Rations with Farm-Raised Grains — 389
My Foray into Formulating Grain Rations — 390
The Importance of Proper Grinding Equipment — 391
Grinder-Mixer Troubleshooting — 393
Putting Together a Ration — 393
Grain Mixes for Dairy Cows — 396
Rations for Poultry — 398
Feeding Farm-Grown Grains to Pigs — 400
Some Last Thoughts on Feeding Your Own Grain — 402

CHAPTER NINETEEN
Where Do We Go from Here? — 403
The Human Consumption Factor — 403
Other Infrastructure Considerations — 406
Weather, Climate, and the Future — 408
Corporations and Seed Sovereignty — 410
The Long Look Back and a Peek into the Future — 411

REFERENCES — 413

INDEX — 417

Foreword

We grew grain the very first year on our farm in Maine. Even though we had begun with wooded land and were working hard to clear enough for the first vegetable crops of what would become our specialty, we made sure to clear enough extra for a small plot of field corn. We were lucky to get seeds from another farmer and grew an old New England heirloom, Longfellow Flint. Polenta and corn bread have always been favorite foods in our house, and flint corn makes the very best.

My standard breakfast is a bowl of oatmeal so a few years later, after locating seed for a hulless oat variety, we grew a field of oats. Unlike the corn, which was simple to hand-harvest and husk, the oats required another technique. By following pictures in the graying pages of an old book, we figured out how to cut the standing grain with sickles, tie the stems in bunches, and stand the bunches in the field for further drying. When the oats were dry, we threshed them on a canvas tarp by beating out the kernels with homemade flails. We winnowed the chaff from the grain over that same canvas tarp on a windy day.

By the time we grew our first field of wheat, we had progressed another few centuries. From drawings in one of the wonderful old Eric Sloane books, *The Seasons of America Past*, we figured out how to add wooden cradle fingers to a scythe. That "cradle scythe" made the cutting and shocking operation go easier. We went halves with another farmer on the purchase of a Japanese CeCoCo thresher. The CeCoCo had a foot-powered, treadle-operated, spinning drum covered with wire loops, which effectively removed the grain from the heads of the wheat shocks when they were held against it. That year we adapted a neighbor's blueberry winnower to make our winnowing more efficient.

But fun as all this may sound in the retelling, we were anything but efficient. In reality we were just blundering along. Ahh, what I would have given back then for a copy of this book. There was hardly any information available at that time, because New England grain growing had been on its way out for many years, especially in our area. The industrial concept of "grow it all in the Midwest and ship it east" had almost completely taken over. Recently, however, the burgeoning interest in local foods has begun to spill over into grain. New England grain growing is again becoming a topic of conversation. But Jack Lazor did not wait for a new movement to inspire him. Jack inspired the movement. Jack began reclaiming the small farm's grain heritage right from the start of his farm many years ago. That is why this book is such a delight. These are the words of someone who has talked to all the old-timers and done it all himself. It is like acquiring hundreds of years of knowledge in one book. And he presents everything in an appealing, storytelling manner that will have you sitting up late reading page after page.

If you ever do get to visit Jack's farm, you will see what determination it has taken to do what he does. Perched on a hill in the far northern reaches of Vermont, Butterworks is a farmer's farm. In fact, the first line on the back of their yogurt container says it all: "Butterworks Farm is a real farm." The grain-cleaning and -bagging equipment in the classically designed granary testifies to ingenuity, both Jack's in putting it all together and that of the old inventors and manufacturers who devised the small-scale machines in the first place. But grain growing is not just machinery. It is also biology. The field corn I presently grow, Abenaki Calais Flint, was reselected and improved by Jack Lazor.

Even if you are not a farmer and have never thought about growing grains, this book is worth reading just to understand how an exceptional farmer's mind works. Remember, these are the folks who are feeding you. If you are a small farmer, this is your world. I give my deepest thanks to Jack for paving the way for the rest of us.

—**ELIOT COLEMAN**
Harborside, Maine
March 2013

Acknowledgments

This book was two years in the making, and during this time I learned how to juggle my normal dairy- and grain-farming duties with my newfound writing regimen. My immediate family—my wife, Anne; daughter, Christine; and son-in-law, Collin Mahoney—went the extra mile for me by taking on a good portion of my workload. Ginny and Ursala Mahoney, my two young granddaughters, continued to bring me joy during the whole process. My two main field hands, Brian Krisch and Ed Champine, worked countless extra hours on tractors, cultivators, and combines to allow me more time for writing. I also would like to thank the rest of my farm and dairy plant crew—Mike Heald, Will Krause, Dann Black, Blane King, and Dean Brouillette—for their dedication and work well done while their leader was hiding away in front of his computer. The entire Butterworks Farm community of family and employees contributed to this writing effort by giving me the time I needed to get the job done.

There are a number of other individuals who help keep our operation running smoothly as well. Charles Capaldi, our bookkeeper, has taught me the ins and outs of Microsoft Word and has helped me numerous times with computer glitches and problems. Brian Dunn, a good neighbor and friend, has kept all of my older tractors and machinery in good repair with his natural mechanical abilities. Local welders and repairmen Danny and Ben Guay have been there for me for the more serious heavy-duty repairs that are often needed on a working farm. Ray delaBruerre, my electrician, has been there every step of the way hooking up electric motors and wiring for all the grain-processing and -cleaning machinery I have installed in my granary over the years. I'm definitely more of a visionary than a fix-it kind of person. The help I have received from these people and many others has made it possible for me to create and maintain the infrastructure to grow and process grain here on my farm for the past thirty years.

The greater grain-growing community of which I am a part has provided me with the support, education, and feedback necessary to do what many have considered impractical and impossible here in northern Vermont. University of Vermont Extension agronomist Heather Darby has been there for all of us grain growers with her amazing knowledge, support, and research. Heather was instrumental in the founding of the Northern Grain Growers Association and the rebirth of grain growing as a viable farming enterprise here in Vermont. Alburgh, Vermont, farmer Roger Rainville has transformed his former dairy farm into a grain research center and has been an incredible help to all of us. Fellow Vermont organic grain farmers Brent Beidler, Ben Gleason, and Ken Van Hazinga have all collaborated with Heather to advance the culture of grain in this corner of the country. Extension assistants Erica Cummings, Susan Monahan, Hannah Harwood, and Amanda Gervais have all worked tirelessly on grain research projects for the benefit of our grain community. We have the beginnings of a small-time grain industry here in Vermont because of the efforts of these individuals.

I certainly wasn't born into grain farming, and I have many mentors to thank for much of the knowledge I have accumulated over the years. I met Francis Angier from Addison, Vermont, at a NOFA conference in the summer of 1977 and have been friends with him ever since. He encouraged me to grow wheat just like he did in the 1950s and

1960s. Clarence Huff from Compton, Québec, was in his late seventies when I met him in 1981, and he taught me all about seed cleaning and combine operation. I met Milton Hammond from Newport Center, Vermont, about the same time and was able to tap into his vast knowledge of millwrighting and grain machinery operation. These two wonderful individuals have since passed on, but have left behind some of their experiences with me. I also have to give credit and thanks to two other now deceased old farmers whom I met in the very early days of my farming career. Robert Warden of Barnet, Vermont, and John Ace of Oregon, Wisconsin, were as different as night and day. Robert was a frugal Yankee hill farmer, and John was a relaxed midwesterner who liked to polka dance. However, both of these men had spent their entire lives farming and provided me with role models that I could emulate.

Over the years, I have had the pleasure of meeting and befriending many Canadians involved in the business of growing grain. These folks live in a farming culture where grain growing is the norm and have had lots to share with me. Wilhelm and Gudren Brand from North Hatley, Québec, are German immigrant biodynamic farmers who have shared considerable inspiration with me. Archie Blankers is a dairy and grain farmer from Huntingdon and has given me endless good advice about organic cropping strategies. Loïc Dewavrin from Les Cèdres, Québec, has helped me press oilseeds and provided me with extra grain in lean years. Our Vermont grain-growing community has developed a unique and wonderful fellowship with our contemporaries on the other side of the international border. These folks and many others have so much knowledge and so many years of experience to offer the rest of us, who have so much to learn.

Last, but not least, I would like to express my appreciation to my editor, Makenna Goodman. She sought me out at a NOFA conference in the summer of 2010 and encouraged me to write a book. Up until this point, I had never even considered doing such a thing. Makenna walked me through the process and helped me every step of the way by providing me with tons of positive reinforcement. We broke the usual literary protocol, and I submitted my writing chapter by chapter. Her constructive comments and suggestions helped me turn a daunting task into a finished product. We had a number of relaxed and informal meetings at Chelsea Green over a two-year period. Margo Baldwin, Pati Stone, and the rest of the staff were wonderful and have made the experience of being a first-time author very rewarding. My deepest thanks go to all.

Introduction

Small-scale agriculture has experienced a rebirth in recent years as more and more people seek out food that has taste and meaning. For this reason, consumers have turned to neighboring farms and farmers for an ever-increasing amount of their food supply. At first, produce, meat, and dairy were the primary staples of this "localvore" diet. As a result, local meat producers, cheesemakers, and vegetable farmers—along with their growing ranks of customers—have developed farmer's markets and community-supported agriculture models. Many more alliances among farmers, consumers, and restaurants have developed in recent years, and support for local food systems has come from various state departments of agriculture and a wide array of organizations, from the Farm Bureau to organic farming organizations like NOFA (Northeast Organic Farming Association) and MOFGA (Maine Organic Farmers and Gardeners Association), to name only a few. The recurring theme throughout this movement has been that many people want food with a story. They want to know their farmer and are willing to pay a premium for sustainably produced local foodstuffs.

Until recently, local grain production in the Northeast has been a weak link in the localvore diet. Bulk bins at food co-ops and health food stores were filled with dry goods from everywhere but here. Flours came from distant mills in faraway places like Kansas and the Dakotas. Dry beans might come from China, Michigan, or the Red River Valley of North Dakota and Minnesota. But change is coming ever so slowly to our local and regional food systems, and local grain products are beginning to make their way to pantries and store shelves throughout our country, and especially in New England, where the number of small-scale organic farms is very much on the rise. A small undercurrent of local grain production has existed for years, producing flours and meals from farm-raised corn and cereal grains; more recently, other products like rolled oats and edible oils from sunflowers and flax have appeared. The hulling and pearling of barley, oats, spelt, and emmer present whole new sets of challenges, however, for farmers in a region like the Northeast, severely lacking in the infrastructure and knowledge of grain processing, and elsewhere where commodity grain is the norm and "small-scale" is undervalued and not made simply. Seed cleaning and saving, on-farm plant breeding of locally adapted varieties of grains, and the micro-malting of barley for local beer making are all subjects in need of further exploration and research as we develop a truly sustainable regional food system.

In my state of Vermont, the interest in local grain production has come from a very diverse group of individuals. Initially, the push came from localvore consumers and the retail outlets supplying them. Studies of demand, consumption, and sales histories were commissioned and undertaken by the Association of Neighborhood Co-ops in Vermont and Massachusetts and the Mad River Valley Localvores of Waitsfield, Vermont. Farmers needed to know if local grains were simply a fad or economically viable alternative crops before they invested the energy into a crop so readily available from farther away. Initial investigations by researchers like Ginger Nickerson demonstrated strong demand for locally produced beans, grains, and oils. And as it turned out, homesteaders and back-to-the-landers also wanted to begin growing small amounts of their own grains on a garden scale for home consumption. High-priced imported organic feed grains have also encouraged small-scale organic dairy farmers to consider growing and storing some of their own grains for homegrown rations. Many different individuals and groups have converged at a single crossroads—and the time is ripe for a rebirth of grain growing on a small scale.

As interest in this subject has grown over the past few years, a small and dedicated group of individuals has worked quietly to lay the foundation for this rebirth of grain culture in the Northeast. University of Vermont Extension agronomist Dr. Heather Darby began organizing field days at Vermont grain farms shortly after she began work in her new position in 2003. Workshops were held at the farms of long-established as well as newer growers of grains, and interest was surprisingly strong; the meetings were very well attended by a wide variety of people interested in the production of grains for both human and animal consumption. This led to the inception of a yearly grain conference featuring experts from all over North America who could speak on a variety of themes pertinent to reestablishing a culture of grain here in Vermont and the rest of the Northeast. We've hosted academics from leading agricultural universities who have spoken about wheat and corn breeding, weed science, and plant diseases, and innovative farmers from grain-producing regions of the United States and Canada have shared their successes and failures with a very eager and attentive group of interested individuals from our region. These workshops and conferences have grown exponentially over the past half decade, to the point where many of us deemed it necessary to found a formal organization to promote this work of encouraging grain production here in the Northeast. Thus, the Northern Grain Growers Association was established in 2008, and interest has continued to grow to the point that we had almost two hundred attendees at the 2012 conference in Burlington, Vermont. Large-, medium-, and small-scale grain farming has reemerged in the Northeast to my delight. It is the purpose of this book to provide practical, hands-on information to help anyone and everyone interested grow grain organically, from the tiny homestead garden plot to the large field of a more commercial operation.

My own personal farming saga began in the early 1970s. While attending Tufts University in politically turbulent times, I somehow decided that a simple life providing for myself from the earth would be more fulfilling to me than political protest. Books like Helen and Scott Nearing's *Living the Good Life* helped me solidify my plan of action, and I even created my own major focusing on the history of agriculture. Summers found me working as a costumed interpreter on the historical farm at Old Sturbridge Village, a living historical museum located in central Massachusetts close to where I grew up. My first exposure to grain was spending the day riding on an old combine harvesting rye on a neighboring farm in 1961, and at Sturbridge Village I got to harvest rye by hand with a curved reaping sickle. The hand-cut grain was tied off into bundles that were stood up to dry in the field. Later I got to thresh this grain out on the barn floor with a handheld flail.

In 1973, I moved to Warden Farm in Barnet, Vermont to work on an old-fashioned dairy operation where there was lots of evidence of a once thriving grain culture. One of the farm's outbuildings was a granary that contained many old grain-harvesting tools like reaping sickles, grain cradles, flails, and winnowing pans. My employer, Robert Warden, told me many stories from his earlier days in the 1920s and '30s when they grew oats and flint corn as a matter of common practice. Circumstances took me to rural southern Wisconsin in 1974 and 1975, where I found a culture of dairy farming with farmers growing their own grain as well as hay for their cows. I spent many hours with my friend and mentor John Ace of Oregon, Wisconsin, learning all about oats and ear corn for the dairy. There was also a thriving Amish community around New Glarus, south of Madison. My partner, Anne, and I went to many of their auction sales and began to learn that the Amish grew much of their own wheat for bread using horse-drawn grain drills, grain binders, and belt-powered stationary threshing machines. This was it. I was hooked. I wanted to return to Vermont and grow grain the old-fashioned way for ourselves and our livestock.

Anne and I began our farming careers as back-to-the-land homesteaders on a small farm in Irasburg, Vermont, in May 1975. We were equipped with a lot of idealism and a truckload of old farm antiques we had brought with us from Wisconsin. My first project was to trial six different varieties of heirloom flint corn that I had obtained from the USDA seed bank. (I found out all about how much raccoons like flint corn that first summer.) Anne and I took many trips to the Eastern Townships of Québec that first summer together in northern Vermont, and we were surprised to see so many fields of golden ripening grain during our travels. In August, there were combines everywhere harvesting the oats and barley. The fact that all of this grain was growing quite well only ten miles away across the border was all the proof I needed that I could do this on my side of the border, too. And so we bought our first sixty acres in Westfield, Vermont, in 1976 and grew our first six acres of wheat, corn, and barley in 1977. Planting was done with an antique $25 horse-drawn wooden grain drill pulled by a 1954 John Deere 40 tractor, and we bought a six-foot John Deere grain binder (reaper) and a Dion threshing machine in Coaticook, Québec, for $250. We spent most of the month of July readying the reaper for the upcoming harvest and procuring new canvases from the Amish in Ohio. We ended up buying a second grain binder from Doug MacKinnon of Barnston, Québec, for extra parts. (Doug had used this machine well into the 1960s to reap his and his neighbors' oats.) Beginner's luck was with us, and we successfully reaped and stooked all six acres of our wheat and barley. We powered the stationary thresher with a sixty-foot endless flat belt attached to the belt pulley of our old Super M Farmall tractor. Five or six old-timers from our area provided us with lots of help, support, and good advice, and the general consensus was that field-cured grain reaped, stooked, and threshed was far superior in taste and color to its modern counterpart direct-cut with a modern combine. The old guys were right—our wheat was quite dry and golden-red in color. Another neighbor, Milton Hammond of Newport Center, let us use his small grain cleaner and buhr mill grinder to make whole wheat flour for our first loaf of homegrown whole wheat bread. We thought that we had arrived, but little did we realize that our grain-growing adventure had just begun.

This all happened thirty-four years ago. We tried selling whole wheat flour to our local food co-op but were quite surprised at how little our very special product was worth. No one seemed to give any value to homegrown whole wheat flour except us and the old-timers who had helped us accomplish this modern-day anachronism! But slowly, demand grew, and we have continued to grow wheat and other grains ever since this very humble beginning. We gave up the grain binder and the threshing machine after six years and increasing acreage, and have upgraded to four different combines over the years. I have managed to collect a pile of old and some new machinery for planting, harvesting, storing, and processing our farm-grown grains, and we have gotten mountains of advice from a large group of older farmers in Vermont and Québec, many of whom were French speaking. Most of these old guys have passed on, but their stories, advice, and goodwill live on with me every day. We have made numerous mistakes and learned many hard lessons over the last thirty-plus years of trying to grow grain in northern Vermont—we have lost crops to bad weather, and flooding has become more prevalent over the last decade. But we have learned a lot about proper grain moisture for long-term storage by watching grain mold because it was harvested too wet and not properly dried. The lessons have been difficult, but the successes have been ever so sweet. Indeed, the feeling and security of producing our own food staples represents true wealth and well-being to us.

Lots of water has passed under the bridge over the last three-plus decades, however, and we are no longer young back-to-the-land types. We remain just as optimistic and idealistic, but we are now commercial organic dairy and grain farmers processing our own milk and growing hay and grain for humans and livestock on four hundred acres of land. Integrating crops, livestock, milk processing, and family into a whole farm organism that restores the earth has become a way of life for us; we still grow just about all of our own food and provide all the grains for our cows, pigs, and chickens. Our crop base has expanded from barley, oats, and wheat to include corn, soybeans, dry beans, field peas, sunflowers, flax, and buckwheat, and in most years we can clean and save seed for just about all of our grain crops. Crop growing continues to be my primary agricultural passion along with grain processing, and I continue to learn about and experiment with elementary farm-based plant breeding of open-pollinated corn and wheat as well as on-farm grain processing of spelt, oats, and emmer. It is my dream that our farm can continue to be a center of hope, innovation, and inspiration for those who want to nurture the earth and grow good food for others. We have certainly held many workshops here at Butterworks Farm over the years and have hosted many visitors who have given us as much as we have given them. As interest in grain growing continues to expand, I feel that it is time for me to chronicle as much of my knowledge and experience as I can muster to further the pursuit of a sustainable and organic culture of grain growing in our region. So many people have been so generous to me with their farming knowledge and wisdom over the years; it is my turn to give a little back. I hope that this book about growing organic grains on a small scale will help people avoid some of the mistakes that I have made and achieve their goals with more ease and less pain.

CHAPTER ONE
A History of Grain Growing and Consumption in the Northeast

A thriving and relatively stable Native American culture had been in existence in northeastern North America long before the arrival of the first Europeans. New England was populated by small bands of Woodland Arenac, a subgroup of the Algonquin peoples; Pequots and Mohegans lived in Connecticut; Wompanoags in southeastern Massachusetts; and the Pocumtucks in the Connecticut River Valley of western Massachusetts. Farther north were the Cowasucks and Sokoki along the Connecticut River and in the Lake Champlain basin (Bain, Manring, & Mathews, 2008). The story of grain growing in the Northeast begins about AD 1000, when these hunter-gatherers first began to supplement their food needs with corn, beans, and squash, and these "three sisters" were cultivated on river floodplains all over the region. Because the more benign climate of southern New England was much more favorable to agriculture, corn culture began there a little earlier than it did farther north, and cultivated foodstuffs were traded with their northern brethren. Archaeological evidence from the Skitchewaug site on the Connecticut River near Springfield, Vermont, reveals that these crops were first cultivated around AD 1100, whereas corn and bean culture came several centuries later to the Champlain Basin (farther north) because game was more plentiful and easier to procure in the wider valley and flatter landscape there. The Donahue site at the mouth of the Winooski River on Lake Champlain saw its first agriculture beginning in 1440 (Sherman, Sessions, & Potash, 2004). Flint corn, kidney beans, and squash were grown together in mounds by the women of the tribe, and when the fertility of a field declined, cultivation was moved to a new spot. At the time of arrival of the first Europeans, the Woodland Indians had created a very stable and peaceful society in harmony with the earth, based on agriculture, hunting, and gathering.

French, Dutch, and English traders plied the coast of northeastern North America for beaver pelts beginning in the late sixteenth and early seventeenth centuries, and by 1600 contacts between Europeans and Native Americans were well established. But make no mistake about it—patterns of exploitation and domination established themselves rather quickly. The English took many individuals into captivity and sold them into slavery upon returning to Europe. Tisquantum (Squanto), the famous Indian who helped the Pilgrims, had originally been captured in 1614 by Englishman Thomas Hunt and sold as a slave to Spanish monks. He was later freed and made his way back to his homeland in 1619. Upon his return, Squanto found that most of his fellow Pawtuxets had perished in a smallpox epidemic the year before. Disease and pestilence were the unforeseen consequences of exposure to the white man. In fact, one of the reasons that the

Pilgrims enjoyed abundant gifts of corn and beans from the Pawtuxets was because a majority of the tribe had been wiped out by disease in the previous season (Bain, Manring, & Mathews, 2008). It is in this context that we must view the arrival of the first Europeans in the New World. The English came to settle, clear the land, and establish farms, but the Native Americans had a much different relationship with the earth that did not include ownership. And while this book is not intended to be a history of European and Native American conflict, it is from this context that we must chronicle the first two centuries of agriculture and grain growing on the North American continent.

Grain Production in Early America

The culture of grain is nothing new in New England. Cereal grains are grasses and extremely well suited to this region's cool climate, and grain crops have been raised here for the past four hundred years—ever since Samuel de Champlain first arrived in New France and what was later to become Vermont. In general, cereals have always played an important part in the regimen of mixed or diversified agriculture, as man cannot live by grain alone, and oats, barley, wheat, rye, and corn work very well in a general farming rotation. And while grain is a natural crop for the Northeast, its importance has increased and diminished in cycles over time due to changing patterns of settlement, economic development, and climate. This chapter aims to put grain growing in perspective from its earliest days in the eighteenth century right through today.

European settlers brought stocks of cereal seeds with them when they first arrived upon the shores of northeastern North America, and, as a result, winter wheat, oats, and barley were familiar staples of the first colonists at places like Plymouth, New Amsterdam, and the Massachusetts Bay Colony. Harsh lessons were learned and starvation ensued, however, as it quickly became apparent that English wheats were not very well suited to the harsher climate of the New World, and the frigid winter temperatures and the ravages of the Hessian fly soon put an end to the first crops of wheat. As stated earlier, Squanto, the Pawtuxet Indian, shared his tribes' stores of corn with the Pilgrims that first winter at Plymouth in 1621, and the following spring he taught these new settlers how to plant maize (Indian corn) in hills each fertilized with a buried fish. Indian corn had long been a staple of the Native American diet in the Northeast, and was native to the continent. These sturdy, well-adapted plants produced long, thin ears with eight rows of very hard-textured or "flinty" kernels. Because of its high proportion of hard starch or endosperm, ground Indian corn had a texture very similar to wheat flour. This fact, coupled with a high level of protein, made ground maize the first choice for America's first settlers. After the failure of early wheat crops, it also became apparent that rye was much better suited to the climate of southern New England, as it was (and still is) a much heartier cereal than wheat.

Agriculture in the New World was primarily a matter of subsistence during the first century of settlement, and from 1620 to 1740, settlers pushed inland and upriver from Boston, Hartford, and New York as well as up the coast to southern New Hampshire and Maine. Slowly, trees were cleared and burned for potash while gardens and crops of rye, corn, and hay were sown between the stumps. The New England farmer and his family directed the majority of their time and energy to production for home needs. Of the thirty or more acres of land that a family typically cleared for cultivation, no more than ten to thirteen was devoted to the growing of crops. The average farmer usually grew five acres of Indian corn and three or four acres of rye as staple crops for the family. Planted the previous fall, winter rye was cut in July with a reaping hook or sickle, bound into sheaves, and stored in the barn for threshing later in the season. The grain was beaten from the sheaves on the barn's threshing floor with a wooden flail, a two-piece tool consisting of a long handle attached by a leather thong to a shorter piece of hardwood that was twirled in a circular motion, striking the grain on the barn floor. Once threshed, the straw was removed by pitchfork, leaving a pile of grain and chaff behind, which was gathered and winnowed in the breezeway or center drive of a barn with a door at either end. Elementary grain cleaning was accomplished by tossing the chaffy mixture into the air with a large semicircular wooden winnowing basket or pan, at which point the wind would carry the lighter material away while the heavier grain would fall back into the two-handled winnowing basket. Rye usually yielded eight or ten bushels to the acre for a year's supply of thirty to forty bushels. As for Indian corn, a farmer and his sons planted kernels at the end of May and harvested mature ears from the stalk in late September. Corn finished drying and was stored in corncribs, which were elevated from the ground and had wooden slatted walls. These little buildings were very distinctively shaped, being narrower at the bottom and wider at the top with outwardly sloping walls. Corn was shelled as needed, and both the rye and corn were taken to local water-powered gristmills for grinding into flour and meal. "Rye and Injun" bread was the staple, baked with a mixture of these two grains.

Settlement of New England marched slowly northward and inland as the seventeenth century progressed. At the time, northern New England was controlled by the French and their Indian allies. Samuel de Champlain had explored the coast of Maine in 1605 and the lake that was to become his namesake in 1609, and the French constructed Fort Ste Anne in 1666 on Isle La Motte at the northern end of Lake Champlain. The northward advances of English settlers were held back by the constant threat of attack by protective Indians. Northern New England settlement took place in the context of constant war between the French and the English during the first half of the eighteenth century. After the Deerfield Massacre of 1714, it became quite apparent that the settlers needed protection. Beginning in 1724 with Fort Dummer near present-day Brattleboro, a series of forts was constructed up the Connecticut River. It wasn't until the victory of Wolfe over Montcalm on the Plains of Abraham at Québec in 1759 and the fall of Montreal in 1760 that relative peace came to northern New England. Great Britain took control of Canada and northern New England with the signing of the Treaty of Paris in 1763. Settlement in the region became much easier after the defeat of the French, and a large portion of the Abenaki moved northward to reside at Odanak where the St. Francis River meets the St. Lawrence (Bain, Manring, & Mathews, 2008). Both Newbury and Bennington, Vermont, saw the first arrival of civilization in 1761. Bangor, Maine, was established in 1772. The settlement of northern New England really began in earnest after 1790 as individuals who had fought in the American Revolution began to travel northward to take advantage of land grants in newly opened-up territories.

Self-sufficient agriculture was the standard as early settlers hacked back the forest and grew their first crops among the tree stumps, and the production of potash from the ashes of burnt trees was a primary source of income for many. People in the North grew the same corn and rye they had cultivated in southern New England, but surprises were many in this new territory. Winters were harsher and growing seasons shorter, and stories abound about early frosts and lost corn crops. Fortunately, the Sokoki Abenakis of the Missisquoi delta in Swanton had varieties of flint corn that grew well in more northerly areas. Another pleasant surprise was that these new soils of the north were fresh and not exhausted like much of the land that had been left behind in southern New England. Whereas wheat had been impossible to grow in Massachusetts and Connecticut, it seemed to thrive on these recently cleared northern tracts. A cooler climate and richer earth added up to wheat yields of twenty to thirty bushels to the acre. Samuel Williams in his 1794 volume *The Natural and Civil History of Vermont* reported that the "first crop of wheat will fully pay him (the farmer) for all the expense he has been at, in clearing up, sowing, and fencing his land" (Sherman, Sessions, & Potash, 2004). There were many reports of farms being paid for in one season with a good crop of wheat. Demand was high in the south and also to the north in Canada, and Lake Champlain made it possible to easily transport bulky grains northward to Montreal. In 1792, farmers in the Champlain Valley region of western Vermont had a thirty-thousand-bushel surplus of wheat to export. The wheat boom was on, especially in this region of Vermont that was blessed with wide-open level expanses, fertile clay soils, and hot dry summers.

Wheat continued as the primary grain crop in every region of northern New England for the next thirty years. Farms on both sides of the Connecticut River also produced large amounts for shipment to Boston. Frederic P. Wells wrote in his 1902 *History of Newbury, Vermont*, "In early days grain was the staple product of the farms and thousands of bushels of corn, wheat, rye, oats, and barley, were exported. Some farmers went to Salem, the wheat market for the export trade, several times in each year" (Wells, 1902). Forty-one thousand bushels of wheat left the Champlain Valley of Vermont for New York City on the newly opened Champlain Canal in 1823. But despite this success, a series of developments in the mid-1820s began to lessen the importance of wheat as a primary crop. For one thing, continual cropping had caused soil fertility and yields to decline, and outbreaks of the Hessian fly and wheat midge resulted in several years of total crop failures. The crowning blow came with the completion of the Erie Canal in 1825. It was reported in the Niles' Register that in 1826, Vermont imported approximately fifteen thousand barrels of flour. Only eight years later in 1834, this number quadrupled to sixty thousand barrels. The center of wheat production had moved to the Genesee Valley of western New York. Indeed, by 1835, Rochester had twenty-one active flour mills and had been nicknamed the Flour City. As for Vermont, wool production began to gain importance after the first merino sheep were imported to the area in 1812, and favorable tariff legislation in 1828 gave American-produced wool a distinct advantage over British imports, making sheep farming much more attractive. As grain fields of depleted fertility grew back to grass, pasture and hay for sheep became much more profitable crops than wheat. Writing in his 1831 *History of the State of Vermont*, Nathan Hoskins stated that "the soil is such, and the seasons are so uncertain, for the production of crops of grain, that grazing is the most sure and profitable branch of agriculture which the farmer of Vermont can attend to with success" (Sherman, Sessions, & Potash, 2004). Thus, Vermont embarked on its next phase of agriculture—the Sheep Craze—which lasted until the early 1850s and experienced a brief renaissance with army uniform production during the Civil War.

So while wheat production in Vermont peaked at 644,000 bushels in 1840, despite the decline in export markets, the production of wheat for local consumption continued throughout the remainder of the nineteenth century. A quick perusal of the accompanying production statistics tables indicates that the total bushels of wheat produced declined by a little more than one hundred thousand bushels in 1849; Census of Agriculture figures for ensuing decades show total wheat production holding steadily through 1879. By the turn of the century, the culture of wheat in Vermont had declined by 85 percent to a mere 34,650 bushels. Dairy farming was quickly becoming a way of life in Vermont during the last decades of the nineteenth century. Wheat crops for home and local consumption continued to be important, however, in some of the more marginal upland areas of the state. My own particular farm in the northern town of Westfield in Orleans

County offers a unique view of a society holding on to older traditions. Thomas Trumpass reported to the 1879 Census of Agriculture three acres of wheat, two of Indian corn, and five of oats. Wheat yields were about fifteen bushels to the acre on the Trumpass farm. The only other crop to experience precipitous declines in this period was buckwheat, which was raised primarily for feeding chickens. When compared with modern times, Vermont's total production of 174,394 bushels of buckwheat in 1909 is nevertheless still quite impressive.

Horses supplanted oxen for draft power during this period as more labor-saving farm machinery like mowers, rakes, and reapers became available and affordable, and oat production for horse feed climbed steadily during this period in Vermont. As dairy production became the leading agricultural activity, the production of corn and barley for livestock rations increased. Labor-saving grain-harvesting and -processing machinery like fanning or winnowing mills, threshing machines, and reapers had all been invented during the 1830s. Jerome Increase Case built his first threshing machine in Racine, Wisconsin, in 1842; Vermonter Alfred W. Gray, at first a manufacturer of corn shellers, began making threshers at his facility in Middletown Springs in 1844; and the Samson Power and Threshing Co. of East Berkshire was also a major manufacturer of threshers and treadmill-type "horse powers" during the latter half of the nineteenth century. Due to serious labor shortages after the Civil War, these new machines came into widespread use after 1865, whereby much of the drudgery associated with hand-harvesting and -threshing of cereals had been eliminated. Although the center of grain production had moved westward to New York and then to the Great Lakes region, farmers in Vermont continued producing limited amounts of grains on their farms.

The history of grain production in Maine, however, differs greatly from the rest of the New England region. In its earliest days, coastal Maine was uniquely positioned as a trading partner to the West Indies, and timber and farm products were exchanged for molasses used for making rum. But driven by new settlement opportunities and cheap, fertile lands in western New York and the Ohio Valley, Maine experienced the same decline in agriculture and subsequent outmigration as the rest of New England. However, Maine had its own untapped frontier in the verdant lands of Aroostook County in its northernmost reaches. Aroostook was larger than Massachusetts and Rhode Island combined, and it had vast stores of virgin timber, limestone-rich alluvial soils, a good climate, and plentiful water to harness for power. Dr. Ezekiel Holmes, a prominent agriculturist of the day, was hired by the Maine Board of Internal Improvements to investigate the potential for new settlements in this far-off northern territory. In his *Report of 1838* he wrote, "The staple crop of the Aroostook farms is, and ever must be wheat. For this climate, and most of the soil, is exceedingly favorable." Aroostook County was Maine's own grain frontier in the late nineteenth and early twentieth centuries, and wheat, oats, rye, and buckwheat were the first crops cultivated before the advent of potato culture. Wheat production climbed to 665,714 bushels in 1879, the highest production ever recorded in New England. At the same time, as starch factories proliferated, potato acreage in Maine tripled from fourteen thousand acres in 1879 to forty-two thousand in 1899. Agricultural land expanded by 60 percent in Aroostook County between 1880 and 1930, and Maine became New England's undisputed leader in grain production, a distinction that it still holds today.

As the twentieth century dawned in New England, the production of grain had diminished to almost inconsequential levels. If canals were responsible for declines sixty years earlier, railroads finished the job of eliminating cereal farming from mainstream New England agriculture. Dairying, along with the subsequent production of hay and forages, had become the leading agricultural activity everywhere in the region except for Aroostook County, Maine. Local mills ground provender, a dairy ration of corn and oats, from imported grains that arrived in railcars. By-product concentrates like hominy feed and corn gluten were shipped in one-hundred-pound burlap sacks from far-off mills in places like Buffalo, Toledo, and Ogdensburg, New York. Bulletins from the Vermont Agricultural Experiment Station provided yearly analysis of these feedstuffs to ensure that farmers were really getting the amounts of protein and carbohydrates reported on the product labels. Corn, buckwheat, oats, and barley were still produced on some small farms, however, as illustrated on the accompanying grain production charts, but the University of Vermont estimated that farm-grown grains only amounted to 2 percent of the total grain consumed on Vermont dairy farms. And while a few farmers still brought grains to the local mill for grinding and mixing with imported concentrates, grain corn had begun to fall out of favor as producers switched from traditional New England flints to higher-yielding midwestern dent corns for silage production. Surprisingly, buckwheat for chicken feed remained a popular crop long after wheat had fallen from favor. Wheat production temporarily spiked, however, between 1917 and 1919 because of a worldwide shortage caused by crop failures in Ukraine and by World War I. During this time, wheat prices soared and production climbed to 176,003 bushels in Vermont and 261,185 bushels in Maine in 1919. This was the same period that saw the southern plains grasslands of Oklahoma being plowed up for wheat, which would lead to the Dust Bowl ten years later. But this boom didn't last long; by 1923, commodity stocks had recovered to surplus levels and farmers everywhere found themselves in the midst of an agricultural economic depression.

Organic grain farming has persisted in the Northeast throughout the past century, albeit in an ever-so-diminished state. A look at the Agricultural Census of 1950 for the state of Vermont reveals some interesting surprises. At that time, corn was still being harvested for grain on 2,021 acres out of a total of 53,746. Almost 400,000 bushels of oats were being raised on 13,158 acres; 524 acres of barley and 907 acres of wheat were raised that year; clover seed was harvested from 167 acres; dry bean production was reported at 6,156 bushels from 667 acres; and 298 farms reported owning 309 combines in 1950. I was surprised to see twenty-eight combines reported for Orleans County where I live, but Addison County led the state with fifty-four machines. This diversity of grain production and activity from sixty years ago was heartening to discover. When I arrived here in Vermont in 1973, I heard many stories from older farmers about the grains they had once raised. Hall Buzzell of Shattuck Hill in Derby had two pull-type McCormick Deering No. 64 machines and did custom combining of oats all over the county through the 1950s and early '60s. He told stories of some really difficult combining in the mid-1960s due to several consecutive wet summers; all of a sudden, people stopped growing oats for grain and began cutting them in the milk stage for hay. Shortly after this, Hall parked his two machines in the woods and ceased his custom harvesting operation. (My first combine was a Model No. 64, and I was thankful for these old rusting hulks in the underbrush as a source for spare parts.) Milton Hammond of Newport also had an identical machine to mine, which he used well into the 1980s to harvest his own wheat and buckwheat. Milton was kind enough to let us use his Clipper grain cleaner and buhr mill grinder to clean our first crop of wheat and process it into flour.

The 1950s were watershed years for agriculture all over the United States. Farms continued to get larger as smaller, more marginal operators left the land. In the Northeast, the few remaining teams of horses were retired to the sugar place as farms completed the transition from horsepower to tractor power, and rural electrification reached the last farmsteads that had not yet hooked up to the grid. Beginning in 1952 milk cans began to be replaced by bulk tanks, and farm mechanization became all-pervasive as balers replaced hay loaders and upright silos sprang up all over the rural landscape. At the beginning of the decade, a very few farmers were still reaping, stooking, and thrashing their oat crops with grain binders and stationary belt-powered threshing machines; by 1960, all grain was being harvested with modern combines. The widespread adoption of commercial fertilizers also occurred in the 1950s. Farming was definitely less a relaxed way of life and more a business at the end of this decade of change.

Oats became the last major grain crop of importance during the 1950s. Oats require very little fertility and are relatively easy to cultivate: They withstand the challenges of wetter soils more than any of the other cereals, they fit very well into the crop rotation system of a dairy farm, and the straw portion of the crop was very valuable as a source of bedding for dairy cows. Very few farmers had the wherewithal to grind their own oats into a dairy ration. The standard practice at the time was to exchange your oat crop with your local mill for partial credit toward a complete dairy ration—although chances were that you didn't get your own oats back when you brought home bagged dairy grain from your local feed mill. Aroostook County, Maine, was and still is the region's leader in oat production. Indeed, oats fit quite well into the standard potato rotation and found a ready market with the New England mills of the H. K. Webster Company, the manufacturer of Blue Seal Feeds.

One other grain crop of importance at this time was bird's-foot trefoil seed grown in the Champlain Valley region of Vermont and New York State. Ray Bender, a local county extension agent, brought in some trefoil seed from the Albany, New York, area for farmers to try in the late 1940s, as trefoil thrives on wetter, clay soils and was a natural fit for the heavy clays of the Champlain Valley. For some unknown reason, possibly the fact that these soils were once under an ocean and had a sodium base, trefoil grew just as well as wheat had 150 years earlier in the valley. Dick Sherman, a farmer from Westport on the New York side of the lake, and a group of fifty to sixty other farmers established the Champlain Valley Seed Growers Cooperative in 1950, and seed production took off like wildfire; a processing plant was constructed in Westport, New York. Trefoil helped many struggling dairy farmers supplement their incomes in a world of volatile milk prices. The crop was swathed and combined from a windrow in the late summer. Production of trefoil began to fade in the 1960s; by the early '70s, a root fungus that attacked the trefoil plant had all but finished off the crop. Combines were parked in sheds, and the trefoil era was over. A few farmer members of the co-op continued to produce certified oat and wheat seed, however, and in the late 1970s the Sherman family began to supply the Mill River Flour and Grain Company of Massachusetts with wheat for its flour products, and some of this flour found its way to the Bouyea Baking Company of Plattsburg, New York, for a "Champlain Valley" loaf aptly named "Old Vermonter Bread." Modern flour milling was born in the region in 1985 when Dick Webb, Sam Sherman, and his sister Anne Moisan bought the old seed growers co-op building and installed the Mill River flour mill. Champlain Valley Milling was off to a great start with its "Champy" brand of flour products, and a roller mill and sifter was installed in 1990. A new era of wheat production was being reborn in the same region where it had once thrived a century and a half earlier.

J. Francis Angier of Addison, Vermont, was another grain-growing pioneer in the Champlain Valley during this same era. Francis, a Purple Heart–decorated World War II fighter pilot, returned home to farm in the Middlebury area right after the war. The injuries that he sustained from being shot down over Nazi Germany prevented him from being able to milk cows. He moved to a new farm just south of Addison Four Corners in 1952 and very soon thereafter became a major grower of trefoil seed for the Champlain Valley Seed Growers Co-op. An additional four hundred acres of land on the lakeshore in West Addison gave Francis the land base required to become a major grower of grain at the time. When the co-op began producing certified wheat seed in the mid-1950s, Francis stepped up to the plate. He left the co-op late in the decade in search of independence and higher prices for his wheat. In 1958, he produced his first flour with the help of Herb Ogden, who owned a water-powered gristmill at Hartland Four Corners, Vermont. Francis would leave Herb's mill, cross the Connecticut River, and deliver fresh flour to the Cross Baking Company in Claremont, New Hampshire. He also delivered flour to restaurants and food co-ops; his motto at the time was "Anything but dairy." Aside from wheat, he grew a wide variety of grains including oats, barley, rye, and triticale. He ground grains for a herd of beef cows and four hundred pigs with one of the first grinder-mixers in the region. Francis was also one of the first people to combine shelled corn in the Champlain Valley. His good fortunes came to an end in 1976, however, when a barn fire destroyed his equipment and grain-processing infrastructure. This setback, coupled with a body crippled from his old wartime injuries, was the deciding factor in Francis's decision to leave the grain business. It was just about this time that I met him for the first time when he was giving a workshop at a NOFA conference in Plymouth, Vermont, and he has been a mentor and a friend to me ever since.

While our country and region modernized throughout the 1950s, a few exceptional individuals bucked the trend toward more convenient processed and packaged foods. Vrest Orton was one of the founders of *Vermont Life* magazine, and proprietor of the Vermont Country Store in Weston. His propensity for all things nostalgic and "old Vermont" propelled him to become a promoter of flint corn, johnnycake, and unrefined whole grains. In 1951, Orton's wife, Mildred Allen Orton, published the now classic *Cooking with Whole Grains*, which encouraged people to return to baking and cooking with whole meals and flours. The Ortons stone-ground a wide variety of whole-grain products and ingredients through their catalog and country store. As mentioned earlier, Herb Ogden of Hartland Four Corners was also a whole-grains enthusiast, and is best remembered for his wheat flour and flint corn meal, which he produced at his water-powered gristmill. Grain processing was nothing new in New England, but over time imported grains had replaced local supplies. The Chittenden Mill on the Browns River in Jericho, Vermont, began producing graham, white, and brown flour as well as buckwheat, rye, and cornmeal on a set of five buhr stones in 1855. In 1885, the mill was modernized and six high-speed roller mills were installed for the production of white flour. Chittenden Mills was the only producer of white flour east of Buffalo, New York, in the late nineteenth century. But it didn't last forever. The flour-milling operation closed in 1916; animal feed continued to be manufactured until the enterprise completely closed in 1946. Another grain processor of note at the time was the Maltex Company of Burlington, Vermont. In 1899, C. F. Van Patten first made a product called Malt Breakfast Food by combining toasted cracked wheat with freshly sprouted barley malt. The cereal became a national brand and was soon renamed Maltex. The ever-popular Maypo cereal was invented by the Maltex Co. in 1953, but shortly thereafter Vermont's beloved cereal company was sold to Heublein Inc., a conglomerate of food companies, and the Pine Street manufacturing plant was closed down and moved out of state. And so dairy farming achieved total prominence in the Northeast in the twentieth century, but the processing of grains for animal feed and human consumption continued to be important throughout the region.

A series of wet summers in the 1960s put a severe strain on the few remaining oat growers in the region. Wet weather at pollination and harvesttime continues to be the greatest challenge to anyone trying to raise a crop of grain, and this was certainly the case in the mid-1960s. As oat crops rotted and sprouted in the fields, and farmers stood by helpless to do anything but wait for dry weather, major discouragement manifested itself. Crop quality declined rapidly, especially if the grain had been flattened by thunderstorms. Farmers were using more nitrogen fertilizer by this time, and overfertilization with nitrogen produced lush tall crops of oats with weak straw. A good thunderstorm with high winds could literally level a field of tall oats in a matter of minutes; once the crop was flat on the ground, it was much more difficult to pick up with the combine. If next year's hay crop of clover and timothy had been planted along with the grain, it could be smothered and killed by a heavy thatch of lodged oats. For this reason, many dairy farmers planted their last crops of oats for grain in the 1960s.

The 1970s were the golden years for dairy farming in the Northeast. Many small farms persisted and were quite profitable because of an extended period of high milk prices, and as a result farmers were able to modernize and scale up to more modern, efficient machinery. Most dairy farms outfitted their hay balers with "kickers" that launched bales into a wagon pulled behind—labor savings was the order of the day. Big blue hermetically sealed silos first made their appearance at this time, as the A. O. Smith Harvestore airtight silo was promoted as the answer to all of farmers' feed problems. These big, tall blue steel and fiberglass tubes complete with the American flag logo began popping up on dairy farms all over the region. The taller models were designed for haylage, a drier form of grass silage, while shorter models were installed for the storage of high-moisture corn, something new on the dairy scene at this time. High-moisture corn was considered revolutionary at the time because the grain crop could be harvested much earlier at a moisture of 28 to 30 percent. (Corn was normally harvested at 20 to 24 percent moisture and artificially dried with propane to 15 percent.) Wet corn was put into the Harvestore silo to ferment in the absence of oxygen, which meant drying costs and late-fall harvests were avoided—saving time and dollars. High-moisture corn also proved to be an excellent feed for high milk production, and so grain corn production increased quite rapidly in the region. A few early adopters like Earl Bessette of New Haven, Vermont, built wire-mesh cribs for storing and air-drying whole ear corn. Farmers in just about every corner of Vermont began combining high-moisture corn and filling blue silos—Gilbert Boucher of Highgate and Fred Boyden of Cambridge were leaders in the usage of high-moisture corn in their respective communities, and many farmers simply bought corn in to fill their silos. The big three suppliers of high-moisture corn to northern Vermont at the time were Robert Mosher of Noyan, Québec, Hallie Thurston of Fryeburg, Maine, and Logan Brothers of Fort Plain, New York. And while Harvestore silos were quite attractive and revolutionary when they were new, their high cost was the eventual death knell for many dairy operations in the 1980s when milk prices dropped and payback became more difficult. These expensive units cost well over a hundred thousand dollars and came to be known as "blue shafts" because they drove many struggling farmers into bankruptcy.

Grain Production in Canada

Grain culture never disappeared or declined across the Canadian border to our north in Ontario, Québec, and the Maritimes. As a matter of fact, diversified, mixed farming endured much longer on the other side of the line, as competition from points west was never a factor in the agricultural development of eastern Canada. Western Ontario near Windsor and Detroit is the only part of Canada that has a climate comparable to the American Midwest, but there was simply not enough territory for the region to become a breadbasket for points east. The prairies of western Canada were settled in the late nineteenth and early twentieth centuries, but the prairie climate was too cool for corn production and the land too distant for easy, cheap transportation of grains eastward. In addition, a quota system for milk production was imposed on Canadian dairy farmers in the early 1960s, and overproduction of milk was avoided because each farm was limited in the amount of milk it could produce. Canadian dairy farmers traditionally milked fewer cows than their American neighbors because of the quota system, but their farms were similar in size to US farms. This freed up more of their land for the production of cereal grains because less hay was needed to feed smaller herds of cows. The typical Eastern Townships Québec dairy farm (across the border from where I live) of the 1970s milked between thirty and forty cows and practiced a crop rotation that included barley, oats, and clover hay, with older hay fields plowed in the fall and seeded to barley or oats the next spring. These grain fields were underseeded with timothy and clover, so that after the grain was harvested in August, next year's hay crop could be found in the straw stubble. Canadian milk quotas proved to be so valuable that many farmers have sold their quotas on a special government-controlled exchange for as much as half a million dollars. Grain production, however, is not controlled by quota in Canada. In the 1970s, many former dairy farmers from the highly productive breadbasket region of the St. Lawrence Valley lowlands sold their milk quotas and ventured into cash grain production. Indeed, corn and soybean production has become the order of the day in Québec's Montérégie region; grain elevator legs and big steel grain silos dot the flat landscape, making it look more like Illinois than northeastern North America. Seed companies began to produce shorter-season varieties of corn and soybeans in the late 1980s and early '90s; as a result, corn and soybean production has moved into shorter-growing-season areas north and east of the St. Lawrence Valley. So, while the climate may be the same, the agriculture is entirely different when you cross the border from northeastern Vermont into the Eastern Townships of Québec. Corn silage, hay, and forage crops predominate on the Vermont side of the border, while grain corn, soybeans, and lesser amounts of oats, barley, and hay crops are found on the Canadian side. This juxtaposition of two entirely different styles of agriculture in the same climate zone gives us hope that more grain could be grown on our side of the border.

The 1980s was the decade of farm auctions as the viability of dairy farming declined all over the Northeast, and farmers everywhere scrambled to cut input costs in order to survive the onslaught of low commodity prices. The National Farmers Organization (NFO) was organized as a cooperative venture in which agricultural producers pooled their resources together for increased power in selling their production or in buying supplies and inputs. (Collective bargaining was the motto of NFO, and a handshake was its logo.) A group of mostly French Canadian farmers led by Joe Rainville of Highgate, Vermont, decided to become more self-sufficient with feed grains, so they bought a small soybean roaster and began roasting soybeans imported from NFO farmers farther west. The roasted beans and other NFO grains were then distributed to farmers, who installed their own storage bins as well as grinding and milling equipment. Many farmers also invested in their own combines and began to plant barley to supplement the diets of their dairy cows. Some northern Vermont owners of combines at the time were Gilbert Boucher of Highgate, Jacques Couture of Westfield, Guy Robillard of Irasburg, and Gus Patenaude of Holland. But NFO farmers had a difficult time maintaining their edge in a marketplace controlled by much larger corporate players, and by the end of the decade many had given up growing and milling their own feed grains.

Infrastructure for the cultivation of grains is and has been very limited, especially in New England. However, a very small group of enterprising and enthusiastic individuals has kept the torch burning in just about every part of the region throughout the 1980s and '90s. In Aroostook County, Maine, potato farmers began switching some of their grain acreage from oats to barley to supply dairy farmers farther south with feed grains. Elsewhere in central Maine, producers like dairy farmer Bussey York of Farmington have moved from dry bean production to corn, soybeans, and cereals. Neil and Vern Crane, potato farmers from Exeter, Maine, began growing about eight hundred acres of corn for grain in the late 1970s, and potatoes, oats, and shelled corn have been common along the Androscoggin River near Bethel, Maine, and the Saco River near Fryeburg, Maine, for years. Harry Records of Exeter, Rhode Island, has continued the tradition of growing Rhode Island Whitecap, an heirloom flint corn for johnnycakes, and the Drumm family, owners of the Kenyon Gristmill in Usquepaugh, Rhode Island, has been milling this same white flint corn into meal since 1696. Pockets of grain growing have persisted in Massachusetts and Connecticut as well; rye cover crop seed and grain corn have always been well suited to the Connecticut River Valley and have been raised there for many years.

Organic Grains in Vermont

The Vermont grain experience in recent years has been very similar to that in the rest of the Northeast. A few individuals have never stopped growing grain as part of their crop rotation, such as potato farmer Burt Peaslee of Guildhall, Vermont, who always grew his own rye seed as well as barley and oats to sell to neighboring livestock producers. Toward the end of his life, Burt experimented with soybeans in the very cool and short growing season of the northern Connecticut River Valley where he farmed. As discussed earlier, a number of enterprising individuals began growing both dry and high-moisture corn for their dairy herds, although lack of harvesting and storage infrastructure has posed problems for many, and only the most serious and dedicated corn growers have endured. Former dairyman Ted Grembawicz of Clarendon, Vermont, has turned his Otter Creek dairy farm into a major producer of dry shell corn, soybeans, and wheat; Bourdeau Brothers, a feed and fertilizer business in Sheldon Junction, Vermont, has become the largest player in the business, planting and harvesting six thousand acres of soybeans and grain corn in northwestern Vermont and northern New York. Giroux Brothers, just across Lake Champlain at Chazy, New York, is another grower of five thousand acres of corn for its large poultry operation, and several other farms in Franklin County have also entered the cash crop realm. In the mid-1990s, Eric Dandurand sold his cows and transformed his Morses Line dairy farm into a cash grain operation and processing center complete with grain elevator legs, drying facilities, and tens of thousands of bushels of grain storage. Eric grinds and mixes feeds for area dairy farms in addition to custom grain drying. Ten miles to the south in St. Albans Bay, Jeff Boisonnault has also installed a large grain dryer and storage bins to handle his eight hundred acres of grain corn. Blessed with good soils and a great climate, the Champlain Valley of northwestern Vermont has become home to several very large-scale commercial grain-growing operations. The size and scope of this very chemically intensive agriculture is impressive in its own right. It is far from organic, but worth mentioning as an example of the potential for local grain production.

Organic grain production in Vermont and the Lake Champlain region traces its roots back to pioneering farmers Francis Angier and Dick Sherman, although of course before the advent of chemical fertilizers: In the late 1940s everything was "organic." In the late 1950s and early '60s, *organic* was a term and practice discussed only in Robert Rodale's *Organic Gardening* magazine. These early practitioners made the choice not to use herbicides and other agricultural chemicals, and Francis Angier intercropped clover with his wheat for fertility and weed control.

The next generation of organic grain growers began as homesteaders and back-to-the-landers. A small group of Northeast Organic Farming Association members started the Amber Waves Grain Cooperative in the Brattleboro, Vermont, area in 1975; plans were made to share land, equipment, and knowledge among participating individuals. John Melquist was harvesting his first cereal grains in Vershire, Vermont, with a horse-drawn reaper at this time, and Jake Guest and the group at the Wooden Shoe Farm commune in Canaan, New Hampshire, grew and marketed flint corn during this same period. The Bouchard family of Morses Line, Vermont, was still harvesting their oats with the grain binder and threshing machine during the mid-1970s, and I grew my first wheat, flint corn, and barley in Westfield in 1977. Elsewhere in Orleans County, Vermonter Milton Hammond was growing wheat and buckwheat on the shores of Lake Memphremagog in Newport Center, George Crane was raising wheat and carrots in Craftsbury, and David Allen was harvesting oats in Greensboro. A few new Champlain Valley growers appeared in the mid- and late 1980s. Ben Gleason began raising wheat on his homestead in Bridport in 1983, and Ken Van Hazinga started cultivating a variety of grain crops in Orwell a few years later. These were the folks who set the stage for a movement that would still be twenty more years in the making. Very little advice was available about organic practices from any source other than earlier generations who had grown up farming in a world without chemical inputs.

Official organic certification did not come to Vermont until 1984. NOFA had begun discussing the issue as far back as 1977, but it wasn't until a chapter of the Organic Crop Improvement Association (OCIA) was founded in 1985 that official organic certification came to the region. The organization very quickly morphed into a completely local group known as Vermont Organic Farmers (VOF), which focused mainly on vegetable production. Certified organic grains were not really in high demand at the time save for a few large-end users like O Bread of Shelburne, Vermont, Borealis Breads of Waldoboro, Maine, and Paramount Bakery of Sutton, Québec—Whitmer wheat from Montana was the most popular wheat at the time. The Champlain Valley Mill in Westport, New York, became a major supplier of whole wheat flour in the mid-1980s, and Robert Beauchemin also began milling his homegrown wheat in Milan, Québec, establishing the "Milanaise" brand that would eventually come to dominate the Québec market. Elsewhere in eastern Canada, two other organic grain mills were just getting started in the 1980s. In 1982, Stu Fleischig established the Speerville Flour Mill in Speerville, New Brunswick to supply the Maritimes and much of Maine with organic flours and meals. Hubert Lacoste restored an old water-powered mill at Ste Claire de Dorchester in Québec's Beauce region at about the same time. Hubert bought his organic wheat by the train carload from western Canada and was an early supplier of whole wheat pastry flour to Fiddler's Green Farm in Belfast, Maine, one of the first small-scale manufacturers of organic breakfast cereals and pancake mixes in the Northeast. Smaller, more localized players could be found throughout the region as well; during the 1980s at Butterworks Farm, we grew an acre or two of wheat each year for ourselves and our local community and about twenty acres of oats and barley for our dairy herd. Ben Gleason began selling his whole wheat flour at the Middlebury Natural Foods Co-op at about the same time, and in Penobscot, Maine, on the Blue Hill Peninsula, both Dennis King and Paul Birdsall were growing small acreages of cereals. My old friend Mark Fulford of Monroe, Maine, affectionately referred to all of us small-scale grain producers as "grainiacs." This we knew: There was something extra special about producing our own bread directly from the earth.

Organic food production went from fringe to mainstream in the 1990s. The turning point came when CBS's *60 Minutes* program highlighted the possible carcinogenic effects of alar, a very popular apple insecticide, at which point organic produce consumption shot up wildly and many newly certified organic vegetable farmers entered the arena. The second major boost to the organic sector came in 1994 when Monsanto released BST, also called rBGH or Bovine Growth Hormone, as a means to increase milk production in dairy cows through regular injection. Holistic consumers balked and began buying organic milk and dairy products en masse. Organic dairy was here to stay, and as a result some very large corporate players began to get into the business. Because organic dairy cows needed organic grain, a new era in organic grain production began in the mid-1990s. Peter Flint of the Organic Cow was instrumental in transforming Windsor County Feeds of South Royalton, Vermont, into the Vermont Organic Grain Company in order to supply dairy rations to his ever-increasing pool of organic dairy farmers. Large amounts of feed grains were transported into Vermont from New York State, just as had been done in the days of the Erie Canal, and New York producers Bob Crowe of Canajoharie and Klaus Martin of Penn Yan became major suppliers of organic corn, soybeans, and cereals; Bob Crowe actually delivered ground rations by the bag to organic livestock producers all over New England. In addition, Klaus Martin established Lakeview Organic Grain and became a major producer, broker, and organic leader in the Finger Lakes region. Les Morrison, a very practical and enterprising individual from Peacham, Vermont, and Monroe, New Hampshire, was beginning his own personal odyssey with grains during this same period. In 1983, Les began growing oats and triticale on some old mountain pastures with very good success. With a couple of grain silos and a tiny electric hammermill, Les began delivering grain to two farm customers, and in 1988 he began operating a very small conventional feed mill on a rail siding in Barnet, Vermont. With one delivery truck and a tiny customer base, Morrison's Custom Feeds was born. A quickly growing feed-milling business soon occupied most of Les's time, but he did find the time to experiment with lupine beans as a protein source before he stopped growing grains in 1993. In 2001, Morrison's became a certified organic processor and a major feed grain supplier to a rapidly growing organic dairy sector in Vermont, New Hampshire, and Maine. One other pioneer in grains at the time was Murray Manley of Crysler, Ontario. I bought roasted soybeans from Murray's Homestead Organics in the early 1990s; Murray's son Tom took over the business shortly thereafter and built his own feed mill in eastern Ontario. Homestead was a major reliable supplier of organic grains to livestock producers until September 11, 2001, when border restrictions were increased. The organic feed grain industry really came into its own in the '90s, and clearly New York and Canada were the main centers of production.

Grain growing in Vermont got its first major push in the 1990s with the establishment of Eric and Andy's Homemade Oats in Cabot. In 1991, Andy Leinoff, a semi-retired investment analyst, and Eric Allen, a carpenter, set to work turning Andy's barn into an oat-processing facility, and with the help of local technological wizard Carl Bielenberg they designed and installed a complete system for cleaning, dehulling, toasting, steaming, and flaking oats. Eric and Andy's progress was impressive; in five months' time, the mill was complete and ready to operate. Oats were elevated forty feet to the top of the building for cleaning and then moved downward by gravity to be hulled. Oatmeal was made on the first floor with the help of an old Wolfe cast-iron roller mill. (Unfortunately, a lot of their machinery had to be replaced with newer, more expensive pieces.) Eric grew quite a few oats in the summer of 1991 on Andy's Cabot Plains farm, and Glen Burkholder, a Mennonite from Wolcott, was enlisted as a custom combine operator to harvest oats for Eric and several other area growers. All went reasonably well that first summer, and Eric and

Andy's Oats debuted right around Christmastime. Vermont's first homegrown oatmeal was packed in small white cloth bags notable for the back-to-back silhouettes of Eric and Andy drawn in the same style as the Smith Brothers' cough drop box. Sales were strong and impressive despite the fact that the oatmeal cost twice as much as regular store-bought fare. The little company got a big break in 1993 when they appeared on the QVC home shopping network. Mail-order sales went through the roof. Shortly thereafter, Northeast Cooperatives, a wholesale natural foods distributor, began selling forty-pound bulk bags of Eric and Andy's Oats. At this point, there was some trepidation about growth. Andy turned down several large supermarket chains because he was worried about supply issues. More local farmers were enlisted to grow oats, and Glen Burkholder stayed pretty busy for a couple of seasons combining oats all over the region. Participating farmers had difficulty producing high-quality, heavy-test-weight oats because knowledge and skill around grain culture was pretty limited at the time. Eric and Andy ended up having to buy organic oats by the train carload from Manitoba, which was frustrating and troublesome since the original idea had been to produce a Vermont product. Production continued for a few more years, but eventually ceased at the end of the decade. In January 2000, the pair decided to call it quits and shut down Eric and Andy's Oats.

Several other Vermont organic growers of grains made notable contributions in the 1990s. Skip Sheldon, a McGill-trained doctor of pathology, began sheep farming on Shelburne Point during the 1980s. Skip first began growing hay and oats for his sheep, but soon became a major producer of soft white winter wheat. Another farmer, Tom Kenyon, sold his dairy cows in Monkton, Vermont, in 1988 and shortly thereafter went to work for Skip as his field man. Within a couple of years, Skip's Aurora Farms was planting and harvesting three hundred acres of pastry wheat a year on rented land in Shelburne, Charlotte, and Ferrisburg. Summer visitors to Shelburne Farms would marvel at golden fields of wheat in late July and early August. Aurora Farms had grain storage bins on Greenbush Road in Charlotte and at the Bostwick Farm in Shelburne; virtually all of the wheat was sold to Champlain Valley Milling across the lake in Westport, New York. In 2003, Skip Sheldon passed on the farm and grain business to Tom Kenyon. Tom had a difficult first year farming on his own when his newly acquired barn burned in May 2004, but has since rebuilt his grain facility and continues to be one of Vermont's major grain producers, cultivating eight hundred acres of hay, wheat, corn, and soybeans.

Jim Geer, a veterinarian from Connecticut, bought his Connecticut River farm in Windsor, Vermont, in 1979. The farm had once belonged to Bob Bartlett, who had become a cattle hoof trimmer after leaving dairy farming, but Jim was fortunate enough to hire Bob as his field man and the two of them grew conventional grain corn throughout the 1980s. Low prices and the soil degradation caused by chemical crop farming convinced Jim to switch to organic farming in 1995, when he began growing hay, barley, triticale, and rye seed. Crop rotation and the addition of composted manure from a nearby goat operation really improved the soils at Jim's Great River Farm. He was able to produce some high-quality food-grade soybeans for export to Japan. Jim has learned to rotate hard red winter wheat and smaller acreages of organic grain corn with hay crops on his beautiful river bottomland. He has experimented with milling wheat flour and has continued farming for the past thirty years.

My own experience in the 1990s was interesting. Until 1990, I raised only cereal crops—barley, wheat, oats, and rye—and I bought whole corn and soybeans from Bob Crowe in New York State. The fact that I had extra hay to sell every year convinced me that I should grow less hay and more grain. In the spring of 1990, I plowed up some hay ground and planted ten acres of very short-season grain corn and five acres of soybeans. To my amazement, both crops matured on my cool hilltop farm. I was able to combine the beans with my 1958 Oliver 25 self-propelled combine. I also found an inexpensive New Idea one-row corn picker and built a long, narrow corncrib to dry and store my harvest of ear corn. The harvest was completed on November 7 that year. (On the very next day, we received two feet of snow and lost our power for five days!) The addition of ear corn and soybeans increased my yearly acreage of grain planting from thirty to fifty acres and made me totally self-sufficient in cattle feed. In 1994, after four years of growing corn on my home farm, I had the chance to rent some land in North Troy, which is ten miles north of my farm and eight hundred feet lower in elevation. It was considerably warmer down along the river, and my corn crop grew like blazes that year; I had found a new and better microclimate less than a fifteen-minute drive from my home farm. In the fall of 1994, I purchased a three-row corn head for my recently acquired John Deere 3300 combine, which would allow me to harvest the crop as shelled corn. I also purchased a small-batch grain dryer from Burt Peaslee and proceeded to install propane tanks for a fuel source. Beginner's luck was with us that first year—we harvested and dried thirty acres of corn and stored it in a hundred-ton metal grain silo that I bought from Jeff Naylor of Waterville, Vermont. But several individuals were instrumental in making this new adventure in corn growing a success—retired dairy farmer Louis Berthiaume from Westfield helped me with field work and silo construction, and Les Morrison loaned us his jacks for erecting the bin and helped us install the aeration floor. We also planted and harvested ten acres of soybeans that same year, and lo and behold, the corn and soybeans grown in the lowlands on alluvial soil yielded much better than they had at our higher-elevation home farm. This discovery of other more productive land with a longer growing season was a revelation to me. My grain acreage began to increase every year throughout the mid- and late 1990s, and I took over other pieces of land and even bought some river bottomland on the Missisquoi River right on the Canadian border. Average yearly grain acreage for Butterworks Farm climbed to between 125 and 150 acres during the decade of the 1990s.

Growing Grains to Offset Feed Costs

The high cost of organic feed grains was and still is a major concern for organic dairy producers, so as the organic dairy sector grew, many farmers began working grains into their crop rotations. Spencer Aitel, from South China, Maine, was one of the first dairymen to do so in his state, and concentrated on cereal production, raising oats, barley, and rye for both seed and feed. In nearby Sydney, Jeff Bragg of Rainbow Valley Farm had always grown grain corn for his conventional herd; after transitioning to organic production, Jeff found that he was able to produce organic grain corn quite successfully. Henry Perkins of Albion bought a combine during the same period and was soon feeding homegrown corn, barley, and soybeans to his cows. In Vermont, dairy farmers were also trying innovative ways of reducing feed costs by raising their own grain. Vince Foy of North Danville contracted with Brian Pillsbury to custom-raise fifty acres of high-moisture ear corn on the Winooski River on land rented from an electric utility, and James Maroney had successfully raised corn, soybeans, and cereals at Oliver Hill Farm in Leicester until fire claimed his dairy barn and put him out of business in 1994. Elsewhere in the Champlain Valley, an informal alliance developed between dairy farmer Joe Hescock of Shoreham and neighbors Ken Van Hazinga and Ben Gleason. Joe was and still is a ready market for wheat, soybeans, and other cereals that don't quite make the grade for human consumption. Joe also began picking and grinding ear corn for his one hundred milkers around 1998, and Green Mountain Mills began operations in Pittsford the same year, where owner Jez Harrington produced a full line of flour products milled mostly from western wheat. This much was obvious: The decade of the 1990s saw tremendous expansion in organic food production, and the foundations for organic grain production had been laid by a core group of individuals who demonstrated to others that it could be done here in our region.

The (Re)Birth of the Local Food Movement

The catchphrase for the first decade of the twenty-first century has been *local food*. In Vermont, renowned ethnobotanist and Arizona author Gary Paul Nabhan was invited to Shelburne Farms to speak about his book *Coming Home to Eat: The Pleasures and Politics of Local Foods* in February 2002. Gary told the story of eating foods sourced from within 250 miles of his home. At about this same time, Carlo Petrini was developing the concept of "slow food" as an alternative to the overprocessed and tasteless fare so popular and pervasive in modern industrial societies. Soon after, local Slow Food groups called "conviviums" began popping up all over the country, and these local food groups would have dinners made exclusively with locally sourced ingredients. The word *localvore* very quickly became part of the lexicon, and localvore challenges became standard fare as groups of people banded together and vowed to eat only food produced within one hundred miles of their home. In 2008, the Vermont Sustainable Agriculture Council released Ginger Nickerson's study, *Vermont's Local Food Landscape: An Inventory and Assessment of Recent Local Food Initiatives*, which was presented to the Vermont legislature with much fanfare. As consumption of local foods began to achieve a newfound importance, farmers were relegated to hero status. Locally produced grains held the same importance as vegetables and dairy products, and the first hint of a strong demand for grain products came at a meeting between localvore groups and a handful of Vermont grain growers held in the basement of the Kellogg Hubbard Library in Montpelier late in the winter of 2004. The meeting was quite educational for both groups; the localvores left with tempered expectations of just what they could expect in the way of a predictable supply of food grains from the growers, and our little group of grain producers made it very clear that we needed a fair price for our production as well as market guarantees. We were also amazed that after all the years of competing with cheaper mass-produced grain imports from the Plains and the Midwest, the local community wanted to buy what we were growing at a price far above the standard commodity level. After this meeting, several of us decided right then and there to increase our spring planted acreages of wheat and other grains for human consumption. This was the beginning of a new era—grain farming had regained its rightful place in Vermont agriculture.

Maine experienced its own grain renaissance during the same period. Way up north in the Aroostook County town of Linneus, Matt Williams was retiring from a career in extension and embarking on a new one as a grain farmer. In 1997, he harvested his first crop of barley and sold it to the Vermont Organic Grain Company. For the next several years, he grew both organic barley and soybeans, selling his crops to newly transitioned organic dairy farmers farther south in central Maine. In 2000, Jim Amaral of Borealis Breads put out a call for Maine wheat to supply his successful and rapidly growing bakery. Matt heard the call and began growing wheat, and he erected some storage bins and organized a group of five farmers to plant two hundred acres. The first year's crop was milled into flour at Morgans Mills in South Union, Maine, and Speerville Flour Mill in New Brunswick did the milling the second year. In 2002, Matt was unable to find anyone with enough time or mill capacity to grind his wheat crops so he decided to build his own grain-milling and -processing facility, a project that took two years to complete and came online in 2003. Matt built a building on his farm and installed a thirty-inch Meadows stone mill. He also put up two more large grain silos, doubling his storage capacity. For the next three years, Matt went it alone, seeding and harvesting 160 acres of wheat in an average year. In 2006, Matt made the decision to become more of a miller and less of a grain farmer, so he recruited six additional growers in the southern Aroostook area and began to develop his Aurora Farms brand of wheat flour and rolled oats. Matt's milling and flaking operation produces five to ten tons per month of finished products from an annual production of 200 acres of oats and 250 acres of wheat.

Perhaps the greatest boon to the rebirth of grain culture in northern New England has been the return of agronomist Heather Darby to her home state of Vermont. Heather grew up on a dairy farm in the town of Alburgh in the Champlain Islands, and as a young girl was always interested in crops. Heather studied agronomy at the University of New Hampshire and University of Wisconsin; she finally received her PhD from Oregon State University in 2002. At the age of twenty-nine, she began working as a crop specialist for University of Vermont Extension in March 2003—and Vermont agriculture has been changing for the better ever since. Heather's extreme passion, intelligence, and ability to get the job done have allowed her to break down old paradigms and promote change on the agricultural scene. Upon her arrival in the spring of 2003, she took an immediate interest in the state's small community of grain growers, both organic and conventional. I met Heather soon after she started her job, and we clicked immediately—here was someone from extension who new something about grains, seed saving, and soil fertility issues. Over the course of the summer and fall of 2003, she visited my farm frequently for consultations and "crop tours." Our very first on-farm grain meeting was held at Ken Van Hazinga's Mount Independence Farm in Orwell in September 2003, when Ken showed us his seed-cleaning operation and his crops. In 2004, Heather organized the first in a series of grain meetings at the Grange Hall in Bridport, Vermont, and in March of that year thirty or more people turned out to hear retired UVM agronomist Winston Way talk about his experiences with grains during his forty-year career as the university agronomist. These yearly grain meetings have continued right up to the present. Grain conferences in 2006 and 2007 were held at the Vermont Chamber of Commerce building in Berlin, but by this time the standing-room-only attendance approaching one hundred people was stretching facilities to the limit. At the second Berlin meeting, a straw poll was taken and it was decided that we would form some sort of an officially sanctioned grain growers organization. We formed the Northern Grain Growers Association before the next year's meeting at Vermont Technical College in Randolph. With Heather's help and support from UVM Extension, our organization has thrived; we have been able to sponsor at least three on-farm grain-growing workshops each season. Wheat culture has been the primary focus, but we have also talked about barley, oats, and spelt, as well as seed cleaning and plant breeding. The most recent Northern Grain Growers Association annual conference was held at the Davis Center at the University of Vermont, and close to two hundred people attended to take part in baking demonstrations as well as workshops on soil carbon, on-farm animal ration manufacture, and flax growing.

A Growth Spurt in Research and Community

Major advances in research and education have also taken place in New England in the last few years. In 2007, I teamed up with Heather Darby and wrote a Sustainable Agriculture Research and Education (SARE) grant to grow, restore, and breed heirloom varieties of spring wheat. I was awarded the grant and began to grow out and increase a very small amount of the seed of nineteen old-time varieties of wheat. Breeding and variety-crossing instructions were provided by Dr. Steve Jones, an organic wheat breeder from Washington State University, and Steve actually helped us cross heirloom varieties with more modern cultivars in the hope of developing wheat varieties that would be more suited to the climate of northern New England. The original SARE project lasted two years, but has taken on a life of its own and continued right up into the present. We have developed several promising new varieties of wheat and restored many of the original wheat heirlooms to the point that larger acreages of these varieties are being planted. After four years of seed increases and small plot work, we were finally able to taste and compare the difference between many of the heirloom varieties; in September 2010, artisan baker John Melquist baked bread samples from five different wheat varieties and presented them for a public bread tasting at City Market in Burlington. Brent Beidler of Randolph, Vermont, also received a SARE grant in 2006 to cultivate and process his own Japanese millet seed for his grazing and forage needs on his dairy farm, which has been a very successful venture as the sale of millet seed to other farmers has provided him with an extra income from his farm. Eli Rogosa, a wheat aficionado from Maine and Massachusetts, also started her Northeast Organic Wheat Project about the same time. Eli has a passionate interest in very old European wheat varieties as well as ancient grains like emmer and einkorn, and worked to provide many farmers with small amounts of heirloom winter wheats for trial and increase. After four years of seed increases, heirloom wheats have reappeared in our region; many of these grains are noticeably superior in flavor and baking qualities.

Perhaps the most important indication that the culture of grains has come of age in the Northeast was the award of a $1.8 million Organic Agriculture Research and Extension Initiative (OREI) grant to the University of Maine and University of Vermont in 2009. Heather Darby teamed up with Ellen Mallory, a sustainable agriculture educator from Maine, to write the proposal, which was titled *Enhancing Farmers' Capacity to Produce High Quality Bread Wheat*. The project was envisioned as a partnership among farmers, millers, bakers, and other end users to foster all aspects of better wheat in our region. Numerous variety and soil fertility treatment trials were planted at Heather's research farm in Alburgh, Vermont, as well as at the University of Maine research farm in Orono, which will be conducted for four years. A group of twenty farmers visited grain producers and toured the Milanaise organic flour mill in Québec in the fall of 2009. The same group (of which I was a member) visited wheat producers and processors in Denmark in late October 2010. The experiences have been amazing, and I really believe that we are already growing higher-quality bread wheat in both Maine and Vermont as a result of the early lessons learned from this endeavor.

One other hotbed of local grain culture and production has been the central Maine region in and around Skowhegan. Longtime Farmington grain farmer Bussey (Herbert) York has led the way by transitioning first to organic dairy and then to organic grain production in the last few years. In early August 2007, a group of food activists in Skowhegan organized the first Kneading Conference in a church parking lot. Four portable wood-fired bake ovens were set up in a grassy park on the banks of the Kennebec River, and seventy-five people attended hands-on baking demonstrations as well as lectures on local grain production and food policy. I attended the conference in 2007 and 2008 as a speaker on farm grain issues and small-scale culture, and the camaraderie was extraordinary—and the food excellent. The Kneading Conference has grown to national prominence, attracting well over two hundred people to the last two conferences. More ovens and workshops have been added, and the venue has been moved to the Skowhegan fairgrounds to accommodate the increased attendance. Amber Lambke and Michael Scholz have been very active in the organization of this wonderful event from the very beginning—their mutual vision has extended well beyond the confines of this conference. Together they have envisioned and planned a local grain mill for grinding processing grains grown in the central Maine region. Recently, they purchased the old Somerset County Jail in downtown Skowhegan with plans of rehabbing the monolithic structure into a food-based community center and grain-processing center. Amber and Michael worked nonstop to obtain the funding to make this project happen, and so far it's been a success. This project exemplifies the possibilities for grain production in local economies all over the region.

Grain farming on both a large scale and the homestead level is nothing new in this part of the country. Although it has materialized and disappeared over the years, the legacy is there. Native Americans and early settlers grew grains out of pure necessity, because they needed to feed themselves and their livestock. But while westward expansion and abundant resources changed everything—to the point that localized grain production for local markets all but vanished—recent economic and cultural circumstances have encouraged a new regionalism and the importance of self-reliance. Finite energy reserves and questions about a centralized industrial food system have pushed food security to the forefront as an important consideration in these times. The rebirth of grain growing in our part of the country is taking place in this context. Not only does it make economic sense, but it's fun and satisfying to produce your own bread, cereals, and livestock feed. My hope is that this book will make the task easier and more straightforward.

CHAPTER TWO
Soil Fertility Considerations

A rich heritage of grain culture exists in our region. There are a few basic considerations and requirements when you're thinking of growing your own grains. Land, machinery, and knowledge are all essential. Couple these ingredients with some good advice, willpower, and beginner's luck, and you are on your way to being a grain grower. First, it's important to consider scale; grain, by its very nature, is bulky and lower in value than other more specialized crops, which gives rise to mass production and large scales of operation. Today's average producer of commodity grain crops farms a thousand acres or more to make the same living that his grandparents made on two hundred acres two generations ago. So-called mixed farming, where farm-raised grains were fed to cattle, hogs, and chickens, has been replaced by the cash grain system, and the rural fabric of much of our country has unraveled as small-scale animal agriculture has given way to gigantic concentrated animal feeding operations (CAFOs). In this industrial system, grain is harvested by massive combines and trucked away to nearby grain elevators owned by multinational corporations; crop diversity is an unwanted burden to an industrial agricultural system that is fueled primarily by corn and soybeans. Ninety-five percent of all soybeans and 85 percent of all corn crops are genetically modified, and seed sovereignty is increasingly in question as fewer and fewer large corporations come to dominate the trade. This sad state of affairs certainly poses challenges to anyone who cares about the environment or is trying to live in harmony with nature. It does, however, present us with a unique opportunity to do something interesting and different in our own particular bioregion. We are blessed in the Northeast for so many reasons. For one, we lack the huge expanses of land that make industrial agriculture so well suited to the Midwest and the plains states. Plus, a very willing and open-minded population inhabits our towns and cities, and locally produced foodstuffs are now more popular than ever. Indeed, small-scale agriculture has been in decline and needs a boost, and people are hungry for meaning as well as food. Organic grain production provides opportunities for both farmers and gardeners to expand their repertoires; dairy farmers can very easily add oats, barley, and wheat to their forage crop rotations, and small plots of any sort of grain can be intensively grown in the home garden.

Information about organic grain growing was pretty sparse when I first became interested in growing my own cereals in the early 1970s. Having come to farming as a romantic with a background in the history of agriculture, old farming textbooks were my first source of information about soils and crops, and I was able to purchase many old agronomy texts from the late nineteenth and early twentieth centuries at used-book stores. Studying farming practices from this particular time period provided me with good sound advice because agriculture was still being practiced in harmony with nature; chemical fertilizers (or manures as they called them) were relatively few, and toxic inputs were not yet developed. By the first decade of the twentieth century, mankind had probably reached the pinnacle of knowledge in crop science that was holistic and friendly to the natural environment. MacMillan and Orange Judd Company, both of New York, were the two primary publishers of agricultural texts at the time, and Liberty Hyde Bailey, dean of agriculture at Cornell, was the editor of the *Rural Science* series, *Cyclopedia of Agriculture* (four volumes), and *Cyclopedia of Horticulture* (six volumes), all published by MacMillan. My favorite volume from the time is *The Cereals in America* published in 1904 by Thomas F. Hunt, a professor of agronomy at Cornell. The book contained everything I needed to know as a novice intent on growing my first crops of grain, and I would recommend this book and all others like it to anyone searching for good solid information on any agronomic subject. Occasionally, you might stumble across one of these volumes in a used-book store, and the libraries of most land grant colleges and universities still have many of these old agronomy books on their shelves—simply go to the 630 section in the Dewey decimal system and you will find all of these old books in one place. Certainly, farming in the early twentieth century has a lot to offer us now, one-hundred-plus years later.

A Brief History of Chemical Fertilizers

In its quest for high levels of production, modern agriculture has forgotten and ignored many of the tenets of basic farming knowledge contained in the agronomy textbooks of yesteryear, as quantity is much more important than quality in these modern times. Ever since German chemist Justus von Liebig discovered chemical nitrogen and its importance as a plant food, soluble minerals have supplanted humus and organic matter as the basis of fertility in agriculture. According to von Liebig, plants need nitrogen in the form of ammonia to grow and synthesize protein, but where the nitrogen comes from is of no consequence according to the von Liebig view. This discovery has led to amazing gains in crop yields over the past 150 years. In the mid-nineteenth century, bat guano imported from South America was the first artificial "manure" applied to crops at the time, and nitrate of soda (Chilean nitrate) was the next big discovery, a salt-like substance mined in the Atacama Desert of northern Chile where the rain never falls; it contains 16 percent nitrogen. The history and development of high-yield-production agriculture is directly coupled with the introduction of stronger and stronger sources of mineral nitrogen—Chilean nitrate was followed by ammonium nitrate, also used in the munitions industry. Modern fertilizer production reached its zenith with the invention of the Haber-Bosch process in which urea could be made with natural gas and anhydrous ammonia for direct soil injection. This process—first used during World War II to harden the ground for airstrips in the jungles of the South Pacific—paved the way for higher crop yields in the 1960s. (Now a standard source of nitrogen for corn, it has also succeeded in hardening the soils of farm fields throughout the Corn Belt of the American Midwest.) In addition, nitrate and salt fertilizer use has increased steadily since the end of World War II, and excess ammonium nitrate from the munitions industry was channeled into fertilizer production, which paved the way for the Green Revolution that followed in the 1950s and 1960s. American agriculture has become totally reliant on the fertilizer bag, but the unforeseen consequence of this over-reliance on salt-based fertilizers has been pollution and environmental degradation. Nitrates have seeped into the water table, making drinking water unsafe throughout much of farm country. Perhaps the largest and best-known example of this pollution problem is the ever-growing dead zone in the Gulf of Mexico where the Mississippi River empties the effluent of America's productive heartland into the sea. We have learned that the increased productivity of modern agriculture has come with a price tag.

The forgotten element in this quest for higher crop yields and so-called increased productivity has been carbon. Justus von Liebig was a chemist, not a soil microbiologist; little did he realize that soil-applied mineral nitrates require the oxidizing action of soil bacteria and presence of carbon to make them available to plant roots. For every one part of nitrate, twenty parts of carbon are needed to make this process happen, and soil carbon is literally burned up and released to the atmosphere as carbon dioxide during the nitrification process. Sadly, the application of soluble nitrate fertilizers has reduced organic matter levels in American soils from a high of 5 to 7 percent one hundred years ago to a current average of less than 1 percent nationwide. But humus and organic matter ensure resiliency in our soils, and for every 1 percent increase in organic matter level a soil will retain an additional 160,000 pounds of water per acre. High-humus soils are like the lungs of the earth, allowing soils to retain moisture in times of drought and wick away excess moisture during rainy wet spells. Good aggregation or soil structure goes hand in hand with high levels of humus. A soil with good aggregation properties will very easily crumble into many small little pieces in your hands. This good "crumb structure" ensures that there will be plenty of pore spaces for air and water storage, and a healthy high-humus soil will have the ability to hold on to more mineral nutrients and be able to resist the forces of wind and water erosion. The carbon cycle is the foundation of organic agriculture, and it is entirely possible to achieve the same yields as the chemical guys in an organic system. We get there by working *with* nature to foster a soil that is alive and teeming with microbial life. To grow nutrient-dense, high-quality crops, we must have an abundant supply of carbon—nature's building block.

The Importance of Soil Fertility

Any exploration of soil fertility will take us into three interdependent and equally important realms—the chemical, the physical, and the biological. The chemical realm is an indication of what kinds of minerals are present in a particular soil and in what sort of concentration. This aspect of the whole picture is probably the easiest to understand and measure with a standard garden or farm soil test, where levels of calcium, magnesium, phosphorous, potassium, and nitrates are indicated in parts per million or pounds per acre. With this information and accompanying recommendations, decisions can be made about how much and what types of soil amendments to apply to a specific plot or field. The physical realm refers to a soil's actual physical structure. What kind of texture does it have? Is it clay, silt, sand, or gravel? Is it poorly or excessively drained? Compaction could be a problem because of low organic matter and poor aggregation. A soil's physical qualities are directly related to its humus content and the cultural practices of its human stewards. Good diverse crop rotations and frequent applications of compost and manure will preserve and enhance the physical structure of a farm's soil. The biological realm of soil fertility is all about microbiology and the billions of little organisms that inhabit our soils. Fungi, bacteria, actinomycetes, nematodes, and algae all work in conjunction with the soil's humus and mineral fractions to feed themselves and the resident plant population. Mycorrhizal fungi colonize plant roots to symbiotically consume plant sugars while they break down soil minerals and make them available to the root zone of the host plant. Soil microbiology is the least understood aspect of soil science; these little beasties make it all happen, and we only vaguely understand their function. What we do know is that microbial populations are more diverse and effective in an environment high in humus and carbon, and a holistic approach to soil fertility that considers the interrelationships among chemical, physical, and biological soil properties is a prerequisite for success in an organic farming system.

Until relatively recently, soil mineral fertility was only charted in terms of three elements—nitrogen, phosphorous, and potassium. This "NPK" approach is still pretty much standard fare in today's world of conventional agriculture. Soil tests indicate deficits in these three major elements. Fertility requirements are determined as pounds to the acre of added fertilizers, and the soil is treated like a bank account to which deposits and withdrawals are made. As an aspiring organic farmer thirty-five years ago, I was confounded by the notion of how was I going to get enough nitrogen, phosphorous, and potassium for my crops from cow manure only; I had visions in my head of walking around the pasture picking up cow flops with a dung fork. It didn't seem practical—and this point was certainly driven home to me by my local extension agent, who really wanted me to use chemical fertilizer. *Organic Gardening* magazine wasn't much help. Several years later, I discovered *The Albrecht Papers* published by Acres USA, a collection of papers and articles written by Dr. William A. Albrecht between 1918 and 1957. Albrecht was chairman of the Plant and Soils Department at the University of Missouri for over thirty years and a keen observer of nature. He noticed that the very highest-quality wheat in terms of protein was grown in an area several hundred miles wide that stretched from the Panhandle of Texas northward to the Dakotas, an area that corresponded directly to the tallgrass prairie that once supported millions of bison. Dr. Albrecht theorized that very high-quality foods could be produced in this region because the soil fertility balance was very close to perfect. With a one-to-one ratio between rainfall and evaporation, soil minerals were neither leached nor overconcentrated, but as you traveled eastward from this imaginary line, rainfall steadily increased and soil acidity climbed. A westward journey from this line would take you into the very arid West, where soils were more naturally alkaline because evaporation exceeded rainfall. Simply put, these high plains soils had a perfect fertility profile and produced crops superior in nutrition and feed value. Dr. Albrecht then analyzed the mineral makeup of these soils. He found that negatively charged minuscule clay soil particles (also called colloids) had the ability to magnetically attract and hold on to positively charged minerals called cations. These ideal soils were "saturated" with 60 to 75 percent calcium, 10 to 20 percent magnesium, 2 to 5 percent potassium, 0.5 to 5 percent sodium, and 7 to 10 percent hydrogen. This concept of base saturation was very new and unfamiliar to me in 1977, but it gave me the hope and inspiration that I, too, could grow highly nutritious grain crops here in the Northeast without dumping NPK fertilizers on the land.

Managing Soil Fertility Organically

Organic soil fertility management for grains or any other crops begins with a holistic view of our total environment. Where is our field or plot located? Is it next to a river that floods often or on a cold windy mountaintop? Does the soil drain well or does it remain waterlogged for long periods of time? Good drainage is essential; some crops can tolerate wet feet more than others. For example, oats will survive in a soggy soil for a little while, but barley will shut down its growth habit in the presence of excess soil moisture. Well-drained loams produce the best grain crops. Clays, which have the highest density of colloids or soil particles, have the highest potential fertility because they have the most sites to attract and bond to calcium, magnesium, and potassium cations. The ability of a soil to magnetically attract and hold on to mineral fertility is called cation exchange capacity or CEC. Clay soils have the highest CEC levels; once charged up with minerals, they are hard to deplete. Sands, on the other hand, are composed of much fewer particles that are larger in size and much coarser in texture. For this reason, they drain well, but do not hold on to minerals very well at all. Silts are somewhere between sand and clay in texture and CEC levels. A silt loam is probably the most ideal soil there is for crop production, as silts drain well but still have enough texture to hold on to moisture and mineral fertility. A basic understanding of the physical structure and the cation exchange capacity of your soil will help you make decisions on how to best manage the mineral fertility of your future grain field. If your soil is quite sandy, you will have to apply soil amendments and rock powders like lime and rock phosphate more often in smaller doses. Sandy soils, by their nature, simply cannot hold on to large amounts of nutrients. Although it may not drain as well, heavier clay can accept much larger applications of mineral fertility. Knowing the physical properties and drainage potential of your soil is the first step in preparing to grow grain on your future field or plot.

The Soil Test

Once you have determined the suitability of a particular piece of earth for grain production, it is time for a soil test. But not all soil tests are created equal—we must remember that a soil test is just an analysis of a very dynamic environment at one moment in time. Begin by walking around your field with a shovel or soil probe and getting as many "plugs" as possible in the most representative manner possible. Avoid low-lying spots and other areas that don't seem characteristic of the rest of the field, and lay out your acreage in a grid fashion or zigzag your way through to obtain your soil profiles. Collect and mix together all of your plugs in a bucket to obtain a soil sample that you would consider representative. Separate fields or plots have different characteristics, so it's better to take three individual soil tests from three distinctly different fields than it is to mix them together for one composite sample. A general rule of thumb is to take at least three or four soil plugs from each acre. Once your sample has been collected and mixed together, it is time to put it in the little sample bag and send it off to the lab of your choice. There are many private soil testing laboratories in addition to services provided by state agricultural universities, and soil tests will vary in the amount and type of information provided. When the lab receives your sample, the first thing they will do is to make an extraction, where a specific amount of soil—say, a tablespoon—is mixed with a prescribed amount of distilled water. A very small amount of an extraction reagent (a few drops) is added to this soil solution and shaken in order to dissolve its minerals and prepare it for further testing. Extraction reagents are always some sort of acid. Acids by their nature release H+ hydrogen ions, and plant roots do the same thing. The acidic nature of the extraction reagent is basically mimicking the roots of plants in dissolving minerals from solution. Soil tests will also vary in relation to the strength of the extraction agent. Mild extractions like acetic acid will not dissolve as many minerals from the soil solution as a stronger reagent like ammonium acetate. Soil testing is a science unto itself and can only be briefly discussed here. Most soil tests done in the Northeast are medium-strength soil extractions like Morgan, Modified Morgan, and Mellich 1, and will provide you with a basic overview of the reserve fertility in your soil. Home and farm test kits like the LaMotte and Simplex outfits rely on much weaker extracts and provide you with more of a picture of what nutrients are available right here and now. Testing your own soil for available mineral fertility right on the farm is a fun and rewarding experience—sometimes that little bird's-eye view of what is happening right here and now is a valuable asset when you are trying to fertilize and plant a crop and don't want to wait a week or two for laboratory results from some far-off place.

SOIL REPORT

Job Name: **BUTTERWORK FARMS**
Company: Butterwork Farms
Date: 2/13/2012
Submitted By: _____

Sample Location			A	C&C	Field	1N	NCAE
Sample ID					1		
Lab Number			1	2	3	4	5
Sample Depth in inches			6	6	6	6	6
Total Exchange Capacity (M.E.)			14.39	13.24	11.67	10.71	13.79
pH of Soil Sample			6.40	6.40	6.30	6.20	5.50
Organic Matter, Percent			7.68	7.42	3.31	6.33	7.32
ANIONS	SULFUR:	p.p.m.	21	17	21	19	22
	MEHLICH III PHOSPHOROUS:	ppm	53	38	45	48	47
EXCHANGEABLE CATIONS	CALCIUM ppm:	Desired Value	1957	1800	1587	1456	1876
		Value Found	2192	1919	1719	1501	1594
		Deficit					−282
	MAGNESIUM ppm:	Desired Value	207	190	168	154	198
		Value Found	124	180	116	136	71
		Deficit	−82	−10	−52	−18	−127
	POTASSIUM ppm:	Desired Value	224	206	182	167	215
		Value Found	112	86	81	60	54
		Deficit	−112	−120	−101	−107	−161
	SODIUM:	ppm	21	18	19	17	17
BASE SATURATION %	Calcium (60 to 70%)		76.17	72.46	73.62	70.10	57.77
	Magnesium (10 to 20%)		7.21	11.30	8.28	10.59	4.29
	Potassium (2 to 5%)		2.00	1.66	1.77	1.44	1.00
	Sodium (.5 to 3%)		0.64	0.58	0.71	0.68	0.53
	Other Bases (Variable)		5.00	5.00	5.10	5.20	6.40
	Exchangeable Hydrogen (10 to 15%)		9.00	9.00	10.50	12.00	30.00
TRACE ELEMENTS	Boron (p.p.m.)		0.79	0.71	0.68	0.84	0.48
	Iron (p.p.m.)		197	206	200	236	186
	Manganese (p.p.m.)		38	23	35	43	14
	Zinc (p.p.m.)		2.52	1.78	1.8	1.92	1.03
	Aluminum (p.p.m.)		6.29	4.86	5.2	4.88	3.66
OTHER	Cobalt		784	790	843	815	921
	Molybdenum		0.41	0.37	0.4	0.52	0.26
	Selenium		0.15	0.07	0.06	0.1	0.04
	Silicon ppm		35.83	31.74	26.64	29.39	15.74
	EC		0.17	0.11	0.07	0.08	0.1

Logan Labs LLC

A complete and thorough soil test will help you with soil fertility management. IMAGE COURTESY OF JACK LAZOR, FROM LOGAN LABS

Reading the Soil Test

Whether you interpret the results yourself or you rely on the advice of a professional, a current soil test will give you the basic information on where to start with a potential grain crop. (A sample test results sheet from Logan Labs appears on page 26.) The first thing to consider is the soil's base saturation of calcium, magnesium, and potassium. If the percentage of calcium is lower than 70, calcitic or high-calcium lime should be added. If magnesium and calcium are both low, then dolomite or high-magnesium lime needs to be added. The calcium-to-magnesium ratio is essential for soil health and proper plant growth as calcium is the element that helps all of the other minerals mobilize. (Well-known soils consultant Gary Zimmer calls calcium "the trucker of all minerals.") Magnesium is more of a catalyst and is responsible for proper cell function. Limestone is slow to break down, so it's best to choose the right kind of lime in the very beginning. Overapplications of dolomite can send magnesium levels off the chart, which will cause "tightness" and poor drainage in a soil. On the other hand, excessive levels of calcium will make a soil too alkaline and prevent the uptake of other minerals. Phosphorous and sulfur are both anions or negatively charged elements. Phosphorous is responsible for root growth and early-season plant vigor. Sulfur should be present at least at the fifty-parts-per-million level. It is essential for the proper synthesis of nitrogen compounds into protein within the plant. Boron is more of a trace element and helps with calcium function. Both boron and sulfur are quite easily leached for most soils, and small amounts should be applied yearly to maintain a healthy soil mineral balance. Soil mineral fertility will be discussed at length throughout this book in relation to the individual grain crops, but at this point it is important to understand that a proper balance of calcium, magnesium, phosphorous, and a wide array of minor minerals is far more essential than the simplistic NPK accounting system practiced in conventional agriculture. Having a complete and comprehensive soil test in hand before putting any seeds in the ground is akin to starting a journey into an unknown territory with a good road map.

Soil Fertility as Part of a Holistic System

Nitrogen function, availability, and delivery are seen in a totally different light when the comparison is made between a conventional and an organic system of farming. Notice that there are no categories on the Logan Labs test for ammonium, nitrite, or nitrate. Total percent of soil organic matter is the only category that even comes close. Plants certainly need nitrogen to grow whether under organic or conventional management. Adequate nitrate nitrogen is supplied to plants through a dynamic relationship among respiring active soil microbial life, carbonaceous humic compounds, and balanced soil minerals all taking place in a friable and aerated soil medium. Crop rotation plays an extremely important role in an organic system as well. Lighter-feeding crops like beans and cereals, for example, should always follow heavy feeders like corn and wheat. Our atmosphere is 80 percent nitrogen, and legumes like beans, clovers, and alfalfa will "fix" atmospheric nitrogen into the soil with the help of rhizobium bacteria. Hay crops with plenty of alfalfa and clover mixed with the grasses play an integral part in a low-input organic crop rotation. A heavy-feeding crop with high nitrogen requirements can be grown without any N inputs (manure, compost, or fertilizer) on a plowed hay crop; while the plowed-down sod decomposes throughout the growing season, small amounts of nitrate are continually released by the soil microlife as they consume the hay crop residue. In an organic system, we do everything possible to understand and work in unison with nature's cycles. We are trying to build fertility while we farm, and we are not getting soluble nitrogen from a fertilizer bag while we consume precious energy and resources. The bottom line is that as we coax food from the land, we are also trying to take as much carbon dioxide as we can out of the sky and lock it back into the earth's crust as stable humus. I believe that as organic farmers, we can be extremely productive and heal the earth at the same time.

CHAPTER THREE
Getting Started with Tillage

You've found a plot of land next to your garden or out in your back pasture. What's next? Make some final determinations as to soil type and drainage potential. A few simple observations will tell you everything you need to know—grab a shovel and dig a small hole, and feel the dirt with your hands. Sandy soils will have a gritty texture that will abrade your skin as you rub them between your fingers. Silts are silk-like and will smear when pressed against a hard surface. Clay soils will be more difficult to penetrate with your shovel and tend to clump up into little balls, especially when they are damp. Clay is quite fragile and easily damaged by working it with tillage equipment when it is wet. The end result will be a collection of hard little balls that are almost impossible to work into a fine seedbed—and once the tilth of a clay soil is destroyed, it can take years to get it back. Tillage rule number one is to never work a field if it is excessively wet. To determine a soil's suitability for tillage, try this simple procedure. Take a handful of soil, squeeze it into a ball, and then relax your grip. If the ball of dirt falls apart, the ground is ready for the disc harrow or the field cultivator. If the ball remains round and intact, wait a little longer until things dry out a bit.

Another way to find out more about your particular soil is to pay a visit to your local county USDA Farm Service Administration (FSA) office. Here you will find the NRCS, also known as the Natural Resources Conservation Service. On staff will be several soil conservationists who can provide you with all sorts of soil information as well as assist you with numerous determinations for slope and erosion potential. Most counties in the Northeast have been surveyed and mapped for soil types and classifications, and there are hundreds of named soils like Rumney, Nicholville, and Marlow. The physical nature of each soil is described in detail for features like hardpans, texture, and the height of the underlying water table. The NRCS has an aerial map of each entire county, which is divided into quadrants delineated on a large wall map. Once you've found your location, a soils map can be printed out for your reference. These maps are quite detailed. It's quite surprising how many different soil types there are in close proximity to one another. The various soils are designated by letter abbreviations. For instance, Rumney might appear as RuA. The *A* designates the slope of a particular soil classification site. A is nearly level, while D is the steepest. The NRCS will also be able to provide you with an accompanying soils sheet for each particular classification on your land. These information sheets describe the physical characteristics and drainage as well as crop suitability for each particular soil type. Your county soil conservationist will be able to interpret this soils sheet and provide you with additional information and advice about growing grain on your particular piece of land. The FSA office is the USDA's outreach arm stretching into your community, and is the same outfit that administers crop subsidies and other federal agricultural programs. I have found lots of help here and have collected my fair share of farm payments, as well, but the paperwork involved is daunting and seems a bit overdone. It's also important to remember that what the government giveth, it also taketh away; every USDA farm payment for crops or conservation practices is taxable income and must be declared. Despite this reality, I would recommend an information-gathering expedition to your local county FSA office.

Choosing Your Crops

Next, let's consider what to grow. Cereals like oats, wheat, barley, triticale, rye, and buckwheat are all seeded in solid blocks, while crops like corn, beans, and sunflowers are planted in rows. Row crops offer an easier starting place for the homesteader or gardener because planting, cultivation, and harvesting can all be performed with hand tools and garden tillage can be accomplished with a rototiller or a spading fork. Seedbed preparation requires only a garden rake; planting is done with a hoe; and weeds can be removed with a rototiller, a hoe, or a hand-push-type cultivator. At harvest, ears of corn can be handpicked into a bushel basket and whole bean plants pulled up by the roots. Cereal grains, on the other hand, are a bit less straightforward to grow. These small grains can certainly be hand-scattered and raked into a well-prepared garden soil, but this process is more tedious and much less exacting than planting corn and beans in rows with the use of a hoe. Some homesteaders plant crops like wheat in very narrow four- to six-inch rows with the help of a hand-pushed garden seeder like a Planet Jr. or an Earthway. Each particular type of grain will be discussed in much more detail in later chapters of this book. Suffice it to say at this point that row-planted crops like corn and beans offer far fewer challenges to the beginner with limited time, tools, and experience.

Cereal grains like wheat, barley, triticale, and spelt can be planted in either the spring or the fall. Varieties planted in March through May are called spring grains, while cultivars seeded in late August through October are labeled as winter grains. Certain grains like oats are only spring-seeded, and rye is usually only fall-planted. There is one particular variety of rye called Gazelle that is spring-planted, but the seed is quite rare and very expensive. Winter wheats are classified as either hard or soft. Hard wheats are usually red, have higher protein and gluten levels, and are primarily used for bread flour; soft wheats can be either red or white in color and have much lower levels of gluten. (Whole wheat pastry flour for pies, cookies, and cakes is made from soft winter wheat.) Winter survival can be an issue with fall-planted cereals. Winter barley will do quite well in the mid-Atlantic region of Pennsylvania and Maryland, but is only a marginal crop when it is planted in the Northeast, because stands can succumb to winterkill if they're blown bare of snow and exposed to cold temperatures. Winter wheat stands a much better chance of survival here in our region, and fall-planted wheat is a natural in the milder climate of southern New England. Improved varieties have been developed in Ontario and Québec that perform quite well in northern New England. If you are dealing with heavy clay soils that do not dry out in early April, winter wheat is definitely a better choice for your farm than spring wheat. For the best success, spring wheat needs to be planted as soon as possible in the very early spring: Early planting ensures good weed suppression and a higher yield. If the ground is wet, however, early seeding is out of the question. Winter wheat usually goes in the ground during the third week of September when the earth is much drier. Weed control is much more of a sure thing with a winter grain. The crop germinates in late September and makes good growth in October and early November. Weeds will germinate, but not survive the winter. The wheat itself grows into dormancy and wakes up the following spring already four to six inches tall and ready to begin growing immediately, which gives it a very favorable jump start in the spring. In fact, crop yields can be double those of spring wheat, and weeds are generally not a problem. Spring-planted wheats do much better on lighter, better-drained soils where planting can take place when there is still snow on the edges of a field. In most cases, soil type and drainage will determine whether a spring or winter grain is sown.

Preparing the Soil for Planting

Tillage is the next step. Whether grains are planted in the spring or the fall, the land has to be prepared to accept seeds, and a nice friable patch of dirt, free of other plants and grasses, is needed in order that grain seeds may germinate, grow, compete with other plants, and finally ripen. If you're starting with a grass sod, you'll need to turn the ground over with a moldboard plow. Depending on your scale, plows can range in width of cut from one furrow up to six. A one- or two-bottom plow can be pulled by the small older tractors found on most homesteads or small farms, and larger four- and six-bottom plows are drawn by bigger high-horsepower tractors. A six-furrow plow with eighteen-inch bottoms turns over nine feet of ground on every pass through the field. Implements like this are commonplace on larger acreages where there is plenty of room to maneuver. I have owned a four-furrow International 720 semi-mounted plow for the past twenty years, and it suits the needs of my medium-sized operation just fine.

Plows by their very nature cut a slice of soil usually eighteen inches wide and flip it bottom-side up. Special cutting moldboards called clay bottoms will stand the furrow on edge instead of flipping it upside down. This will allow a clay soil to drain off excess water so the land will dry out faster for secondary tillage. The unique thing about moldboard plowing is that the soil within the flipped furrow remains intact in its inverted position. When I first started traveling around the Eastern Townships of Québec, I was amazed to see thousands of acres of land being turned over in October and November, and I soon learned that fall plowing was standard fare in an agricultural system where barley, oats, and wheat were seeded in the early spring. My first thought was that all of this bare dirt would erode throughout the winter, but it quickly became apparent that these plowed furrows were quite stable and would not wash away in the winter season because the soil's actual internal structure had not been disturbed. The surface of the plowed furrow would weather and break down a bit over the winter, making the soil quite friable and workable the following spring.

Fall plowing certainly gives the sod a head start on decomposition the following spring, and I found out the hard way about the benefits of fall plowing when I grew my first crop of wheat and barley in 1977. We purchased an old McCormick three-bottom trailer plow that spring and set out to turn over our six-acre field with the help of our old Farmall Super M tractor. These plows had a breakaway hitch that "tripped" every time the plow hit a rock. It seemed like we encountered a rock every twenty feet. When this happened, the plows would become disconnected from the tractor, and we would have to stop, back up, and rehook to the plows. Progress was pretty slow. It took weeks to finish the field, and we were late getting our wheat planted. We also noticed that lots of grass hay seemed to invade our wheat field after it germinated and the crop began to grow; this happened because we plowed in the spring instead of the previous fall. It was much more difficult to set the sod back when the soil hadn't had time to digest and mellow over the previous winter. This initial experience taught us quite a bit, and we soon became proponents of fall plowing.

Spring plowing certainly has its place, especially if you're planting a row crop. Corn, soybeans, and sunflowers are usually seeded in mid- to late May after the ground has warmed up, whereas dry beans need really warm soil temperatures and oftentimes are planted in the first two weeks of June. April plowing works quite well for these crops because there is a six-or-more-week interval between plowing and seeding, and during this period a plowed field can digest a good bit of its sod with help of secondary tillage equipment like discs and field cultivators. Row crops also have the advantage over cereal grains because they are planted in thirty-inch rows and receive two to four row-crop cultivations during their early growth period; any emerging grasses or annual weeds can be removed from the row crop in this manner. Once a cereal grain has emerged from the ground, any major intervention with cultivation equipment is no longer possible.

The Benefit of Cover Crops

Fall plowing for cereals doesn't work everywhere, as some soils are more prone to erosion than others. In a situation where the ground really needs winter cover, a cover crop can be sown in the late summer; we begin by plowing and planting our field in late July or early August. Cover crops have numerous benefits. For one, a rapidly growing broad-leafed cereal like oats is a great choice because it can gobble up left-over soluble mineral fertility and accumulate it in its plant tissue. Oats will grow well into November and early December, producing lots of plant biomass as stems and leaves aboveground and roots belowground. Eventually, the oat plant will be winterkilled, leaving the earth covered with a brown mat of dead plant mass and the soil profile filled with root biomass. The following spring, this carpet of winterkilled oats is tilled up, and grain is planted in much the same manner as the traditional bare-ground fall-plowed system. When working with a cover crop preceding a spring-planted cereal, it is very important to plant something that will winterkill. If a crop like winter rye is planted instead of oats, it will survive the winter and be extremely hard to eliminate from an early-planted spring cereal. Little is worse than having a bunch of volunteer rye in a spring wheat seeding.

The Dangers of Tillage

There is no doubt that tillage is destructive to the soil. Oxidation is the end result of most traditional tillage practices, because air is being injected into the ground, which promotes microbial activity. Soil life increases in activity in the presence of increased air in much the same manner as a fire does when air is provided, and the end result of all of this increased microbial activity is that some of the soil's carbon fraction or organic matter is burned up in the process. The by-products of this reaction are carbon dioxide, which is released to the atmosphere, and nitrate nitrogen, which is used by the plants for growth. As stewards of the earth, we need to seriously consider the consequences of our tillage actions. We certainly need not overtill the land, and replacing the spent organic matter with more crop residue for future reactions is essential and goes without saying. The addition of cover crops and well-made compost will increase organic matter and soil fertility over time as well.

Tilling Your Grain Plot

The moldboard plow is a bit impractical for primary tillage in a garden-sized plot. Rototillers are the most common tillage tool for small areas. Bearing in mind the potential destructive nature of the tillage process, exercise extra care not to overtill your tiny grain plot; there is no need to whip up garden soil into a fluffy whipped-cream state. Start early the previous summer and rototill only enough to get a shallow yet firm seedbed. Plant buckwheat as a cover crop and incorporate it before it sets seed; follow the buckwheat with oats planted in early to mid-August. The following spring, lightly till in the mat of winterkilled oats for a seedbed into which you can plant your early cereals. A series of light tillings and cover crops will ensure success with a garden-planted grain plot.

Secondary Tillage

A plowed field is rough and bumpy and not yet in any condition to accept seed. We are now ready for secondary tillage—also known as the fitting process—in which we smooth and dress the plowed furrows into a mellow, yet firm seedbed with as few tractor passes as possible. The first tillage tool of choice for this job is the tandem disc harrow. The disc is a butterfly-shaped device equipped with four rows of conical round blades that rotate on an axle or gang bolt. Given its butterfly shape, the discs are pulled through the earth, cutting up the inverted sod at a slight angle to the implement's direction of forward travel. The disc's two front gangs slice through the earth, throwing it outward, while the two rear gangs recut and throw the dirt back inward. All tandem disc harrows do the same thing whether pulled by oxen, horses, or tractor. Modern transport discs are heavier and wider than their older counterparts, and these newer implements are equipped with a set of transport wheels and a hydraulic cylinder so they can be lifted up and moved from field to field. We still have our first disc that we used from 1977 until 1990; it is still intact and sits in a position of honor atop one of our stone walls. This machine was a McCormick Deering, just like our trailer plows, and it was seven feet wide with no extra set of wheels for transport—if you wanted to go down the road with it, you pulled on a rope from the tractor seat and backed up. This action closed the angle of the tandem gangs, which allowed the conical discs to roll forward without cutting into the ground. If you wanted to cut through the sod, you pulled on the same rope while you drove the tractor forward, moving the gangs back to their butterfly position. It's amazing to think that these old plows and harrows worked so well and were at least thirty years old at the time. We paid a hundred dollars apiece for these fine old pieces of farm machinery.

The first pass over plowed ground with the disc harrow is always done in the same direction as the furrows to avoid having the living daylights rattled out of the tractor operator. In order to more completely smooth out the furrows, each pass with the disc is half lapped over the previous pass, which is known as double discing. After you've covered the field once, it is time to change the direction of travel by forty-five or ninety degrees. Generally, two double discings will nicely fit a plowed piece. Sometimes a third pass is necessary to chop up the remaining little pieces of sod on a spring-plowed field. The fewer trips made over the field with the disc harrow and tractor, the more petroleum and organic matter we can conserve.

The Final Fitting

Our future grain field will need a light and final fitting to get it ready for planting. This final pass should be done just prior to planting to ensure that any little annual weeds that are sprouting are uprooted and eliminated—we need to give our future cereal crop every chance possible to beat the weeds. The implements of choice for this project are the spring-tooth harrow, the field cultivator, and the spike-tooth drag. When we first started growing grain, all we had was a two-section drag or smoothing harrow. This was a very simple tool, consisting of two steel frames, each with four rows of pegs or spike teeth. The two frames were attached with special hooks to a hardwood four-by-four, and a steel cable was attached to the front of this wooden beam and then hooked to the back of the disc on its final pass through the field. Each section of the drag was outfitted with a handle to adjust the angle of the spike teeth, making it more or less aggressive. Pulling these two implements through the field in unison was a great saver of time and tractor fuel.

In 1986, we bought a twelve-foot Glencoe field cultivator for the purpose of preparing the seedbeds of grain fields, and we still operate this machine today (see the color insert for a photo). This particular implement has a set of transport wheels and a hydraulic cylinder in addition to twenty-three spring-loaded vibrating C shanks to which are bolted triangular cutting sweeps. Adding the field cultivator to the tillage equipment lineup was a major step forward for me. As this machine was drawn through the earth, the spring-loaded shanks vibrated and stirred the soil, while the six-inch-wide sweeps sliced off weeds under the surface. The faster I drove with the cultivator, the better it worked, and I very quickly added another section to the spike drag and hooked it to the rear of the cultivator. Grain crops seemed to get a better start when the ground was prepared with the cultivator, and the increased aeration increased the amount of soil nitrogen to the grain crop while the sweeps eliminated more weeds. With the field cultivator, I was able to eliminate one pass with the tandem disc. Eventually, I bought a five-bar tine-tooth harrow that attached right to the back of the cultivator frame. The spring-tooth harrow basically performs the same type of "scratch" tillage as the field cultivator. It is much simpler and has no transport wheels, and the spring teeth are raised and lowered by a long steel handle similar to the teeth of the spike drag. These implements complement and enhance the rotary and chopping action of the disc harrow with a pulling and stirring action that puts the finishing touches on a grain field.

Preparing the Seedbed

Seedbeds must be level, clean, and firm to ensure good seed-to-soil contact. Germination is often reduced in fluffy and soft ground. Occasionally, the ground needs to be tilled and firmed at the same time. This is a job for the culti-mulcher. This implement consists of two heavy cast-iron packing rolls attached to the front and back of a transport frame. Between these rolls are a number of cultivator shanks with spikes for aerating and stirring the soil. As the culti-mulcher is pulled across the field, the front roll levels, smooths, and crushes clods. The earth is then cultivated and finally leveled and firmed one more time by the rear roller. The culti-mulcher works best in very finely pulverized loose soils and is really only necessary to break up bothersome little balls of clay.

Grain crops do not always follow plowed sod. Many times, a cereal crop will follow corn, soybeans, or another cereal, and in these cases the moldboard plow is not needed for primary tillage because the sod is absent. A wide variety of tillage equipment is available for use in these situations, and the heavy offset disc harrow is often employed to incorporate the residue from the previous crop. Chisel plows, with their larger, deeper working shanks and twisted shovels, will really cut through light crop residue for an alternative form of primary tillage, and some farmers like to use rototillers and power harrows for this work. We have recently begun using a German rotary-harrow-like piece of equipment called the Lemken Rubin 9 for incorporating crop residue in advance of a new grain seeding. They all have their advantages and shortcomings, but we do know that we are trying to get the best seedbed possible with the fewest rips over the field.

CHAPTER FOUR
Sourcing and Planting Seeds

Now that we've balanced our fertility by adding compost and minerals, and our soil has been worked and the seedbed prepared, it's time to plant the crop. But sourcing where to buy your seed can be difficult. The first order of business is to select a variety of grain that is well adapted and will flourish in a colder region. This task is easier said than done because there is hardly any commercial seed production in the Northeast; most of the corn and soybean seed used here is produced in Corn Belt states like Michigan, Illinois, Iowa, and Minnesota. Row crops are particularly climate-sensitive and are categorized by maturity ratings, so it's important to find out how many heat units or growing-degree days you have in your particular location and match your variety choice to the local growing conditions. It's better to be a little conservative and choose a slightly shorter-season variety than to select something that pushes your growing season to the maximum because it needs every available heat unit to mature and make grain. An early frost or a cool, damp growing season can spell disaster and crop failure for longer-season varieties of corn and soybeans. (A more detailed discussion of seed and variety selection for row crops will follow in the individual corn, soybean, and sunflower chapters.)

The Value of Canadian Varieties

Cereal grains like oats, wheat, barley, and rye, on the other hand, pose fewer complications than their row-crop brethren, but much of the seed for these crops is also produced in faraway places like the high plains of North Dakota and northern Minnesota. University of Vermont Extension agronomist Heather Darby has tested many of these varieties in her research plots, finding that some of these cultivars have flourished while others had trouble adapting to our wetter and more humid climate. But our Canadian neighbors in Ontario, Québec, and the Maritimes have been planting, breeding, and selecting wheat, oats, and barley for a very long time, which is particularly advantageous to us here in the Northeast because we have such a similar climate. I have been using Canadian varieties that I have purchased just across the border from me in Québec for over thirty years. Many of these cereal varieties are prefixed with the letters *AC* or *OAC*—*AC* stands for "Agriculture Canada," while *OAC* represents "Ontario Agricultural College," now known as the University of Guelph. There are numerous cereal-breeding programs in eastern Canada, and new varieties have been released by the Agriculture Canada Experiment Station in Ottawa and Laval University in Québec City. Private seed companies like Semican and Prograin in Québec and C&M Seeds in Ontario have also released many of their own excellent grain varieties. But aside from the public breeding programs at Cornell in Ithaca, New York, plant breeding in the Northeast is virtually nonexistent at both the university and the private level. Some of us here in Vermont have been doing some preliminary selection of wheat crosses on our farms, but this pales in comparison with the potential of a more serious university-based cereal breeding program, and we are extremely fortunate to have a good selection of Canadian and some domestic cereal varieties from which to choose cultivars that will thrive in our region.

However, in recent years, it has become increasingly difficult to import seed from Canada. In the not-too-distant past, anyone could cross the border and buy farm seed to bring home and plant, but things have changed since September 11, 2001. Protocols have been put in place to ensure the quality of imported seed stock and to protect our seed supply from disease. You must now make prior arrangements with your Canadian seed supplier to test a representative sample of the seed shipment at an approved seed laboratory. Once the lab determines that the seed has a clean bill of health, a Seed Quality Certificate (US Customs Form Number 940) is issued to accompany the seed shipment, and with this piece of paper in hand, border crossing and seed importation are painless. The only other consideration is the actual dollar amount of your seed purchase, as any importation valued at more than two thousand dollars must go through a customs broker. Canadian cereal varieties can also be purchased from seed dealers on the US side of the border, which is probably the best option if you don't live close to the border or don't want to import your own seed.

Producing Your Own Seed

Seed doesn't necessarily need to be purchased from an outside source. The ultimate beauty in the culture of cereals is that the grain we produce is also our seed. Farm-produced seed has many advantages. For one, it ensures our independence from off-farm suppliers; it is a tremendous money saver as well. Oftentimes, you will find a variety of wheat or barley that works extremely well on your farm, but not be able to buy the seed for the next season because the variety has been discontinued and replaced with something new. To save seed, set aside a high-quality sample of grain and process it with the help of a fanning mill air-screen-type cleaner. Farm seed cleaning and saving is a subject unto itself and will be discussed at great length in a later chapter.

But there are a few caveats. For example, whatever grain is to be used for seed should be dry, in good physical condition, and come from a field with good soil fertility. When combining a crop, I always choose my "seed piece" and set this grain aside in a separate wagon. It is also essential to know if the variety with which you are working is in the public or private domain. Chances are that if you're growing something that has been recently released, it's a "PVP" variety—this stands for "Plant Variety Protection." Plant breeders' rights are a legal institution put in place to protect the breeding investment of a seed company in much the same way that a copyright protects an author. As an individual grower, you have the right to save your own seed from a PVP variety, but you cannot sell the seed or share it with others. Farmers have saved their own seed since the beginnings of agriculture, and if you've got the grain quality and the wherewithal to clean and store seed, it's the ultimate in farm independence and self-reliance.

However, there are as many reasons to buy new seed each crop season as there are to save it yourself. Many grain growers like the assurance of purchasing and planting "certified" seed. Certified seed will have a blue tag sewn onto the top of the bag, indicating the name of the variety and the state or provincial seed certification program that inspected its production. Commercial seed growers produce certified seed plant "foundation stock seed" in order to ensure purity and protection from disease. These foundation seed plots are meticulously tended and constantly rouged for off types and diseased plants. When you buy certified seed, you are thus assured of starting the season off with a high-quality product guaranteed to be free of disease and to have excellent germination. Although it is considerably more expensive, certified seed is a must for the beginning grain farmer, even if the goal is to produce your own seed down the line.

Germination Potential

You only get one chance to put in a grain crop each season, which makes planting a seed an act of faith, and you hope and pray everything is perfect as you wait for your newly planted grain crop to germinate. It's heartbreaking to see a poorly emerged crop with spotty germination, and replanting isn't always an option. Moreover, even if you are in fact able to replant it can still result in a major economic loss, because yields are usually much lower in a later-planted crop. Adverse weather conditions of extreme drought or excessive rainfall can result in poor germination. A fluffy, unpacked seedbed will reduce soil-to-seed contact and lower germination as well, and poor seeding performance can also be caused by improperly adjusted or operated planting machinery. The most common cause for marginal stand establishment, however, is reduced germ in the seed being planted. This problem can be totally avoided with a little planning and forethought. First, find out the rate of germination for the seed that you intend to plant long before you plant it. If the germ is 85 percent or lower, develop a contingency plan, sometimes, it's as simple as upping the seeding rate in proportion to the reduced germ. Seed replacement is another option, especially if the germ drops below 80 percent. All certified seed will have the germination rate displayed on the special blue tag stitched onto the bag, but it's a good idea to verify the date of the germ test to make sure it's fairly current. If it's year-old seed, you might want to do another test. Farm-grown and -cleaned seed should also be tested for germ; you can do your own germ test right in your kitchen or send a small sample to a qualified seed lab. Many state land grant experiment stations will do seed germination tests. There are, for example, many different types of germination tests performed to simulate wet, cold, or warm soil conditions, and the true vigor of a seed will definitely be revealed when it is placed in a cold soil situation. To do a home germination test, simply count out two hundred seeds of whatever grain you want to test. Place the seeds between two moist paper towels on a plate. Maintain at room temperature and water often. After five or six days, peel the paper towels from the seed, count the number of sprouted seeds, and divide this number by two hundred to determine the percentage of germ. Knowing the germination potential of your seed is essential before you can even think about establishing an excellent stand of grain.

Seedborne disease is something else that can stop a growing crop of grain right in its tracks. Most cereal grain diseases are fungal in nature and are caused by airborne spores that infect cereal grain plants at flowering. Fusarium head blight and "take-all" fit into this category and will be explored in more detail in the disease sections of each plant chapter. Loose smut (*Ustilago nuda*) is a fungus that infects perfectly good seeds; fungal spores enter the flower of a grain plant during the flowering period, becoming systemic in the entire plant. Eventually, the spores get right into the embryo of the seed, where they will lie dormant until the right conditions trigger an outbreak of the fungus on the developing seeds of the plant. These little black "smuts" or fungal bodies grow all over the head of the plant, eventually consuming all the grain kernels. A few individual grains may survive, but an outbreak of loose smut usually spells disaster for an infected crop. This disease is treatable, however, with common anti-fungal seed treatments like captan and Formalin, which are used in conventional agriculture. Thirty-five years ago when I first began growing cereals, I couldn't understand why every grain grower I met in Québec insisted on using treated seed. I had no idea how serious an outbreak of loose smut could be. I got to see this situation firsthand two summers ago when some invisibly infected organic wheat seed came in from Minnesota and was planted in Heather Darby's trial plots in Alburgh, Vermont. Little black fungus bodies grew all over the heads of these wheat plants, leaving nothing but a few shriveled kernels behind. Some organic seed treatments are being developed using steam and ultrasound, but this work is only in its infancy. If organic farmers have a loose smut problem with a particular variety that they want to preserve, the best thing to do is to give the seed to conventional neighbors to grow out using standard nonorganic seed treatments. After one generation of conventional management, the loose smut infection of the seed will have been eliminated and the seed will be ready to plant again without treatment. A good laboratory test is highly recommended and can detect the presence of loose smut and other seedborne diseases.

The Mechanics of the Seed-Planting Process

Now that we have made sure that our seed is free of disease and has a good germ, it's time to plant it. The proverbial scene of the farmer of yesteryear walking through his field scattering seed by hand from a cloth sack may come to mind. This system of seed broadcasting works just as well in today's smallholdings as it did back in the nineteenth century. Cereal grain seeds evenly scattered on freshly prepared earth were covered by pulling a small tree over the field with a team of oxen or a horse, and sometimes a field was rolled again after this process to promote better soil-to-seed contact.

Today you can polish your hand-seeding skills or choose from a number of mechanical seed broadcasters. The most common tool for this job is the Cyclone-brand hand-cranked broadcaster. The seed is placed in a little cloth bag suspended from your shoulder. At the bottom of the bag is a hand-controlled gate valve to control the seeding rate, as well as a hand-cranked spinning disc that distributes the seed evenly over the field. To sow the grain, you simply walk along slowly turning the seed distribution crank. This same concept of spin seeding has been taken to the next level by the Herd Seeder Company of Shelbyville, Indiana. Herd manufactures very inexpensive, lightweight seed broadcasters that are driven electrically by battery power or mechanically by the power-takeoff shaft of a small tractor. Herd seeders can be mounted on just about anything from an ATV to a truck to a tractor, but when using one of these devices start by reading the operator's manual very carefully, and follow the directions and measurements for opening the seed distribution gate as well as the suggested rate of travel for the machine. When broadcasting seed, the spread pattern is sometimes heavier right under the spinner and lighter at its outside extremities. Experienced seed broadcasters recommend going over the field a second time, taking another pass between each of the original traveled routes. This methodology ensures that a heavy coat of seed is applied everywhere in the field. Machine-broadcasted seed can be covered with any number of lightweight tillage tools. A spike-tooth drag or a disc set to just barely penetrate the ground

Frost-seeding legumes into a winter cereal field in early spring.
PHOTO COURTESY OF SID BOSWORTH

will nicely cover the seeds. When all else fails, grab a large branch from a nearby tree; one pass with any sort of an earth roller will put the finishing touches on this low-key grain-seeding project.

Sowing Seeds in the Home or Research Plot

The rate at which grain seed is sown is just as important to the success of a newly planted crop as the quality and germ of the seed. If planted too thick, the seedlings will choke one another out as they compete for limited soil nutrients. On the other hand, if the seeding rate is too low and there are not enough plants to fill the available space, weeds could very well take over and dominate the new seeding. Recommended seeding rates have climbed over the past several decades because many of the newer, more modern cereal grain varieties have been bred not to tiller, which means there are fewer side shoots. When I planted my first crops of oats, wheat, and barley in the mid-1970s, the standard grain seeding rate was one hundred pounds to the acre. Today we think nothing of planting 150 to 175 pounds of wheat to the acre. Whatever seeding rate you decide to use, you don't need a whole lot of seed when you are planting small garden plots of one hundred square feet. Given the proportion of 100 square feet to the 43,560 square feet found in an acre, you only need half a pound (or a little less) seed to plant a ten-by-ten-foot garden plot of wheat. Lay out and figure the total square feet in your little garden grain plot before you begin, then predetermine how much seed you will need and pre-measure it by the coffee can or yogurt cup. A little measurement and calibration is a good prerequisite to field seeding a cereal crop.

Mechanically planted grain is generally sown in rows six to seven inches apart with the help of the modern grain drill. For really small research plots, cereal grain seeds can be hand-sprinkled into little furrows made with a garden hoe or other hand tool.

A good compromise between the use of the grain drill and the broadcast method is to plant grain seeds with a hand-pushed mechanical garden seeder like an Earthway or a Planet Jr. Will Bonsall of the Scatterseed Project in Industry, Maine, is the master of this method, and grows hundreds of small grain test plots each season, seeding everything with his antique Planet Jr. hand seeder. He uses the beet and chard seed plate and likes to drop his seeds one to the inch in rows that are eight inches apart. A little row marker extends eight inches off to the side of the hand-push seeder and makes a small mark in the dirt. The next row is seeded by following this mark. Will says that if he is not satisfied that he is laying down enough seed, he will turn around at the end of a row and double back over what he has just planted. The little garden seeder is an ideal tool for this job because it not only plants and covers the seed, but also packs it with a little iron wheel that follows the seeding mechanism. This is probably the most precise and exact method there is for hand-planting cereal grains.

The Grain Drill

The grain drill was invented by English nobleman Jethro Tull in 1701, though it wasn't widely adopted by farmers until well into the nineteenth century. Apparently Tull came up with the idea of seeds sliding down tubes as he gazed at a pipe organ while in church. The grain drill drops seeds from a mounted box down through a series of equally spaced flexible tubes as it travels on wheels across a field; the bottom of each flexible tube is positioned next to a disc opener that slices a furrow through the soil into which seeds drop. Grain drills are classified as either single or double disc. A single disc opener drills and buries the seed with one concave disc that is usually ten inches in diameter. The flexible seed tube on a double disc drill makes its way down between two twelve-inch-diameter flat discs that are touching in the front and two inches apart in the rear. Both types of drills have covering chains that trail along behind the openers to help cover the seed. (Most older seed drills have single disc openers, while models made in the last twenty to thirty years have double disc openers.) All of these seeding machines, old and new, can be equipped with a grass seed attachment. This sweet little unit is a much smaller seed box that's bolted to the front or rear of the main seed box for the purpose of sprinkling tiny little hay seeds on the ground while seeding the cereal crop. This two-crop capability is quite unique because it lets you "seed down" by planting your future hay crop right along with your annual crop of cereal grain. Both the grass seed and main seed boxes contain a drill shaft that is chain-driven by the machine's wheels. Attached to this turning square shaft are little fluted gears positioned over each hole atop the seed tube. Seeding rates for the grain drill are adjusted by moving a lever, which pulls the drill shaft and its seed gears closer to or farther away from the seed outlet hole; more seed will fall down the hole when these little gears are pulled farther over the hole. I purchased my first grain drill at an auction in 1975 for twenty-five dollars—it was made completely out of wood and had large-diameter old-fashioned wooden spoked wheels. The machine was six feet wide. It was an old horse-drawn model whose long pole had been shortened to accommodate being pulled by a tractor, and painted on the wood in very fancy letters was the word HOOSIER. The drill had been manufactured by the American Seeding Machine Company of New Richmond, Indiana, in 1910, and stood me well for another fifteen years until I retired it in 1989 when I purchased an eight-foot John Deere rubber-tired model. These old wooden seed drills were commonplace on just about every New England farm well into the 1960s and '70s. They still exist today and can be found stored in many old barns and sheds. If you are just starting out growing grain crops on small acreages, an affordable old wooden-wheeled grain drill may be just what you need.

Farm machinery manufacturers have increased the width, seed hopper size, and weight of grain drills over the past half century. I thought that I had made major progress in 1989 when I replaced my old wooden-wheeled drill with a John Deere FB. The machine had rubber tires and seed boxes constructed of galvanized metal. The biggest step up with my new drill was its lift system for positioning the seeding discs in and out of the soil. On the older antique model, the seeding discs were raised and lowered by a hand-operated lever in the rear where the teamster stood. The John Deere FB, which had been made in the 1950s, had two wheel-driven ratchet-and-pawl-style lift boxes, one for each side of the drill. All you had to do was pull the ropes attached to the lift-box handles to engage the gearboxes and lift the seeding discs out of the ground into a neutral position. When you were ready to resume seeding, a quick tug on the ropes would drop the discs back down and grain would begin falling down the tubes to be "drilled" into the earth by the single disc openers. With two lifting mechanisms, each side of the drill could be raised and lowered independently. The ability to plant with half the width of an eight-foot seeder had many advantages, especially in tight little spots or when finishing up a field. The mechanical lift system employed on this grain drill was also used to raise and lower plows and other trailing-type farm machines before the advent of hydraulic lift cylinders.

Another wonderful feature on my new drill was an acre counter, a little gear-driven device that would mesh with a corresponding driving gear attached to the drill shaft. Whenever the drill shaft was turning and the machine seeding, the little acre clock would be engaged and ticking off the acres planted in tenths. Calibration of seeding rates was much easier if you had an idea of exactly how much land you were covering. Most grain drills have a seeding chart on the inside of the cover. If, for instance, you want to seed wheat at a rate of 150 pounds to the acre, you look for wheat on the chart to find the corresponding setting for the desired rate. If the chart indicates setting number nineteen for the 150-pound rate, you simply slide the drill shaft adjustment handle over to the number nineteen position and lock it in place with a wing nut. To verify if this is the correct setting, I put the 150 pounds of seed in the drill box, reset the counter to zero, and go out and plant. I was extremely pleased to find out that the seeding rate chart under the lid of my John Deere FB grain drill was right on for accuracy. In earlier times, I calibrated the old wooden drill by hoisting it up in the air and turning the big wooden wheels by hand. I measured the circumference of the wheels and found it to be twelve feet. After suspending the drill in the doorway of my toolshed, I put a tarp on the ground below the seeding discs to catch the seed. I put some grain in the seed box and then began to turn the big wooden wheel by hand. After turning the wheel one hundred revolutions, I stopped and made some calculations. One hundred revolutions multiplied by the circumference of twelve feet came out to twelve hundred feet of travel. I then multiplied twelve hundred by the six-foot width of the grain drill to determine that we had just seeded the equivalent of seventy-two hundred square feet. Seventy-two hundred square feet divided by the 43,560 square feet found in an acre came out to 0.165, somewhere between a tenth and two-tenths of an acre. My next step was to gather up the seed that had fallen onto the tarp and weigh it. I knew that the drill was set just about right when the weight of the planted seed was twenty-five pounds, which was the proper proportion of the 150-pound rate. Indeed, calibration of any piece of seeding machinery can be a fun and creative process; it gives us a chance to dig up our mathematics training from our elementary school days.

I finally caught up to the contemporary age of modern farming when I purchased an International 510 grain drill from a neighbor in 1993. This machine was twelve feet wide and could be raised and lowered from the tractor seat with the flick of a hydraulic control lever. My particular model was classified as a 21×7, which meant there were twenty-one seed tubes each positioned seven inches apart across the entire width of the machine. The seed box held close to fifteen hundred pounds of seed, which meant ten acres could be planted between refills. This larger machine rolled along on twenty-inch truck-style tires that protruded a foot and a half on each side of its twelve-foot main frame, which put the total width of my new grain drill at close to fifteen feet—wider than anything I had ever operated before. When seeding, I had to learn where to line it up in relation to the last pass so as not to leave an unseeded space (sometimes referred to as a "rabbit run") in the field. Since the wheel marks were a good foot to a foot and a half outside the planted area, I very quickly learned to run the outside double disc at least six inches inside the tire mark of the last pass. This larger machine was considerably less maneuverable than my earlier grain drills, so I had to navigate the field in an entirely new manner. John Ace, my Wisconsin mentor, had taught me to seed a piece by driving the drill in the back-and-forth style. This was quite easy with my two older grain drills pulled by a tiny old tricycle-style John Deere 40; the tractor could turn on a dime, and there was plenty of room at the end of each pass to turn 180 degrees and double back right next to where I'd just seeded. Seeding in this manner meant I didn't ever have to lift the drill up—when the field was completed, I seeded the headlands where I'd turned the drill back and forth several times in the opposite direction. This geometrical approach to laying out a field for planting surely wasn't going to work with this new piece of equipment that was fifteen feet wide, carried fifteen hundred pounds of seed, and was pulled by a much larger tractor. Larger, newer grain drills are driven around the perimeter of a grain field either clockwise or counterclockwise, depending on what direction suits you. For some unknown reason (maybe because I'm left-handed), I prefer the counterclockwise approach. After you've planted around and around a piece five or six times, you begin to notice that the corners start to get a little tight and are much more difficult to negotiate in a manner where all the ground is being evenly covered with seed. When you take a sharp turn with a grain drill while seeding, the inside wheel revolves more slowly than the wheel on the outside of the turning radius. Since the two-piece drill shaft is powered by both wheels, the inside shaft revolves much less than its outside counterpart, planting much less seed. This uneven distribution of seed on sharp corners has the potential to affect the final stand because the seed will be laid down too heavily on the outside of the turn and not heavily enough on the inside. To avoid this situation, drive straight ahead instead of making the ninety-degree turn with your grain drill. Lift your drill up into the neutral nonplanting position when it is on top of what has already been planted. Continue driving straight ahead and then begin a broad, wide turn in the opposite direction to the one you're planting, eventually coming around 360 degrees or full circle until you are lined up with the next quadrant of the field. This will be at a ninety-degree angle to where you just were. For example, if you are planting in a counterclockwise pattern, turn your drill clockwise on these corners. You will find you have turned ninety degrees after coming around full circle in the opposite direction. At this point, lower your drill back down into the seeding position and continue merrily on your way. I call this the "loopdy-loop" method. This 360-degree turn system will permit you to negotiate a very tight turn in an easy, gentle manner. You will be able to very accurately plant around square corners without having to crank your tractor and grain drill around in tight quarters. As the area left to seed shrinks and the sides of the square get shorter and shorter, you can skip every other pass and just drive over to the opposite side to continue seeding. Eventually the two long sides will come together and you will be finished planting your piece. View your field as an elementary geometry problem. Bigger grain drills can cover more acres in much less time, and they are easy to negotiate.

In 2000, I traded up to a newer and even more modern grain drill, when a farm machinery salesman from a large dealership in Québec stopped by one day and caught me in a weak moment. I traded my International 510 for a Case International Model 5100 Soybean Special model. The newer grain drill was equipped with a few more bells and whistles that allowed it to do a better job of putting seed in the ground. Each double disc opener on the 5100 model was outfitted with a trailing rubber press wheel, which rolled over the newly planted seed to better pack it into the seedbed. By turning a little D handle on each press wheel attachment, more or less pressure could be applied to the seed. This adjustment allowed me to match the machine to the existing field conditions. I noticed better stands of grain the first year I used my new drill with press wheels. Seed-planting depth is also essential to achieving a productive stand of cereal grain. If you are seeding along and notice kernels of grain on top of the ground, you are losing yield potential. Sometimes the ground is just plain too hard or tight for proper penetration by the double disc openers. The 5100 drill has two features to make sure that seeds are properly buried. Each row unit has an adjustable down-pressure spring on the rod that connects the opener to the lift arm and rock shaft. All grain drills are equipped with this feature. Compressing these springs will put more down pressure on the double discs, which in turn will push them farther into the soil. Many times it is necessary to apply more pressure only to the disc openers that follow in the wheel tracks of the tractor pulling the drill. These wheel tracks can be hard to penetrate, especially in more easily compacted clay soils. One other nifty feature on the 5100 grain drill is an adjustable lift rod on the hydraulic cylinder that lifts the seeding discs in and out of the ground. The top of this rod is equipped with a large nut that can be turned either inward or outward to change the length of stroke of the hydraulic cylinder. This in turn will control the depth of the double disc openers. Small-seeded crops like flax only need to be buried an inch or less deep, so for shallow planting the cylinder nut is backed off to shorten the stroke of the hydraulic cylinder, which will then exert only partial force on the seeding mechanism. For deeper planting, tighten down the adjustable nut; this will extend the cylinder's stroke and pull the seeding discs down to their maximum depth. Precise control of seed-planting depth coupled with just the right amount of seedbed packing are the icing on the cake when it comes to planting that perfect crop of grain.

The Case IH 5100 grain drill has one more new feature that can improve seeding performance. The Soybean Special attachment allows you to halve the speed of the drill shaft by changing the driven end gears from a seven-tooth sprocket to a fourteen-tooth sprocket. These sprockets are quick and easy to change. With the drill shaft turning at half speed, seeds can be accurately applied at extremely low rates. This is very helpful when you're planting soybeans in narrow rows with a grain drill. The slower-speed drill shaft allows the soybeans to be dropped at the rate of three to six per foot, instead of the usual eight to twelve when the full-speed drill shaft is used. Properly spaced soybeans planted in seven-inch rows and packed in with press wheels have almost the same yield potential as soybeans seeded with a corn planter. The slower-turning drill shaft also helps to improve the placement and planting accuracy for other small seeds like sorghum sudan and flax. Newer grain drills also have three-position seed cups that help to gently deliver seeds as they are metered out by the fluted rollers and dropped down the flexible seed tubes. Larger seeds like field peas and soybeans need more room as they roll out of the seed-metering device, and dropping the seed cups down to a lower position will allow these seeds to pass through the system without being ground up and destroyed by the fluted seed rollers.

Caring for Your Grain Drill

Before we finish this rather extensive discussion of seeding and seeders, we would do well to attend to some "housekeeping" concerns. First, whenever possible, always park your grain drill back in the shed immediately after using it. If you have to leave it out in the field, cover it with a tarp or a sheet of plastic. Even though most machines have seed compartments with tightly fitting covers, moisture will still manage to find its way inside through little cracks and openings. Left-over seed is a wonderful medium to attract, collect, and absorb atmospheric humidity or a stray raindrop. Left unattended, the remaining seed in a drill will swell and eventually sprout, which can be bad news—this sprouted seed can severely plug the same interstices through which seed must flow to get to the soil. There's little worse than hauling your drill out of the shed to plant next spring and finding the seed gears, cups, and tubes totally filled with dried-up and hardened sprouted seed. This sort of situation can take hours to remedy with a shop vacuum and a piece of wire for poking and cleaning. The best time to use the shop vacuum is right when you finish planting. Clean your grain drill seed boxes (both grass seed and grain seed) immediately. (This cleaning process is also highly recommended when changing to another variety or type of grain.) As you vacuum out your seed compartment, you will be surprised how many seeds are left down in the hole next to the seed gear. You can't see them, but they are there, and stray seeds can contaminate your next grain crop—you don't want rye in your wheat or oats in your barley. The grain drill is a marvel of agricultural engineering and functional design; as operator of this amazing piece of equipment, you must do your part to keep it clean, lubricated, and maintained.

Although newer-model grain drills work by the exact same principles as their older predecessors, many of the little add-on features developed in the last several decades have allowed grain growers to fine-tune their planting craft. Grain seeding isn't as precise a project as you would think; once you've learned the basic principles of planting and mechanical seeding of cereal grains, you just have to try it and learn from your mistakes. This is the only way to learn as far as I am concerned. Everyone makes mistakes—I certainly have made plenty. But I have one last piece of advice before we move on to discuss crop growth. *Never back up with a seeder that's in the ground.* Backing a drill or a row-crop planter up while it is in planting mode will plug your disc openers with packed soil, which will impede the flow of seed and stop you dead in your tracks. Depending on how long it takes to discover this problem, you might find it difficult to know exactly where to begin replanting once you've got your drill or planter working again. In this chapter, I have concentrated on planting cereal grains and have purposely omitted row-crop planter adjustment and operation. There is just as much if not more to know about planting corn, beans, sunflowers, and other row crops, but the mechanics of row-crop planting will be discussed in great detail in the row-crop chapters of this book. Meanwhile, have fun sowing those cereal grains.

CHAPTER FIVE
Early Growth and Weed Control

Once you've completed field preparation and planting, you must wait for the crop to emerge from the soil. Spring cereals, especially wheat, should be planted absolutely as early in the season as weather conditions allow—wheat seeds will germinate at temperatures as low as thirty-eight degrees. This is probably the one single most important grain-growing principle there is, and it also brings the most doubts to every new grower of spring cereals. You might wonder, "Won't my wheat be damaged by late-spring frost or snow?" Wheat is tougher than you might think. If the earth is thawed and dry enough, it's time to plant seeds; many years it is early April when I plant wheat on my northern Vermont hilltop farm, and there are still unmelted snowdrifts on the edges of the fields. Extra-early seeding gives the cereal crop ample time to make vegetative growth while the daylight is increasing through the summer solstice. Your crop will also get the jump on the weeds, because most of the problem ones like lamb's-quarter wait until May to germinate when the soil warms up. I have seen many an April snowstorm blanket a field of recently germinated oats or wheat. Sometimes I wonder if this little shot of "poor man's fertilizer" doesn't actually give the new cereal seeding a boost. Low temperatures and heavy frost are also of little consequence to a newly planted crop of grain.

Frost seeding is another lesser-known grain-planting strategy developed and practiced by farmers in Québec and Ontario. This practice involves seeding wheat down into frozen ground with a heavy modern minimum-tillage style of grain drill. The window of opportunity for successfully frost seeding a spring wheat crop is very narrow. The snow should be mostly melted away, but the ground still frozen. A very cold morning followed by warm afternoon thawing action is ideal, and a heavier no-till style of drill with good seed disc penetration will help to get the seed down into the frozen earth; the thawing afternoon soil will finish the job. Spring frost-seeded wheat seeds basically lie dormant until they are awakened by warmer spring temperatures.

The First Signs of Growth

Optimal germination temperature for most cereal seeds is sixty-eight degrees. Grain planted in early to mid-May will germinate and push through the soil in four to five days. Early- to mid-April-planted crops will take seven to ten days to emerge. The appearance of the primary root, also called the radicle, is the first indication of germination in the recently planted seed; the coleoptile, or first leaf, will push out of the same end of the seed shortly thereafter. It is important to remember that cereal grains are grasses and basically have the same growth habits. Cereals have seminal as well as crown or nodal roots, and the radicle forms from cells present in the seed and is a seminal root. Very soon after the appearance of the radicle and the coleoptile, the first node of the infant cereal seedling appears. The

A wheat sprout—coleoptile and radicle roots.
PHOTO COURTESY OF SID BOSWORTH

crown roots, the main stem, and the first leaf all begin at the first node, and more leaves, roots, and tillers soon follow.

A tiller is a side shoot that develops from buds at the base of the main stem. Several tillers will appear in conjunction with first leaves. Tillers develop their own crowns and individual root systems, which gives them the potential to become separate stalks of grain. Factors like light intensity, plant density, and the availability of soil nitrogen affect the number and vigor of tillers produced by an individual seed. The tendency toward high seeding rates and increased stand density so commonplace in modern high-production agriculture seem to limit the amount of tillering found in most contemporary grain fields.

Dealing with Weeds

Weed competition is perhaps the biggest challenge to growing productive crops of cereal grains. This is especially true for spring grains. As discussed earlier, fall-planted winter grains already have five to ten leaves and some tillers by the time the snow melts and the growing season begins, at which point the soil is quickly shaded and weeds cannot really get a foothold. Spring cereals don't have this sort of advantage. Row crops, on the other hand, are planted thirty inches apart in rows that can be mechanically cultivated. Some experimentation, however, has been done in Denmark and Northern Europe with growing cereal grains in wider twelve- to sixteen-inch rows, and a special steerage hoe called the Schmotzer is driven through these wide rows to mechanically remove weeds. Both the University of Maine and University of Vermont have Schmotzer hoes and are conducting trials to determine the efficacy of wide-row cereal planting in northern New England. In the meantime, the rest of us have to figure out some other sort of strategy with our six- and seven-inch rows of wheat, barley, and oats. My own personal nightmare weed that afflicts my spring planted cereals is common mustard. Mustard will germinate just as early as that super-early-planted wheat that I strive so hard to make happen. My old-timer neighbors told me numerous stories about pulling kale (which is our local variety of yellow mustard). Sam Pion, who grew up on the farm where I now live, told me that his father used to spray the oats with something called "blue vitrol," which as near as I can tell is copper sulfate, and he said it would kill the kale dead. (I wondered what kind of copper toxicity this had left behind in my soil.) Thirty years after hearing about blue vitrol, however, I have learned that mustard is a sulfur deficiency indicator. Sulfur is certainly a necessary element for protein synthesis in plants. I have been using very trace amounts of copper, zinc, and manganese sulfates on my hay crops for years, so it may be time to try this fertilization as a foliar spray on cereal grains.

As my first crops of wheat, barley, and oats grew up and turned bright yellow thanks to mustard flowers, I got a lot of pressure to spray my grain fields with herbicides. My old friend and mentor, seventy-five-year-old Clarence Huff from Compton, Québec, showed me his five-gallon pails of MCPA 48 potassium salts—a mere ounce or two of the stuff mixed with water and applied to an acre would make those pesky mustard plants disappear. My own brother-in-law, Ted Welch, was working for the Velsicol Chemical Company of Chicago, Illinois, at the time. Velsicol's primary product was Banvel, a 2-4-D herbicide for use on cereal grains, and Ted kept asking us if we wanted a free five-gallon pail. We knew that dioxin, which was one of the deadliest chemicals on the earth, was a by-product of 2-4-D, and there was no way we were having any of that stuff on our farm. This was in 1977 and 1978, and the concepts of organic farming were just beginning to be understood at the time. And so we simply learned to live with the mustard. Fortunately for us, it turned out to be not such a bad weed after all. Mustard flowers quickly and ripens to make seed quite early in the season. Unsprayed grain fields will turn completely yellow with mustard flowers when the cereal crop is twelve to sixteen inches tall and still in the vegetative stage. Then all of a sudden, the yellow color is gone and grain heads begin to pop up. The mustard is still there in the bottom of the grain crop in its reproductive state, forming and filling its tiny little seedpods. The good news is that mustard will make dry mature seed several weeks before the grain is ready to harvest. So at least the mustard seeds don't pose a threat to the ripe kernels of grain once the crop has been through the combine. Other later-maturing weeds like red root pigweed and lamb's-quarters, however, are still quite moist and green at the time of the grain harvest and will impart moisture to an otherwise dry grain sample. If these soggy wet weed seeds are not removed immediately, the newly harvested grain will heat, develop mold, and spoil. Other unwanted plants like ragweed not only impart moisture into the newly harvested grain crop, but also flavor the grain with undesirable pungent smells. A trained nose can very easily detect the sharp essence of ragweed in wheat berries even after they have been cleaned. After considering all of these possibilities for disaster, a dry mustard seed that is basically inert and stable doesn't seem like such a bad thing to have mixed into your grain sample. It's also quite interesting that the clean grain outlets of most older combines and threshing machines made before 1960 were all equipped with little rotary screen cleaning devices known as scourkleens. The tiny round black mustard seeds would fall through the screen into a separate bag while clean kernels would continue on to the combine's grain tank. Farm equipment manufacturers eventually did away with scourkleens in the 1960s because herbicide use on cereal crops became universal. But in my early days of grain growing, I was fortunate enough to have harvesting machines with these built-in cleaners.

Although mustard may seem like a "dream weed," we would still do well to eliminate it from our grain crops, because those little dry seedpods shatter and split open quite readily during harvest. As the bats of the combine reel hit the crop during the cutting and gathering process, viable mustard seed is spilled and scattered all over the ground below. Weeds have evolved to survive, and this shattering process ensures that there will be plenty of mustard seed in the soil seed bank for years to come. As I became familiar with the problems of mustard during my first few years as a small-time grain grower, I was told by numerous people that mustard seeds were viable for at least eighty years, and I found out firsthand that farmers before me on my farm must have also been plagued by the stuff. I would plow up an old piece of pasture or field that had not been turned over for thirty or forty years, and still encounter mustard in my grain crop. As the years passed by, my mustard situation didn't seem to improve much, and there was certainly lots of free advice to be had for the asking. This is when my neighbor Sam Pion told me about spraying the blue vitrol on the grain fields. Several other old-timers mentioned running a roller over a recently planted field of a grain for mustard control. The strategy was to wait until the grain crop was six to eight inches tall. Early on a cool crispy morning in late May, you flatten the entire field of grain and mustard with the same roller you would use to pack a newly seeded field. Since the mustard would be stemmy and brittle and the grain, grass-like and supple, the grain crop was supposed to spring back while the mustard stems would actually break off. I must say that I have never tried this approach, either. The one thing I did try was to cut off the mustard that towered over the grain crop with a mowing machine. This set the mustard back a bit, but it left wheel tracks in the field—and the mustard still managed to flower and seed a week or so later. Throughout this whole process, I was thankful to have the seed-cleaning machinery necessary to clean black mustard seed from my wheat, barley, and oats.

The Tine Weeder

Sometime after 1980, I began hearing about tine weeders from some of my Canadian friends. These devices were all made in Europe at the time. They were lightweight little harrow frames mounted to the hydraulic lift arms of a tractor. Attached to this mounted framework were three or four rows of long narrow tines made of spring steel, which were eighteen inches long and in theory vibrated their way through a new grain seeding plucking out those pesky little weeds. The theory was that the grain crop would be well rooted enough to withstand the vibrating scratching motion of the tines, while the smaller ungerminated weeds in the white hair stages of early growth would be pulled up and exposed to the baking action of the sun. Of course, I had to have one of these miracle machines to solve all of my problems. The first thing I did was send away for the glossy literature to leave in the bathroom for viewing and subsequent temptation. The only model of tine weeder widely available at the time was manufactured by the Dutch farm equipment company Lely, and these machines were quite expensive for their size—I seem to remember that the four-meter thirteen-foot model sold for close to twenty-eight hundred dollars. After dealing with a year or more of "weeder envy," I decided to buy one for the 1984 crop season. I was able to borrow the money from a philanthropic farm advocacy organization based here in Vermont called the Working Land Fund. Lely's North American distribution center was located in Burlington, Ontario, about an hour west of Toronto, and I drove to my friend and former apprentice Alex English's farm two hours east of Toronto the first day. The next day we battled five lanes of snarled traffic back and forth through Toronto on the 401 to pick up the weeder. It seemed odd fighting city traffic to obtain this piece of equipment that was going to make its way back to a peaceful rural setting. Nevertheless, I managed to get home with my new prize. It was totally in pieces and unassembled, and took several more days to bolt the thing together and get it ready to use. By this time, my crop of barley was almost six inches tall, and I was ready to tine weed my crop.

My first tine-weeding experience was scary and unpleasant; it seemed like I was flattening the crop more than I was eliminating any weeds. I attached the Lely weeder to my little narrow-front-end John Deere 40 tractor. The weeding job looked particularly bad where the wheels of the tractor flattened out the grain in advance of the weeding tines. The wheel-flattened six-inch-tall barley plants got pretty badly covered with loose soil. Another problem I had not anticipated was that the aggressive tine action of the weeding device also pulled small rocks up from the seedbed. I finished weeding the ten-acre field of barley wondering if I had done more harm than good, but the flattened soil-covered plants did reemerge and survive. However, several weeks later, the barley field still turned its normal mustard yellow color, because mustard grows with a little taproot that makes it difficult to remove mechanically if it has progressed too far in its growth habit. I vowed to be ready to hit my grain field a little earlier the next season.

I learned many more lessons the next year with my new Lely weeder. I tried to get out in the grain field at the three- or four-leaf stage in 1985. That time, I encountered a new problem—the field was infested with quack grass roots. As I progressed across the field with my shiny red machine, the tines began grabbing the long quack grass rhizomes and very quickly balled up. This slowed the weeding process down quite a bit. Once the tines filled up with roots, I had to lift up the weeder, drive it over to the edge, and remove the roots by hand. This turned into a long, tedious process, but it was worth it in the long run because the rhizomes were removed from the grain crop where they could have rerooted and sprouted to outcompete and choke everything. The other lesson that I learned that season was that tine weeding and underseeding grain crops with clover, timothy, and alfalfa seeds for a future hay crop after grain harvest were two incompatible operations. If you sowed grass seed at the same time as your cereal crop, the weeder had the potential to wipe out the newly sprouted hay seeds. So here I was ten years into my grain-growing adventure with more knowledge and experience, but still bewildered by weeds. I had to repay the loan for the Lely weeder, and I couldn't really say that it had solved my mustard problem as I had hoped it would, but I learned that farming is a fluid business where nothing stands still. Flexibility, good observation skills, and a willingness to try something new every year were the qualities I needed to survive as a newcomer to this business of growing grains.

Kovar tine weeder at work in young corn. PHOTO COURTESY OF SID BOSWORTH

Lely stopped manufacturing weeders in the 1990s. My machine had many bent tines that were quite expensive to replace. It had never really served me that well, so I sold it to a vegetable producer for use on his carrot beds. By this time, tine weeding was becoming a much more widely used practice for weed control in all sorts of organic crops. The Cadillac of tine weeders was and still is the Austrian-made Einbock, and all of my Québec organic grain-grower friends were buying these machines; the Einbock is also a rear-mounted, hydraulically lifted machine. Its main frame is equipped with five rows of pipe-mounted spring-steel weeding tines. You can adjust the angle of these tines in relation to the soil below to five different positions by means of a top-mounted control lever, which allows you to weed a field more or less aggressively. The Lely weeder was also adjustable, but each individual spring tine had to be manipulated in its holding bracket. This process took over an hour, but the same adjustment on the Einbock took five seconds. Tine quality is also much approved on the Austrian model—there are more tines and they are positioned much closer together, whereas the soil end of the weeding tines on the Lely machine was bent at a ninety-degree angle to the upper part, making it an almost too-aggressive puller of weeds. Einbock tines are much gentler on the crop because they are only bent at a forty-five-degree angle on the bottom.

The Einbock had only one drawback, and that was its price. I thought twenty-eight hundred dollars for the thirteen-foot Lely weeder was outrageous. I looked at a fifteen-foot model so I could also use it to weed six rows of corn on each pass. This wonderful piece of Austrian technology retailed for close to five grand at the time. Wider models for larger-scale grain growers were also available in thirty- and forty-foot widths, and these machines cost between ten and twenty thousand dollars. My introduction to European farm machinery was complete. I simply couldn't afford this kind of money for something that only got used one week a year.

It wasn't too long after this that I had a conversation with Finger Lakes organic grain farmer Klaas Martens. Klaas had discovered a little-known Minnesota-based manufacturer of tillage accessories called Jack Kovar Sales. The Kovar Company had been around since the 1930s and fabricated just about anything needed for field finishing and post-tillage. Chain harrows and spike-tooth harrow finisher attachments for field cultivators had been the mainstay of their business. Kovar also made a weeder very similar to the Einbock, but for about a quarter of the price. I put a call in to Pete Kovar in Anoka, Minnesota, and found that the fifteen-foot model I needed cost a little less than fifteen hundred dollars. Pete is a wonderful and very trusting person, and a true friend to the organic grain farmer, always available by phone and full of good advice and suggestions. Needless to say, I ordered a Kovar mechanical weeder and had it several weeks later in time to weed my cereal grain and row crops for the 1997 season. The Kovar weeder is set up like the Einbock with five rows of pipe-mounted tines that can also be adjusted for ground angle by means of a top-mounted lever. My fifteen-foot model consisted of three individual five-foot gangs hung from a foldable main frame by small chains. The tines were made of slightly larger spring-steel round stock, and they were straight all the way to the bottom. These straight tines vibrated along the top of the ground in a circular motion and were fairly aggressive. They seemed to work best in taprooted crops like beans and sunflowers. Pete Kovar worked quite closely with Klaas Martens to develop a forty-five-degree-angled tooth similar to the Einbock's, and after several years I replaced the straight teeth with angled ones purchased directly from Pete. Bob LaFrancois of Organic Equipment Technology in Byron, New York, has also been instrumental in spreading the knowledge and the good word about grain-crop weeding. Bob improved on the Kovar's chain suspension system by developing what he calls "wish bone" mounting brackets. Instead of hanging by chains, each individual five-foot gang is attached to a four-inch round pipe on which it can rotate freely and better follow the contours of the ground below.

Tine weeders and their operation could be the subject of an entire book. The best advice that I can give to any organic grower of crops is to seek out other farmers with experience for questions of what to buy and how to operate a weeder. Einbock and Kovar are the two most popular brands in this part of the country, and German farm machinery manufacturers Hatzenbichler and Rabher also offer weeding equipment. The Rabher employs spring-loaded knife-like pieces of steel instead of round tines, and the Hatzenbichler is quite similar to the Einbock in appearance and operation. These two lesser-known brands of weeders are quite popular in Canada among farmers of German descent. If you have questions about weeders, how to use them, or where to buy them, Bob LaFrancois from Organic Equipment Technology is the man to speak to.

How to Operate a Tine Weeder

A weed is simply a plant out of place. When we plant wheat, barley, or oats in six- or seven-inch rows, we leave plenty of space between these rows for other unwanted species to germinate. Usually, our little seedling cereal plants are clearly discernible in their evenly spaced rows; a cereal seedling may be two or three inches tall at the two- or three-leaf stage of development. A close inspection of the soil surface between the rows of grain plants will reveal many little weed sprouts just popping up. The grain still has the upper hand at this point since the weeds are relatively minuscule, but the race is on. Ideally, we would like to see the grain seedlings grow quickly and shade the spaces between the rows, depriving the weeds of light and soil nutrients. Unfortunately, this is not often the case at this stage of cereal development. Once the weeds have established themselves, they can quickly catch up to and eventually dominate the new grain seeding.

Pre-emergence weeding, also known as blind harrowing, is the best practice we can employ to ensure that our grains sprout into a seedbed free of competition from any other species of plants. There is no better time to remove weeds than the period between the initial planting and the first emergence of a newly seeded crop. We can scratch the soil surface as aggressively as we deem necessary as long as the sprouted seed remains below ground level. Once the tender little coleoptiles have pushed through the seedbed, we risk damaging our future cereal crop if we attack it with the weeder at this point, so close monitoring of the germination progress of our newly planted seeds is the order of the day. Check your grain field daily by digging up a little soil to see how close the seedling is getting to the surface. Once you've got your hands down in the dirt, you can also monitor the development and germination of our friends the weeds. We're looking for little weed sprouts in what we call the white hair stage; such hairs look like little white threads. The main principle here is to expose these white threads to the drying rays of the sun. Timing is everything at this stage of the game. It's best to run the weeder over the newly planted seeding on a sunny, warm day just before the new grain shoots are ready to break through the ground. Ideally you would also like the weed sprouts to still be in their infantile white hair stage. A little forethought and gambler's luck is a key ingredient to success when making a management decision about when to weed a not-yet-germinated crop. If your grain seed is only halfway up to the surface and you find little white weed threads everywhere, you might just want to go out and weed if the weather is going to be sunny. (Mechanical weeding is totally ineffective if there is no sun or heat to cook the exposed weed seedlings.) Once a weed has developed roots and poked through the surface, it is more difficult to scratch out mechanically, and the rate of weed kill is much higher when weeding is performed sooner rather than later. Mechanical weed control in organic cereal culture can be a nerve-racking and worrisome process—vigilance and decisive action are the keys to success.

Let's imagine you head to the field with your weeder. It's about nine o'clock in the morning, the weather is clear, and the sun is beginning to heat things up. You've dug down into the soil and determined that your recently seeded grain crop is a quarter to half an inch below ground level. The weeder is attached to the three-point hitch of the tractor. Lower the machine down into weeding position with the three-point hitch control lever, and get back off the tractor and stand behind the machine to see if it is level from side to side. The machine can be leveled by shortening or lengthening the three-point hitch arm support linkage, which is usually done by turning a little crank. Once you have made this adjustment, check the weeder's fore and aft level. You want the front and rear ranks of tines touching the ground at the same time. This adjustment is made by changing the length of the tractor's top link. Lengthening the third arm will drop the rear tines lower; shortening the top link will lower the front tines of the weeder. This adjustment is a little tricky and most times must be made while the machine is in use. Because of the springiness of the tines, the weeder will act a bit differently once it is being drawn over the ground. Since the tendency is for the weeder to dig deeper in the front, I have found that the top link needs to be lengthened considerably so that the rear tines can do their job. This bit of fine-tuning has been a struggle over the years. With a fore and aft setting like this, the weeder will not have much ground clearance when it is raised into a transport position. I have recently been able to remedy this problem by installing a hydraulically controlled top link to my fifteen-foot Kovar model—it's basically a heavy-duty hydraulic cylinder that replaces the manual top link—which I bought for about $350 from Bob LaFrancois of Organic Equipment Technology. This is European ingenuity at its best and it was well worth the investment; I can now change the front-to-back angle of my weeder on the go. If the rear tines are dragging in transport position, a flick of the hydraulic lever picks them up. When I need to resume weeding, the hydraulic top link has a stop mechanism that you can preset to return to the same position every time. Proper leveling is essential for maximum mechanical weeding performance.

The next thing you need to determine is just how aggressive you want to be out there weeding your field of grain that has not yet emerged from the seedbed, which might take a little trial and error and experimentation. I will assume that tines with forty-five-degree tips are being used, as the tip of a weeder tine will stir more soil when its angle of penetration is decreased. If you pull the adjustment handle backward toward the rear of the machine, the weeder tines will move forward to a position that is more perpendicular to the ground below. This will in turn allow the forty-five-degree tine tips to enter the soil at a more acute angle. Returning the adjustment handle to its closed position, however, will have the opposite effect, and the weeding action will be much gentler. You want to be as aggressive as possible without damaging the grain crop, as a lot of this is shooting in the dark. You should at least get off the tractor, dig around in the soil, and have a little look. If you see broken-off grain seed shoots or whole seeds up on the surface, you know that you have to lighten your touch a little. Chances of injuring young grain seedlings are much less at this stage of crop development.

The manner in which you approach the grain field has some relevance. When working in a field that has not yet germinated, I like to weed perpendicular to the how the grain was seeded. This approach is not essential to good weeder performance and is probably more about my sense of geometry than anything else. It is important, however, to operate a weeder in as straight a line as is practical and possible; as with all implements attached to the three-point hitch system of a tractor, a slight turn of the steering wheel will increase the sideways travel of a rear-mounted machine a much greater distance than you would expect. When you turn a tractor to the left or the right while weeding, the weeder tines will be moving both forward and sideways simultaneously, which doubles or triples the soil disturbance and has the potential to injure the crop you are trying to preserve. For this reason, I would recommend lifting the weeder out of the soil at the completion of each pass. If you are operating the machine back and forth, lift the machine and spin the tractor right around so that you can head back to the left or right of your last trip across the field. When the job is completed, run the weeder in the opposite direction along the headlands where you did your turning. It is also important to avoid traveling over the same spot too many times with a weeder. Too much tine action can overdisturb the grain crop growing underneath. You want to weed just enough to uproot the weeds, not the crop.

Understanding Your Crop

It is also possible to underseed a grain crop with alfalfa, clover, and timothy seed before or during the weeding process, and Loïc Dewavrin of Les Fermes Longprés in Les Cèdres, Québec, has been a pioneer in this process. Loïc and his brothers plant a clover cover crop into two hundred acres of organic wheat each season. The Dewavrin brothers are talented fabricators of agricultural equipment specially modified and adapted for their particular grain operation, and they mounted an inexpensive air seeder right to the frame of their large Einbock tine weeder. Grass seed is propelled by a hydraulically driven blower through a series of tubes to equally spaced points across the main frame of their weeder. Depending on soil conditions and how much seed coverage is required, they have the option of moving the seed delivery system either forward or backward on the weeder frame. Air seeding of grasses and cereals is much more commonly practiced north of the border than it is here in the States, and many organic grain growers in Québec have outfitted their tine weeders with air seeders for the purpose of underseeding a hay or cover crop. I have achieved this same result by mounting an electrically driven Herd grass seed broadcaster to the front of a tractor with a tine weeder mounted on the back. Performing two tasks at once is certainly more efficient than making multiple trips over a field, but I have found it better to concentrate on laying down the seed first and going back over the piece with the tine weeder. As discussed earlier in the planting chapter, seed broadcasters do not distribute grass seed evenly when seeding width exceeds eight feet. The fifteen-foot working width of the Kovar tine weeder isn't quite congruent with the eight- or ten-foot coverage of my low-tech, inexpensive Herd grass seed broadcaster. It takes me a little longer to lay down my grass seed, but at least I know that I am taking the time to do a thorough and complete job.

When to Weed Again?

Once the grain seeding is up and growing, it's time to stay out of the field for a while. Cereal seedlings are quite tender and easily damaged when the coleoptiles have just poked through the earth's crust. Hopefully you've been able to set the weeds back a bit with pre-emergence weeding, because you will need to wait a week or two until the plants have begun to tiller and are showing five or six leaves. The effectiveness of a weeding operation in a field of cereal grain that has been growing for several weeks is a toss-up at this point. Experienced grain producers like Loïc recommend not going back in to mechanically weed after cereals have emerged. Studies have been done in Denmark comparing the weed removal advantage in a mechanically weeded grain crop with the amount of plant damage and potential yield loss incurred in the process. In most cases, the process was a draw. Mechanical weeding reduced plant density by 10 percent or more, which was just about the same as the potential for yield increases by knocking back the weeds. This is not to say that tine weeding doesn't have its place in an established cereal grain crop; if the weeds are coming on fast and furious between the rows of grain, it might just be best to attack them one last time, as it would be the last chance you get before the grain gets too tall. The choice and the crop belong to us, the growers. This is where the fun comes into our farming endeavors—we try something and it might or might not work. If we don't do the experiment and stick our necks out, we will never know if a particular management decision had the potential to improve our situation. If you go ahead and run the tine weeder through a standing crop, always leave yourself a control strip or section where a particular area is untouched by the weeder. You can look at your little gambling experiment in true scientific fashion this way, comparing the final results of your different approaches. Then you need to remember what you did or didn't do. My advice (and I should do better to heed it) is to write things down in a diary or crop journal.

Weeds as Soil Messengers

Those pesky little weeds that you are trying to set back and eliminate with the tine weeder are more than just plants out of place; they are also a reflection of the health and well-being of the soil of which we are stewards as organic farmers. Weed scientists like to talk about the weed seed bank that is present in any particular soil. Most academics equate the amount of weed pressure experienced by a crop to the level of weed seeds present in the growing medium. I know that my own particular seed bank, for example, is definitely full of mustard seed. But why is it that some years weed pressure is high and other years the problem is minimal? My mustard problem has definitely declined in the last decade since I have begun applying my own finished compost and trace minerals to my grain crops.

Perhaps the two most common soil conditions responsible for unbridled weed growth are compaction and the presence of excessive soluble nitrates. Compacted soils tend to be low in oxygen and drain poorly, and weeds like foxtail and velvet leaf thrive in these sorts of conditions. Large leafy, thick-stemmed weeds like lamb's-quarters and redroot pig are actually beneficial cultivars because they metabolize excess nitrates present in the soil, but perennial weeds like quack grass have been associated with soils lacking the biology and sugars to feed soil microbes. A little application of molasses, for example, might just do wonders in a quack-grass-infested field. We have already discussed the importance of soil testing and subsequent soil mineralization, and while mechanical weeding is certainly one approach to weed control in cereal crops, you also would do well to learn what each particular weed is telling you about your soil and your past management practices. Other soil amendments like chelated minerals and micronized humates sprayed in tiny amounts onto our soils and crops at the right moment could help us create the conditions under which weeds might just disappear. A whole-systems approach based on balanced soil fertility, good soil structure, diverse soil biology, and timely mechanical weed control will be the ticket to success.

CHAPTER SIX
The Growth Cycle and Harvest of Cereal Grains

At this point, you've planted your crop and weeded your grain once or twice. The plants have five or six leaves, and some tiller shoots have appeared. Once the crop has gotten five or six inches tall, driving machinery through it is no longer an option—the plants will not spring back easily and may stay permanently flattened—so it's time to let the grain grow. Growers who depend on foliar feeding their cereals will still drive spraying equipment through a field at this stage, the logic for this process being that two little wheel tracks left through standing grain is small price to pay for the increases in crop yield and quality. Most sprayers are equipped with twenty- to thirty-foot booms. A tractor with eighteen-inch-wide tires will knock down the equivalent of three feet of grain on each pass. If a thirty-foot spray boom is being used, 10 percent of the crop is flattened. Cereal grains fields in Europe are laid out with "tram" lines every thirty feet or so for the purpose of making multiple passes through standing crops with a variety of inputs. These permanently established wheel tracks seem to be standard fare in just about all of the grain fields I have seen in England, Denmark, Italy, and France. Perhaps tram lines will become more commonplace in organic grain fields on this side of the Atlantic as we develop more sophisticated approaches to managing the fertility needs of our cereal crops. We do know that supplemental nitrogen applied to wheat at tillering will increase both the yield and protein level of the subsequent crop. As your grain crop begins its growth cycle in the month of May, you can leave it to its own devices.

The Vegetative Growth Period

The vegetative growth period for cereal grain plants is amazingly rapid. Don't blink for too long because things change quickly from day to day; your wheat, barley, or oats may be five or six inches tall today and well over a foot in height by the next week. Leaf sheaths appear as tillers continue to emerge from the crown node during the first several weeks of growth. Stem elongation, also called jointing, begins with the appearance of the first internode of the stem. Tillering is complete at this point, and stems grow as a series of internodes that increase in length from the base to the top of the plant. Wheat has six distinct internodes, each defined by a leaf and a circular joint in the stem, and once the jointing process is complete, a flag leaf will emerge from the uppermost node. This signifies that a plant's vegetative growth phase is complete. Stem height is predetermined by the genetic code of the particular variety of cereal. Some modern dwarf varieties of wheat barely reach knee height, for example, while older heirloom varieties like Red Fife and Marquis are between waist and chest height when the flag leaf first appears. Six to eight weeks pass by very quickly, and that grain that was just a few inches tall in late April or early May is now ready to reproduce.

The Reproductive Period

The flag leaf is the last leaf to appear on the uppermost node of a grain plant, and the reproductive phase has begun when the grain head begins to swell in the flag leaf sheath. Cereal grains, like all grasses, go through this "boot" stage just prior to the emergence of the seed head where the top of the plant appears swollen. Peel back this extra-wide flag leaf and you will find the seed head inside complete with spikelets. Each spikelet will produce two to five florets, each of which has the potential to produce a single seed of grain. Final yield is directly proportionate to the number of spikelets formed in a head of grain. High light intensity and optimal nitrogen availability will increase the number of spikelets expressed on the grain head that is still in the boot, while low temperatures and excessive moisture can have an adverse effect on spikelet development. Finally the head will emerge from the flag leaf sheath, and your field of tall-looking grass will be transformed into a genuine grain field complete with a sea of heads waving in the breeze.

Pollination

You've been waiting for the heading stage for that little bit of positive reinforcement that you're really growing a field of grain. Pollination must take place for a plant to produce viable seed, as cereals are self pollinating. Once the head has emerged, the anthers inside each floret begin to elongate, allowing the pistils (female) and stamens (male) to express themselves. Like any flower, four stamens surround each pistil. Anthesis or pollination of a cereal flower begins in the midsection of the head and works its way toward the top and bottom over the course of about one week. Pollen granules are quite visible hanging from the spikelets of each head of grain. I became quite familiar with the pollination process during my little heirloom wheat cross breeding project, which involved the emasculation of wheat heads for breeding purposes. The tops of the florets are snipped off to expose the anthers. With the help of intense magnification and a very fine pair of tweezers, the four stamens surrounding each stamen are plucked out, turning the flower into a unisex female. Then pollen shed from a different variety is dusted onto the emasculated flower, and the male and female heads are clipped into a special little paper bag, where hopefully the cross will be successful. A discerning nose will pick up the malty, almost sweet smell of a field of pollinating cereal grain. The smell is comparable to a cornfield during silking and tasseling. This period of anthesis is a critical time in the development of a cereal plant, as grain heads are susceptible to fungal infection from airborne spores when flowers are opening to accept pollen. A prolonged wet and damp period during pollination week, for example, could increase the load of fusarium spores present in the environment, and once these are present in the flower of a cereal plant, fusarium mold develops right along with the seed. This could spell disaster for a grain crop three to four weeks later at harvesttime. Fusarium has the potential to produce the mycotoxin deoxynivalenol, commonly known by its acronym, DON, which is also known as a vomatoxin and as one might expect causes vomiting symptoms in animals and people. All ripe grains, for this reason, should be laboratory-tested for mycotoxins. Wheat with DON levels that exceed one part per million (ppm) is considered unsafe for human consumption. All livestock except pigs may be fed grains with DON levels up to ten ppm. The best you can hope for is warm, dry, and sunny weather during pollination.

Grain Filling

Grain fill commences once pollination is complete. Starch and protein are metabolized in each kernel during the grain-filling process. Grain kernels begin to swell as they fill with a white or cream-colored substance that is simply called the milk. The plant is still very green looking while it is in the milk stage. As maturation progresses, the forces of plant growth transform this very liquid white inner core to a more solid, dough-like state. The soft dough stage is your first hint that a grain kernel is actually developing into something that you'll eventually harvest. You can see the form of the kernel, but it is still soft enough to be crushed or smeared between two fingers. The whole plant appearance will change to a splotchy green color once the soft dough stage has been reached, and chlorophyll in the leaves and stem will begin to fade as the seed develops, giving your once grass-like plant the golden qualities of straw. Leaves will turn yellow first, followed by the stem of the plant, and the soft doughy interior of the ripening grain kernel will harden up to the point where it can be dented with your fingernail, but no longer easily flattened between your thumb and forefinger. Cereal grains are physiologically mature once this hard dough stage has been reached. At this point, the plant's life is over. The seed is fully developed, but grain moisture may still be as high as 25 to 35 percent. Plant stems may still be a bit green, but they are now hollow in the center like a drinking straw. Whole kernel moisture must drop to the 13 to 16 percent level for safe harvest and storage, and the dry down process has begun. Let's hope for hot, sunny, dry conditions to speed this process along.

The Harvest

The dry-down period can be the most nerve-racking part of growing a crop of grain. Once the hard dough stage of development has been reached, it's hurry up and wait. You need a string of hot, sunny days to complete the drying process. You just might find yourself going to the field every day and sampling a few kernels of grain between your teeth for hardness. When the grain begins to crack between your teeth, you know it is close to ready. Once the kernels of grain are physiologically mature, the stalks below begin to weaken, and thunderstorms and wind-driven rain can lay a field flat in a matter of minutes. Once a crop is down or "lodged," a whole new series of problems presents itself. For one thing, the drying process is impeded when grain sits directly on the ground; heads of grain are much more susceptible to sprouting when they can't dry out after a rainstorm. Lodged grain is also difficult to harvest because the combine cutter bar cannot get entirely underneath the crop. Sometimes the only way to harvest a field of flat grain is to operate the combine in one direction toward the grain heads, which makes it take twice as long as normal. Hail is another consideration—a late-July hailstorm can strip all of the ripe heads from a field of grain in a matter of minutes. In addition, an extended rainy period can keep standing grain permanently wet and cause ripe heads to sprout right on the stalk. In a worst-case scenario, little green shoots will become visible as they emerge from the head. Frequent wetting and drying of ripe grain will encourage unseen internal sprouting of seeds in the head. When this happens, grain quality is reduced because starches are transformed into sugars during this process. In wheat, this means that less endosperm will be available for the baking process. Levels of the amylase enzyme will be high because the seed, which is still attached to the plant, has basically woken up and begun to grow within itself. The level of amylase activity within a sample of grain can be measured with a falling number test. Good falling numbers are necessary to make a quality loaf of bread. Whatever the weather situation may be, a timely grain harvest is the best assurance there is for a high-quality crop.

Binding and Reaping

The grain-harvesting methods of yesteryear have the potential to provide us with higher-quality grain samples than the modern combine. When grain is reaped, by either sickle, scythe, or mechanical grain binder, the straw is cut when the kernels are still in the hard dough stage. (Lodging is not an issue, because stalks are still quite strong at this point.) Hand-reaped grain is generally gathered and tied into a bundle about eight inches in diameter, and bundles can be tied with loose straw stems or string. The reaper or grain binder was invented by Cyrus McCormick in 1835 and perfected throughout the late nineteenth and early twentieth centuries. This machine revolutionized grain farming because it not only mowed the grain, but also packed it and tied it into a bundle. One horse-drawn grain binder could do the work of many men. I have very fond memories of operating my seven-foot John Deere grain binder in the late 1970s and early '80s. It was fun traveling around the back roads of Québec's Eastern Townships trying to speak French and looking for a *moissoneuse*. I found one in Coaticook for $250; it was a later PTO-driven model made in the 1950s. Earlier binders were chain-driven by a massive iron bull wheel. I heard many stories about binders not working well because the metal drive wheel would lose traction and slide along on clay soil. The PTO drive solved this problem and made the machine truly "modern." We pulled our binder with a little twenty-horsepower John Deere 40 tractor of the same vintage.

All grain binders are outfitted with a mowing-machine-style cutter bar, a wooden reel with bats, and a platform and two elevator canvases. Conveyor canvases are made with very heavy tarpaulin cloth and have attached wooden slats about a foot apart. We bought new canvases from the Amish of Holmes County, Ohio, and new canvas is still available for these machines today. Grain binder platforms are operated about a foot off the ground. As the machine is pulled forward, the reel lays segments of cut grain (heads back butts forward) onto the platform canvas. Grain is then carried sideways and drawn upward between the elevator canvases toward a packing arm. When the bundle is the proper size, another cast-iron arm revolves around, tying a binder twine knot, and chucking the newly tied bundle onto the ground or an attached carrying unit. My mechanical skills were in their very early stages of development when I set about resurrecting my old machine. Fortunately, I had the little paperback owner's manual. The most complicated task was measuring the diagonals between the ends of the wooden canvas drive rollers to square up the conveying units so the canvases would track properly. After that, the only other daunting obstacle was adjusting the knotter to tie properly, and I got a great lesson about knotters from one of my older neighbors. I highly recommend grain binder harvest technology to any new small-scale farmer starting out with more time and energy than money. These machines are still around hidden away in sheds, and once in a while one will appear at a consignment auction. If you have the wherewithal to take a trip to the Great Plains or the prairie provinces of western Canada, these machines are everywhere.

Stooking Grain

Let's get back to the harvest. Whatever method you use, you're going to do your best to gather up all the standing grain into bundles before it gets completely ripe and totally dry. Hopefully, this will give you a little advantage over having to leave the grain standing out in the field to dry down for combining at 13 percent. Dry-down can take as long as two more weeks, and a lot could happen to a crop during that time. Traditionally, once the grain was cut and bundled, it was stood up in groups of four bundles called stooks, which is still a great method for home-scale growers. Stooking grain is backbreaking work, and many a young boy was ecstatic when the combine became the harvest machine of choice. A bundle of grain is picked up in each hand and the two bundles are simultaneously driven butt-end-first down into the straw stubble so that they lean against each other. A little wiggle action during this process lets the grain heads from the two bundles nestle into each other to form a tight and somewhat waterproof mat. This process is repeated with two more bundles at a ninety-degree angle to form a stook of four mutually supported and intermingled sheaves of grain.

Once a stalk of grain in the hard dough stage has been severed from its connection to the plant's root system, the speed of the ripening process is increased. The last remaining bits of energy and nutrients left in the straw will be translocated up through the stem into the ripening head of grain. Five or six different old-timers told me that the color and appearance of grain that had been reaped and stooked was far superior to combined grain that had stayed in the field for several more weeks. I have to admit that the old fellows were right; my first crop of wheat that had been harvested in this manner had a golden red glow.

The hard dough harvest of cereal grains offers even more advantages to the mini grower. Small garden-sized plots of grains can be quite easily cut down with an old reaping sickle similar in size and shape to the instrument on the old Soviet flag. These tools have a serrated edge. To hand-harvest grain, you'll need to bend over at the waist and cradle a small amount of standing straw in one arm, then draw the reaping sickle backward across the cradled grain with a sawing motion. Repeat this process several times and you will have the makings of a bundle of grain. The beauty of a small patch of grain is that the reaped material can be better protected from the elements. If you have only a few stooks of grain, they can be covered with a tarp when rain threatens. Bundles of grain can be brought onto a covered porch or into an airy barn to cure and ripen. But using this method on larger acreages would be out of the question, so incessant rains can sometimes be a problem for a field of stooked grain. For instance, we experienced a very wet harvest season in 1979 which necessitated that we go out to the grain field and turn all the shocked grain inside out for better drying.

Swathing Grain

Swathing grain is the modern-day equivalent to reaping and stooking the crop at the hard dough stage. Grain swathers have reels and canvases and operate much like a grain binder without the knotting and bundling features. These machines are either self-propelled or pulled through the field by a tractor. I have a very old Massey Ferguson eighteen-foot swather, which transports sideways down the road, but must be opened up to harvest a field of standing grain. An eighteen-foot lightweight steel reel lays the cut grain onto two platform conveyor canvases that each revolve toward an open space near the middle of the cutter bar. In the same manner as the binder, the grain is cut eight to twelve inches off the ground to leave lots of stubble, and as the swather makes its way through the field, cut grain is conveyed toward the middle of the machine and deposited in a crisscross fashion on top of foot-high stubble. All of the heads are pointed in the same direction. The grain remains dry and off the ground, so that air can get underneath the windrow. Swathed cereals cut in the hard dough stage finish the ripening process in much the same manner as reaped and stooked crops—kernels of grain harden up and absorb the remaining goodness from the straw.

Swathing works wonders on weedy crops because it allows the green weeds to dry down along with the primary cereal, which makes combining much less problematic. Also, normal rainfall won't bother a windrow of swathed grain, but excessive heavy rain and other wild weather conditions can wreak havoc in this system. In 1998, I had twenty acres of beautiful swathed barley that got pounded by a hailstorm. The hail drove the windrows of barley right down into the stubble, and I ended up losing half the crop. My 20/20 hindsight probably doesn't serve me well here because had we left the crop standing for the combine, it would have been knocked flat and totally ruined by the hail anyway. If you do plan to swath grain, it would be advisable to look at the long-range weather forecast and hope that the weatherman is right. If all the conditions are favorable, swathed grain can totally dry out in the field. The harvesting process is completed with a combine that has been equipped with a special windrow-pickup attachment.

Harvesting in a Wet Season

After three decades of growing grain crops, I have begun to experience a lot more wet weather during my window of harvest, which begins in late July and continues through August. I have come to understand how much this rainfall can impact the quality of my grain crops. Harvesting a wheat crop that has begun to sprout internally and has low falling numbers can either make or break me financially, as there's a big difference in the prices paid for feed wheat versus those paid for high-quality milling wheat. The same holds true for barley and oats: If falling numbers of barley are low, the crop won't make it for malting, and oats that are subjected to constant wetting darken as they lose the nice white color that is so necessary to make attractive rolled oats.

Once again, our Canadian neighbors to the north have developed a system to preserve grain quality in wet harvest seasons. On several of our grain field trips to Québec, our Vermont group of grain farmers has observed the importance of pushing the envelope on the timely harvest of cereals at moisture contents higher than we ever deemed possible. I had always thought that if a safe storage moisture for cereals was 13 or 14 percent, grains harvested at 16 to 18 percent moisture and bin-dried with forced air would be the limit of possibility. But many Québec producers of bread wheat have now begun to harvest at 25 percent moisture if wet weather threatens their crop. Tremendous volumes of air and well-maintained drying silos are needed to keep grain that is this moist from spoiling in the bin. My own personal comfort level for bin-drying wet grain has now moved up to the 20 percent level. Several years ago, I installed a squirrel-cage-style fan on one of my grain storage silos. This particular unit, which is also known as an air pump, has allowed me to harvest and bin-dry grains at much higher levels of moisture. It is important to realize that all of this so-called quality advantage comes with a cost, as large electrically powered bin-drying fans gobble up the kilowatts. When I run my seven-and-a-half-horsepower air pump day and night for a week to ten days to finish drying fifteen to twenty tons of wheat, it adds hundreds of dollars of energy costs to my total harvesting expenses. This makes machine- or hand-reaped naturally air-dried grain look pretty darn sustainable. Nevertheless, those of us cultivating larger acreages of grains can take solace in the fact that a wet year doesn't necessarily mean we will have a lower-quality harvest. We have been so fortunate to have had the opportunity to learn about these increased options for grain quality from our friends in Québec.

Threshing

As described in the first chapter of this book, the handheld flail and the threshing floor were standard fare for removing kernels of grain from straw for centuries. It all began in the Fertile Crescent of ancient Samaria between the Tigris and Euphrates Rivers when cattle and other hoofed animals were driven back and forth across piles of harvested grain stalks. Ripe grain heads need some sort of applied forceful abrasive action to give up their tightly held kernels, and when we consider the pounding force of a piece of three-inch-diameter hardwood being twirled around and around from a five-foot handle smashing into a pile of grain on the floor, we very quickly understand why this process is also known as "thrashing." It's all about the grain being hit by something in order that we may extract the precious kernels from the whole grain plant. The French use the verb *battre* (to hit) to describe the threshing

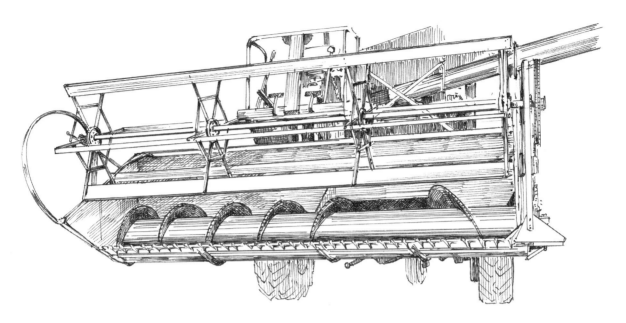

Typical grain header with auger feed intake.

process, and the threshing machine is known as a *moulin à battre* (a mill to hit) in French. Threshing is akin to batting or whacking something with a wooden object.

Threshing Machines

Although the flail and the threshing floor were employed well into the nineteenth century, the first threshing machines were developed in the previous century. Scottish mechanical engineer Andrew Meikle is credited as the inventor of the first working machine in 1784, which was a pretty crude tool consisting of a large cast-iron cylinder with spike teeth rotating inside a wooden box. The spike teeth meshed with corresponding stationary teeth through which the grain passed. The friction between the intermeshing metal teeth rubbed the kernels of grain from the heads, and steam power provided the motive force. These very early machines performed only one function, which was to dislodge the grain from the stalk. Chaff, grain, and straw all ended up in the same pile, necessitating further separation and winnowing. The first US threshing machine was patented in 1837 by Avery and Pitts. As the nineteenth century progressed, many improvements and modifications were made to these amazing machines by large national firms like McCormick Deering and J. I. Case as well as smaller regional outfits. A. W. Gray of Middletown Springs, Vermont, manufactured wooden threshers from 1857 until almost 1920, and the Samson Machine Company of East Berkshire, Vermont, made similar machines during the same period. These early threshing machines were fairly compact units designed for indoor use. The base unit was constructed of select-grade hardwood and measured four feet wide, four feet high, and up to twelve feet long. Motive power was provided by a flat canvas belt that connected a pulley on the side of the thresher to the larger pulley of a horse-powered treadmill. These power units were called "horsepowers" and were also manufactured by small shops like Samson and Gray. The heart of these threshers was a revolving spike-tooth cylinder and corresponding concave below. The concave could be raised and lowered to provide increased or decreased clearance for the harvested grain that passed between it and the cylinder above. Kernels of grain rubbed off in the threshing process dropped through the concave, down a little wooden chute, and into a burlap bag. The remainder of the threshed straw and loose kernels of grain continued through the thresher on a series of reciprocating wooden racks called the straw walkers. The straw walkers were powered by a bell crank connected to the main drive pulley of the thresher. These louvered wooden racks shook back and forth with a four-inch stroke, allowing the remainder of the threshed grain to fall through to the bottom while the straw made its way to the outlet end of the machine. Meanwhile, a belt-driven wooden paddle fan located at the bottom of the thresher blasted air up through the straw walkers to blow and separate the lighter-weight material from the clean grain. These old wooden-style threshing machines were quite commonplace, mostly as permanently installed fixtures in barns all over northern New England; one is still demonstrated every fall at the Tunbridge Fair in central Vermont. Grain threshing was dirty, dangerous, and extremely dusty. Human limbs and fingers were put at risk because they were at such close proximity to the whirring spike-tooth cylinder when grain was hand-fed into the throat of the machine. Threshed straw had to be pulled away from the other end of the machine with a pitchfork. Eyes, ears, and lungs filled with dust and chaff while clothes became filthy and bodies itched. Many old-time farmers have told me that they were glad to be able to buy grain off the railcar because threshing was one of the worst jobs on the farm. This seemingly crude technology still beat pounding the grain out on the barn floor with a wooden flail.

Larger threshers made of galvanized steel began to appear after 1880. These machines were manufactured and distributed nationally by larger farm equipment companies like International Harvester, John Deere, and J. I. Case and were first adopted on the Great Plains and in western Canada. These more modern "tin" threshing machines were equipped with many improvements and deluxe features like self-feeders, clean grain elevators, and straw blowers. Hand-feeding of grain was no longer necessary because whole bundles could be thrown onto the self-feeding conveyor chain. Reciprocating knives at the throat of the thresher cut the binder twine around the bundles instead of the human hand. Grain separation was much improved with longer straw walkers, better blower fans, and the invention of the cleaning shoe, which consisted of two vibrating cleaning screens stacked on top of each other located below and at the end of the shaking straw racks. The top screen sifted grain from chaff by letting the grain kernels drop through while the coarser material made its way out the back of the machine. The bottom screen, also known as the sieve, had smaller openings to give the threshed grain one final sifting. Kernels of grain that passed through the sieve went out as clean grain, and anything that passed over the back of the sieve was returned to the threshing cylinder by means of a return or tailings elevator. The cleaning shoe was truly revolutionary, and chaffers and sieves are still used on modern combines.

The backbreaking job of pitchforking an avalanche of threshed straw away from the outlet end of the thresher was eliminated by the introduction of the straw blower. Threshed straw tumbling off the back of the straw walkers was sucked up by a rather large four-bladed metal fan and blown into a pile through a twenty-foot-long, one-foot-diameter galvanized metal pipe. These larger machines were meant to be used outdoors in the wide-open spaces. They were powered by endless flat belts attached to a large driving pulley on the side of a steam-powered traction engine. As the twentieth century dawned, more and more of these larger, more "modern" threshing machines found their way into New England. Usually one farmer who owned a thresher would travel around to neighboring farms at grain harvesttime. The work was communal. Families helped one another by sharing labor at "thrashing" time, and stories abound about noon-time feasts prepared for threshing crews by farm women. Threshing bees continued into the 1950s in northern Vermont and into the 1960s in the Eastern Townships of Québec, where cereal grains were even more commonplace.

Threshing at Butterworks Farm

I bought one of these old tin monsters in Coaticook, Québec, in the fall of 1976 for $250. It was a Dion 22×32 model manufactured in Ste Thérèse de Terrebonne in the late 1950s. The two numbers indicated that the threshing cylinder was twenty-two inches in diameter by thirty-two inches wide. This was a medium-sized machine, and came with all of the previously described bells and whistles as well as a Hart clean elevator equipped with a mechanical bushel counter and a bagging unit attachment. In August 1977, after our wheat was reaped and stooked, we enlisted the help of three older farmers ranging in age from seventy-five to ninety years old. The machine hadn't been used for a decade, so there was work to do. Chains and bearings were lubricated. Flat belts were repaired and tightened. We pulled the thresher to the edge of the wheat field, leveled it, and secured it to the ground. The next step was to position our old Super M Farmall tractor and hook up the ten-inch-wide thirty-foot flat belt to the big pulley on the side of the thresher. We started up the tractor and flat belt and were surprised that the old piece of equipment didn't really make any more noise than the soothing whir of the straw blower fan and the slight squeak of the self-feeder conveyor chain. Our friend Milton Hammond brought along his Starrett indicator tachometer to determine and adjust the speed of the threshing cylinder. Milton also brought along his ninety-year-old father, who had hopped on a train just across the border in Highwater, Québec, in 1917 to travel out west and join a threshing crew in Saskatchewan. Our first threshing experience went without a hitch. We had been told by our elders that a stationary thresher had the potential to do a better job than a modern combine because you could be right there next to it monitoring its performance. Modern combines weren't operating under perfectly level conditions, and harvesting efficiency was hard to check because the machine was constantly moving through the field and the operator was stuck behind the steering wheel. The Dion threshing machine, however, had a little side inspection door at the end of the straw walker right in front of the straw blower, and it was easy to tell if grain was going out with the straw. You simply opened the door, reached into the cascading straw with your forearm, and checked for kernels bouncing off your skin. Golden wheat poured into burlap bags and mustard seed from the cleaning shoe trickled into a separate bag. As long as someone else was willing to pitch bundles of grain onto the feed conveyor, I could walk all around the operating machine to make sure that all systems were go. If something overloaded or jammed up, a flat belt would fly off a pulley somewhere on the side of the big tin giant, but unclogging a machine and putting a belt back on a pulley was an easy fix. Back then, grain threshing was such an amazing process for someone like me who so far had only dreamed of producing my own wheat.

Working with Old Machinery

I'll be the first to admit that there is a lot of romance and nostalgia in restoring, reviving, and reusing an old threshing machine. Does an old piece of equipment like this have any merit to someone wanting to do a small-scale grain harvest today? I did a yearly harvest of six to eight acres of grain for six years before I retired the old thresher. I stored it away in a shed and finally gave it away to a homesteader from New Hampshire in late November 1994. We put the thresher on his homemade trailer, and he secured it with chains and binders and headed south with his pickup truck, trailer, and new prize. I just happened to pick up a copy of the newspaper the next day—and there on the front page was my old thresher lying on its side blocking traffic on Interstate 91 in St. Johnsbury. The poor fellow had hit a patch of black ice on the road, and all of a sudden his trailer—which had no brakes—decided to go faster than his truck. He ended up sideways in the road and rolled over. Luckily, no one was injured in the mishap. This was, however, a sad end to a machine that I loved dearly.

Although there are fewer threshing machines around now than there were thirty-five years ago, they still can be found. Occasionally, someone will want to clear an old wooden one out of a barn and will offer it up free for the taking. Old threshers have become collectors' items and may sell for upward of a thousand dollars. This might be too much to pay when you can buy an old pull-type combine for half to a quarter of this amount. There are still quite a few metal threshers left in Canada, however. If for some reason you must really have one, the Dakotas are loaded with them. It rains a whole lot less there, so machines that have been left out in the weather don't degrade as fast as they would in our wetter climate. If you want to run a thresher, you will also need a tractor with a belt pulley attachment. This means you will need a thirty-five-horsepower or better tractor from the 1950s or early '60s. If you are as determined as I was to process your grain in this manner, make sure you get a machine in excellent condition. Check the bearings on all of the shafts and make sure that the internal wooden parts aren't rotten. You can't just order up parts for these things from an equipment dealer, so you might consider getting a second identical machine for replacements. Another word to the wise: Give yourself plenty of time before the harvest to get your old relic running smoothly and flawlessly. If you're really smart, you'll start your restoration project a year in advance. These machines are nowhere near as complicated as a modern combine. Someone like me with only average mechanical abilities should be able to tackle a threshing machine.

Threshing on a Home Scale

There are also a number of options in the mini threshing machine for the ultra-small-scale garden grower of grains. Perhaps the most interesting recent development in small grain threshing is the battery-powered Minibatt, made in France. This little handheld unit sells for around six hundred dollars and operates on rechargeable battery packs much like a cordless drill. It is constructed of durable hard plastic and has a cylinder and adjustable concave for threshing just about any type of grain. The Japanese have been making small threshing units powered by Honda motors for years. These machines do everything that a larger model does, but are quite compact and transportable; they generally measure about eighteen inches wide by two feet long and three feet high. Japanese threshers are generally used for processing rice but can be easily adapted to all sorts of grains. They are quite expensive, usually selling for several thousand dollars. This would be a modest investment if you're totally dedicated to growing all your own food. The final and most interesting sort of small-scale thresher is the one you make yourself. A quick little tour through YouTube on the Internet will transport you to an almost unlimited collection of home-built threshing machines. Most are small plywood boxes equipped with homemade cylinders and concaves powered by very small electric motors. Some inventive folks have retrofitted gasoline-powered leaf shredders into workable threshers. The basic cylinder and concave is pretty easy to simulate in a homemade small-scale machine. The separation technology of straw walkers and the grain shoe are a bit more of a challenge to simulate on a small scale. Whatever your bent, threshing grain can be accomplished in any number of ways that will thrill and excite you when you are holding that finished grain in your hands.

The Combine

The combine harvester gets its name from the fact that it "combines" the reaping and threshing of a grain crop into one continuous process. The very first combines were massive wooden affairs pulled by as many as fifty horses up and down the rolling hills of eastern Washington and in California's Central Valley. There are many archived pictures (often seen in draft-horse books and magazines) showing an early Holt combine harvester in the golden wheat fields of California. These massive pieces of equipment were little more than a large grain binder attached to the side of a threshing machine. The first more modern-appearing machines fabricated from steel weren't developed until a few decades later, when the Curtis brothers of Kansas founded the Gleaner Manufacturing Company and began marketing a much more compact combine harvester in 1923. Gleaners are still a popular brand of machine today. Massey Harris, a Canadian company, released a very lightweight versatile combine in 1940, and by the late 1940s most of the major farm machinery manufacturers were offering a complete line of grain combines to the North American farmer. John Deere's first self-propelled combine was the Model 40, with an eight-foot cutting width; it was manufactured right up until 1961. Larger models like the Number 55 and 95 were made through the 1960s. Other companies had their own particular versions of these early self-propelled machines, like International Harvester's McCormick No. 141. Grain combines were still pretty modest in size through the 1950s and into the '60s, and eight-foot cutting widths were standard. Some of the larger machines had ten- and twelve-foot cutting platforms, which seem pretty tiny by today's standards.

Pull-type machines were much more common in New England where fields were smaller and grain a less important crop. Thousands of John Deere 12As and McCormick 64s were sold in this part of the country, and farmers had several choices when it came to outfitting a new combine. (Pull-type machines could be powered by a PTO shaft coming from the tractor or an independent gasoline engine mounted on the combine drawbar, for instance.) The predominance of small low-horsepower tractors and hilly terrain found in the Northeast gave farmers plenty of reasons to choose independent motor-driven models instead of the PTO-driven machines, however. Pull-type combines could also be equipped with self-unloading grain tanks or a bagging station complete with carrying platform and bench seat. All of these combines had "scourkleen" rotary grain cleaners located between the clean grain elevator and the grain tank or bagging unit. They were farmer-friendly, built with the smaller operator in mind.

Widespread adoption of combine harvesting came first to the drier regions of the country where grain growing was the predominant agricultural activity. What a relief it must have been not having to hand-stook all those thousands of acres of grain. Gone were the threshers and the huge crews of men needed to gather grain bundles. But grain has to be "dead ripe" to be successfully harvested by a combine, which means that the crop must be left in the field for a week or two more than it if was reaped in the hard dough stage with the old-fashioned grain binder. This practice was much more easily done in places like Kansas or North Dakota where the annual rainfall is oftentimes less than fifteen inches, but rainfall levels in the Northeast are about four times that amount. Combine harvesting may have come to the Great Plains in the 1940s, but it wasn't until the next decade that it became commonplace in our part of the country. This "new" type of harvesting came along with a learning curve for many farmers, especially when they realized that the huge pile of oats they'd just put in the wooden bin wasn't quite dry enough. It was time to jam as many dry wooden fence posts as possible down into the grain to suck up as much moisture as possible to keep the crop from heating. Spoiled grain that has heated turns moldy. It can happen to anyone trying to beat the weather and harvest a crop. I have learned quite a few hard lessons in this department. I cannot reiterate the old adage enough: *Check the moisture before combining a crop of grain.* If it is over 14 percent and you don't have aeration or a dryer, leave the crop in the field until it is dry enough. As stated earlier in this chapter, it is possible to combine grain at higher moistures, but you must be prepared to condition it in a manner that allows the crop to dry during storage. Access to a grain moisture tester is essential in this business. If you don't live close to a feed mill or someone who can test your grain moisture levels, you will have to get your own tester, which can cost anywhere from two hundred to a thousand dollars, but they are a good investment. All you need to do is save one crop from spoiling, and you've made your money back. Dry grain will crack between your teeth, but it's nice to have a precise moisture measurement before you put the combine in the field.

How the Combine Works

The modern grain combine must first mow the grain in the field. There are numerous functions involved in cutting down the grain crop and delivering it to the threshing unit, and some of these features have been borrowed from the grain binder of yesteryear. Standing grain is mowed down with a standard reciprocating cutter bar, which is part of the header at the front of the machine. A revolving reel, similar to the one on a grain binder, guides the standing grain into the cutting mechanism. Earlier combines had the same wooden batt style of reel as the binder, but the Hume reel was developed in the 1960s when some enterprising and observant individual added five-inch-long spring steel teeth to the reel batts. It now became much easier to pick up down and tangled crops. Within a very short time, all makes and models were using the Hume reel. Reel height much be watched carefully, however; if a Hume reel is operated in too low a position, the metal teeth can get caught between the knives of the cutter bar section, causing all sorts of breakage. For this reason, specially molded plastic teeth soon replaced the spring-steel variety for even better crop handling at the header. On earlier pull-type machines, cut grain made its way into the threshing chamber on a wooden slatted elevator canvas very similar to what was used on the grain binder. The canvas system of grain transfer left quite a bit to be desired in the design and functionality department. For one thing, the metal driving rolls on which the canvas rode had to be square in relation to one another for the canvas to track properly, and if they were out of adjustment, the canvas would ride to one side or the other and get stuck in place. Things had to be really tight for the canvas to get proper traction on the rolls. My first combine was a 1954 International McCormick Model 64, which was equipped with a six-foot-wide canvas feeder pickup. Every time I started up the machine, I had to risk life and limb by leaning over the side of the header to yank on the canvas to get it to revolve properly.

Farm machinery manufacturers soon replaced canvas pickups with an auger/feeder house system. This worked a whole lot better for getting cut grain into the threshing chamber, and it has been standard fare on combines ever since. Cut grain is conveyed to the center of the grain platform by a sixteen-inch-diameter revolving metal auger located right behind the cutter bar. As this large tube turns, the attached spiral flighting delivers the grain to a series of special metal pegs that protrude from its very center. These spinning short metal fingers grab the grain and propel it into the throat of the combine—called the feeder house—which is basically a long upwardly sloping rectangular metal box that connects the cutting platform to the main body of the combine. It contains an endless conveyer chain of crosswise metal slats that pulls the grain upward and into the threshing chamber. This conveyor system is known as the raddle chain. Feeder house design hasn't changed much over the years except to get larger and wider as machines have grown in size.

Space is quite tight inside the feeder house; wads of grain can get jammed in there and bring everything to a screeching halt, especially in green or weedy conditions. Combines work much better with bone-dry material. If you plug your feeder house, you must stop the machine, grab a pipe wrench, and turn the main drive shaft backward with as much brute force and ignorance as you can muster, and I speak from experience about this particular predicament. Combine manufacturers began installing automatically reversing feeder houses on machines during the 1980s, and I bought one in the late nineties—as soon as I could afford it.

Once the grain has traveled up through the feeder house and into the front of the combine, the threshing process begins. The workings of a modern grain combine harvester are very similar to those of its predecessor, the threshing machine, with some added refinements and improvements. Once the grain enters the machine, it travels over a deep pocket called the rock trap before it makes contact with the threshing cylinder. Rocks and other debris that get sucked up through the feeder house can be very damaging to the internal components of a combine. Rock traps are a wonderful feature, and if you are working in stony conditions and have to run the header low to the ground to pick up lodged grain, it is advisable to check and empty the rock trap often. To do this, you must lift the header as high as it will go, crawl underneath, and open a little trapdoor. Remember to drop down the little safety stop around the hydraulic cylinder rod anytime you must go underneath this part of the machine. It only takes a second, but it could save your life if the header decides to drop when you are below it. Once it's past the rock trap, harvested grain passes between the threshing cylinder and the corresponding semicircular concave grate below it. Hard metal rasp bars bolted crosswise around the cylinder rub the grain against the concave to complete the threshing process exactly as it would happen in a threshing machine. The actual rasps on the rasp bars are specially cast little pieces of metal that are slightly raised up and on a diagonal.

The roughness of these little spurs is what removes the kernels of grain, by means of the friction that develops as the straw and all passes between cylinder and concave. Rasp bars and concaves do wear over time and eventually need replacement to ensure complete and efficient threshing of crops. There is a whole lot more potential for fine-tuning and adjustment when you're threshing with a modern combine. Cylinder speed is infinitely variable, from four hundred RPM for tender crops like peas and beans to over twelve hundred for tougher-threshing crops like wheat and flax. It can be dialed in right from the cab and accomplished through a special set of variable sheave pulleys and rubber V drive belts on the side of the machine. Variable sheave pulleys change in diameter, which allows the drive belt to travel either faster or slower in much the same manner as the different sprockets on a ten-speed bicycle will allow it to move faster or slower. Threshing adjustment is much more precise because combine concaves also have the ability to be raised and lowered on both the front and back sides of the cylinder. Threshed kernels of grain drop through the concave, whereupon they begin to make their way to the shoe for final cleaning.

The cleaning and grain-separation area of the modern combine also operates much as it did on the threshers of yesteryear, although a few of the cleaning processes have been improved with some technological refinements. Grain movement in the very bottom of the combine has been improved by the addition of a series of parallel steel augers that run the entire length of the machine in semicircular troughs. Once the threshed kernels drop through the concave grate, they are carried along by these side-by-side augers to the front of the cleaning shoe. As much as 80 percent of the grain will drop directly through the concave; the remaining 20 percent is mixed with threshed straw, chaff, and other material and is separated out in the rest of the machine. After the grain goes between the cylinder and concave, it immediately passes by a lighter-weight revolving steel drum called the beater. The beater applies a little more friction to the threshed straw, slowing down its rate of passage through the machine and allowing more kernels of grain to fall through to the augers below. After this point, the coarser straw finds its way to the straw walkers while the finer material ends up at the front of the cleaning screen. The internal workings of the whole back end of the combine are shaking madly back and forth. Meanwhile a fan located underneath the main body of the machine is pushing a blast of air up from underneath. Cleaning fan speed is also adjustable—higher speeds are needed for heavier seeds like wheat and soybeans while lower fan speeds are required for oats, flax, and other light-seeded crops. The air blast is necessary for good separation because it keeps all of the threshed material suspended so that the heavier kernels of grain will drop out. All of this shaking, vibrating, and blasting air is quite a sight to see and experience. It's a wonder the whole business doesn't shake itself to death. (These bell cranks, the reciprocating arms, and the bearings that carry them do wear out by the way, and should be greased, inspected, and maintained often.)

All the threshed grain must make its way back to the center of the machine to be scooped up and carried up to the grain tank by the clean grain elevator. The augers that run along the bottom of the machine are designed with both left-hand and right-hand flighting so they can move grain from underneath the concave to the middle as well as from the rear of the machine forward to the clean grain elevator. This little feature—having an auger that turns only one way while it moves grain in two different directions—is absolutely ingenious. Grain separation takes place in several different areas. As the straw walkers seesaw back and forth in the top of the separating chamber, kernels of grain drop through slots into troughs below, and the straw is carried rearward and dropped out the back of the machine. Grain slides down these troughs and is deposited right in front of the cleaning shoe, which is the heart of the combine's cleaning department. Two screens with little louvered metal fingers are

Bird's-eye view of grain passing through a combine harvester.

positioned one above the other in a frame that is being madly rocked back and forth, and these little louvers can be tipped up or closed down by means of an adjustment lever. Different settings for various grains allow for more or less space between the louvers. The top screen, called the chaffer, performs the rough sorting of the mixture of grain and chaff that is moving through the lower section of the combine. Any grain that doesn't quite make it to the front of the chaffer or that falls off the front is dumped directly onto a small cross auger and taken across the bottom of the combine to the clean grain elevator. Chaff and coarse material are blown off the back of the chaffer onto the ground. Kernels of grain that drop through the chaffer fall onto the sieve directly below. Sieve louvers are usually closed up tighter than the chaffer above it, and clean grain that falls through the sieve is delivered directly to the clean grain elevator. Material that falls off the rear of the sieve is delivered by another cross auger to the tailings elevator; this takes it up the side of the combine all the way to the front of the machine and back to the cylinder for rethreshing. This is a highly complex process that is quite difficult to explain with only words.

Combine Troubleshooting

It's difficult to know just how many different processes are happening at the same time when a combine is out there in the grain field doing its job. Every moving part of the machine is getting its motive power from one engine, which is situated in its own compartment next to the driver's seat, and power is transmitted to various parts of the combine through an array of pulleys and rubber V belts. All the systems for cutting, threshing, separation, and cleaning must be working in unison, and if one particular part of the process fails or is compromised for some reason, the machine cannot do its job. Therefore it is necessary to be totally vigilant and alert while operating, as you never know when a bearing will fail or a belt will break. It's hard to tell what's actually going on while you are in the driver's seat looking forward and down at the cutting platform and standing grain in front of you, but there are many clues. A change in sound or in the pitch of the general racket of the machine is usually a good sign to stop and check on things, and a quick peek out the rear window of the cab will tell you if grain is still streaming into the tank. The rattling sound of a slip clutch will tell you that an auger or an elevator somewhere is plugged or jammed. The smell of burnt rubber will indicate that a belt is slipping or broken. Frequent glances at the instrument panel will alert you to potential problems as well. If the RPMs on the cylinder tachometer begin dropping, something is wrong in the threshing chamber. Engine temperature is also well worth monitoring because dust and chaff will plug a radiator quite quickly, causing poor cooling and overheating. Most manufacturers began installing slow-shaft indicators and alarms in combines during the 1980s, and if these systems are maintained, they can help you shut the machine off early before a small problem becomes a large one. Like any piece of farm equipment, a combine needs to be maintained. Bearings should be greased often; engine and hydraulic oil levels should be checked often and kept full at all times. It's best to study your operator's manual and follow all of its recommendations. A well-maintained cutting platform with sharp knives and good guards is essential for smooth combining. But this is just an introduction to the world of combine harvesting, and there is no substitute for the experience of actually running a machine yourself. With a good basic knowledge of how your combine works and a few of these details and considerations in mind, you will be well on your way to harvesting your own crops of grain with your own machine.

Operating a Combine

Once you understand how your machine works, you can begin to make operating adjustments. For a general idea of how to set your combine, consult your operator's manual. Threshing efficiency is determined by the speed of the cylinder and the height of the concave, and both of these adjustments can be made from the cab with levers and cranks while the machine is running. The tachometer will tell you the speed of the cylinder, and a little dial and pointer on the right side of the combine will tell you how much space there is between the cylinder and concave. Generally speaking the faster the cylinder speed and the closer the concave, the more aggressive the threshing will be. These settings are quite important because overthreshing will damage grain by cracking and breaking it. So be only as aggressive as is needed—cracked grain will not keep as well and is worth a whole lot less. Consult your manual for chaffer and sieve settings. The chaffer is usually opened up almost twice as wide as the sieve underneath it; if your chaffer is five-eighths or three-quarters of an inch open for wheat, the sieve might be open half or three-eighths of an inch. Setting these little louvers can be difficult with a common tape measure because it's hard to fit the measure down in there. Try fitting a bolt of the proper diameter down between the louvers to determine the setting; once you have begun harvesting, small adjustments to the shoe can help you get a cleaner sample in the grain tank. If the grain is on the dirty side, close down the chaffer a little. This will clean up the grain sample, but can also send more grain out the back of the machine. If the sieve is closed down too much, the tailings elevator can be overloaded and you may hear the slip clutch chattering. The tailings elevator in most machines can be inspected through a little door on the floor of the cab right by your feet. This is the grain that is coming back to the cylinder for rethreshing. You only want a moderate flow of grain coming up the tailing elevator; too much tailings grain is a sign that the grain is not getting threshed enough on its first pass through the combine. The last major combine adjustment is the speed of the cleaning fan. This is done by turning a little knob that changes the diameters of a couple of variable sheave pulleys. Like the cylinder speed, this adjustment must be made while the combine is running. Go with the fan speed listed in the book for starters, and once you're in the field, get someone to walk behind the operating combine and catch some of the material blowing off the chaffer screen in their cap. Empty the cap slowly, letting the chaff fall into your hand. The presence of grain kernels indicates grain is going out the back and the wind needs to be turned down. On the other hand, you might want to crank up the wind for better separation. The last small adjustment to make once you are out in the field is up front on the cutting and feeding end. Most machines made from the 1970s onward have hydraulically driven variable-speed reels. First find a ground speed that allows for even feeding of grain up into the feeder house; believe it or not, faster is sometimes better than slower. You definitely don't want slug feeding where a pile of grain builds up behind the cutter bar and then is sucked into the machine all of a sudden. Lower the reel in relation to the platform to evenly guide the grain to the cross auger, and then adjust the reel speed so that it is turning about 20 percent faster than your ground speed. If the reel turns too fast, it can "shatter" the kernels right off the stalk, which means that grain ends up on the ground instead of in the machine. Just as there are many systems to maintain and observe on a combine, there are many adjustments that must be made all in relation to one another. The best thing to do is educate yourself, get some help from an experienced mentor, and go at it.

Combine Use at Butterworks Farm

I've owned five combines over the past thirty years and every one of them has been a great learning experience. My first machine after reaping and threshing for six years was a 1954 McCormick Model 64 pull type made by International Harvester. This was what they called a "straight through" combine—the cylinder was six feet wide, just like the cutting head and elevator canvas. Later machines had a much narrower threshing cylinder in relation to the cutting platform. Wide cylinders like this were somewhat problematic because they were weaker in construction and took more of a beating from the grain passing through, and I had to replace bearings on the cylinder shaft. My combine was driven by an auxiliary motor that was situated right next to the bench for the bagging unit. If the person running the bagger wasn't inhaling grain chaff, they were breathing in exhaust from the engine that was two feet away. The canvas that brought the grain up into the threshing chamber was constantly slipping or running crooked, and sometimes the grain was too slippery to travel up the canvas. I struggled with this machine for several years, but can't say that I ever enjoyed running it as much as my old thresher.

In 1987, I purchased my first self-propelled combine from my old friend and mentor Clarence Huff of Moes River, Québec. I got to know Clarence because I bought seed and feed oats from him; he was almost eighty years old at the time and had just parked his old Oliver Model 25 for good. He sold it to me for three thousand dollars, and he came along with it for the first year. This was a step up from the old pull-type machine, and Clarence came down two days in a row and ran his old combine through twenty acres of barley for me. I was amazed at how easy it was, especially since he operated the combine—all I had to do was haul grain away with my old farm truck! The old Oliver had a ten-foot cut and had been made in 1958. It was powered by a Continental flathead six-cylinder gas engine. The combine's only weakness was its raddle chain, which pulled the grain up through the feeder house. Clarence had babied it and was constantly replacing weak links, and it wasn't until the next year when I was totally on my own that I got to experience breakdowns and the pleasure of repairing my own piece of equipment. This machine was so old that parts were not readily available, but I was able to buy a new raddle chain from a Mennonite repair shop in central Ontario. My repair experiences went pretty well, and I started building my confidence. However, in late September 1988 a very hard freeze descended upon us and froze and ruined the block of the Continental engine. Clarence had always drained the block instead of using antifreeze, but I had waited too long to do the job. I was devastated, but determined to find a new motor. We had a friend in Ontario put a want ad in the *Ontario Farmer* and found a block right off, five hundred miles away in Collingwood, Ontario, near Georgian Bay; it belonged to a farmer who was just about as old as Clarence. It took a week or more to install the "new" old engine, but we got the job done in time to harvest some really nice wheat and barley in August 1989. I learned even more about operating and adjusting my Oliver combine in 1990 when I began growing soybeans, where I had to slow the cylinder down and change some other settings. The old Oliver 25 served me quite well and trained me in the process. I sold it to the Massachusetts Audubon Society's Drumlin Farm in Lincoln, Massachusetts, in 1994 for a thousand dollars. This was a happy ending, and I was glad to see it used on another small farm.

There weren't too many other farmers growing grain in Orleans County, Vermont, in the late 1980s and early '90s. Jacques Couture from my town of Westfield and Guy Robillard just over the mountain to the east in Irasburg were both growing substantial acreages of barley to feed their dairy herds. Guy gave up raising barley in 1991 and switched to buying in high-moisture corn. He parked his John Deere 3300 combine in the shed, and I bought it in 1992 for five thousand dollars. The 3300 was the smallest of John Deere's 100 series of combines, made in the 1970s. It was my first machine with a cab, and it seemed gigantic to me. The cutting platform was thirteen feet wide, and I was able to breeze through my entire twenty-acre field of barley in one day that summer. I consider the John Deere 3300 to be my first grain combine from the modern era. I could order any part that I needed from my local dealer and it would be delivered by UPS the very next day. This was paradise. As my grain acreage and diversity of crops increased through the 1990s, I had my trusty combine, and after using the machine for several years, I replaced the grain platform with a dual-purpose "flex" head for cutting soybeans as well as cereal grains. Removal of a few bolts along the front of the cutting platform allowed the whole head to be lowered right down to ground level and follow the contours of the land quite precisely. This so-called shaving action was needed in crops like soybeans whose pods hang right down to within an inch or two of ground level. When it was time to harvest barley, oats, or wheat, the platform could be made rigid again by reinstalling the bolts, and I bought a three-row corn head for my little 3300 in 1994 when I combined my first crop of grain corn. Every time I bought a new head or add-on for my combine for two to three thousand dollars, I realized what a great deal Guy had given me on the machine. But then again, Guy must have known that I was going to have to repair the thing, because those dollars added up, too. Once my grain acreage reached the 150 mark, I began to realize that the 3300 was indeed a small machine. The thirteen-foot grain head was really a bit too large for the combine, especially in weedy and green conditions where the feeder house would plug and the combine would come to a screeching halt. Then it was time to jump out of the cab with a pipe wrench, latch onto the feeder house drive shaft, and lever it backward with all my might. This wasn't a speedy process. It could take up to half an hour to unplug things enough to restart the combine; knuckles would get skinned and epithets would fly. The worst thing about this whole situation was that it could happen again five minutes later fifty feet down the field. Time losses at harvest started to add up, and I began to think about getting something a little bigger with more capacity.

My next move in the world of combine acquisition will hopefully act as a lesson in what not to do for anyone out there considering buying a machine.

I met a fellow at an Acres USA Conference out in the Midwest who had a friend with an older John Deere 6600 combine that was for sale cheap. Since the 6600 model was two sizes bigger than my machine, I assumed it would be everything that I ever needed. I sent the thirty-five-hundred-dollar check out to Illinois, and my combine arrived on a flatbed truck about a month later. It came along with a six-row 653A row-crop head for harvesting sunflowers and soybeans, and yet when the truck pulled into my yard, my jaw just dropped. The combine was a hulking pile of rust. I hoped that its insides would be in better shape than its outside. Unfortunately, that wasn't the case. The thing had sat out in the weather for years about sixty miles east of St. Louis. I figured that chemical plant fallout from companies like Monsanto might have been partially responsible for some of its corrosion. To make a long story short, I ended up repairing everything from the combine's motor right down to its very guts. Once we got the entire machine restored and repaired, we had gained an intimate understanding of the inner workings of a John Deere 100-series combine and we were ten grand out of pocket. The saddest thing about this whole adventure was that this machine was underpowered for its size. There was a lot of combine to carry around with a pretty small motor. When it was all said and done, the 6600 didn't work any better for us than the little 3300. I loaned my "folly" machine to a friend in New Hampshire and eventually sold it at a great loss to another beginning grain farmer in the Connecticut River Valley of Vermont. Sight-unseen purchases from individuals you don't really know are to be avoided at all costs. My pocketbook and my pride have the scars to prove it.

I began to look around for another combine. This time I decided to step up to the 1980s, when John Deere went from the 100 series to the 20 series. Instead of looking for a 6600, I now thought a 6620 would be a better choice. Once again, I started looking in the Midwest, but this time I stuck with reputable dealers. I checked out a few machines in Wisconsin and Illinois, and actually paid a knowledgeable third party to go and inspect these combines. The reports weren't glowing, so I didn't buy. Finally, I found a combine advertised in the monthly *Fastline* used-farm-machinery magazine, and I bought a John Deere 6620 and grain head from Reitzel Brothers of Edon, Ohio, for the 1999 harvest season. The price was eighteen thousand dollars for the basic combine and twenty-five hundred for the grain head. My new combine was a Titan 2, which meant that it was one of the last of the 20 series made in 1988. I was so glad to have purchased a piece of equipment that was only a little over ten years old. There were only twenty-five hundred hours on the tach, and it ran and performed like a top. This was the first hydrostatically driven machine that I had ever owned or operated. There was no more clutch or any variable-speed pulleys and belts on the side to propel the thing along, and ground speed was infinitely variable through a series of hydraulic pumps and hydraulically driven wheel motors. Best of all, the feeder house was reversible and could be unplugged with the push of a lever from within the cab. The pipe wrench days were over. Instrumentation in the cab was much more advanced as well, with digital readouts for ground, cylinder, and cleaning fan speeds. Warning lights and buzzers for various slow shaft speeds and combining malfunctions were standard equipment. This was indeed the machine of my dreams, and it was well worth the interest I had to pay on the banknote I took out in order to purchase something this new and expensive. The very best thing about the new combine was that it didn't break down for a couple of years; it was all paid for before I had to start sinking more cash into it.

More than ten years have passed, and I still have my 6620. It has allowed me to efficiently harvest even more grain, and we have had well over two hundred acres in some years. The machine has broken down and been repaired many times, but breakdowns on combines are simply a way of life. There are so many moving parts that are wearing every second the combine is operating. Bearings seem to go the most often, and there is usually more of a problem removing belts, pulleys, shields, and elevators on the side of the machine to access the bearing than there is to actually change it. Small hydraulic leaks can be perplexing and difficult to find as well. Several years ago, we had to replace the engine, which we did right out in the field during harvest. Most recently, we went entirely through its inner workings, replacing rasp bars, a concave, and a straw walker. Bigger, newer machines are nice, but are tremendously expensive to maintain and operate. The 6620, for example, has 160 horsepower and can suck down more than fifty gallons of diesel fuel in just one day of grain harvesting. It's easy to understand why many custom combine operators charge a hundred dollars per hour. All in all, though, I have found my machine to be quite versatile across a wide array of crops and harvesting conditions. We have used our combine for cereals like oats, wheat, barley, spelt, and rye as well as for row crops like corn, sunflowers, soybeans, and dry beans. We have had corn heads, grain heads, row-crop heads, and windrow-pickup units on the front of our machine. At this point, I have no intention of getting anything bigger or newer, as I think we have found our sweet spot. We will keep operating and repairing this machine because we know it so intimately and we own it free and clear.

Choosing the Right Combine for You

There are many considerations when outfitting your farm or operation with a combine. What sorts of crops do you expect to harvest? How much land do you expect to cover? What kind of terrain are you dealing with? If you are only doing a few acres and there is an available machine nearby, it is definitely wiser to hire the job done. If there is no one you can depend on, an older pull-type combine will probably be more than adequate. Always look for a machine that has been stored inside, as rubber belting and canvases don't last long out in the elements. Buy a canvas conveyor-type machine if you must, but an auger-style pickup is much more trouble-free. The McCormick Model 64 from the early 1950s had the canvas pickup, while the Model 92 from the late '50s had the auger style. The same is true with the older John Deere pull types. The Model 12A from the late '40s and early '50s had the canvas while the Model 30 from 1957 had the auger feed. Probably the most versatile of all the pull-type combines was the Allis Chalmers All-Crop. Over three hundred thousand of these machines were made between 1935 and 1969. They are distinguishable by their bright orange color and sideways delivery of threshed straw. The All-Crop combine was unique in that it had a rubber-fingered cylinder and rubberized concave, which allowed it to do a much better threshing job in tender crops like dry beans. The All-Crop can work as well in clover and canola as it can larger seeded cereals like wheat, oats, and barley. These machines are highly prized by small-scale homestead farmers. Prices for pull-type machines are all over the map; antique farm machinery collecting has become quite popular, and sometimes this pushes prices upward. Sometimes the opposite is true, however, when children or grandchildren want to clean out the barn or shed and are quite willing to part with Grandpa's old combine quite reasonably. Prices for an old pull-type machine can range from two hundred dollars to two thousand. There are fewer of these machines around than there used to be, but they still exist and can be put to work on your farm or homestead.

The most amazing thing about finding, acquiring, and reviving older combines is how little they cost compared to their original purchase price. The first owners of most machines paid very dearly for them in dollars that were worth a whole lot more then than they are now, as farm machinery values decline very rapidly with time and usage. This is one of the reasons why I recommend acquiring a small self-propelled model from yesteryear over a pull type. A John Deere 40 on an International Harvester 101 self-propelled machine from the late 1950s or early '60s can sell for the same low price as a pull-type model from the same era. Since they can't be towed around by an antique tractor, these old self-propelled hulks aren't nearly the collector's items that the smaller units are. Combines from the late '60s and early '70s are equally cheap, as collectors don't want them and they are too old and too tiny to be put to work in present-day large-scale production agriculture. These machines don't sell for much more than two grand. My John Deere 3300 combine that sold for five thousand dollars in 1992 cost pretty close to thirty thousand when it was new in 1977. Combines from the 1980s might be a few thousand dollars more, but are still a great value for the money if they are in reasonably good condition. The trick is to get a good one. Some machines have been better maintained than others over the years. General appearance is a good place to start gauging quality. Has the machine been cleaned of old grain? Do the lights work? Is the sheet metal all dinged and dented? The pride of the former owner is often reflected in the machine. Every once in a while you might get a bad machine with a new paint job, but this is much more common in something like a tractor that is less trouble to paint.

When looking for combines, it's always a good idea to take someone along with you who knows more about equipment and general mechanics than you do. If you can get a parked combine to start up and run, this is all for the better. Sometimes this isn't possible if the old combine has been sitting for many years. If you can get the machine to run, listen for bearing squeals, thumps, and banging. These are all clues to be investigated. If something is inexpensive enough, it might pay to fix it if it's only a minor problem. Look at the hour meter if the combine has one—two thousand hours seems to be the cutoff line for not having to do too many repairs on a combine. One last consideration about buying a combine from the Midwest is the cost of transporting the thing to the East. They are a dime a dozen out there, but it is over a thousand miles away. Sometimes the trucking can cost more than the combine. Still, as I've said before, machinery breakdowns are inevitable and a fact of life. Once you get to know and understand your new combine, a good preventive maintenance schedule and careful observation of all the moving parts can go a long way toward avoiding a breakdown during harvest. If you can have some common spare parts like V belts and bearings on hand, you might stand a better chance of getting back to the field sooner if it does break. Hopefully, this long treatise about ripe grain and the mechanics of its harvest will help you harvest your future grain crops as happily and easily as possible.

CHAPTER SEVEN
From Field to Storage

After and during the harvest, grain must be stored and transported in some sort of a vessel. Depending on the scale of the endeavor, anything from a grain sack to a large truck can be used.

A very tiny garden harvest can be easily stashed away in bags, pails, or trash cans, but rodents and wildlife can make quick work of eating or damaging your stored harvest, so a sealed vessel is best. Mice especially have an uncanny ability to crawl between sacks of stacked grain and burrow into bags that you can't see from the outside of the pile, and once they're installed in these cozy confines the little critters take up residence. They have all the food and shelter they need, but unfortunately, they don't leave to attend to their bodily functions; urine and feces begin accumulating, accompanied by the telltale odor of mouse pee. Once this distinctive smell has been imparted to the stored grain, contamination is a done deal. This odor won't go away and often will permeate a much larger amount of material that surrounds the nesting area.

Storage Container Options

If bags are your only option for storage, try to store them out of reach of rodents. A defunct old chest freezer placed somewhere in a dry leak-proof building makes a good storage cabinet for grain bags. In recent years, shipping containers both large and small have become commercially available at fairly reasonable prices. These watertight metal boxes, which are twenty or forty feet long, are ideal for keeping mice and rats away from grain stored in bags. However, you must be vigilant and keep the doors closed to avoid problems with marauding unwanted guests, as raccoons and squirrels also want to sample your hard-won harvest. Electronic plug-in rodent repellers are also an option in storage situations, but are not as sure a thing as actually preventing physical access to that very tempting source of food and shelter.

Trash Cans and Metal Drums

Trash cans and metal drums are the next storage choice after the grain bag, and you have numerous options in this department. A pretty nice rodent-proof storage container can be had for twenty-five dollars or less. The cheapest and easiest thing to do is to purchase a galvanized steel thirty-gallon trash can and corresponding lid from your local hardware store. Food-grade fifty-five-gallon drums are often had free for the taking from food manufacturers that receive ingredients in them, and these barrels generally have tightly fitting lids that are secured with a special metal band closure. I once purchased brand-new unused fifty-five-gallon plastic toxic waste storage drums for the purpose of keeping corn seed safe from rodents, and these barrels have stood up well over the years. They are made of extra-heavy plastic and come with a metal lid secured by a special metal band. Precautions must be taken, however, to keep barrel-stored grains away from excessive heat and moisture. Cool, dry, shady places are best, and direct sunlight should be avoided because grains that have stored up the sun's warmth can condense when the ambient outside temperature drops. Condensation is the worst enemy of stored grain and will be discussed at length later in this chapter.

The Bulk Bag

The bulk bag or "tote" is probably the most versatile unit of storage for the producer who has more grain than will fit comfortably into a few barrels, but not enough to warrant investing in a larger fixed grain bin. Totes are large flat-bottomed square sacks constructed of heavy-duty woven plastic that measure three feet by three feet, and these bulk bags stand anywhere from waist to shoulder height, depending on their capacity. They are meant to sit on top of a standard wooden pallet and are generally moved around by forks on a tractor front-end loader or standard forklift. A waist-high bulk bag will hold between five hundred and a thousand pounds of grain, while the taller five- or six-foot model can store a ton or more. All of these units are equipped with four very strong nylon loops, which are strategically sewn to each of the top four corners. These loops receive the pallet forks and allow the bag to be suspended in the air for filling and emptying. The flat bottom of the tote bag has a special opening chute that can be cinched closed by means of a nylon drawstring. To discharge the grain from one of these storage sacks, tractor forks are inserted in the top loops and the bag is raised up into the air. First, an outside draw rope is released, which allows the inner exhaust chute to drop down. This little tunnel of material hangs down about a foot, is somewhere around eight inches in diameter, and is tied on the bottom with a very flexible lightweight nylon rope. When the knot is undone, gravity empties the grain from the bag at breakneck speed.

Safety and forethought are two highly encouraged considerations when dealing with totes that are suspended from forks. Remember that there is a lot of weight up there over your head. To avoid crushing injury, stay out from underneath the suspended bag. Untie the unloading neck carefully and deliberately. Once the grain comes streaming out, there is no stopping it, so make sure it has a place to go where it can be contained. One little secret for emptying bulk bags is to suspend them so that the foot long unloading neck sticks down into another vessel like an auger hopper or some other sort of container. When the level of grain below reaches the bottom of the nylon chute, grain falling down the tube will shut itself off until more room is made for it. This same process can be accomplished by lowering the boom and the bulk bag in order to shut off the flow of grain from the bottom. These very same storage bags (especially the very tall ones) can also be quite top-heavy sitting there on a pallet. Tumbled-over totes are quite common as they are being transported in trucks. Sometimes they slowly lean over and eventually collapse just sitting there in a shed. It's a good idea to relift stored bags by their straps every once in a while and reset them onto the pallets below. Once a bag falls onto its side, grain will most likely spill from the top.

Automatic Filling

Small bags and barrels can certainly be filled with grain being delivered from a stationary threshing operation. My old Dion threshing machine had an elevator and a bagging attachment, and smaller garden-scale threshers are designed to empty into a low receptacle like a washtub. From here, grain can be hand-scooped to bag or barrel. Some of the older self-propelled combines like my McCormick Model 64 had bagging stations: As grain bags were filled and tied off, they were stacked on a special carrying platform. When a load of bags had been accumulated, you kicked a little lever and dumped the bags in a heap for later pickup out of the field. Combines with grain tanks are a different story. A larger hauling container is needed to accommodate greater quantities of grain that empty quite quickly from the combine's unloading auger, however. An old dump truck or a retrofitted wagon will work. Still, it's important to realize that grain is heavy and flows to the place of least resistance. For this reason, all holes must be patched or plugged and sideboards must be reinforced when an old farm truck or wagon is being used as a grain hauler. Large cardboard appliance boxes and a staple gun will work to seal corners with gaping holes, and plastic sheeting or inexpensive blue tarps make good floor liners. The last thing you want to see is your precious grain running out a hole in a side or the floor of your makeshift grain hauler. It's also a good idea to run a chain between the two sides and around the back of the sideboards on a farm truck or wagon. Pull the chain tight with a hand-operated binder and your load will be more secure as the weight of the grain pushes outward against the sideboards. Check the tires of your truck or wagon for adequate pressure; a normal-looking tire can appear pretty flat once a load has been put in place. Last but not least, make some sort of provision to unload your wheeled grain receptacle. Official grain bodies will dump hydraulically and have a coal door in back through which grain can flow to its next destination. You can cut your own grain unloading door in the back of your homemade grain body and block it with a piece of wood or heavy cardboard. If you don't have the luxury of a hydraulic dump body, get ready to do some shoveling and pushing of the load off the truck or wagon. This all sounds like quite a lot of work, but it sure is worth it, especially when you are harvesting and transporting your first crop of grain. Every scoop shovel full of sunlit kernels that you must move several times is your golden reward for a risk taken and a job well done.

Gravity Boxes

As your grain acreage and harvested quantities increase, hand-shoveling begins to lose its glory. It might be time to consider getting some equipment that was actually designed for the job of hauling grain from the field. Farm trucks with well-made dump bodies and rear outlet chutes will work just fine, but the tractor-drawn gravity-flow wagon is much simpler, cheaper, and better suited to the job. "Gravity boxes" are self-unloading metal wagons specifically designed for hauling grain from the field. They are rectangular in shape and have a floor that slopes upward at a forty-five-degree angle from a high wall on one side to a short wall on the other. A large unloading door with accompanying grain pan is situated at the bottom of the high side wall, and grain will naturally flow down the sloping floor and out the door when it is opened by turning a hand wheel. Gravity box capacities are measured in bushels, and they come in sizes from one to five hundred bushels. (A bushel is basically one and a quarter cubic feet.)

After struggling for several years with wooden-slatted hay wagons and stapled cardboard sides, I bought my first gravity box in 1991. It was basically a piece of junk, but not to me. It held 125 bushels and did not tow straight, but, at the time, I felt like it was good value for my three hundred dollars because I didn't have to shovel grain by hand anymore. Average-sized gravity wagons range between 250 and 350 bushels, and a 300-bushel wagon fully loaded with a heavy crop like wheat or soybeans weighs in excess of nine tons. That's a lot of weight to pull out of a field, especially if you have to climb any inclines. I've managed to accumulate six more of these medium-sized wagons over the years, and I've learned some painful lessons. For one thing, because of the inclined bottom design, these units carry much of their weight up high, and this elevated center of gravity makes them quite top-heavy and prone to roll over. I remember pulling nine-plus tons of wet corn up an icy road out of a river bottom field in late November 1995. We had two tractors hitched one in front of the other to get enough traction to make the hill, and we were moving right along and almost to the top when the gravity box began to sway and rock wildly. In a matter of seconds, it was lying on its side with its load of corn all over the ground. The lesson to be learned from this is to make sure you have a heavy-enough tractor with adequate traction to pull a fully loaded grain wagon through difficult and challenging terrain. The same holds true for descending a steep slope, because a heavy wagon full of grain can push a tractor down a hill and into a potentially disastrous situation. Another little word to the wise about using a gravity box is to always make sure the unloading door is fully closed before pulling the unloading lever on the combine. There's nothing more frustrating than watching grain pour all over the ground from up in the combine cab. The wagon chassis, also called the running gear, under the gravity wagon must be in good condition to safely bear all the weight being carried. Good tires, well-lubricated wheel bearings, and a wagon hitch and turning apparatus with tight tie-rod ends are all essential to keep a heavy grain wagon safely on the road. The last thing you want is an out-of-control wagon heading for an oncoming car. The very last consideration when buying a used gravity box is the condition of the box itself—older used wagons are bound to be somewhat rusty, but should be free of holes. Out in the Corn Belt, gravity wagons have been pretty much replaced by large thousand-bushel grain carts that combines unload right into while on the go; for this reason, there are plenty of these smaller units for sale. Prices range from five hundred to two thousand dollars and are generally much higher as the harvest season approaches. Several good gravity-flow grain wagons can really lend a good helping hand even in a small grain operation.

Grain Moisture

Now that you have a container full of recently harvested grain, it doesn't matter if the harvest is contained in a sack, a trash can, or a gravity box; it must be sufficiently dry or it will not keep. Stored grains have varying tolerances for water content, but 13 percent moisture seems to be the magic number at which most crops will store safely—although for some reason grain corn stabilizes at 15.5 percent moisture. However, if you are planning on holding on to your corn harvest for a year or more, 13 percent is still the recommended level. Commercial growers of grain strive to sell their crops at the highest moisture possible because they want to sell as much water along with the grain as is legally permissible, and they also want to avoid "shrink"—a term for grain that has gotten too dry. Commercial corn, for instance, is always sold at the 15.5 percent level because it takes less corn to make a bushel at that level than it does if the corn was at 11 or 12 percent in moisture. It's all a matter of density—drier grain is denser because there is less water contained within. When considering total grain moisture levels for long-term storage, it is always better to err on the side of caution. Wheat or barley at 11 and 12 percent moisture levels will store much better than the same crop at the recommended levels of 13 and 14 percent. Slightly lower storage moisture levels also provide a cushion against unforeseen incidences of bin condensation and prolonged periods of high-humidity weather. Oilseeds like flax, sunflowers, and canola do best at the 8 to 10 percent level.

Harvested grain that is exceedingly wet will heat up and spoil almost immediately, but this sort of situation is easily detected by inserting your hand into the barrel or wagon. If there is heat there, it's time to take action and do something about it. Another indicator of high moisture in a harvested grain crop is poor flow ability—wet grain hangs up everywhere from the combine tank to the wagon. Recently harvested grain will only begin to heat if it is full of green weed seeds or is 20 percent or higher in moisture. Grains between 15 and 20 percent can be deceiving because they can appear normal and may be stable for a week or more. However, slightly wet grain is a ticking time bomb. Respiration will begin slowly at first and then increase. Grain at this in-between stage may not heat up and produce mold like really wet material, but it will develop a distinctly musty smell that will render it unusable for human consumption and compromised for animal feeding. At this point, you're probably thinking it seems as if there are peril and pitfall around every corner in this business! While it's true that I have had moldy and musty grain countless times, I only have my own complacency and lack of vigilance to thank for it. The bottom line is that we need to very quickly find out how wet or dry our recently harvested crop is and do something about it if we need to.

Determining Moisture Content

So how do we determine the moisture content of our harvested grain? The easiest and simplest thing to do is take a sample to a neighbor or local feed mill with a moisture tester. Unfortunately, moisture testers are few and far between in this part of the country because very little infrastructure for grain production exists. It is possible, however, to measure grain moisture yourself using a low-temperature oven and a small lab scale that can measure five hundred grams or less. Begin by selecting a plate or some other vessel to contain your grain sample. Determine the weight of the plate in grams, and then add a hundred grams of grain to the plate. If your plate weighs three hundred grams, in other words, place it on the scale and add grain until the total weight is four hundred grams. Then place your plate full of grain in a very low-temperature oven of 200 to 250 degrees. Slow drying is best, as you don't want to burn up the grain. If there is a Koster forage moisture tester nearby, use the special Koster oven to dry out your grain sample. If you are doing this procedure in an ordinary kitchen stove, remove the plate often and stir things up to facilitate more uniform drying of the grain kernels. After twelve hours, weigh the sample each time it is removed from the oven. When the total weight drops only one gram or less between weighings, the drying process is complete. Subtract the three-hundred-gram weight of the plate from the total weight to determine the weight of your sample. Subtract this number from the original hundred grams to determine the difference in weight before and after drying. Then divide this number by one hundred to find out the moisture of the original sample. If your sample weighed sixty grams after drying, the difference between sixty and one hundred is forty. Forty divided by one hundred indicates that the original one hundred grams of field run grain was 40 percent in moisture. This is a time-consuming and somewhat tedious process, but well worth doing especially if grain growing is a fairly new venture for you. We need all the available information that we can muster to make intelligent decisions about what to do next with our harvest.

It won't take too many oven-drying adventures before you will be ready to purchase your own moisture tester or cooperate with others to own one communally. Most feed mills and grain establishments have rather large, expensive grain analyzers, and most of these units are manufactured by the DICKEY-john Company of Auburn, Illinois. The standard analyzer is a box-like instrument into which a grain sample is poured. The type of grain being tested is punched into a little keypad; seconds later a digital reading for grain temperature, test weight, and moisture pops up on a screen. This is much fancier than you will ever need on your farm, but it's a great place to bring your farm tester for calibration. DICKEY-john also manufactures a complete line of farm moisture testers called the GAC line. Some of these units will provide test weight and grain temperature, but most just indicate grain moisture. I am most familiar with the Dole 400B moisture tester made by the Dole Radson Company of Chatsworth, California. I bought a Dole moisture tester more than fifteen years ago, and it still works quite well. The Dole, like most other brands, measures the resistance of an electrical current as it passes through a specifically weighted grain sample. Five ounces of grain is measured out by filling a counterweighted cup that sits on top of the unit. When the little rectangular cup tips up the counterweight, the sample is poured down an opening. Once the grain is inside the tester's inner compartment, a little button is pushed that sends current from a nine-volt battery through the sample, and while the button is being pressed, a large circular dial on the front of the tester is spun around until a little red needle lines up with a dot smack dab in the middle of a special scale just above the button. Once the needle is lined up, the button is let go and the corresponding grain moisture can be read against a line on the circular dial. There are numerous numbered moisture scales for all sorts of grain on the front of the Doles moisture tester. Scale A and Scale B can be read to provide coefficients for other less common grains and beans. For example, after pressing the button and turning the dial, Scale A may indicate the number forty for a sample of black turtle beans. A quick look in the manual under black turtle beans will tell you that forty corresponds to 13 percent moisture.

Moisture testing is pretty straightforward, but there are a few rules and basic procedures to follow to ensure accuracy. Take grain from several parts of a wagon, truck body, or bag to make sure your sample is representative, and repeat the test at least two more times and compare the results. Measure the temperature of your grain sample with a small thermometer. (The Dole tester comes with its own easily broken glass thermometer.) Moisture test results are dead-on accurate at eighty degrees. Cold grain is always lower in moisture than the tester will indicate, while the opposite is true for warm grain that has just been discharged from a dryer. The thermometer as well as a chart in the manual will tell you how many points to add or subtract depending on the grain temperature. For example, 105-degree grain that measures 15 percent on the indicator dial is actually closer to 13 percent in moisture. Careful attention to grain temperature is needed for accurate moisture testing. One last little fact about moisture testers is that as of February 2008, the Dole is no longer being made. It must have been such a good unit that people were not buying enough new ones, but there are plenty of used Dole testers out there for sale on eBay and at other venues. Luckily, Agri-Tronix Corporation of Indianapolis, Indiana, is still providing parts, repair service, and calibration for these reliable moisture testers.

Measuring Grain Density

Each of the grains has its own standard specific density or test weight. Corn and rye for example, both measure fifty-six pounds to the bushel, while soybeans and wheat weigh sixty pounds. A bushel is a volume measurement for one and a quarter cubic feet, and standard bushel test weights for other lower-density grains are considerably lighter. Oats weigh thirty-two pounds to the bushel, whereas barley comes in at forty-eight pounds. Grain density is a measure of crop quality and can be measured with a handy-dandy little plastic test weight measuring scale. This small unit consists of a standard-sized cup attached to one end of a counterweighted ruler scale. The cup is filled level with a grain sample and then the ruler-and-cup is slid one way or another until it balances on the pivot point. Once balance is achieved, a little arrow on the pivot point will correspond to the bushel weight indicated on the ruler scale. Knowing the test weight of your particular sample is a good thing, but it's not the end of the world if you are a few pounds underweight. Wet and cloudy weather conditions could have been responsible for a lighter-weight crop, for example, and you'll have something to strive for next time. Sometimes you might be pleasantly surprised and find that your harvested sample might be running a few pounds higher than the normal weight. This happens often with oats—a thirty-eight-pound test weight will provide us with just the right grain for hulling and flaking. These little test weight measuring devices sell for ten dollars and are well worth it.

Grain Precleaning

Most field-run harvested grain needs a bit of conditioning before it goes into long-term storage. As discussed in a previous chapter, dry weed seeds from plants like mustard aren't a problem in a freshly harvested crop; it's the wetter seeds from weeds like lamb's-quarters and ragweed that are a real threat because they can impart moisture and scents to an otherwise acceptable load of grain. These weed seeds need to be removed immediately before they do damage. The pungent smell of ragweed that quickly migrates onto the surface of neighboring wheat kernels will most certainly turn up in the resultant flour, and green clover leaves and blossoms can impart a cinnamon flavor to wheat kernels. Indeed, when high-moisture foreign material is present in harvested grain, it's time to act quickly.

The rotary screen cleaner will remove small weed seed as well as coarser trash from harvested grain. These screen-covered revolving drums are available in many sizes, but most are three to four feet in diameter and five to seven feet in length. Most of these units seem to have been manufactured in the Grain Belt—Nebraska, Iowa, Manitoba. Some of the more common brands are DMC, Snowco, Farm King, and NECO. The standard rotary cleaner is basically a long cylinder that is covered with a fine-mesh screen. The cleaning drum is supported by a center shaft that allows it to rotate within a steel frame. An electric motor provides power to the spinning drum, which must be higher at one end to allow the grain to pass through it. The frame of the rotary cleaner has adjustable legs to vary the angle of cleaning. Steeper angles provide for quicker passage of grain through the revolving drum. Most of these machines also have a smaller-diameter interior drum that is outfitted with a larger mesh screen for removal of coarse material like bits of straw, stems, and unthreshed heads. Rotary cleaners are intended for high-speed cleaning of larger volumes of grain. (The finer and more meticulous conditioning of grain for seed and human consumption will be explained in the next chapter.) Prior preparation and setup of the rotary screen before the combine starts rolling is highly recommended, because you need to be ready to deal with the gravity boxes as they arrive from the field. You will also need an additional receptacle like a bin, grain dryer, or extra gravity wagon to receive the cleaned grain as it exits the rotary cleaner.

Rotary screen cleaners can be purchased brand new for between two and three thousand dollars. Watch for them at auctions and in used-farm-machinery advertisements, where they can sell for as little as five hundred dollars. Here are a few hints and considerations about the operation of rotary screen cleaners. First, begin by selecting the proper inner and outer screen sizes for the grain you intend to put through the machine. This information can be found in the owner's manual or through a quick call to the manufacturer. Screen mesh size is calculated by the number of squares to the inch; smaller-seeded crops like flax may take an 8×8 or 10×10 outer screen while larger grains like corn use a 3×3. Make sure the inner screen is large enough to let the majority of the grain drop through it. If the mesh is too fine, too much usable grain will come out mixed with the coarse stuff. Swing the cleaner's loading auger out under the discharge pan of the gravity wagon and begin delivering a light flow of grain to the revolving drum. At this point, you can adjust the angle of the drum as well as the flow of grain being delivered to it. It's best to use the full capacity of the cleaner, but not overload it. If the grain doesn't seem to be working its way through the rotating screen quickly enough, increase the incline a bit. Once the process has begun, you will see weed seeds dropping through the screen. Check the grain at the outlet for cleanliness and slow down the flow if too many weed seeds are sneaking through. If the weed seeds and grain traveling through the cleaner are excessively moist, a slower cleaning speed will be necessary to do a complete job. You will also need an auger and receiving pan at the outlet of the cleaner to carry the cleaned grain away to its next destination.

The Grain Auger

Another piece of necessary equipment that has already been mentioned numerous times is the grain auger. Archimedes is credited with the invention of the screw in ancient Greece, which is basically an electrically or engine-powered turning screw inside of a tube. Grain augers are made of steel and are available in all diameters and lengths. Just about everyone who grows grain, except for someone with a few trash cans full, needs an auger to move the harvest from here to there and back. Small four-inch-diameter models are used to empty hopper bottom bins in addition to moving lower volumes of grain whenever and wherever needed. These smaller models are quite inexpensive and are usually powered by a one-third- or one-half-horsepower electric motor attached to the upper outlet end. If you are using an auger to load grain into a bulk bag or some other container, you will need to suspend the auger from its motor end, which can be done by hanging it with a chain or cable from an overhead beam or an elevated front-end loader. A basket or pan is attached to the lower inlet end of the auger to contain the flow of grain and guide it into the tube. The spiraled interior of the auger, called the flighting, is constructed of lightweight steel and is a bit tender, so it's important to protect flighting from getting bent or otherwise damaged. Augers are basically metal turning inside of metal; for this reason, they shouldn't be run empty for overly long periods of time. As augers age, the flighting becomes thinner and thinner, to the point where it gets quite sharp and can easily cut your hands as well as the bottom of the plastic basket at its base, so protect yourself and the rest of your equipment. (A thin little piece of wood placed between the flighting and the pan, for example, will prevent the flighting from ripping a hole in the plastic.)

Augers are also built larger in diameter and longer in length for bigger grain-moving jobs like filling grain silos. I have a fifty-one-foot, ten-inch-diameter auger that is mounted on a wheeled transport frame. Augers of this size are powered by the PTO shaft of a farm tractor and have a capacity that is measured in tons per minute instead of tons per hour. With the turn of a crank, these larger augers can be winched up quite high in the air to reach the center filling hole on the roof of a large steel grain bin. Those of you familiar with physics know that the steeper inclines require more power to propel the grain up the tube, and auger capacity declines in direct proportion to the angle of steepness. These bigger augers can also be powered by larger electric motors. I own two eight-inch-diameter electric augers—one is sixteen feet long and the other is thirty. The sixteen-footer is quite versatile, but very heavy, and it takes two people to move it from job to job. The thirty-foot auger is mounted on a steel transport frame and can be wheeled anywhere. Like most other transport augers, this model can be raised or lowered with the turn of a crank. Since a tractor and PTO shaft are not required for operation, motorized augers will fit into some very tight places. If you end up with an electric-motor-powered unit and must leave it outdoors, be sure to protect the motor with some kind of waterproof covering. Once you have one auger on your farm, you'll want more to be able to perform tasks that require several augers to be working simultaneously. Look for these helpful tools at farm sales and in used-equipment flyers, because a new Westfield brand fifty-one-foot rubber-tired transport auger can cost as much as six thousand dollars. Used augers usually sell for five hundred to two thousand. When you're buying a used machine, check the flighting and the tube for signs of excessive wear. I would also recommend hooking a tractor to the PTO shaft and running the thing; if it vibrates wildly and seems as it is going to rattle itself apart, it's best to keep looking for another one.

Grain augers are to be treated with respect. Steer clear of rotating flighting, because it can grab you or your clothing and suck you in. The bottom of an auger tube should be caged where the flighting emerges. If for some reason you must sweep grain toward the rotating flighting, do it with a paddle instead of your bare hands. There are plenty of other rotating protrusions on an auger, from the PTO shaft to V belting, and loose-fitting clothing can wind around a shaft in a split second. I speak from experience here—I have lost two pairs of pants to PTO shafts in the last thirty years. So please, be careful. Another major hazard with long-transport augers is electrocution from a nearby power line. Many times an auger must be cranked quite high up in the air to deliver grain to a high place. The temptation is to just scoot the thing across the yard without letting it back down. If you do this and hit an electric line, your chances for survival are pretty minimal. Augers left cranked way up in the air are also quite tippy. A big gust of wind can come along and push the thing over. Again, I speak from experience—I've rolled two augers in my farming career. Granted, I live in a violently windy location, but had I not left these augers halfway up in the air, I wouldn't have found them collapsed on their sides the next morning. Farm safety and common sense go hand in hand. A few extra minutes to put something away properly will save dollars and agony.

Grain Storage and Elementary Drying

New England weather conditions are rarely dry enough to permit grain harvest moisture to drop down to the 13 percent level. It's so wet here, if we go without rain for a week or more, we start to feel parched. The forty-five inches of annual rainfall we receive here in the Northeast keeps our fields, trees, and lawns green all through the summer months. In drier grain-growing regions like the Dakotas or eastern Washington, most of the vegetation is already brown and dormant by late July or early August, and crops grown in these regions are usually dry enough to go from the combine right into long-term storage. Our abundant rainfall certainly promotes good hay and pasture production, but it impedes the complete dry-down of a grain crop, and many times the best we can do is harvest grain at 16 to 18 percent moisture. One interesting little fact is that grain will actually increase 2 to 4 points in moisture as it gets thrashed and shaken through the combine, so you may hand-sample a ripe grain field at 14 percent moisture, but find it testing closer to eighteen percent out of the combine's grain tank. Let's say the grain has been cleaned by the combine's scourkleen or run through some kind of rotary screen and is still testing in the high teens for moisture. Some kind of drying or conditioning is necessary or this precious harvest will begin to degrade shortly—but all is not lost. There are numerous options for bringing the grain down to a moisture level that will allow for safe storage.

The extremely small garden producer of grain will probably have the easiest time drying the mini crop. An empty greenhouse and a couple of fans to move the air through will suffice nicely. Lay down some tarps or plastic sheeting either on the benches or the floor, and simply spread your one hundred to five hundred pounds of harvested grain out as thinly as possible. Stir it around several times daily to constantly expose more kernels to the sun. Grain moisture must have a place to escape, so leave the greenhouse doors at least slightly ajar. If you have access to some fine-mesh window screening, you can build some rugged frames upon which grain can be spread to dry. Just make sure to lift the screens off the floor by putting legs under them or placing them between two tables—elevated grain will dry even faster, because it is exposed to air from the top and the bottom. Grain keeps quite well in cloth, especially burlap. The coarse weave of burlap will allow for the passage of air through the grain contained within the bag. If you have five or more burlap bags full of grain, you can use an eighteen-inch box fan to power an "in the bag" drying system. Choose a nice dry location like an enclosed porch or an unused room. Stand the fan up with a bag of grain on each side. Next, lay a bag horizontally across the top of the two standing grain bags. Now block the hole between the two standing grain bags with two more bags placed in the upright position. The idea here is to build an airtight tunnel of sorts around the box fan. Nestle the bags into each other so there are no holes through which blowing air can escape. Turn on the fan and check for air leaks. You want to feel a small flow of air coming through the burlap. It is very important to seal any leaks between bags in order to build up some back pressure on the fan. Move the box fan and bags into a greenhouse setting for improved drying efficiency. Although this may seem like grain drying at its most basic and fundamental level, it really is all you need to know. All of the more sophisticated grain-drying systems that will be described in the next few paragraphs operate on the same principle. When air of lower humidity and higher temperature is pushed through wet grain, it removes the moisture.

The Screw-In Grain Aerator

The next level of small-scale grain drying is the screw-in grain aerator. This simple and inexpensive little unit is ideal if you have a tote bag or gravity wagon full of grain that needs conditioning. The screw-in aerator is a two-piece tool consisting of a six-foot piece of five-inch-diameter steel pipe and a small one-twelfth-horsepower twelve-inch electric fan. The bottom three feet of the pipe is perforated while the top three feet is solid, and the tip of the pipe is pointed and equipped with spiral flighting. This pipe is literally turned into the grain by means of a small cross handle slipped through its top. When the pipe is screwed in far enough that the perforations are totally buried in the grain, the cross handle is removed and the fan unit is mounted on top. Once the pipe and the fan are connected together, the aerator is plugged in and the drying process begins. You would think that the fan would blow air down the tube and out the perforations into the grain, but this very clever invention works in just the opposite manner: The top-mounted fan actually pulls the air through the grain and up through the pipe. When you think about it, this method pulls moisture from the grain in a much more uniform manner. As far as I know, there is only one US manufacturer of this unique aerator. The B and W Manufacturing Co. of Columbus, Nebraska, has been making its Model 128 Grain Aerator and two larger units since 1958. These units can be purchased from catalogs and some farm supply stores for about $240; used ones sell for around $100. This is cheap insurance. These aerators can be screwed into a gravity wagon or standing tote bag, but if the gravity box is only half full, screw the thing in at an angle. All you need to do is to make sure the perforations in the tube are totally buried. I have been amazed at how well these simple little things work. When we are really busy during the harvest season, we end up with an odd gravity wagon full of some moist grain. There isn't sufficient time to really deal with the grain since we have to head to another field and harvest the next crop. In situations like this, we will screw in two of these aerators and then park the wagon in the shed. When we come back to it a week or ten days later, the grain is stable and dry. The twelve-inch fans that do the work are rated at two amps, which is only a tiny amount of electricity. For many small-time grain growers out there, two or three screw-in grain aerators will handle any grain drying they need to do.

Drying Grain in Silos

Larger amounts of wet grain pose more challenges. When you've got twenty to forty tons of wheat at 18 percent moisture, you've got to be ready move quickly. In larger-scale situations, grain can be dried and conditioned right in the bin. Galvanized-steel grain silos are a rare sight here in New England, but they do exist, and as more farmers rediscover growing grains, more of these specially designed storage bins will appear. On-farm grain storage silos are larger in diameter and much shorter in stature than the common upright silo found on dairy farms. Indeed, most grain bins are equal in height and diameter. An eighteen-foot-diameter structure is usually eighteen to twenty-four feet tall. (For some unknown reason, silo diameters are measured in three-foot increments.) Fifteen feet is considered a small silo, while thirty feet is the industrial size found at large cash grain operations and feed mills.

Constructing a Silo

The walls of a grain storage silo are built by bolting together many sheets of galvanized steel, each individual sheet slightly curved and ribbed for extra strength. Standard bin panels measure thirty to thirty-six inches in width and eight to ten feet in length. Smaller-diameter silos may use five or six sheets per ring, while larger structures require ten or twelve. The standard grain silo sits on a permanent circular cement base that is usually two feet wider in diameter. The bottom ring of a grain silo is sealed and attached to the cement foundation. This bottom layer is constructed of heavier-gauge steel to withstand the increased side pressure from the weight of many tons of grain above. Six or seven of these rings of panels stacked, sealed, and bolted together will form the walls of a finished bin. A side access door is usually placed between the second and third ring. Special pie-shaped panels are bolted together around a center fill opening to form a roof for the structure. The center fill spout is protected by a hinged cover that can be opened to accommodate an auger for bin filling. Contrary to what we might think of as normal construction methodology where buildings are put together from the bottom up, grain silos are built from the top down. Roof panels are bolted together first right on top of the circular cement foundation. Once the roof section is completely assembled, it is raised several feet in the air by special jacks placed around its circumference. The top ring is then bolted in place right underneath and attached to the roof. Once the roof and top ring are joined, cables from the jacks are attached to special brackets and the whole business is elevated once more to accommodate the next ring of side sheet panels. This jacking and bolting process continues until the bottom ring is attached and the storage silo is complete. I built two grain silos in the 1990s here on my farm and learned quite a bit in the process. My one piece of advice is to develop a very detailed and specific marking system when dismantling a bin. With the right information, each modular sheet of galvanized steel can be returned to its very same original position during the re-erection process. Side panel sheets are supposed to be identical, but they do bend and change a little over time, which can make it difficult to line things up for bolt placement. If each sheet is returned to its original position, holes will line up and bolts will fit in easily. Little marks and notes on the steel done with an indelible marker or a paint stick can save a lot of time and aggravation later on in the process.

Drying Grain in a Silo

Now that you've built a bin and installed a working aeration floor, it's time to use it. There are only two fundamental principles to understand in grain drying. First, moisture must migrate from the interior to the surface of each individual kernel of grain. Second, this moisture must be evaporated from the surface of the mass of grain into the surrounding air. Several other factors must be considered when trying to remove moisture from grains. If you're pushing high volumes of air through contained grain, the air must be lower in relative humidity and higher in temperature than the grain you're trying to dry. The general velocity of the air being blown up through the grain will also influence how much and how quickly moisture is removed. Grain drying follows the laws of thermodynamics that apply to the transfer of energy states and matter. In search of equilibrium, heat will migrate to colder states and moisture will migrate to drier states. So if you're blowing warmer, drier air through grain, heat and moisture will eventually be removed until a balance is achieved between the grain sample and the ambient temperature and humidity.

But drying grain in the bin is more than turning on the fan and walking away. First, you must make sure the grain in the silo is level. You want all the air coming up through the mass of grain to travel the same distance, as forced air will definitely take the path of least resistance. If there is a low spot in the silo, more air will exit there, which will leave higher levels of grain untouched and undried. The temperature and humidity of the outside air will also influence the speed at which moisture is pushed out of the grain. Warmer and drier summer conditions certainly allow for better drying efficiency, whereas leaving the fan on during a rainstorm or during excessively humid weather might have the opposite effect of adding more moisture to the grain. When doing static in-bin drying, the fan should be left running twenty-four hours a day whenever possible. The grain contained within will dry from the bottom toward the top as a moisture front is pushed upward through the entire mass of material. When the fan is turned off, the moisture front will begin to recede downward. It is also quite important to have plenty of vents in the roof of the silo to provide an escape route for air and moisture. Most grain storage structures come with at least one J vent in the roof. Bin drying can be a nerve-racking and patience-testing business. Axial fans running twenty-four hours a day for weeks on end can drive you crazy because they literally scream, and if the noise doesn't send you around the bend, it may disturb your neighbors. As time passes during this process, you will be constantly climbing into your silo and grabbing grain samples off the top for moisture testing. Test after test will indicate that the grain moisture is unchanged or maybe even a bit higher than it was when you put the grain into the silo. This is where you need to have faith. The moisture front will arrive at the top of the grain at the very end. As long as there is sufficient air traveling up through the grain, heat and the potential for spoilage are removed. Eventually, everything will stabilize. The grain will dry and you can turn off the fan.

Once a grain silo has been constructed and made weathertight, it can be filled with grain. It's what you install in the bin's interior that will allow you to dry and safely store your crop for long periods of time. The addition of a perforated drying floor will give you the ability to pump volumes of air up through stored grain. Standard aeration floors sit twelve to sixteen inches above the concrete foundation in the bottom of a grain silo. Floor sections are made of long narrow sheets of heavy-gauge galvanized stock that snap together in an interlocking pattern. These pieces of flooring are completely covered with a geometric pattern of very tiny holes to permit the through passage of forced air; they're supported by numerous support legs and designed to support the weight of a bin full of grain. Grain bin drying and aeration is accomplished by forcing large amounts of outside air up through the floor and grain with a very powerful fan. Fan size and capacity are determined by diameter and the horsepower of the electric driving motor. An eighteen- to twenty-four-inch axial fan powered by a three-horsepower motor will supply more than enough air to an eighteen-foot-diameter bin containing a hundred tons of grain. Blown air finds its way under the floor through a piece of hardware known as the transition, and the large-diameter fan is sealed and bolted onto a square plate at the inlet end of the transition piece. A transition designed to hold an eighteen-inch fan may be two-foot square on the inlet end. This rather large piece of galvanized metal tapers down to a two-foot-wide by one-foot rectangle at its outlet end where it bolts to the side of the grain bin. The transition is what allows the air from a large fan to get into the small twelve- to sixteen-inch space between the cement foundation and the perorated floor of the grain bin. Good air seals are essential for efficient bin-drying and aeration, so for this reason everything must be gasketed and caulked with silicone or some other good sealant. The most important joint is where the bottom grain bin ring sits down on the cement foundation, and special anchor bolts and brackets are used to pull the bottom of the bin down onto a rubber gasket. Over the years this seal can be breached as the gasket shrinks and the metal flexes and moves ever so slightly. A grain bin supplier from Québec once advised me to use a light application of hydraulic cement around the base of the bin. This product—which is intended for underwater use—expands to fill in crack and holes, but it will crack over time and needs to be renewed every few years. Metal seams along bin walls should also be inspected for air leaks and remedied with sealant. A simple test for air leaks can be performed by turning on the fan. You'll be able to feel the air as it escapes. Mark these places with a paint stick. Turn off the fan and go to work with the sealant. Make sure you give the caulk time to harden up and stabilize before you start the fan again.

Removing Your Grain from a Silo

The aeration floor in a flat-bottomed grain bin is generally equipped with an unloading auger that is located between the perforated floor and the underlying cement. A special open-topped metal box measuring a foot square and ten inches deep is fitted flush into a corresponding hole cut into the very center of the aeration floor, and an eight-inch-diameter hole in one side of this floor well accommodates the inlet end of the auger tube and its protruding flighting. From here the unloading auger runs underneath the perforated floor and out through a perfectly cut round hole in the side wall of the grain bin. When it is time to remove grain from the bin, gravity takes it to the lowest point in the center well, and the horizontal auger delivers it to the outside. Most center wells are equipped with a sliding top plate for opening and closing. This little flat piece of steel slides in special grooves and is attached to a length of steel pipe that runs between the auger and the floor. This pipe handle also fits through a little hole in the grain bin wall. Pulling the handle from the outside will control the opening size and the amount of grain that feeds the unloading auger. Pushing the handle all the way in shuts off the grain flow completely. When you are finished getting what grain you need from the bin, close the little sliding door and let the auger run until the flow of grain stops. The auger and grain well are clean and empty and ready for the next time unloading needs to take place. Leaving the unloading mechanism full of grain might make it difficult to start the auger the next time it is used. The sliding door shutoff apparatus should be checked and maintained at the beginning of every crop season, as anything that moves back and forth often will eventually wear and break. If the pipe becomes detached from the sliding door, it will be impossible to open it and remove grain from the bin. I've had the opposite problem happen to me where the pipe handle became detached from the sliding gate while it was in the open position. This certainly wasn't the end of the world, but it was nevertheless unnerving. For proper aeration it is also necessary to seal the auger and pipe handle where they come through the wall of the bin. This prevents air loss at these points. The electric motor that drives the auger is mounted to its outlet end and sits out in the weather; when it's not in use, it should be protected with an upside-down bucket or washtub. A basic understanding of the workings of an aeration floor and a good maintenance program will ensure trouble-free operation for years to come.

Adding Supplemental Heat

Added heat can be the missing link in a static grain-drying system. When you raise the ambient air temperature moving through the grain, moisture will outmigrate much more quickly. Of course, there is a trade-off here. Supplemental heat usually comes in the form of propane, and increased energy costs translate into higher drying costs. Axial grain bin fans can be purchased with supplemental propane burners. These burner kits are installed right in front of the fan and will vary in BTU output. To preserve grain quality, choose a burner attachment that will not let the grain being dried climb above a hundred degrees. High drying temperature is a sure way to turn good-quality grain into a poor-quality product very quickly: High heat robs grain of quality by cracking kernel seed coats, which allows oxidation of the endosperm. Vitality is reduced as well because the germ (which is the life center of the seed) basically gets cooked. It's worth climbing into your bin with a thermometer probe to determine grain temperature as well as moisture. It's also possible to provide some added agitation to the mass of grain inside the silo while it is undergoing the heating process. The handy outfit called a "Stir-Ator," is made by DMC (David Manufacturing Company) and available for purchase from any grain bin sales and installation company. One to three vertical electrically driven auger flightings are suspended from a steel I-beam frame that is attached to the roof and the uppermost ring of the grain bin. These little units act like corkscrews, constantly lifting grain upward in the bin. The metal frame from which these vertical screws hang rotates 360 degrees in a circle to fluff and agitate the entire contents of the bin. Stir-Ators move very slowly; one entire cycle around the whole circumference of a bin takes several days, but the plus side is that drying time can be reduced by as much as 50 percent with this sort of system. Since the grain is constantly being moved, static pressure is reduced and the warm air traveling upward through the grain has a much easier time reaching the surface. Grain quality is improved as well because overdrying—which occurs when a moisture front is moved upward through a bin—is avoided. This type of technology and grain infrastructure is for the serious grower only, however, as Stir-Ators cost thousands of dollars and are mostly intended for drying shelled corn in October or November. They will work in August for bin-drying cereals without supplemental heat. Equipment of this nature can sometimes be purchased quite reasonably at auction when a large cash grain operation is selling out, and entire grain bins and their contents are put on the auction block. It's not uncommon to see a twenty-four-foot-diameter bin with drying floor, propane fan, and stirring system sell for less than five thousand dollars. To get a deal like this, you might have to journey down to the southern part of New Jersey, where grain farms are being turned into housing developments. I have seen many of these farm sales advertised in agricultural papers and am always tempted to attend one. However, my common sense and my wife always prevail and I never go, because when a bin is purchased at a farm auction, the buyer is always responsible for dismantling and removal. If we can find something a little closer to home, we are a whole lot better off even if the price is a bit higher.

Pressure-Cure Drying

Over the years I have made friends with many farmers actually farming in regions where grains are the predominant crop. These are my go-to people for organic seed, specialized equipment, advice, and, ultimately, friendship. One such individual is Rodney Graham of Hunt, New York. Rodney grows dry beans, corn, and cereals on eight hundred acres in the Genesee Valley about fifty miles south of Rochester, and I've purchased black turtle bean seed from him several times. I was describing my grain-drying difficulties to Rodney in one of our phone conversations when he asked me if I had ever heard of pressure-cure drying. He went on to describe how he had installed special high-powered "air pump" fans in several of his large grain silos. These squirrel-cage-style blowers produce a much more powerful blast of air than axial models of similar horsepower, and they run more quietly, too. Axial fans provide a very rapid initial blast of air, but airflow slows down as it experiences back pressure on its way up through the mass of grain. It's like driving a car up a steep hill in high gear; after a while you start to lose power and speed. Air pump fans have more torque and the ability to maintain good positive pressure from the bottom all the way to the top of a bin full of grain. Pressure-cure drying, on the other hand, has become quite popular because supplemental heat is not necessary. This system of drying was developed and is sold by CMC (Custom Marketing Corporation) of West Fargo, North Dakota. The company sells special grain bins, drying floors, and air pumps, and is a bit of a pyramid scheme—which might make anyone skeptical. (You get a discount on your next purchase if you talk someone else into buying their equipment.) I liked the idea of drying my grain in the bin without using heat, so I ordered a seven-and-a-half-horsepower air pump unit from CMC.

There are, however, a few caveats to this sort of drying system. It can take weeks, even months to push a moisture front up through a silo full of grain. My first experiment with this new squirrel-cage fan was trying to dry ear corn in my 18 × 18 grain silo. Little did I realize that as the outside air cooled off in November and December, the progress of the drying front up through the silo slowed to a crawl. We ran this fan day and night for six weeks from mid-November until early January, and while it may have screamed less than a horizontal axial fan, it still made plenty of noise. I was also advised to leave the fill opening cover off the very top of the roof, as apparently the air pump moved so much air that the roof could be damaged if there wasn't a place for it to escape easily. Rain and snow could get into the bin, which was supposed to be just fine: Moisture would be evaporated before it even touched the grain inside because the airflow was so powerful. This was partially true; if you climbed up on the roof and peered down into the bin, the escaping air would indeed blow your hat off. After six weeks of listening to this unit howl, I finally turned it off in early January. There was some spoilage on top of the grain, mostly from weather that had made its way inside. The corn had dried for the most part. Sure, the quality was probably higher. It didn't shrink like it would have if it was dried with heat. The worst thing about it was my electric bill: Pulling fifty or more amps of power with a seven-and-a-half-horsepower motor for six weeks cost more than a thousand dollars. What I failed to realize was that these units are probably intended to do their pressure-cure thing in places like North Dakota where the general relative humidity is much lower than here in damp New England. I also failed to realize that the added heat I had used in the past for drying late in the fall helped to offset the ambient cool and humid conditions we have in this part of the country. The CMC sales literature touted the ease of pressure-cure drying for grain of up to 25 percent moisture. My corn was probably closer to 30 percent in moisture. Herbert "Bussy" York of Farmington, Maine, had the same experience with a CMC drying system for his grain corn harvest, and he ran his fans most of the winter. In his first year with the system, he spent much more on electricity drying his corn than he would have on the equivalent amount of propane.

Every cloud has a silver lining. While pressure-cure drying in the late fall doesn't work as well here in the Northeast as it does in North Dakota, it does work quite well in the summer months when the air is warmer and drier. Due to rainy weather, I have had to harvest wheat at 20 percent moisture several times over the last few seasons. During times like these, I have sent my harvested wheat through the rotary screen cleaner and then directly into the bin with the air pump fan. After leveling the grain and leaving a door open, the big fan was turned on and the grain was left to cure. Even in very humid conditions, the wheat dropped down to 13 percent in ten days or less; after it was dry, the bin was emptied and the next crop was put in for another round of drying. I have found this system to be quite workable. I have also dried smaller amounts of shelled corn in October this way. I now provide supplemental heat by placing a gas-powered heating unit in front of the intake end of the air pump. I recommend this type of aeration very highly for certain applications here in our eastern climate.

Drying Small Amounts of Grain

To the smaller grower with a garden plot or an acre or two, all of this equipment and infrastructure probably seems daunting and unattainable. Don't worry: There is a scale of storage and drying that makes sense for much smaller amounts of grain. The first thought might be to obtain a retired hopper-bottomed grain bin from a former dairy farm. Drive down any country road where former dairy barns persist and you will see these forlorn galvanized bins constructed of the same ribbed steel panels used on larger flat-bottomed bins. These were (and still are) the silos where the grain company delivered feed rations to dairy farms. These bins stand on four angle-iron legs and have a bottom section that tapers to a point not far from the ground. A small four- or five-inch-diameter auger emerges from this very bottom point to deliver the grain ration through the wall of the barn to the cow stable within. Sitting on top of this hopper bottom are anywhere from two to five ringed sections, each of which comprises two bolted panels and measures six feet in diameter. A two-ring hopper-bottomed bin will hold at least two tons of ground dairy ration while a six-ring unit will hold six tons. The first thought might be to simply put a small grain crop in one of these bins for storage, but a closer inspection of one of these older grain bins will usually reveal corrosion, weak metal, and sometimes actual holes in the bottom of the tapered hopper section. These bins have spent their entire lives out in the elements, and water from rain and melted snow always finds its way into the bottom. The salt in dairy feed also contributes to rust and subsequent metal breakdown. A silo like this might work if it is moved inside another building. It is possible to attach an axial fan to the side of the hopper, but it will not perform properly because there is no perforated floor or inside air space to ensure an even distribution of forced air up through the grain. A hopper bin is a poor substitute for a flat-bottomed grain storage silo.

The six-foot rings that form the top section of a hopper-bottomed bin are the big prize. Often an old farm delivery silo with a spoiled bottom can be had for the dismantling and the taking. The hopper bottom may be shot, but the rings above are generally in great shape because all the water went to the bottom to do its damage. A six-foot-diameter ring section unbolted from the larger silo can be used as a temporary or permanent storage enclosure anywhere there is enough room to set it up. This ring section will sit nicely on a floor anywhere inside a building. It can be filled and unloaded using a five-gallon pail. One six-foot-diameter ring section will stand between thirty and thirty-six inches tall and hold up to fifteen hundred pounds of grain. A B and W screw-in aerator can be inserted if drying needs to take place, and if you need more storage, simply bolt another section on above. To protect your inexpensive grain bin from rodents and debris, cover the top with corrugated steel roofing or hardware cloth mesh. If you want to get really fancy, situate this grain bin ring on an upstairs wooden floor and install a simple four-inch sliding gate valve in the floor before filling takes place. Connect a piece of four-inch plastic drainage pipe to the valve and you have a gravity-flow system to the ground floor below. Small indoor wooden grain bins can also be constructed with boards or plywood. I've done

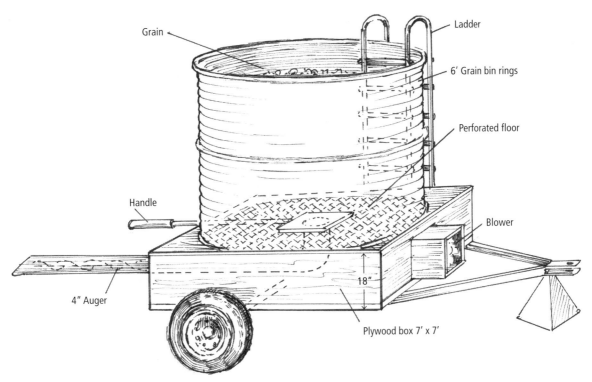

Homemade grain dryer for smaller amounts of grain

it both ways and prefer using the metal sections because rodent access is eliminated.

These very same six-foot round grain bin rings can be used to build a miniature version of a flat-bottomed grain silo. An aeration floor is just as easily installed in a six-foot-diameter grain bin as in a larger-diameter structure, and the unloading auger under the floor can be scaled down to a four-inch model to fit the smaller bin. About ten years ago, I found myself with a variety of small harvests that needed conditioning and drying. Five hundred pounds of one particular variety of dry beans or emmer was too small an amount to cover an aeration floor or fill a stationary grain dryer, so to accommodate my small-scale drying needs, I decided to build a miniature drying bin on wheels. My very talented neighbor, Brian Dunn, followed my design and fabricated the thing in less than a week.

We started out by welding together a seven-foot-by-seven-foot steel underframe with an extended triangular channel iron hitch in the front. An old car axle with rims and tires was welded to the bottom of the frame for mobility. Next we built a seven-foot-square plywood box with eighteen-inch-high sides. This plywood plenum was completely reinforced on the inside with a web of steel angle iron to give it strength and durability. We then mounted the blower box to the metal frame and transport axle. The next step was to center and cut a five-foot, eleven-inch hole in the top of the plywood box. In this hole we fastened scrap pieces of perforated grain bin flooring to construct a tiny aeration floor. We cut a grain well into the center of the floor and installed a four-inch auger underneath for bin unloading. Then two old six-foot grain bin rings were fastened to the top of the plywood box, a medium-sized axial fan was attached to the front behind the hitch—and the job was finished. We caulked and sealed all the joints and connections to provide an airtight plenum underneath the floor. Brian also attached a ladder to the outside and inside walls of the little silo for easy access and cleaning. We were able to build this little unit for less than a thousand dollars. But ours is only a suggested design; certain aspects can be adapted to meet anyone's particular needs. Creativity and the ability to source and recycle other people's castoffs will determine how much you might have to spend on a miniature small-scale drying bin like this.

We have found our little mini bin on wheels to be quite versatile, and we have saved many a wet crop with this unit. One hundred pounds of dry beans will cover the floor to a depth of several inches, while a full dryer will hold close to two tons of wheat or black beans. Once the grain harvest begins, we park our little dryer in our shop to receive small loads of grain. Supplemental warm air can be added to the mix by running a two-foot-diameter cardboard "sona-tube" between the side-mounted axial fan and a wall-mounted Modine steam heater. This little bit of extra warmth makes all the difference when it comes to drying time. I don't know how we ever got along without our homemade mini dryer—we use it from early August right through Christmas some years, and each fall we fill it full to the top with open-pollinated ear corn that we must dry slowly and gently for seed use. There are times when I wish we had more than one of these units. Detailed plans for this type of unit can be found on the Northern Grain Growers Association website: www.northerngraingrowers.org.

Stationary Grain Dryers

In the busy harvest season, the flow of grain coming out of the field can exceed the drying capacity of a flat-bottomed grain silo. This is especially true if two or more types of grain are being harvested and there is only one drying bin, and sometimes harvest moisture is just plain too high for efficient bin-drying. We find ourselves in this situation when shell corn is being combined in November, ambient air temperatures have dropped into the thirties, and relative humidity has begun to climb as winter approaches. If you are racing against the onset of winter and have fifty to one hundred acres of corn to combine, a propane-fired, standalone grain dryer is almost a necessity.

I encountered a similar problem during barley harvest in the summer of 1992, which was cool and cloudy. After waiting what seemed like weeks, we went out and harvested twelve tons of barley. (This was two years before I had an aeration grain bin on the farm.) I knew the grain was damp and didn't want to repeat my past mistakes. The previous summer, I had spread out a similar amount of barley on a concrete floor in a toolshed only to find that it did not dry enough and developed a musty smell. So I jumped in my truck and drove sixty miles east and a little south to Guildhall, Vermont, to ask potato and grain farmer Bert Peaslee what to do. Bert suggested that I take home one of his stationary batch-type grain dryers and use it to circulate and aerate my damp barley. I paid him five hundred dollars for it, hooked it behind my truck, and made the long, slow journey home. It didn't weigh much, but it was quite large and unwieldy. The dryer looked like a large upright funnel made of lightweight perforated sheet steel, drawn along on two little bald rubber tires. It stood about twelve feet tall and measured nine feet in diameter at the top. The sides dropped straight down for about six feet and then tapered to about a foot in diameter at the bottom. I definitely got some looks in passing on the way home. I found out that many of the potato farmers who had grown some grain years earlier in my area had these round batch dryers on their farms. Most of them were no longer in use and had been left outside to rust away, but Bert had taken good care of his dryer, lucky for me.

These were the early days in my acquisition of grain infrastructure. I didn't know a thing about grain dryers, but I sure did learn fast, although it was a bit of trial by fire. The brand name of my new prize was Meyer-Morton. It held 250 bushels of grain. The company had long since gone out of business, and the machine was supposedly obsolete, but it sure worked well for me because it saved my barley crop. Once I was home with the thing, I cleaned it, greased the bearings, oiled the chains, and hooked it to the PTO shaft of the tractor. The machine was much more than a big perforated freestanding basket. Inside, there was another self-contained perforated cylinder that formed a large air space. This plenum was connected to a two-foot-diameter axial fan driven by the PTO shaft. Coming up through the very center of the plenum section was a solid-steel twelve-inch-diameter pipe that contained upright auger flighting. The old dryer was sort of like an onion with three skins—a center vertical auger, a six-foot-diameter plenum, and a nine-foot-diameter outer wall. Once filled with grain and in operation, it was an amazing piece of equipment to behold. Grain was hauled up through the central auger and thrown outward at the top of the tank. From here the whole mass of grain would gradually fall back to the bottom, traveling between the plenum and the outer wall. Meanwhile, a very powerful air blast produced by the fan made its way into the plenum and outward through the eighteen-inch "sandwich" of falling grain. The dryer was elementary and ingenious at the same time. I had the good fortune of several hot, sunny, low-humidity days when I ran my barley through the new dryer that year. It took a day or more of the tractor running full-tilt and the dryer fan screaming to lower the moisture of my barley into the safe zone. I came to appreciate the drying power of a beautiful day. The old machine held together, and I felt like putting a feather in my cap.

Additional Drying Techniques

I learned more about dryers over the next few seasons. In 1994, I combined my first crop of shelled corn and needed to be able to dry it with heat. Before this time, I had only used the power of the atmosphere on a nice day to do my grain drying. The learning curve was steep; I knew nothing about propane except that it was potentially explosive and dangerous. Thank God for the gas man—I ordered a five-hundred-gallon tank and special hose to run between the tank and the dryer. The propane we use in our homes for cooking and hot-water heaters is in the vapor state and is drawn from the top of a gas tank, but grain-dryer burners require so much gas that they consume liquid propane off the bottom of a gas tank. Once the liquid enters the dryer, it passes through a special pipe coil and gets preheated and vaporized by the warmth of the burner element. Grain-dryer controls are run by a six- or twelve-volt electric line connected to the battery of the tractor. I opened the control box on my dryer and found that I had to learn how to adjust four different thermostats to dry the grain safely and properly. There were settings for grain temperature, plenum temperature, and a high limit switch that shut the gas off if things got overheated. These thermostats were all connected to various solenoid valves that controlled the flow of gas coming through the machine. This was literally trial by fire, but I had plenty of help figuring things out. I did have an operator's manual by this time and was sure glad of it. I dried all of my corn that year with very few problems, a good example of beginner's luck.

I've moved on to bigger and newer dryers since these earlier days. I did get Bert Peaslee's second Meyer-Morton in 1995 so that I could speed up the corn harvest. The continual hard running of a two- or three-week stretch of corn drying took its toll on these old machines. They began to get tired and break down fairly often. In 1998, I bought an abandoned 570-bushel M and W brand dryer for two hundred dollars from a farmer just outside of Middlebury, Vermont. It was so inexpensive because it needed some sheet-metal repair, but after I replaced the bottom clean-out canister, the thing was as good as new and ready for service. It was bigger than my two original dryers put together and less prone to breakdown because it had fewer moving parts and a much simpler V-belt system of power delivery. This thing is a real monster. It's twelve feet in diameter and more than fourteen feet in height. It has, however, held up quite well over the years, and I'm happy to say the M and W 570 is still in use on my farm today.

Drying grain with propane in a freestanding batch dryer is serious business and not to be taken lightly. I can't give you a complete education here in a few paragraphs, but I will try to touch on the high points of safe operation of this amazing grain-saving tool. First of all, make sure the dryer is supported by its extra metal legs and that it is dead level. After your dryer is hooked to the gas tank, check all connections for propane leaks with a soap bubble solution. If gas is escaping, you will see it in the bubbles. Turn the valve at the bottom of the tank very slowly and deliberately to activate the gas supply. Most tanks have an anti-slugging device that automatically closes an internal safety valve if gas rushes out of the tank too quickly, and all dryers have a sail switch in the fan that must be activated for gas to flow and the burner to fire. If you start everything up and the burner won't fire, it could be a stiff sail switch. Conversely, the sail switch will shut off the gas if the tractor powering the dryer runs out of fuel. Study your manual and set all of the controls as recommended. It is very important to keep the grain temperature as low as possible to maintain quality. One hundred degrees is the maximum allowable temperature to maintain germ, and a properly set grain temperature thermostat will ensure that the grain doesn't get too hot. When drying grain, check grain moisture and temperature often, as overdrying can be just as detrimental to grain quality as not drying it enough. Remember that grain loses moisture as the temperature drops. One-hundred-ten-degree corn with a moisture level of 16 percent will stabilize at 13 percent or less when it cools. Once you understand the moisture and temperature relationship, you can turn the gas off earlier and let the grain cool down in the dryer.

Grain should never be put in the bin hot, and you should never leave a grain dryer unattended for very long, as there is too much that could go wrong. If the thing catches on fire because of a faulty sail switch, it could take a nearby building with it, which is especially true if you are drying a high-oil crop like sunflower seeds. Follow the manufacturer's recommended setbacks from buildings, people, and livestock. Also, extremely wet grain (like corn over 25 percent moisture) will sometimes dry unevenly and poorly because it tends to stick to the tapered walls at the bottom of the dryer. So be right there with the moisture tester when the dryer empties out to make sure your grain has dried uniformly. Remember that a grain dryer must be full of enough grain to cover the entire top of the internal plenum. Any exposed metal will allow precious heated air to escape to the atmosphere instead of traveling through the grain. Also be aware that as a batch of grain dries, it will shrink in volume and could eventually bare the top of the plenum. Keep your eye on the gauge at the propane tank; a batch of wet corn can take up to five hours in the dryer and will use an inordinate amount of gas. If you have a lot of drying to do, you might want to have a second tank or a really good gas man who will keep you well supplied with propane.

Propane-powered grain drying is out of reach for most novice and small-scale growers of grain, and we have barely scratched the surface when it comes to discussing this practice. These old round batch dryers have pretty much given way to newer, higher-capacity continuous-flow systems, and it's getting to the point now where very few modern commercial cash grain farmers bother with old batch dryers. For this reason, they are incredibly cheap. Newer rectangular column dryers have a diamond-shaped internal plenum. Moist grain comes from the field and is stored in a hopper-bottomed "wet tank." From there it enters the top of the dryer and very slowly makes its way to the bottom. Basically, the grain goes in the top wet and comes out the bottom dry, and there is no circulation of the grain. These systems cost hundreds of thousands of dollars and are intended for people with rather large acreages. My advice is that if you find a reasonably priced old batch dryer, study it carefully and buy it well enough ahead of time that you have time to get it fully operational before the grain harvest season begins.

CHAPTER EIGHT
Preparing Grain for Storage, Sale, or End Use

Once a crop of grain has been properly conditioned and dried, the storage phase begins. It doesn't matter if grain is kept in a bag, barrel, or large silo, constant vigilance is necessary to keep it in good condition during long-term storage. For one thing, grain kernels can be hydroscopic and absorb moisture from the surrounding environment over time, and bags of stored grain left on a cement floor will wick humidity directly from the porous surface below. Earlier in my farming career, I stacked some sacks of seemingly dry barley on a cement slab and promptly forgot about them. Six months later, I returned to find the barley had a very musty odor, and lo and behold, mold had invaded the bottom third of each bag. After this incident, I began to take better care of my stored grains. A wooden floor is much more stable than a cement slab, which will transmit moisture from the atmosphere as well as the earth below. If you must store grain on cement, install a protective layer of boards between the grain and the floor.

Outdoor galvanized-steel grain silos present their own unique set of problems, as the metal walls will certainly hold grain in and weather out, but heat and cold are easily transmitted through this very thin steel membrane. This can be especially problematic during the spring warm-up period. The mass of grain contained in the silo cools off through the winter months to the ambient outdoor temperature, but when warmer spring days arrive, the sun's rays beat down onto the south-facing wall of a grain silo and warm the metal to the point that it is almost too hot to touch. The frying-pan-like heat begins to migrate into the cold grain on the other side of the bin wall, and condensation is the by-product of this encounter between warm metal and cold grain. Of course, condensation produces moisture, and grain spoilage is the result. This scenario is most intense in stored shell corn because common practice is to dry it only to the maximum allowed moisture level of 15.5 percent. This situation is easily remedied, however, by warming up the contents of a grain bin every spring. When the sun comes out and the temperature begins to climb, that's the time to start the aeration fans. This process can take a few days or up to a week depending on the size of the silo and the amount of grain contained within. Professional cash grain farmers (I am certainly not in this category) monitor their bins with temperature sensors and alarms that will notify them when there is a ten-degree temperature difference between stored grain and the outside air. These automatic systems will actually turn aeration fans off and on when ventilation is needed. The rest of us, however, can happily perform this little task by hand. I was reminded how important this procedure is in 1995. I was grinding a batch of cow grain in early June when I noticed green dust wafting out of the corn as it poured from the bin unloading auger. All of a sudden I got that sinking feeling in my stomach that something had gone horribly wrong, so I climbed the ladder up the side of the corn silo and opened the inspection hatch. The beautiful yellow corn had crusted over and turned an odd shade of yellow-gray, and a pungent sour smell burned my nostrils. The problem was definitely most pronounced along the interior south wall of the bin where the grain had warmed and condensed the most. I could only think of all the time and work that had been invested in that remaining fifty tons, and I was devastated. The corn was not completely ruined, but its quality had been seriously reduced through my own ignorance and lack of attention to detail. So please, don't let this happen to you! Turn those fans on every spring.

The Butterworks Granary

I built my post-and-beam granary in 1990, but the project began a few years earlier when I erected a twenty-six-foot-tall large hopper-bottomed bin to store organic shell corn purchased from Bob Crowe of Canajoharie, New York. The bin stood on legs and was nine feet in diameter, and held a tractor-trailer-load of corn—twenty-five tons. Like all bins with lower walls that taper down to a bottom-mounted unloading auger, this particular silo was fraught with problems once the ice and snow arrived. The bottom auger boot was constructed with a special metal plate that slid in and out of a narrow opening. The gravity flow of the grain from above could be shut off by pushing this plate all the way into the slot, which allowed access to the auger and the bottom of the bin for repair and maintenance. It turned out that the little slot for this sliding shutoff gate was also an ideal inlet route for rainwater and melting snow, and the water that entered the bottom of the bin during the warmer months would soak the corn and cause spoilage. In the colder months, the wet corn would freeze solid, bind up the auger flighting, and hinder the unloading process. After several years of struggling with spoiled and frozen corn, we decided to construct a building around the bin. Perhaps for romantic reasons, I copied the design of an old defunct grain mill located thirty miles to the north in Ayers Cliff, Québec. The old Dupuis mill was quite tall and had an extra monitor-style roof that ran its length. I liked the looks of the building and found out that the specially designed extra roof had been installed to accommodate large internal upstairs grain bins that were fabricated from wood. I didn't know much at the time, but I did know that this was what I wanted. It seemed like style and function could work well together in this situation. Our granary went up in August 1990. The building measured twenty-eight feet wide by thirty-eight feet long and close to forty feet to the peak of the monitor roof. Over the next few years, we sheathed the granary and constructed several massive overhead bins, which was my first foray into bulk grain storage aside from a few little plywood affairs slapped up in the corner of an equipment shed. These bins integrated perfectly into the heavy support timbers of the building. The first step in putting together indoor grain storage was to build rectangular frames of two-by-eight planks, eight-inch sides flat, overlapped and bolted at the corners. These frames were spaced a foot apart at the very bottom, two feet apart in the midsection, and thirty inches apart at the top. The logic behind this design was that the support framing needed to be closer together at the bottom of the twenty-five-foot-tall bin because of increased weight and side pressure located there. Vertical one-inch boards were then nailed to this frame to form an internal skin, which was finally covered with sheets of galvanized steel; a four-inch sliding gate valve was cut into the floor for gravity flow emptying of stored grain. Each of my newly constructed overhead bins held between twenty-five and thirty tons of grain. The building was divided in half with bins on the west side and a drive-through passage on the east. This layout allowed us to back a tractor and feed grinder-mixer into the granary and supply it with grain from above by gravity, and the monitor roof above proved to be quite advantageous because we could roll our fifty-one-foot auger right up to a window opening to fill the internal overhead grain bins from above. This whole process taught me a lot about the advantages of inside storage bins. Warming and cooling of stored grains was not necessary, because a bin inside a building offered double protection from the vagaries of the outdoors. All that was necessary for the safe storage of a crop was to make sure the crop was dry enough before it went up the auger and into the bin. The problem of moisture infiltration into the hopper-bottomed bin was also eliminated once it had the added protection of the granary walls and roof.

Preparation for Cleaning

Ordinary precleaning by rotary screen or other methods is adequate for grains intended for animal consumption. Sometimes this step can be skipped entirely, because a certain amount of weed seed can be tolerated in almost any livestock ration. However, it's a good idea to be moderate in the weed seed department. Mustard seed is quite high in both protein and oil, but it is spicy and hot tasting. Too many of these little round black seeds will flavor ground grain, making it unpalatable. Stored grain intended for seed to be replanted or for human consumption will require further cleaning and conditioning. Lots of forethought should be put into seed production. Above all else, do everything possible to preserve germ, including timely harvesting and gentle handling of crops. Apply some sort of aeration to the grain as soon as possible to avoid heating. It may be as simple as screwing an aerator into a gravity wagon. Choose your "seed piece" before the harvest begins. There will usually be at least one standout field of a particular variety whose quality shines above everything else. Look for the field with plumper kernels, good standability, fewer weeds, and the best overall appearance, and check for rogue plants. You don't want rye sticking up in your winter wheat or oats growing in with your barley. Combine your specially selected field into a separate wagon or other suitable grain receptacle. This grain will need to be kept separate through the entire post-harvest process. The standard of perfection for grain destined for flour milling, hulling, flaking, or malting is also high, but can be one tiny notch lower than seed grain. Weed seeds, dirt, chaff, and other foreign material must be removed from the stored crop before you can proceed any farther in the process of adding value to it.

The Basics of Cleaning and Grading

Grain cleaning is not complicated. There are hundreds of specialized pieces of equipment for the task, but the process only involves two basic physical principles—shaking and blowing. Seed grading is a shaking process accomplished by sifting grain kernels through oscillating sieves. Several other types of grading are practiced, but the sifting of grain with reciprocating screens is by far the most common and effective way to remove over 90 percent of foreign material from the product. Air aspiration is the other method. Here lighter material is actually blown from a grain sample with a controlled blast of air. This is the same winnowing process practiced by farmers who first settled the land here in our region. People stood with a winnowing basket in the barn's center alley and actually tossed grain into the windy cross draft to blow out the lighter material from their harvest. We can refine this process for modern use by adding a few more features that weren't available 150 years ago. If you need to clean small amounts of grain from your home garden harvest, a couple of five-gallon plastic pails and a three-speed home box fan will suffice quite nicely. A shop vacuum with a blower attachment will also work well. Situate the box fan on a chair and place an empty pail on the floor in front of it. Continue by turning on the fan and pouring a fine stream of grain directly in front of it from the upper pail to the lower pail. The wind from the fan will carry away the lighter material, while the heavier better-quality kernels will drop to the pail below. This will take some trial and error, as well as experimentation with the fan speed and the height from which the grain is dropped. Remember to spread out a tarp before you begin dumping grain from pail to pail, as this will allow you to sweep up and collect your mistakes during experimentation. My one other piece of advice in setting up this home grain aspiration system is to wear a mask to protect your respiratory system from the dust and dirt flying around in the air.

Small amounts of grain also can be readily sifted through homemade screens, which can be inexpensively fabricated from wooden lath and hardware cloth. Begin by nailing together three-quarter-inch by inch-and-a-half wooden strips into a twelve-by-twelve-inch frame. You can be as simple or fancy as you like, with plain butted corners or sturdier mitered corners. Obtain an assortment of wire-mesh hardware cloth and tack it to the screens. Standard sizes run from one-eighth-inch to one-inch squares, but window screening will suffice for smaller seeds like flax. Most grain cleaner manufacturers sell wide assortments of hand screens; these are used to sift grain samples to determine which sizes of screens should be used for a cleaning job. A good collection of these hand screens will allow for some precision home cleaning. Screen grading of grain is usually a two-step process. Grain is first passed through a screen with holes just large enough for it to fall through. Larger material like dirt clods, chaff, and plant parts is retained by the screen and falls off the end, while the desirable kernels fall through. This method of removing the larger unwanted material is called the "scalping" process. The grading process comes next, where a screen with holes smaller than the grain is employed. The good grain is retained by the screen, while the smaller material like weed seeds falls through. A basic understanding of grading concepts and proficiency with hand-sifting and aspiration will get your small quantities of grain properly cleaned and give you a good foundation of understanding to take the next step toward machine cleaning of grains.

Air Screen Cleaners

After years of trying to mechanize the winnowing process, the hand-cranked fanning mill appeared on the scene around 1880.

Many of the same artisans and small shops that made threshing machines began to manufacture winnowing mills at this time. These early cleaners measured three feet wide by four feet long by four feet tall and were constructed of furniture-grade hardwood. A rounded housing built of curved pieces of wood was attached to one end of the machine, and this half-barrel-like protrusion contained a very simple fan constructed of rectangular wooden paddles rotating on a steel center shaft. Motive power from a side-mounted hand crank was transmitted to the fan through a series of gears that stepped up its revolutions enough to achieve a good blast of air. All fanning mills were built with a square top-mounted hopper and a reciprocating screen directly below. The back-and-forth motion of the cleaning screen was achieved by means of a connecting arm and bell crank driven from the main hand-turned crankshaft. These early winnowers were usually supplied with several interchangeable screens to match the crop being cleaned. To operate a fanning mill, one person poured a steady stream of grain into the hopper while the other turned the crank as evenly and steadily as possible. Fan speed was controlled by the person turning the crank on the side of the machine. Simple wooden dampers were installed inside second- and third-generation machines to better direct and control the velocity of the air blast.

Fanning mills from the late nineteenth century seem like pretty crude affairs to us today, but they were an incredible step up for farmers who had been hand-winnowing their grains in the breezy center drive of the barn. These old devices are still pretty common out there in the countryside. In a moment of dreamy idealism, I bought one at an antiques shop for twenty-five dollars in 1973. It was sturdily constructed of maple and painted a beautiful shade of blue, which had faded over the years. I had no farm or access to land at the time, but it sure seemed like a good idea to stock up on grain equipment. I was able to use my prize a few years later on some dry beans that were in need of cleaning. It definitely was a step up from standing in a windy doorway pouring material from bucket to bucket. If you are dealing with small amounts of garden-grown grains, I would recommend acquiring a fanning mill if one comes your way for free or cheap. And while acquiring an antique fanning mill can be a good introduction to the grain-cleaning process, it's definitely not worth paying exorbitant antique prices for something like this; you might as well get a more modern motorized cleaner in that case.

The aspiration, scalping, and grading processes are all combined together in the workings of the modern air screen cleaner. Picture yourself swirling a hand screen around and around in front of a blowing box fan. Add a second screen underneath and a grain delivery system and you have the beginnings of a mechanical grain cleaner. It didn't take American ingenuity long to improve upon the design of the first hand-cranked fanning mills, and the job of grain cleaning was vastly improved by the addition of a second lower screen. This new added feature, which was positioned directly underneath the top screen, had much smaller holes for finer sifting. Once cleaning performance got better, manufacturers began building larger machines with small cup elevators and bagging attachments, which allowed for much easier delivery of bulk grain to the machine as well as the simple takeaway of clean seed in bags. The late nineteenth century was also the age of motor power. It didn't take long before the hand crank was replaced by an early kerosene or gasoline engine. Many grain mills that had been powered by water installed turbines and electric generators. The A. T. Ferrell Company of Saginaw, Michigan, began making its Clipper line of grain cleaners as early as 1869, and hundreds of thousands of these machines were used on farms and in grain establishments all over the country. Clipper is still making seed cleaners in Bluffton, Indiana, today.

The Forano 150

I bought my first grain cleaner in 1983. My old friend and mentor Clarence Huff from Compton, Québec, had a Forano 150 to clean and size the seed oats he grew and sold. Every time I visited Clarence, I marveled at the big wooden hulk of a machine with its belts and pulleys. On one of my visits to his farm, Clarence told me about another machine that might be for sale. It was located in old Dupuis mill in Ayers Cliff, which was only ten miles north of the Canadian border. I drove over and visited Newton Blake, who had bought the building and turned it into a Laundromat and storage facility for boats and old cars, while his wife ran a small beauty parlor next to their living quarters in the back. We dickered back and forth and finally came up with the price of two hundred dollars, which was a considerable amount of money for me at that time. We used rollers and planks to get the thing loaded in the back of my truck. Once loaded, it took up all the room I had in the bed of the truck; it was so tall that it towered three feet over the top of the truck's cab. The load was quite top-heavy, so it was a slow drive home. It must have been quite a sight riding along in the back of my truck. When I passed Canadian customs in Rock Island, Québec, they sounded the siren because they wanted to inspect my "export" from their country. Buying this particular grain cleaner must have been in the stars for me, as I had admired the Dupuis mill for many years every time I had driven by, and buying the Forano 150 from Newton Blake gave me free access to explore and inspect every nook and cranny of this monitor-roofed grain building that intrigued me so much. I got to know old Newt quite well, and he was touched when I patterned my new granary after his building seven years later.

When I arrived home with my new Forano 150 seed cleaner, a new chapter in my grain education began. I set the machine up in an open pole shed. It wasn't an ideal location because it was open to the weather on the east side, but it was the only spot I had. A bit of head scratching had to take place because the cleaner had already been dismantled when I picked it up. Two cup elevators and a grain-receiving hopper had to be integrated to the main frame, and belts and pulleys had to realigned and installed. I called Clarence a few times for advice. The machine was made entirely of high-grade maple. It came with a collection of thirty or more screens that were all numbered. Penciled onto many of the screen frames were indications for their use, like "avoine haut" (which meant "oats upper") or "blé bas" (which told us it was the lower screen for wheat). Also written on the wooden sides of the cleaner were the names of farmers along with a count of how many bags of oats or barley they had had cleaned. Much to my liking, this old machine was full of history. Forano was a very well-known Québec foundry and manufacturer of equipment for sawmills, grain mills, mines, and farms. The company had been located in the town of Plessisville, halfway between Sherbrooke and Québec City. The modern-day Forano of 1983 was more of a welding shop—a mere shadow of its former self. My cleaner (or "crible" as it was known in French) had been built in the 1920s. Needless to say, you were on your own when it came to getting replacement parts. The beauty in all of this was that it didn't need new parts because it was extremely well built. At this time, during the early 1980s, every small town in Québec still had an active feed mill that actually ground grains for local farmers, and every mill had a Forano 150 for cleaning farm-grown oats for replanting. Because of this, a knowledge and parts base for these old machines had persisted. Shortly after this time, the trend in grain milling north of the border began to move toward a more centralized industrial-style model. The little "Shurgain" and "Co-op Féderée" mills were all closed down by 1995. I was glad to have bought one of these old screen cleaners when I did.

Stenciled onto one side of the cleaner was "200 RPM." This machine was not meant to turn fast. To attain the proper operating speed, I bolted an old twenty-four-inch flat pulley from my junk collection onto the main cross shaft of the machine, and connected a six-inch flat belt from the rear pulley of my tiny John Deere 40 tractor to the larger pulley on the cleaner. Once again, beginner's luck prevailed: The little old twenty-horsepower two-cylinder putt-putt ran the seed cleaner at just the right speed. Grain was dumped into a tin-lined receiving hopper and then elevated twelve feet straight up, where it trickled into the inlet end of the cleaner. This was my first experience with a cup or bucket elevator. This very simple contraption was an upright long wooden box about eight inches square and twelve feet in height. Sprockets were mounted in the top and bottom of this special box; running around and around on these sprockets was an endless square-linked chain that had a very small steel cup attached to every fifth link.

These little buckets scooped up grain, whose flow was controlled by a sliding door at the bottom of the receiving hopper. Once a bucket arrived at the top of the elevator, it flipped over the top sprocket, dumping its little load of grain down a feeder pipe and into the cleaner. The empty cup would then make its upside-down way down the back side of the elevator box. Once it reached the bottom, the process would begin all over again. A properly working cup elevator, set up in tight quarters, can deliver a large amount of grain quite quickly.

East Tyrolian Austrian horizontal stone mill. PHOTO COURTESY OF JOHN MELQUIST

The first stop for grain elevated by my Forano 150 was the debearder. This device was an enclosed metal cage made of very sturdy fine-mesh hardware cloth. Little steel fingers propelled the grain horizontally through this cage from one end to the other across the top of the machine. As the grain traveled the three feet, the little steel fingers rubbed off any awns or beards that were still attached. At the same time, very fine seeds like those of mustard dropped through the fine wire mesh to a cloth bag hanging below. This was a precleaner as well as a debearder. I remember trying to empty the bag while the machine was running. It weighed fifty pounds and was over my head. Once it was removed, all of this mustard seed would still be dropping down right down my neck and under my shirt. (We eventually replaced the bag with a catch tank and a pipe to gravity-deliver the seed to a bag waiting below.)

Once the precleaned grain dropped out of the debearder, it then fell into a specially designed distribution box situated over the vibrating screens. This was a very simple design with a slowly turning auger inside the box and a slowly revolving wooden drum below. The grain was pulled back across the box by the auger, whereupon it trickled through a small opening between the revolving drum and the edge of the box. This opening in the bottom of the distribution box was adjustable for different sizes of grain kernels, and if the machine was being fed properly the grain being augered across this special box would make it all the way across to the other side. This would ensure that the vibrating screen below would receive an even covering of grain. My first order of business with my new seed cleaner was to learn how to feed it properly: Just the right amount of grain had to be going up the elevator, and the small opening through which the grain dropped out of the distribution box had to be set just right. If the opening was too big, the screen below would not be uniformly covered with grain. If too much grain was traveling up the elevator or the opening in the bottom of the box was too small, the distribution box would overflow and make a mess. Luckily there was a little glass window in the side of the distribution box to allow for inspection from the ground seven feet below. I hadn't even started the real cleaning process yet and had already learned that there were a ton of little adjustments to make to this beautiful old piece of equipment. The rate at which to supply a cleaner with grain is the first lesson to be learned when you're trying to master grain cleaner operation. Fortunately, the worst thing that can happen if you make a feed supply mistake is a little spilled grain or a flat belt flying off a pulley.

The "shoe" is the heart of any air screen cleaner. This is where the screens that do the shaking and sifting are located. The shoe on the Forano 150 was a three-sided box suspended from the top of the hardwood main frame by four flexible supports. Its fourth side was open on the end to allow for the placement of two interchangeable sieves. The entire shoe, frame and screens, was attached to a couple of reciprocating shaft-driven arms that shook it back and forth in a two-inch stroke. Underneath each screen, a long brush traveled back and forth from one side to the other and back to keep the seed moving and sifting properly. I was so impressed by the ingenious technology employed by the designers of this very old piece of equipment.

I cleaned barley on my initial run with the Forano. I followed the directions on the list of screens, installing a number fourteen screen in the top slot of the shoe and a number nine in the bottom. Once the machine began to operate, I came to a better understanding of how the thing actually worked. Barley dropped from the distribution box onto the top reciprocating screen. Most of the barley dropped through the number fourteen holes, while the chaff, unthreshed heads, and small pieces of straw slid off the end of the screen into a bag. This was the scalping process. Once the barley hit the lower number nine screen, the small stuff (weed seeds and dirt) dropped through, leaving the retained quality grain to drop off the back side. This was the sifting process. Next came the coolest part of the whole operation: A rather large steel paddle fan situated in the bottom of the cleaner provided a blast of air upward toward the grain tumbling like a waterfall off the lower end of the bottom screen. (Fan speed could be controlled by turning a wheel on the side of the machine.) This controlled air blast blew the last of the chaff and light material out of the falling grain and up through a vent to a catchment hopper. The heavier, high-quality barley dropped right by the fan to a collection area in the very bottom of the cleaner. From there it entered another smaller cup elevator, which also served as a double bag-filling unit. To my utter surprise and amazement, you dumped grain into the intake hopper and it came out of the machine in five different places: (1) mustard seed from the debearder; (2) scalped material off the top screen; (3) weed seed and other fine material from underneath the bottom screen; (4) number two grain blown by the fan into a large hopper in the rear; (5) clean number one seed-quality grain that made its way past the air blast of the fan. I didn't know much about what I was doing that first year with the cleaner, but I was able to make the thing function. This was a real confidence builder for me. I was able to transform barley that was full of weed seed and coarse material into a high-quality finished product.

Additional Notes on Cleaning Infrastructure

Just about every piece of grain equipment you will ever use will need to be adjusted to work properly. Developing a clear understanding of how something works is the first step. After that, it's time to give it a try and see what happens. You can make refinements while on the job. Mistakes will most certainly happen, but it is part of the learning process. Let common sense be your guide and make safety your number one concern—keep hands and fingers away from moving parts, and watch those loose sleeves and pant legs. My adventure with the Forano 150 had only just begun that summer of 1983. I eventually moved my old cleaner into the bottom floor of our new granary building eight years later. Sitting on a dirt floor in an open shed began to take its toll on the underpinnings of the machine's wooden frame. I had to rebuild some of the wooden understructure. Somewhere along the way, I procured an ancient electric motor with a flat belt pulley keyed to its shaft. So we switched from tractor power to electric motor power. This was much handier. The machine could now be operated with the flick of a switch. The quarters were much drier and more protected from the elements. If grain spilled (and it always does), cleanup was simple. We could now sweep the spillage from a cement floor instead of the damp earth. There was more room to stack and store bags of cleaned grain. Feeding the machine was also much more streamlined because a gravity wagon could be backed in next to it and unloaded by a small electric auger. We had to use a bucket brigade to fill the intake hopper when we were operating in the open shed. This is the story of gradual improvement in my own infrastructure. It won't be any different for any other fledgling or wannabe grower of grain. You have to go at your own pace and make do with what you have. If you happen across a reasonably priced old grain cleaner with a good collection of screens, buy it. The finished building can come later.

As I have built and improved my infrastructure over the years, I have also increased my understanding of the sifting process and the screens that do the job. In the very beginning, I followed screen size recommendations to a T. It didn't take long to realize, however, that every kernel of barley, oats, or wheat was not totally uniform in size and shape. This was especially evident with the top scalping screen. Sometimes lots of good kernels of grain would roll off with all the chaff and other coarse material. Stepping up one size from a number thirteen to a number fourteen screen would solve this problem, and soon I got accustomed to shutting down the cleaner and experimenting with different screens. I found out after a while that the numbers on each screen referred to hole size measurements in sixty-fourths of an inch. There were slotted screens in addition to the round-hole ones. Slot width was also measured in sixty-fourths along with slot length in fractions of an inch. For example, a 9 × ¾ screen had slots nine sixty-fourths of an inch wide by three-quarters of an inch long. In some instances, slotted screens worked best for the initial scalping action on the top, while round-hole screens did a better job sifting down below. Oats with their hulls and boat-like shape cleaned much better with slotted screens. I also found out that slight differences in bottom-screen hole size could make a big difference in how much grain went for number one versus how much ended up in the weed seed bag. Dropping down to a smaller screen size number put more grain in the number one bag. If the weed seed contained too many good quality mini kernels, I simply ran it through the machine again with different screens. As I added other crops to my repertoire over the years, I found that I needed additional screens that didn't come with the original collection. These screens were no longer available from Forano, and while I was able to get a few here and there from various individuals in southern Québec, I couldn't find everything I needed, especially numbers sixteen through twenty-five. I needed these larger hole sizes for cleaning beans, peas, and corn. A bit of research took me to the A. T. Ferrell Company, the manufacturer of the Clipper line of seed cleaners. It turns out that my Forano 150 must have been a Canadian copy of a Clipper grain cleaner because Clipper screens were just a fraction of an inch smaller than mine, and they fit perfectly into the shoe of my machine. I have since bought quite a few screens from A. T. Ferrell. There are over 150 different sizes available in just about any configuration imaginable. Clipper screens cost close to two hundred dollars apiece and come with round holes, triangles, slots, or wire mesh. It's hard to believe that buying a screen today costs the same as I paid for my cleaner in 1983.

Over the years, I have also become much more familiar with the nuances of the fan adjustment on my machine. When I first began cleaning grain, I noticed that sometimes there would be just as much number two grain falling into the bag below the catch hopper as there was number one grain being produced. It didn't take me long to realize that if I backed off the wind, more grain would drop past the fan and into the number one compartment. Once I began putting plumper, heavier grains like corn and beans through the machine, I began to appreciate the advantages of increased airflow from the fan. These grains need a much stronger blast of air than the cereals do. (I have actually been able to blow splits out of black beans with the cleaning fan.) On the other hand, lighter crops like flax and grass seed need only a little whisper of wind. I have also found out that the air system on my old Forano is ancient and somewhat limited in its ability to provide the cleanest grain sample possible, and newer cleaners have a much more advanced air system. Whereas my machine provides an air blast from below to aspirate grain falling off the lower screen, more sophisticated models lift lighter material right off the tops of the reciprocating screens. Fans are positioned higher up in relation to the shoe; air is conducted to the upper side of the screens through a vent system. This same vent also serves as a conduit to remove the lighter material once it is aspirated. These more advanced air screen cleaners have three and sometimes four screens. Three-screen machines do a very complete job of scalping and sifting because the screens are triple-stacked in the shoe frame. A four-screen cleaner is equipped with two shoes. Grain is scalped and sifted once, then dropped onto the lower shoe for a second run-through. Three- and four-screen cleaners are more commonly used by professional high-volume seed cleaners.

Grain and seed cleaning needn't be a stumbling block or hindrance in the process of learning how to produce homegrown grains on any sort of scale. There are still plenty of old Clipper 2Bs around the countryside, especially in regions where grains culture persisted longer than here in the Northeast. These old machines pop up at auctions all the time out in Iowa, Illinois, and Ohio, and since large-scale corn and soybean farming is the order of the day in these places, an old seed cleaner doesn't usually sell for much more than a hundred dollars. Of course, it's hard to go to the Midwest to attend an auction, but there are a number of people who make it their business to buy equipment in the Midwest and bring it back to New England. Their little cast-off stuff is just the right size for us out east. Commodity Traders International sounds like some huge multi-national corporation, but it is a father-and-son outfit based in Trilla, Illinois, that specializes in finding, reconditioning, and reselling old grain-processing equipment. I have purchased some really neat old stuff from these two salt-of-the-earth individuals—Charles Stodden Sr. and Charles Stodden Jr. You will always pay more for something if it comes from a dealer or a middleman, but in most cases it's worth it: They can find exactly what you want because they live there and they have the contacts and the connections. A used grain cleaner might cost you five hundred to a thousand dollars. This still isn't much money when you consider that a new Clipper cleaner from A. T. Ferrell may sell for over three grand. The most important thing is to get your grain cleaned however you can. It may be as simple as packing up your small number of bags and journeying somewhere not too distant to use someone else's equipment. Or you can buy the hand screens you need from A. T. Ferrell and do the job right at home. Most old cleaners are made of wood and are quite easily restored and repaired. Just remember to set your machine up level wherever you decide to use it. Keep an old cleaner protected from the elements, and it will give you many more years of good service.

The Gravity Table Separator

I happily cleaned all of my many different grains for years with my Forano 150, and had no idea that there was anything else out there besides the common fanning mill. I found out differently in the early 1990s when I began experimenting with dry beans. I purchased seed from Tony Neves of the Freedom Bean Company in Maine and began to learn what else was out there in the world of seed cleaning. Dry beans have their drawbacks in that they tend to split in half during the threshing process, not to mention that excess rain and ground moisture can stain bean seed. Tony had decided to shrink the size of his operation by selling all of his extra bean equipment. He showed me how he was able to remove splits and lighter-weight stained bean seed with a machine he called a gravity table. He wanted to sell it, so I paid him a thousand dollars and brought the thing home on a trailer. I bolted it down to the cement floor in my granary and wired up its ten-horsepower electric motor. The machine had been manufactured by Forsbergs Inc. of Thief River Falls, Minnesota, and its design was rather simple. The gravity table separator was basically a large wooden box that contained a powerful squirrel-cage fan to blow a strong blast of air up through a top-mounted vibrating screened deck. The model I purchased measured four feet high by six feet long by three feet deep. The upper deck was shaped like a trapezoid, longer in the front and angled in on the sides. This deck could be tilted up in the back and on its right side. The fan speed was controlled by a set of variable sheave pulleys adjusted with a hand crank. Air intake was governed by opening or closing a sliding door on the back of the unit. I installed a small intake hopper over the left side of the gravity table and tried it out on some black beans. I didn't change any of Tony's settings. The machine worked beautifully as long as I followed a few basic rules. A constant flow of beans was necessary to keep the wire-mesh deck covered with beans at all times. If the layer of beans was too thin or had mesh showing, the gravity table would not function properly. Once I turned the thing on and supplied it with the proper flow of beans, its performance was flawless, and the fan below hummed and whirred while the deck vibrated the beans gently across from one end to the other. The table was tipped up slightly on its right-hand side. Magic seemed to be happening as the beans dropped from the hopper above down onto the vibrating table and air stream. The sea of vibrating beans seemed to miraculously separate right before my eyes. Lighter off-color and split beans migrated to the left and fell into a hanging grain bag. The heavier, shinier beans seemed to walk uphill to the right and deposit themselves in two more hanging sacks. Small stones and pebbles ended up piled up against a barrier at the extreme right-hand side of the deck. This was truly an amazing piece of equipment. My black beans looked a whole lot better once they had been run across the top of the gravity table.

Whereas air screen cleaners grade grains mostly by size, gravity tables separate materials by density. Two identical kernels of grain may be the exact same size, but one may weigh more than the other. The kernel with the highest specific gravity (density) will have more potential as a seed or processed product because it contains more and better quality-material within itself. This takes us back to our discussion of grain test weights. Corn is supposed to weigh fifty-six pounds per bushel, but if the bushel weighs fifty-eight or sixty pounds, it is higher-quality corn. The gravity table separates out these heavier kernels of grain for you. Everything that makes it to the far right-hand side of the vibrating deck is high-density, quality grain. For a number of years, I only used my gravity table for beans. I finally decided to try it out on some wheat with shriveled kernels that could not be removed by my other cleaner. Much to my surprise, the gravity table did an even better job on the wheat than it did on the beans. Since then, we have begun to use it quite often as the second and final step in our grain-cleaning process. To expedite things, we have installed a small cup elevator leg to feed grain to the overhead hopper above the gravity table. When we're really cooking, we can auger clean grain from the air screen cleaner directly over to this elevator. We certainly aren't professional high-volume cleaners of grain, but I do know that we have come a long way since we first hooked a flat belt between the Forano 150 and the old tractor.

Although I am quite proud of the advances we have made in grain cleaning over the years, I have come to realize that my gravity table, just like my old Forano 150, is a bit ancient and outdated. I've called Forsbergs several times with questions and a request for an operator's manual. When I told them that I had a Model 10 gravity table separator with a serial number of 940, the phone went dead for an instant. Finally, Gary in the parts department told me that my machine had been built in 1946. "Don't break anything," he proclaimed. "There are no parts available for that machine." Newer gravity tables have features that allow a greater degree of separation to take place from one end of the deck to the other. The Oliver Manufacturing Company of Rocky Ford, Colorado, makes gravity tables that provide four individual air blasts to the deck. Each blast of air is independently adjusted and regulated to achieve an enhanced degree of separation. Apparently, Oliver machines are capable of achieving close to a 20 percent difference in material density across the top of the deck. Gustafson is another popular make of gravity table. Each particular brand of gravity table has its own set of strengths and weaknesses. An Oliver may have better airflow, but a newer Forsbergs has the advantage of a better eccentric system for improved table movement. Many of the newer models (twenty years old and younger) are equipped with metal hoods and an auxiliary vacuum system to suck up airborne chaff and dust. I will have to be happy with my ancient gravity table for a while because newer models are pretty expensive. A brand-new Forsbergs or Oliver machine can cost as much as twenty to thirty thousand dollars. Good used gravity tables sell in the five- to ten-thousand-dollar range. If you're in the market for a gravity table and can afford to wait a while, be patient. Something will come along sooner or later for a reasonable price.

Other Types of Cleaning Equipment

I've had the good fortune to meet many people involved in the grain business from all over the United States and Canada. Crops and climates differ from one geographic region to the next, but out on the Great Plains and in the Corn Belt of the Midwest, grain is king. When I travel, I am utterly amazed at how many different kinds of equipment exist in the world of grain, and the same holds true with grain-cleaning machinery. While a good air screen cleaner and gravity table can take care of 95 percent of anyone's cleaning requirements almost anywhere, there are also a number of very specialized machines tailored to very specific cleaning tasks. I don't have a lot of direct experience with most of this cleaning equipment, but I do understand what each machine is designed to do and how it works. I would find some of this equipment helpful on a rare occasion, and am not planning on procuring any of these items anytime soon. A brief description of a few of the more common pieces of seed-cleaning equipment follows.

Debearder

Barley, rye, and some varieties of wheat have little elongated spikelets protruding from the top of the grain head. These awns, or beards as they are often called, are attached to the upper end of each individual kernel of grain in the head. Many times the beard will remain attached even after the crop has been harvested. Debearders are a longitudinal cylinder equipped with a number of shaft-mounted metal paddles that apply friction to the bearded grain kernels moving through. The abrasion process literally rubs the spikelets from the kernels. Grain that retains its awns is just fine for ground livestock feed, but not human consumption. Grain destined for malting especially needs debearding.

Destoner

Crops that bear their fruit close to the ground can end up with small stones in the finished grain sample. This is especially common in dry beans and soybeans. Soybeans are harvested with a special flex-head cutting attachment on the combine. The flex head floats along right on top of the ground, shaving the stems at ground level to retain the lowest-hanging pods on the plants. Often a little ridge of dirt will cross the cutter bar, allowing soil to enter the combine. This is how little rocks can get into the soybeans. Dry beans are "pulled" from the ground by special undercutting knives, and are then windrowed and harvested with a special bean combine. Lots of soil always ends up mixed with dry beans. Rocks will rattle the moving metal parts of a feed grinder as well as crack the teeth of people who unsuspectingly consume beans with small stones. Destoners do work, however, and look exactly like gravity tables. (Forsbergs is famous for its destoners.) Grain passes over the destoner in much the same way it does a gravity table. The only difference is that when the stones hit the farthest wall, they are collected. The grain or beans proceed on their merry way while the stones are separated and left behind.

Aspirator

This piece of equipment is used to apply secondary air to a flowing stream of grain loaded with light material. Aspirators are commonly used to blow loose hulls from oats, spelt, or barley that has just been through the dehulling process. These can be stand-alone machines or combined with a multi-functional grain processor like a dehuller.

The Carter Disc Cleaner

If you want to grade grain kernels by size and shape, this is the tool for the job. As many as thirty three-quarter-inch-thick, twenty-inch-diameter cast-iron discs are positioned two inches apart on a rotating shaft. The surface of each individual disc is pockmarked with little indentations. These little indents are actually sized in sixty-fourths of an inch, just like cleaner screens. The shaft and discs rotate inside of a metal housing that is designed so that grain can flow through the bottom in and around the rotating discs. This is accomplished with the help of a small auger that runs the entire length of the machine underneath the discs. The shaft and discs rotate through the grain in the bottom at sixty revolutions per minute. As the discs pass through the grain, kernels of the desired length are picked up by the indents and tossed over into another compartment as everything rotates upward. The Carter disc is especially useful when trying to separate shorter-grain kernels from longer ones. This machine will very readily separate stray rye kernels from wheat. There are many different indent sizes for the discs on this machine. Professional grain cleaners will most often have a different Carter disc cleaner for each individual grain-cleaning job.

Grain cleaning is a necessary skill needed by just about everyone who harvests a crop and wants to take it to the next level. The job can be performed on a very small scale with hand screens and a box fan. A very small investment in some basic equipment will take you to a more intermediate level of grain cleaning. It is important to remember that other infrastructure is required for cleaning and storage to be successful. You will need something like a small four-inch auger to move grain from here to there and back. If you want to get really fancy, you can find and install some small cup elevator legs. A small leg has the advantage of fitting into limited space—it will take your grain straight up in the air and drop it into a nearby vessel at a forty-five-degree angle. If you want to take grain eight feet away, you need to lift it eight feet up to drop it the proper distance at the forty-five-degree angle. Augers will usually suffice for moving grain sideways, but they can be cumbersome and get in the way of human movement in a cleaning situation. It seems like we are always tripping over augers in our granary. Elevator legs keep everything out of the way above your head. We have discussed the necessity for clean, dry, and well-aerated storage. The last point to emphasize is the importance of being ready to jump into action when the grain is harvested and ready to be cleaned. Designing and building some kind of a functional cleaning operation of any size or scale is a good activity for the off season.

Unless otherwise noted, photos courtesy of Jack Lazor

The author's farmstead from the lower pasture.

Looking north to Québec. PHOTO COURTESY OF BETHANY DUNBAR

Mixed grains: oats, peas, and wheat. PHOTO CREDIT UNKNOWN

Wisconsin farmer John Ace, 1974. PHOTO COURTESY OF ANNE LAZOR

Feeding wheat into a threshing machine, 1977. PHOTO COURTESY OF ANNE LAZOR

Planting wheat in early April on the home farm, 1998.

Reaping our first crop of wheat. PHOTO COURTESY OF JIM REHRMAN

Detail of a threshing machine. PHOTO COURTESY OF ANNE LAZOR

Reaped wheat stooked for field-drying. PHOTO COURTESY OF ANNE LAZOR

A field of grain corn half in Vermont and half in Québec.

Applying a light coat of compost in the early spring.

Processing farm compost with a Sittler windrow turner.

Straw-based compost medium.

Windrows of compost at Butterworks Farm.

Clover interseeded into standing corn.

Last pass before planting with a field cultivator.

Discing down the furrows in a spring grain field.

Spring tillage with a moldboard plow.

Plowed furrows.

Fall plowing for a spring cereal crop.

The Lemken Rubin 9, a German one-pass tillage tool for cereal grain planting.

The offest disc, a primary tillage tool in lieu of moldboard plowing.

Spring trip shanks and tillage sweeps on a field cultivator.

Planting wheat with an end-wheel grain drill with press wheels.

Planting Red Fife wheat in April at Butterworks Farm.

An old-fashioned 1960s-era mid-mount row-crop cultivator.

First pass with the tine weeder after spring wheat emergence.

The work of the weeder.

Application of organic soil conditioner just before wheat emergence.

Tine weeding in soybeans.

A vigorous seeding of spring wheat is hardly impacted by tine weeding.

A Danish tine three-point-hitch row-crop cultivator.

Tine weeding a row crop to eliminate weeds at the white hair stage.

Antique rotary hoe. PHOTO CREDIT UNKNOWN

Harvesting swathed wheat with a combine.

Hard spring wheat glowing red at harvesttime.

Combining a beautiful crop of hard red spring wheat.

Unloading dried cereal grain.

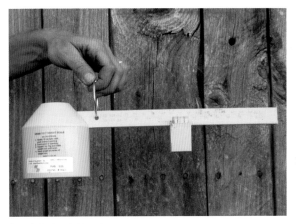
A grain test-weight scale.

Dried barley going into a gravity wagon.

A screw-in grain aerator.

Testing for grain moisture.

Companion-planted red clover in grain stubble.

Aeration of grain in a batch dryer.

Rotary screen grain cleaner atop separated weed seed prior to storage.

Unloading a gravity wagon into an auger.

From the rotary screen to the dryer.

Close-up of grain ready for the bin.

Recirculating grain.

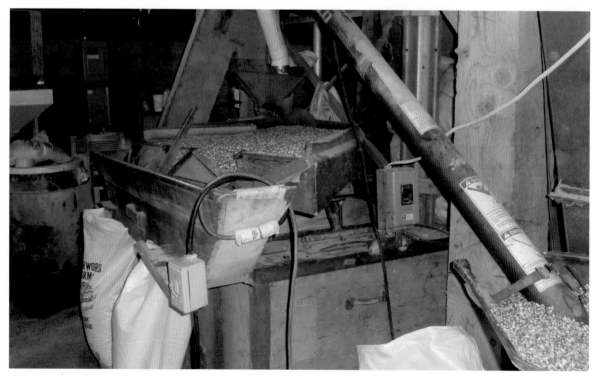
Grading seed corn for density on a gravity table.

The fanning mill grain cleaner at work.

Unloading a tote into the seed cleaner.

Weighing a tote on an antique grain scale.

Finished processed grain products ready for market.

Top deck of a Forsberg gravity table separator.

An antique Forano 150 fanning mill grain cleaner.

CHAPTER NINE
Corn

Corn is one of the few major grain crops that originated in the Americas. It is believed to have descended from the wild grass teosinte over ten thousand years ago in the Balsas River Valley of southeastern Mexico. Teosinte is a wild grass with a tassel very similar to the corn plant's, and as with its descendant, each individual seed of the teosinte plant is enclosed by its own little husk and is borne in the axil of each leaf. Archaeologists and ethnobotanists have found evidence of corn dating as far back as 8700 BC, but major progress in domestication really began in earnest around 1100 BC. In the first millennium AD, maize culture spread rapidly northward into the area that would eventually become the American Southwest, and corn arrived in the Northeast sometime after AD 1000. (The vast fields of corn cultivated by New England Indians have been described in chapter 1.) While history and archaeology don't provide us with all the details about the early culture of *Zea mays*, we do know one thing for certain: The plant as we know it today would not have made any genetic progress without the help and intervention of humankind. Corn cannot survive as a wild plant without help from us; with its seeds encased in a tightly fitting husk, it can only reproduce if someone husks and replants it. If a ripe ear of maize was to fall onto the ground, its kernels would rot before they could sprout in most circumstances, so some kind soul had to remove that husk and the seed from the cob before germination could take place. This is where we entered the picture.

Corn is the ideal grain for a small grower with limited equipment, land, and expertise. Everything from soil preparation and planting to harvesting can be accomplished quite easily by hand on a very small scale, and breeding selections are always made by hand. Because no specialized machinery is required aside from a hoe, a spade, and an able body, a small plot of corn is much easier to manage than an equal-sized parcel of barley, wheat, or oats. If the Native Americans could raise corn with no more than a hoe or a digging stick, we should be able to do it with a few more sophisticated tools.

Growing Corn in Colder Climates

Corn is an amazingly productive plant. One small seed planted in May has the potential to produce a towering twelve-foot-tall plant with several ears of ripe grain by the end of September. This is due in part to the fact that corn is slightly more efficient at photosynthesis than most of the other major species of grain plants. *Photosynthesis* refers to the interactions among light, carbon dioxide, and enzymes that together produce sugar, starches, and ultimately plant biomass. Oats, wheat, barley, beans, and oilseeds are all C3 plants; corn, sorghum, millet, and sugarcane are C4 grasses. C4 plants convert more of the sun's rays and atmospheric carbon dioxide into plant material than their C3 brethren because less photorespiration (the process whereby oxygen and carbon are lost to the surrounding atmosphere and soil during photosynthesis) takes place during photosynthesis. Corn is certainly a full-season plant. It remains vegetative much longer into the growing season than the cereal grains. Oats, wheat, and barley enter their reproductive stage of growth in June when grain heads begin to appear, but corn waits until late July or early August to show silk and tassels. This extra four to six weeks of vegetative growth provides the corn plant with the stature and leaf area to produce a much higher volume of grain per acre than cereals. Corn is just beginning to develop and fill ears of grain as the small grains are being harvested. This situation has its advantages as well as its drawbacks, however. Corn bears and ripens its fruit at the end of the growing season when days are getting shorter and temperatures are cooling off. Yields are certainly much higher, but the ripening process can be slowed or curtailed by frost and rainy weather. Small grains field-dry well in the heat and sunshine of July and August, but autumn weather, especially in northern New England, is anyone's guess. Where I live on "the rooftop of New England," it's not uncommon to have a little snowstorm in October. The early snow always melts quickly, but it is a stark reminder of what's to come sometime in the next few weeks. So while corn dries down in the field quite well out in the Corn Belt of the Midwest—the combines start rolling in late September and early October, when the grain is already down to 15 percent moisture—it's a different story here. If our corn has dropped to the 25 percent moisture level by the end of the first week of October, we are having a good year. But despite all of these reasons to not grow corn, many of us continue to do so with great relish. It's addictive to plant seeds in May on a bare field and watch a jungle develop over the next several months, and the potential for an abundant harvest of golden grain on a beautifully sunny autumn day keeps us all going year after year.

The Basic Biology of Corn

Corn has the uncanny ability to evolve quickly as it adapts to new and changing environments. The extremely short-season varieties that we grow in far northern New England and southern Canada come from the same parent plant found in southern Mexico. Corn can be classified as one of three distinct types—flour, flint, and dent—depending on the starch makeup of the kernel. Each kernel has an embryo or germ at its base. The germ harbors the future prospect and genetic potential for another generation because it contains a minuscule corn plant complete with roots and stalk. The embryo comprises the scutellum, plumule, and radicle. The tip cap at the very bottom of the kernel protects the germ, and what little oil a corn plant produces is found in the germ.

Corn kernels are protected by two outer layers—the pericap and the aleurone. The outermost layer, the pericap, is clear and papery thin. The aleurone is thicker and forms the actual skin around the kernel; if the corn is colored, this is where the pigment is contained. The remainder of the kernel consists of starch or endosperm. There are basically two kinds of starch—hard or flinty endosperm and soft or floury endosperm. The ratio of hard horny starch to soft floury starch as well as their location in the kernel determine if the corn is flint, flour, or dent. Flour corn primarily comprises soft starch surrounded by a thin outer layer of horny starch. Flour corns grind quite easily into a fine fluffy corn flour. The Indians of the American Southwest and the Great Plains were major users of flour corns. Flint corns are on the opposite side of the spectrum, containing mostly hard starch. The flint corn kernel has only a tiny center core of soft endosperm. Dent corns have a relatively large inner core of floury endosperm surrounded by a thin ring of flint-like hard starch. The juxtaposition of a soft center and a hard outer shell is what causes this type of corn to "dent" or dimple in the center during the ripening process. The soft interior dries much more quickly than the hard exterior. The uneven drying of the two distinct starches is responsible for the center shrinkage and resulting dent. Popcorn is a type of flint corn that consists completely of horny endosperm, and the resultant makeup of 100 percent hard starch is what causes this type of corn to pop when heated. Sweet corn can be either flint or dent in origin. There is such a high proportion of sugary endosperm in a kernel of sweet corn that it shrivels and shrinks when dried. Just about every corn out there is a flint, dent, or flour type.

Flour Corn

The original maize grown by Native Americans was either flint or flour corn. Flour varieties were commonly cultivated in southern regions and northward through the High Plains into the Dakotas. Due to the softness of the starch, kernels of flour corn were easily pulverized into a very fine, silky-textured flour with very simple implements. Small grinding holes cut into native rock are a common sight in most Indian ruins of the Southwest. Masa harina is a type of ancient corn flour used to make tortillas. It was (and still is) made by steeping and cooking kernels of flour corn in a caustic wash of water infused with lime or wood ashes, in a process called nixtamalization. The alkalinity of the soaking solution loosens and removes the pericap and aleurone outer skin of the kernel, leaving the soft endosperm behind. After cooking, the outer skin is rubbed off and floats to the top. The remaining soft starch is then ground while still in a wet state. The resulting flour, masa, can then be dried or immediately pressed into tortillas. In addition to removing the outer layer of the corn kernel, the slaked lime used in the process also serves the beneficial functions of adding calcium to the diet as well as liberating the amino acid niacin in the corn; combining lye- or lime-treated corn with some beans makes for a complete protein. This process was developed out of sheer nutritional necessity in Mexico over two thousand years ago.

By virtue of its extremely soft starch, flour corn can absorb water from the surrounding environment as it finishes its drying process in the fall, and prolonged periods of wet weather can foster the development of molds in these corns. For this reason, flour maize has remained native to southern regions and the semi-arid plains where dry autumn weather is standard.

Flint Corn

Harder-textured flint corns acclimatized to the cool, rainy, humid weather of the eastern United States and Canada. These are the same long, slender eight-rowed ears of corn that the first European colonists found upon arriving in the New World. Just about every tribe of Native Americans cultivated their own regionally adapted variety of flint corn. This is where we see the amazing ability of corn to adapt to a wide cross section of climatic conditions. Three-foot-tall plants of Gaspé yellow flint make ripe grain in the very short growing season of northern Québec, while ten- and twelve-foot plants do the same thing twenty-five hundred miles to the south in Georgia and the Carolinas. Flint, or Indian corn as it was called, was universally adopted in North America by the white man. Because of their hard texture, flints mill with a bit more difficulty into a much grittier-textured meal that's better for steaming than it is for baking. Flint corn makes the best polentas, and johnnycake—a favorite food of early New England settlers—was a very moist mixture of ground flint corn and animal fat steamed on a hot griddle. There were literally thousands of varieties of flint corn grown all over the eastern United States and the northern plains. Unfortunately, most of these are now extinct, but old varieties including Longfellow, Rhode Island Whitecap, and King Philip have persisted into modern times, and many heirloom varieties of flint corn are available from Seed Savers Exchange in Decorah, Iowa. Flint corn is prized by many for its unique sweet flavor and gritty texture.

Dent Corn

Dent corn is a relatively new phenomenon in the grand scheme of things. It may dominate today's world of corn, but it did not exist in early America. In fact, when New Englanders began to migrate in the 1820s and 1830s to the "Northwest" of Ohio, Indiana, and Illinois, they brought their local varieties of flint corn with them. These varieties crossed with some of the flour corns that had also been brought to the region by settlers from points farther south like Virginia. Once again, corn demonstrated its ability to adapt to new environments and growing conditions: The combination of flint and flour genetics produced larger, girthier ears of corn. These ears yielded increased amounts of grain that made much better animal feed. Dent corn was born in the region to which it was best suited. The high humidity, warm nights, and fertile black soil of what was to become the Corn Belt all combined to further the popularity of this newly adapted miracle of cross pollination.

Dent corn was conspicuous by its multiple rows of puckered "dented" kernels. Numerous local varieties became native to each little region of this newly settled territory. After 1850, named varieties like Reid Yellow Dent, Boone County White, Leaming, and Hickory King began to emerge in different areas of the newly settled territory. Dent corn soon became king in the region that was to become the American Midwest, and a new mixed farming model of wheat, corn, hogs, and cattle grew exponentially because of the bounty of dent corn. Farmers were the first breeders of this miracle plant, selecting and saving seed from their most productive specimens. Then land grant colleges got in on the act after the Civil War with the development of state experiment stations. Varieties like Minnesota 13 and Cornell 25 were developed by some of the first public corn breeders.

Ensilage

Flint corn began to fall out of favor in New England in the late nineteenth and early twentieth centuries. As dairy farming became the predominant form of agriculture, farmers began to grow corn for silage. "Ensilage" was made by cutting down the whole corn plant when the grain was only half ripe. Cut and bundled corn plants were then run through a special belt-driven ensilage blower that chopped everything into small pieces and blew it up a pipe into a silo. This mixture of chopped stalks, leaves, cobs, and milky kernels was then packed and left to ferment. Corn silage was the sweet-smelling, very palatable product of this fermentation process. Corn silage revolutionized northeastern dairy farming in much the same way dent corn had changed midwestern farming fifty years earlier. Midwestern dent corn varieties were much better suited to corn silage culture than the native New England flints, because they yielded more. As a result, seed corn from dent varieties began moving eastward as dairy farmers began using midwestern corn for silage production. On-farm seed saving of locally adapted flint corn varieties met its demise during this same era, because as New England farmers became specialists in milk production, seed corn very quickly became a purchased input. The closest area of seed corn production was the Finger Lakes region of central New York, and Stanford Seed of Buffalo, New York, was a major purveyor of seed to the Northeast. Western and central New York had a climate very similar to that of the American Midwest. Flint corn was still prized as a meal for human consumption there. Locally adapted flint corns still took most of the prizes at county fairs all over the region. Harold Putnam, a ninety-six-year-old retired dairy farmer from Cambridge, Vermont, told me about a northern Vermont variety named Davis Flint that took first prize for many years in a row right up until the 1930s at the Champlain Valley Fair. I report with sadness that seed for Davis Flint has not been available since World War II.

Breeding for Productivity

Up until the 1930s, all the corn that was grown was open-pollinated, and the many different varieties available produced seeds that were true unto themselves. When you planted Davis Flint or Reid Yellow Dent, you harvested seeds identical to the ones you planted. The hundreds of varieties available to farmers in 1920 were the culmination of fifty or more years of breeding and selection efforts by farmers and university plant breeders. But while the initial crosses between eastern flints and southern flour corns had produced these higher-yielding dent varieties, and average corn yields had climbed from fifteen bushels to the acre in 1850 to close to twenty-five bushels by 1900, yields hit a plateau at this point.

In 1920, farmers were still only averaging twenty to twenty-five bushels of corn to the acre. So, late in the nineteenth century, university scientists began to apply the genetic and heredity work of Charles Darwin and Gregor Mendel to the world of corn. In 1906, George H. Shull began experimenting with inherited traits in corn at the Bureau of Plant Industry in Cold Spring Harbor, New York. Shull first worked to develop inbred lines with very specific traits. This was accomplished by selfing individual corn plants. *Selfing* refers to the process of breeding a plant only to itself in order to accentuate certain desirable traits like yield or grain quality. This was accomplished by dusting the silk of a particular corn plant with the pollen from its own tassel above. Once pollinated, the future ear was enclosed in a special paper bag to prevent any other pollen entering. Inbreds are the product of inbreeding, and although these plants may have the ability to transmit a specific inheritable trait, they are weak and spindly and not viable for commercial production. In 1908, Shull then took two inbreds and crossed them together. He was delighted to find that seed from the offspring produced a corn plant far better than either of its two inbred parents or any other mass-selected variety of open-pollinated corn. He had set out to prove Darwin's theory of heterosis, also known as hybrid vigor.

Other scientists followed suit and continued to isolate and develop workable inbred varieties for corn breeding and research. Researchers soon found that crossing inbreds produced even better breeding material; soon after, the "double cross" became the norm in corn breeding. A corn breeder would start out with inbreds A, B, C, and D. A and B would be crossed to make AB and C and D would be crossed to make CD. The next growing season, AB would be crossed with CD to make hybrid ABCD. D. F. Jones of the Connecticut Agricultural Experimentation Station produced his first double-cross hybrid from four inbred parents in 1918; the first commercial double-cross hybrid, Burr-Leaming, was released by the Connecticut station in 1921.

Hybrid corn was a little slow off the mark because commercial seed production still needed to be developed. Iowan Henry Wallace was the first person to figure it out, in the 1920s. Wallace was also an influential farm editor and leader. He had written several books on corn, and his family had published *Wallace's Farmer* out of Des Moines, Iowa, for many years. (Henry Wallace was Roosevelt's secretary of agriculture from 1933 to 1939 and vice president from 1940 to 1944.) Wallace developed a seed production method whereby two inbreds were planted in side-by-side rows. The variety that was to be used for seed production was detasseled by hand; then pollen from the neighboring row would fertilize the ear on the detasseled stalk. The male row that provided the pollen would be discarded while the ear on the female detasseled plant would provide the hybrid seed. Henry Wallace's Hi-Bred Seed Company (soon to become Pioneer Hi-Bred) marketed its first hybrid corn seed in 1926. Hybrid corn not only yielded 15 to 20 percent more grain, but also remained standing late into the fall because of improved stalk health. Unlike most other new agricultural developments, hybrid corn took the farming community by storm: By 1930, over half the corn grown in the United States was hybrid. By 1940, open-pollinated corn was a thing of the past, and in the space of fifteen years farmers went from savers of seed to buyers of it. A whole new seed corn industry had been born in the American Midwest.

The Growth of Corn Agriculture

No other domesticated plant in American agriculture has experienced as much change in the last ninety years as corn, and yields have climbed exponentially over the years. Thirty- to thirty-five-bushel yields were commonplace just before the beginning of the Second World War. Commercial fertilizers became more widely used after the war because excess ammonium nitrate left over from the production of munitions was channeled into agricultural production. By 1950, many farmers began to see averages of fifty bushels to the acre. In 1965, the average national yield for corn had climbed to sixty-five bushels an acre. Increased yields were also due to the fact that hybrids could be planted much closer together than their open-pollinated ancestors; when corn was planted three or four plants to a hill forty inches apart, total plant population was about twelve thousand to the acre. Modern corn planted six or seven inches apart in thirty-inch rows has a population of thirty-three thousand to the acre. In 1930, hybrid corn was a totally new plant compared with the open-pollinated (OP) varieties of yesteryear. OP corn was conspicuous by its diversity, with tall, short, spindly, and stocky plants all growing in the same field, whereas hybrids were all about uniformity; each stalk of a particular hybrid corn variety was (and still is) identical. When you look down a row of hybrid corn, every ear will be attached to its stalk at exactly the same height. Sometimes it almost reminds you of a field of little clones.

Indeed, we have traded the plant diversity of the old-fashioned OP corns for the increased yields and narrow genetics of modern corn. But this lack of genetic diversity didn't seem like much of a problem until an outbreak of southern corn leaf blight, a fungal disease, nearly wiped out the nation's corn crop in 1970. Diverse genetic material was quickly brought in from Latin America to produce leaf-blight-resistant inbreds for the next growing season. It was also about this time that corn breeders began to abandon the production of double-cross hybrids in favor of varieties bred from improved single crosses.

Corn yields have continued to climb, to the point where the national average has reached 165 bushels to the acre. Breeding has certainly been partially responsible for this miracle of productivity, but we mustn't forget that artificial inputs in the form of herbicides, pesticides, and chemical fertilizers have played a major part in all of this "progress." High-yield corn farming is all about petroleum usage. Critics of modern high-yield agriculture like to mention the fact that the energy equation in terms of petroleum calories in and corn calories out is just about even, if not somewhat tipped in favor of the oil well. Even though we may look askance at all of these bushels per acre coming from millions of acres of hybrid corn, there is no substitute for watching an excellent crop of golden yellow corn flow from the combine into the grain wagon. As organic farmers, we partner with sun, soil, and the environment to help this miracle along without dependence on artificial inputs.

The Rise of Genetic Modification

A farmer from 1920 transported to a modern-day cornfield by time machine would not know what to make of today's corn. In addition to the previously described petroleum inputs of fertilizers and chemicals, today's corn is also genetically modified. DNA research began in earnest during the late 1970s and early '80s. Once molecular biologists began to understand the genomes of particular cells, the next step, in their minds, was to manipulate DNA for the betterment of humankind. This was especially true in the medical field, where laboratory researchers used this technology to find cures for disease. Somehow, this newfound field of knowledge very quickly found its way beyond the walls of the laboratory and into the greater realm of agriculture, and biotechnology was born. In 1996, the first transgenic varieties of corn were released with the Bt (*Bacillus thuringiensis*) gene inserted right into the plant. "Yieldguard" corn was the first of many biotech varieties released by the Monsanto Company in which genes from another totally unrelated organism were placed into the DNA of common agricultural plants. This nifty little genetic insertion protected the corn plant from the European corn borer because it filled the entire plant with Bt toxin. The *B. thuringiensis* fungus has long been the pesticide of choice for organic farmers for use on worms in the larval stage—and now every cell of the plant contained this pesticide.

Soon after, scientists at Monsanto isolated a specific protein that would make plants resistant to the herbicide Roundup. "Roundup Ready" soybeans hit the market in 1996, followed by Roundup Ready corn in 1998. Roundup is a nonselective post-emergence herbicide that kills all plant life upon contact; Roundup Ready crops are resistant to it. There are plenty of places aside from a book on organic grain production to read more about genetically modified grain crops and their influence on modern agriculture, but suffice it to say that things have never been the same since these crops were widely adopted by American agriculture. In 1996, GM corn didn't seem like such a big deal. We were all more worried about insect pests becoming resistant to Bt. But none of us in the organic sector had any idea that GMO corn and soy would become so all-pervasive in today's modern farming world. Presently, 93 percent of the soybeans and 87 percent of the corn grown in the United States have GMO traits, but in the late 1990s, conventionally bred hybrids still dominated the scene. GMO corn varieties were few and far between, and were more concentrated in the Corn Belt. There just wasn't enough corn grown in the Northeast to warrant the development of genetically modified varieties to suit the cooler climate and shorter growing season of our region. This situation didn't last long, however. *Biotech* became the buzzword, and by 2005 it was increasingly obvious that the entire US seed industry was heading in that direction. If you thumb through a seed catalog from one of the large corn seed companies like Pioneer or DeKalb, you may only find one or two conventionally bred hybrids these days out of a total of twenty varieties that are offered. These varieties sport names like Liberty Link and Herculex for resistance to the herbicide Liberty and for protection from the corn rootworm. Many of these modern-day GMO varieties are triple-stacked, which means that multiple traits have been inserted into their genomes. You may know by now that corn reproduces by pollen, which is shed by the tassel and absorbed by the silk lower down the plant. You may also know that pollen is light and windborne. Herein lies a challenge: protecting a field of organically grown non-GMO corn from the pollen drift from neighboring GMO corn. Unfortunately, the technology is here to stay for the time being. We can pass judgment on this type of agriculture for being in defiance of the natural order and for being diametrically opposed to our principles, but if we want to grow corn, we still must farm in the presence of this challenge. It is my hope that if we can intimately know the culture of corn and grow higher-quality organic crops that yield better than what Monsanto has to offer us, we can turn the tide of the biotech monster.

My Personal Corn Odyssey 1975–2011

My first experience with corn growing was at Old Sturbridge Village Museum in Massachusetts, where I worked as a farmhand on the living historical farm from 1971 to 1973. We planted Rhode Island Whitecap flint corn, four plants to the hill along with beans and squash. The corn was handpicked, dried in a wooden-slatted crib, and baked into johnnycake in the farm kitchen. The whole business of raising seed and locally adapted varieties of flint corn for human consumption has intrigued me ever since. I got my next taste of corn culture when I spent a year in southern Wisconsin in 1974. Round wire corncribs were to be found everywhere, and every dairy farm picked ear corn and ground it into meal for dairy rations. I was very fortunate to get to know dairy farmer John Ace of Oregon, Wisconsin, who taught me, an eastern kid who wanted to be a farmer, all about midwestern mixed farming. In October, I got to help run the corn picker and haul corn to the crib. John fed his pigs by pulling a gravity wagon around the pasture with its unloading door open just enough to dribble a stream of ear corn on the ground. The pigs would come running to devour the corn. He also loved to tell stories about the past, and I listened intently to him reminisce about selecting ears of corn for seed and planting corn in hills with a horse-drawn planter and a check wire. We would drive around the countryside and go to farm auctions together. This was ten miles south of Madison in an area where lots of hybrid seed corn was being grown by two small seed companies named Blaney's and Renk's. I remember seeing fields where four out of every six rows of corn had no tassels. John explained to me that hybrid corn was the product of two different "corn parents," and that the corn had been detasseled by college kids the previous July. I also learned that corn had been harvested with the pull-type corn binder before the debut of the mechanical corn picker. Bundled corn was stood up to dry in the field in large shocks; there were still Amish in the area doing it the old-fashioned way. When we went to farm auctions, John would always take me over to the antique farm machinery and tell me all about it. I remember being fascinated by the Rosenthal corn shredder. These were rather large contraptions, similar in size and shape to the threshing machines of yesteryear. Corn shredders were belt-driven, and bundles of dry corn were fed into one end of the machine. Ear corn was delivered up an elevator to a crib, loose shelled corn ended up in a sack, and the remaining corn stover was shredded and blown into a large stack for winter feed and bedding material. No one wanted these old relics, and it was so sad to see them go for ten to twenty dollars. I was certainly young and impressionable at the time, but I was intent on farming this way, so I listened carefully to every story and anecdote. I feel quite lucky to have spent almost an entire year under the spell of John Ace in 1974 and early 1975, as he had a reverence for the past and he had lived it as well. This was also the end of an era when small mixed farms with cows, pigs, chickens, corn, oats, and hay were predominant in the American Midwest.

Our Move Back to the Northeast

Armed with knowledge of and experience in small-scale diversified Wisconsin agriculture, Anne and I returned to Vermont in May 1975 to give it a try in the Northeast. I wanted to grow grain corn on our little rented homestead, so before leaving Wisconsin, I contacted Dr Willis H. Skrdla at the North Central Regional Plant Introduction Station in Ames, Iowa, to see if he had any old varieties of corn that would do well in the cool climate of the mountains of northern Vermont. Much to my surprise, Dr. Skrdla provided me with eight little manila envelopes that each contained fifty seeds of some very old heirloom varieties of short-season corn. Aside from putting in a large garden and milking a family cow, my little corn project consumed most of my time that first summer in Irasburg, Vermont. I convinced a neighboring farmer to plow up and manure four small parcels of land for my project, and I planted the corn by hand and tended it through the summer. Most of the varieties were flints like Gaspé Yellow, New York, and Canada, but there were a few dents like Minnesota 13. The corn grew quite well, but became quite popular with the raccoons when it got to the milk stage. I tried traps and red pepper on the silks, but in the end the wildlife got the better part of my crop. Since there weren't many ears left to harvest, it was difficult to tell what would have been the varieties best suited for my particular growing region, but I do remember that the Gaspé Flint got about three feet tall and produced miniature ears of ripe grain by the middle of August. This inauspicious beginning to corn culture taught me that my crops were food for more than just me. If I was going to succeed in this business, I needed to find a way to outsmart the raccoons.

Despite the fact that my first attempt at corn growing was a bust, I was undaunted. There definitely wasn't any grain corn being grown in these parts during the mid-1970s. This didn't stop me from asking every old-timer I met if they remembered growing corn for anything besides silage. Surprisingly, there were old folks who remembered husking bees still happening in the 1930s. One neighbor, Daisy Orne—from whom we bought an old hand-crank cream separator—told us that if a boy got a red ear at a husking bee, he got to kiss the girl of his choice. The other common theme from these conversations with elder members of the community was that the summers were hotter back then and the corn grew better because of the heat. I'm not sure if it was my desire to replicate the idyllic self-sufficient farm of yesteryear or my love of history, but I liked to look backward as much as I did forward in those first days of my farming adventure. I was struck by the number of old granaries and wooden-slatted corncribs that hung from the second or third story of old barns. This was evidence enough that corn had been picked and cribbed in our region in days past. Of course, I had to root around and explore the interiors of these old corncribs that hadn't been used in fifty years. Sure enough, there were long slender cobs left on the floors of these structures, but the grain had long since been eaten by some really well-fed rodents. These were flint corncobs, and it seemed like such a shame to me that this corn had once been so common, but was now extinct. No one had cared enough to keep these varieties going, but I was more determined than ever to unearth an ancient strain of flint corn and bring it back to fruition with my tender loving care.

Flint Corn: The First Success

It wasn't long before I had a breakthrough in my search for the elusive northern Vermont flint corn of yesteryear. Anne and I were visiting with UVM agronomist Winston Way in January 1976 when I happened to mention my interest in flint corn to him. Win went over to a file cabinet, opened a drawer, and pulled out two beautiful ears of eight-rowed yellow flint corn, which had been given to him by another extension agent in Lancaster, New Hampshire. They had been grown by an elderly gentleman and were ours to plant and increase the seed stock. We planted our kernels of flint corn on a friend's farm in Albany, Vermont, in May 1976. Our little corn patch was located in an old cow feeding yard where the soil was quite rich; the corn grew like blazes and the raccoons left it alone. We handpicked the corn in late October with snow flurries falling down around us. To finish drying the corn, we stuck a nail in the butt end of every cob and suspended the cobs on ropes strung across the ceiling over our wood cook stove. After we shelled the corn with a small hand-cranked corn sheller, we stored the seed away in a metal can to protect it from mice. The harvest was good in 1976, and we ended up with close to forty pounds of seed—enough to plant two acres of corn.

In May 1977, we took our half of the seed and planted an acre of our precious flint corn on our recently purchased farm in Westfield. I hardly knew a thing about this process, but I did manage to borrow a two-row corn planter to seed my first acre of corn ever. Beginner's luck was with us: The corn germinated well, and the field looked respectable. We used a little front-mounted two-row cultivator on our little John Deere 40 tricycle tractor to control the weeds, and I pampered and babied my acre of corn with lots of hand-weeding. In July, I thought the corn needed a little boost, so I mixed up fish emulsion and seaweed extract and walked through the rows with a hand sprayer to foliar-feed the crop. September came along, and expectations ran high that the corn would start to dry down and ripen. There was also the worry of early frost killing the corn before it got totally ripe. September 1976 was a very cool, wet month. We escaped the frost because it was cloudy and overcast for most of the month. However, the corn did not ripen, because there were no warm, sunny days to promote the process. The first killing frost finally came on the fifth of October, when the corn was still awfully wet and not ready for harvest. It was another hard-earned lesson, to tend and nurture a corn crop through an entire season only to see it killed by frost a week before it gets ripe. Farming is certainly full of joy and sorrow, but if we can't take the good with the bad, we might as well not even try—there will always be disappointment in agriculture. I was doubly sorry that I had planted all of my seed for this old variety of flint corn. My wheat had done well that summer, however, so I decided to postpone my flint corn restoration project to concentrate on grain crops that were more reliable.

Experimenting with Hybrids

I had several more adventures with corn during the 1980s. In 1981, I bought some ultra-short-season hybrid dent corn from Johnny's Selected Seeds. It was time to join the modern age. We salvaged an old two-row horse-drawn corn planter off a stone pile and put it to work. I was determined to get this corn ripe, so we planted it in early May amid snow flurries. It was a battle keeping this corn weed-free, but we did manage to get it ripe before the arrival of the first frost. We left the corn standing in the field into the month of October for drying. Just before it was ready to pick, we noticed that there weren't many stalks left standing in the field. Black bears had found my corn the previous evening and rolled it flat. I could now add bear damage to my repertoire of corn disappointments, and I must say that at this point, I was ready to give up on growing corn. It seemed like it just wasn't in the stars for me. In fact, Anne, my wife, made me promise never to try to grow field corn again. But in 1986, we planted five pounds of Garland Flint (purchased from Johnny's) in our garden. This time we just barely got it ripe; we had to finish drying the ears in our kitchen. I was beginning to come to the conclusion that we lived in too cold and wild a spot to reliably grow corn.

It's amazing how short memory can be. By 1990, I was ready to break my corn promise and give it another try. We were selling hay to other farmers and buying tractor-trailer-loads of organic shelled corn from New York State. Selling hay seemed like a good way to export our soil fertility to other farms, and I figured that we had about ten more acres in hay than we needed for our cows. We plowed ten acres and planted it to a very short-season Canadian hybrid corn called King 127. Once again, I was flying by the seat of my pants. I bought an International 56 two-row pull-type corn planter for three hundred dollars. Thank God the thing came with an operator's manual, and I figured out how to set the plant spacing. The chart on the corn seed bag told me what plates to use for the small round seed that I had purchased, and three bags of seed planted the ten acres with just a little left over at the end. Weeds were a bit of a problem for me that year because I waited a little too long to run the cultivator through the crop, but we ended up getting a whole school-bus-load of people from a nearby evangelical Christian commune to come and spend a day pulling out weeds from the field. These folks saved the day for us. One other stroke of good luck for us that year was that the frost held off until the fifteenth of October, allowing the corn to get good and ripe. Our two interns, Joe Bauch and Simon Dennis, helped me construct a long, narrow corncrib from long cedar poles and page wire. We finished the crib in late October. It measured four feet wide, sixteen feet high, and forty feet long, complete with a galvanized-metal roof. We were working right down to the wire. I bought a used one-row New Idea corn picker the last week of October, and we didn't have any gravity wagons to pick the corn into, so we stapled multiple layers of cardboard to the wooden-slatted sides of my three hay kicker wagons. Last but not least, I bought a forty-foot solid-bottomed elevator to load the harvested ear corn into the crib. We began picking the corn on the third of November and finished on the seventh, and someone had to ride in the hay wagon behind the picker to keep the corn shoveled to the back. I had no idea how heavy ear corn was; the hay wagons definitely were not designed for the amount of weight we were subjecting them to. When we finished on the seventh of November, it started to snow as we picked the last couple of rows. It snowed all through the night and into the next day. Temperatures dropped into the low teens, and the wind began to howl. Eighteen inches of snow had fallen by the next afternoon, and the electricity went out for four days as a bit of mid-January took hold in early November. The corncrib filled up with snow, and we didn't have time to unload the last three wagons of corn because we were so busy being in survival mode with a herd of cows and no power for milking or water pumping. After everything calmed down and the power was restored, we finished filling the corncrib: Our ten acres of corn filled it full. We weren't able to start grinding the ear corn for our cows until later in January after the wet ear corn had had some time to dry in the winter winds. I was a bit surprised at how much more effort it took to run whole-ear corn through my grinder-mixer. The shelled corn we'd been using flowed easily out of an auger into the feed grinder. Ear corn, on the other hand, had to be shoveled into the feed grinder by hand. Whole ears of corn also made considerably more noise as they were being smashed up inside the hammermill. I did find out that the cob portion of an ear of corn contained valuable fiber and very digestible forms of energy, which made for a superior dairy ration for our cows.

Lessons from Québec

The 1990s were the years when I really learned how to grow corn, and it looked like hybrids were what worked best in our short-growing-season location. In the fall of 1991, I took a little excursion to the Eastern Townships in Québec just north of the border to buy some straw from some friends. I hadn't been to Canada for quite a while, and as I drove up the Stanstead Highway to Hatley, I was literally amazed to see field after field of corn that had been harvested for grain. This was the same territory where I had discovered vast expanses of barley and oats growing fifteen years earlier. Grain corn was now the rage just twenty miles away to the north, but it was still unheard of on this side of the border. I ended up stopping at the Bonnieburn Farm in Hatley to inquire about what varieties they were growing. There I met Keith MacDonald, who was trying his hand at cash cropping at the time. Keith explained to me that most farmers, including him, were growing low-heat-unit short-season varieties supplied by Pioneer Hi-Bred Limited of Chatham, Ontario. These Canadian varieties seemed much better suited to the cool climate of northern Vermont than what I had been buying from American seed companies, so I made the switch to Pioneer corn from Canada. In 1993, I planted ten acres of Pioneer 3967, a seventy-seven-day corn with twenty-four hundred Canadian heat units. Much to my surprise, this variety ripened beautifully for me. It stood only six feet tall, but it carried a giant ear of corn. Canadian plant breeding had allowed me to grow very productive varieties of wheat, oats, and barley, and it wasn't any different for corn. These were the golden years of corn breeding, before the advent of genetically modified varieties.

Québec has continued to be my go-to place for everything pertaining to the culture of grain, and the same holds true for corn. Cash crop corn farming didn't really come to the province until the mid- to late 1970s. In 1976, the newly elected Parti Québecois government of Rene Lévesque decided that a separatist Québec needed to be totally self-reliant in the production of feed grains. As a result, the agricultural sector received tremendous subsidies for land drainage, grain silos, dryers, and field enlargement. Farmers in the warmer regions like the St. Lawrence Valley sold their dairy herds to take up cash crop farming. By the early 1980s, the Montéregie region between Montréal and Granby began to take on the appearance of Illinois. Huge fields of corn, giant storage silos, and elevator legs proliferated everywhere. A good portion of this grain grown in southern Québec found its way southward into Vermont and the remainder of New England as the primary ingredient in dairy rations. This change in agriculture was all about agribusiness and corporate expansion. The Québec government grain stabilization program (fondly known as *le stab*) provided farmers with two hundred dollars for each acre of corn planted, which more than covered the cost of seed, fertilizers, and chemicals. As corn acreage increased, the seed companies came to the rescue with higher-yielding varieties that performed well in the cooler-season regions that had once been off limits to grain corn.

The fact that grain corn production had moved eastward into the Eastern Townships of Québec was an incredible boon to a little organic wannabe corn producer like me. I wasn't interested in all the fertilizers, herbicides, and pesticides, but I sure liked the fact that companies like Pioneer were breeding and producing varieties that would mature just south of the border on my farm. I got to know Keith MacDonald quite well over the next few years as he became a more modern-day mentor for me. Despite the fact that Keith was growing five hundred acres of corn with lots of artificial inputs, he still had plenty of good tips for me on my ten to twenty acres. His primary and most important recommendation was for timely and early planting. For each week earlier in the spring the corn was seeded, two weeks could be gained on maturity in the fall. The race was on. Common logic had always told me to wait and plant corn when the ground warmed up to fifty degrees or better. This really wasn't necessary when planting seeds covered with pink antifungal chemical seed treatment, however. Treated seed was accepted by the organic standards at the time, and I used plenty of it. We aimed to plant our corn by the first week of May come hell or high water, and sure enough the corn matured earlier the following October.

It was also about this time that I began growing corn along the Missisquoi River, eight hundred feet lower in elevation than my hilltop farm. In my discussion of grain dryers, I have already described what a revelation it was to grow corn on a riverbank. This is where the heat is, and corn thrives on heat. In 1994, my corn acreage tripled when I moved this part of my grain operation ten miles north to some valley land in North Troy. All of a sudden, yields began to climb to more than one hundred bushels to the acre. Keith recommended that I try growing Pioneer 3921, a hard-textured flint/dent cross hybrid. This particular variety had a very high test weight and matured reliably, and Pioneer 3921 became my variety of choice for the next twelve years until it was discontinued by the company about five years ago. I grew my corn in rotation with cereal grains and soybeans, and for a number of years I was caught up in the glory of it all. By this time we were combining and drying shelled corn, and operating the machinery was as much fun as growing the crop; Gobbling up row after row of grain corn with the combine at harvesttime made me feel pretty important—truly I'm not sure if I was more enamored with the growing process or the machinery. It was easy to forget about the trials and tribulations of years past. At that time, I didn't seem to mind that I was totally reliant on a rather large, distant corporation for my seed. After all those early years of failure, I was simply glad to be harvesting an ample crop that I could grind up and mix with barley to feed to my cows. I had a hero image of Pioneer Hi-Bred International as being the only large seed company out there that was still independent and not owned by a big chemical company like Monsanto or Novartis. I thought that the Pioneer 3921 was the best corn going because it was so dense and so orange in color, and I must confess that I sported a Pioneer jacket and baseball cap. At least this decade of success with commercial hybrid corn provided me with the knowledge and experience to polish my corn-growing skills.

My Return to Vermont Flint

Luckily, the corporate stupor began to fade after a while. My interest in and passion for flint corn and the seed saving culture of yesteryear was still there. In 1997, I finally found the old variety of native flint that I had been searching for twenty years earlier. Tom Stearns, an idealistic seed saver just out of college, gave me a couple of pounds of Roy's Calais Flint seeds to plant. The seed had originally come from Roy Fair, an old-time hill farmer who had lived a nineteenth-century life in the twentieth century in Calais, Vermont. Roy had died in 1981 and left his legacy to two bachelor brothers just up the road, Mike and Doug Guy. Tom salvaged a bunch of twenty-year-old ears from the rafters above their kitchen and planted them out in 1996. I was so happy that there was a variety of northern Vermont flint corn that wasn't extinct. I planted an eighth-acre patch in 1997 with great success. The raccoons left the corn alone this time. The ears were dry and dead ripe by the second week of September. The corn had been cross-pollinated by garden corn because there were a few sweet corn kernels present in some ears, but even so I ended up with thirty pounds of viable native flint corn seed in September 1997. I planted progressively larger patches of this old corn each season for the next few years, and made an effort to select the best plants with the best ears, rouging out anything with sweet corn germplasm present. This corn stood about six feet tall and produced both golden-yellow and beautiful maroon eight-rowed ears of true Indian maize. I've grown Calais Flint just about every year since, and have produced seed for Tom's company, High Mowing Seeds, as well as for Fedco Seeds of Waterville, Maine. And while the earliness of Calais Flint is remarkable, it does have a few disadvantages. Once the ear is ripe, the stalk gets weak and has a tendency to break when subjected to autumn winds and rain. For this reason, ripe ears must be harvested early before the corn has a chance to fall over. Also, the plant never really develops an extensive root system; brace roots, those little support roots that emerge from the stalk just above ground level, are also missing. But even though the Calais Flint lacks the "agronomics" of modern corn, it makes up for it with a superior sweet grain flavor and early enough maturity to save my own seed. I have to thank Tom Stearns for his persistence in ferreting out this old corn and bringing it back from near extinction.

The Search for Open-Pollinated Varieties

My good experiences growing, selecting, and saving seed from Roy's Calais Flint corn gave me the desire to look beyond the hybrid corn on which I had become so dependent for feeding my livestock. I began to look around for varieties of open-pollinated dent corn, which I hoped might eventually replace modern hybrids on my farm. I knew that the old standards like Reid Yellow Dent and Minnesota 13 required more growing season than I had, so these old standards were out of the picture.

Beginning in 1998, I tried one bag of something new and out of the ordinary each growing season. Pioneer 3921 was still my standby, but I really wanted to find an OP corn that would mature nicely and allow me to save my own seed. Commercial seed corn was over one hundred dollars per bag even then. The first corn I tried came from the private collection of Rob Johnston (the owner of Johnny's Selected Seeds). It was called Matheson Dent, and was a very light yellow-colored girthy midwestern dent corn from the northern reaches of Lower Michigan. Needless to say, Matheson Dent didn't think much of our northern Vermont climate. Its maturity was marginal, and some stalks never even developed ears. Right after this, I met Victor Kucyk, a small-scale corn breeder and seed producer from Dublin, Ontario. Victor called me up because he wanted some Calais Flint to breed into some of his lines for earliness; he had many open-pollinated corns and was very willing to let me try them in my locale. So began a four-year stint of planting Victor's varieties. I tried Krugs, Wapsie Valley, and a cross between Reid's and Wapsie. The Wapsie corn had the most promise, and in 2003, I very foolishly decided to plant twenty acres. Wapsie was a big, tall ninety-day corn with massive ears. It grew quite well, but just barely made maturity about the middle of October. Sadly, it had the same weak stalks as the Calais Flint, and as October worked its way into November and the corn began to dry down, it also began to fall down. When we finally went to harvest it with the combine in November, nearly half the stalks had broken and fallen to the ground. This was extremely frustrating because all of these nice ears were never harvested. (The wild turkeys did quite well that season, however.) Open-pollinated corn had quite the reputation for poor standability. I found that out the hard way.

After my twenty-acre fiasco with Wapsie Valley, it seemed like I was stuck with commercial hybrid corn forever. I thought, *At least I can still produce corn, even if I'm dependent on the big seed corporations for my seed.* This was just about the time that I became friends with Frank Kutka. I had first heard of Frank from Walter Goldstein, an organic and biodynamic corn breeder at the Michael Field Institute in East Troy, Wisconsin. I went to visit Walter because he had successfully bred his own open-pollinated corn called Nokomis Gold. After I had explained my predicament of not being able to find a suitable OP corn for my climate, Walter told me about Frank,

who was breeding corn in northern Minnesota, not too far from the shores of Lake Superior. Every three months, Frank published a little pamphlet called *Corn Culture* (www.cornculture.info) dedicated to OP corn. I contacted Frank to get on his mailing list, and the world of OP corn opened up to me at once, as there were people all over the country interested in the same thing. Frank and I got to be pretty good telephone friends. A year later, Frank and his family moved to Ithaca, New York, so that he could enroll in the plant breeding PhD program at Cornell under Margaret Smith. Now he was only three hundred miles away, and Frank even made several trips to visit us in Vermont. He looked at my Calais Flint and Wapsie Valley corn plots and made many suggestions for selection and breeding improvements. My knowledge about corn breeding and its possibilities increased by leaps and bounds. Frank thought that OP corn had been stagnating since hybrid corn had come to total predominance in 1940. He thought that if as much breeding work had been done on OP corn as had been done on hybrids in the last fifty years, OP corn would be every bit as productive as hybrids are today.

Frank set about breeding what he called composite varieties. He would take as many as five different synthetic lines and cross them all together in one breeding plot. Synthetics are used by breeders, each of which consists of as many as ten different inbred lines all crossed with one another. When Frank crossed five different synthetics, he ended up with the genetic diversity of fifty different strains of corn in one variety. One of his first releases combined three very early synthetics from the University of Guelph in Ontario with two from the American Midwest. Klaas Martens, a very well-known organic grain farmer and industry leader from Penn Yan, New York, grew out Frank's first composite variety with great success in 2003. Frank and Klaas decided to call this new composite variety Early Riser because it had such great early-season vigor. I planted a bag of Early Riser in 2004 to try it out, and I was amazed by how well it did. The corn literally blew out of the ground; it had large sturdy stalks very similar to hybrid corn, and it matured quickly in the fall. Klaas encouraged me to save the seed so that the variety could acclimate to Vermont. He made arrangements with New York Certified Seed (a division of Cornell) to process my ear corn into finished seed for sale to other Vermont farmers. In early October, I called up my extension agronomist, Heather Darby, to come over and inspect the field for seed quality. The corn was beautiful and ripe enough to pick. The three-acre piece was right along a busy highway, and once we walked back into the field five or six rows, we found that the raccoons had been having a field day in there. These little creatures had demolished more than half the field. There was hybrid dent corn for silage nearby, but the wildlife skipped it to concentrate on the OP stuff—the raccoons knew quality. We very quickly picked the remainder of the field into one gravity box, and the ears were sorted by hand and trucked out to Ithaca for seed processing. I had waited years for this, as Early Riser corn seemed particularly well suited to my growing conditions. It was beginning to look like seed independence might be a possibility after all.

Frank's Early Riser: Here to Stay

Sometimes corn breeding is just as dependent on luck as it is on skill. Breeders may cross many different combinations of inbreds before they come up with a hybrid they like. Early Riser just happened to be one of these lucky combinations. According to Frank and Klaas, the Early Riser composite variety was not very well suited to the climate of the Finger Lakes region. It was, however, a natural for my cold little corner of northern Vermont. So Frank's Early Riser became my personal corn, and I have been selecting it, saving seed, and replanting it ever since. The corn was not entirely without problems. There were standability issues for the first year. I did my best to choose seed ears from the very best stalks.

Early Riser may have had the rugged stalks of hybrid corn, but it did not have the uniform appearance and agronomics of modern corn; my newfound corn variety was all about genetic diversity. Some plants were tall while others were short, and ear expression was all over the map. Most of the plants bore deep orange-yellow ears with hard-textured flinty kernels; some ears dented while other did not. There were also a few lighter yellow traditional dent ears with very soft grain. The smallest flint-style ears were earlier in maturity than the larger dent ears, and husk coverage varied quite a bit from one plant to the next.

The greatest thing, however, about Early Riser was that there was so much genetic diversity from its five original parents. Dr. Margaret Smith, Cornell's corn breeder and Frank's mentor, came out to our farm and did a workshop on corn breeding and selection. Her advice was to let it stay diverse. Margaret thought the worst thing we could do was to "narrow up" Early Riser by selecting for only one type of plant. So we have limited our selection criteria to standability, overall plant health, husk coverage, and earliness. I feel so thankful to have an open-pollinated corn that works so well for us. A number of other farmers in Massachusetts, Maine, and Vermont have begun to grow Early Riser, and some of them are saving their own seed and adapting it to their own particular climatic conditions. This has been a dream come true for me. Twenty years ago, it didn't seem possible, but it's so nice to know that there is a very small group of individuals out there who want something different from the standard fare that modern agriculture has to offer. We are choosing genetic diversity and seed independence over the narrow and limited choices offered to us by the corporate seed sector.

A lot has happened in the world of corn growing since I first began growing Early Riser corn in 2004. Almost 90 percent of the corn grown in North America is now genetically modified to withstand the application of herbicides like

Roundup and Liberty. Several different Bt genes have been inserted into corn plants to deal with everything from the European corn borer to rootworms. Many corn varieties are now triple-stacked with several different genetic insertions, and there is plenty of GM pollen drifting around corn country. This makes it more important than ever to have our own seed that is free and independent of what the corporate seed industry has to offer. Vigilance to avoid cross pollination is crucial as well. I feel extremely lucky to live in an area where corn is only a minor crop, because it means I have been able to isolate my Early Riser corn to keep it away from pollen drift. I have lots of very well-protected fields. I continue to grow some Calais Flint seed and do battle with raccoons. (Unfortunately, these little masked bandits made their way through an electric fence and consumed most of my 2010 seed crop.) I have also been growing a little longer-season flint corn called Garland, a variety that was developed in the 1960s and '70s by George Garland of Acworth, New Hampshire. George crossed Longfellow Flint with King Philip Flint to come up with a variety with very good standability and yield. My fun with corn continues, and I have still not totally weaned myself from the yield advantages of modern hybrid corn. Early Riser will need a whole lot more breeding and selection work before it is competitive with what the big companies have to offer.

I mustn't forget to mention that most OP corns are much higher in protein and flavor than their hybrid counterparts. You can't beat a corn like Early Riser for making superb polenta or corn bread. OPs also offer many advantages in livestock feeds because they seem to contain more minerals than the hybrids. My personal corn story has gone on and on because it has been such a big part of my life. There were many ups and downs. I felt like quitting the quest many times. I am so glad that I stuck with it and that I was able to meet and learn from so many amazingly gifted people.

Planting Corn

Corn has the potential to produce more bushels and tons to the acre than any other commonly grown grain crop. Whereas fifty-bushel wheat and eighty-bushel barley are considered excellent yields, a one-hundred-bushel corn crop is mediocre. (My top yields have been in the range of 130 bushels to the acre, which is over seven thousand pounds of grain.) Yield advantage alone makes corn the first choice for many growers of grain. But there is a downside to all of this: All of those tons and bushels come at a cost. Corn is a heavy-feeding crop, and it has the potential to remove quite a bit from the land. Nitrate nitrogen (NO_3) is primarily responsible, and as discussed in the earlier chapter on soil fertility, this must be metabolized through the carbon or organic matter fraction of the soil. Continual bombardment of the soil with soluble nitrogen will eventually begin to deplete the organic matter reserves of the soil—and once the soil's humus is degraded, soil structure and water-holding properties are also reduced. The next step is soil erosion. All of this is simply the long way to tell you that corn must be grown in rotation with other crops that are regenerative and soil building, as continuous corn culture is a ticket to disaster.

Cover Crops and Rotation

It is always best to plow sod or a heavy cover crop of winter rye for corn. The typical midwestern rotation for the past 150 years always began with plowing a hay field down for corn, because as the microbial soil life consumed the turned-under sod, nitrates were slowly released to the growing crop of corn. It's also common knowledge that a turned-under field of sod will feed a crop of corn without the addition of any other nitrogen inputs like chemical fertilizer, manure, or compost. Weed pressure will be minimal if corn follows sod, too. Indeed, the plowed-down grass and legumes of a hay field are candy to all the tiny soil beasties that are invisible to our eye. We have already discussed the old-fashioned three- to four-year crop rotation, but it is worth mentioning here again. Plow down a field of clover and timothy either in the spring or the previous fall. Plant corn in year one; oats are seeded down the second year, along with clover and timothy seed. The oats are harvested and the straw baled, leaving behind a thick bottom of timothy and clover of hay for the following year. Years three, four, and sometimes five are devoted to hay and soil building. As the clover and other legumes in the hay crop begin to run out, the field is plowed again for another round beginning with corn. There are many variations on this basic rotation. One thing does remain constant, however. There is always a cycle of renewal that takes place as a heavy-feeding crop like corn is followed by plants that contribute to renewal. This is just plain old good farming practice—something that's been forgotten by much of today's high-production agriculture.

Field Preparation for Corn

Moldboard plowing is the easiest and most effective way to deal with sod. Fall plowing is essential for the establishment of a cereal crop, but not quite as important for corn, and some farmers in longer-season areas will take an early cutting of hay and then plow the field for corn. Plowing down green sod in the spring, however, provides even more living food for the soil life below. A couple of passes with the disc should cut up and smooth out the plowed sod. This is the time to apply manure or compost. Some farmers even like to spread the manure on top of the sod before they plow, as they figure that plowing the manure under with the sod will make the manure available in July when the corn roots get down a little deeper. This is when corn goes through its wildest vegetative growth spurt, and the manure will be right there to feed the crop when it is needed. Either way, a coating of manure or compost is a nice complement to the rotting sod in the root zone environment. A little soil smoothing with a field cultivator, spring-tooth harrow, or spike-tooth drag is all that is necessary to prepare the field for the corn planter.

Seedbed preparation for corn is a much less fussy affair than it would be for a hay or cereal grain crop. The row-crop cultivator and other weed-control machinery will be passing through this same field in a few weeks, so it need not be golf-course smooth. If you are using a field or spring-tooth cultivator to finish the seedbed, however, be careful not to run it too deep. Cultivator shanks run too deeply can pull chunks of sod (as well as rocks if you have them) up from below, which will make it more difficult to plant and cultivate the corn crop. My last little bit of advice for field preparation with corn is to do some sort of light tillage pass just prior to planting, as this helps to set the weeds back. Remember that it might take the corn a week to ten days to germinate and break through the soil; every little bit of weed control that you can perform at the very beginning of the season is absolutely essential. If you get your field all ready to plant corn and find that you must delay planting for two or three days due to weather or other circumstances, give the field a quick light touch with a finishing harrow or field cultivator.

Choosing Your Varieties

Variety selection is just as important as soil temperature and planting date. It's important to keep in mind that more than any other grain crop, corn needs water, heat, and humidity to grow and mature; yield potential increases in direct proportion to the amount of heat available to the corn plant. We talked earlier of corn's amazing ability to adapt to climate as it moved northward from Mexico, but keep in mind there are corn varieties that will do well in Texas, but not in New England and vice versa. Open up any corn seed catalog and you will see varieties described in terms of "days to maturity." The shortest-season corns are the 75-day varieties, while the longest-season ones are described in terms of 120 days, but this method of classifying corn maturity is a bit deceptive and not really accurate. A ninety-day corn doesn't make grain in ninety days. If this was the case, corn planted in the middle of May should be dead ripe by the middle or the end of August. Sometimes you will see corn described as "90 days CCB." This little acronym stands for "Central Corn Belt," which means the corn might mature in that amount of time in Illinois or Iowa where the growing conditions are perfect. Corn maturity ratings are probably much better categorized in terms of the number of accumulated growing degree days (GDDs) or heat units necessary for a plant to reach maturity, which is how the Canadians do it. GDDs are determined by taking the average daily temperature over fifty degrees Fahrenheit or ten degrees Celsius. If, for example, the high for the day was seventy-five degrees and the low was fifty, the average between these two temperatures would be twelve or thirteen degrees, which would give you the number of GDDs accumulated for that particular day; you can add up all the days to determine the yearly average number of heat units. Since I live so close to Canada and have primarily grown Canadian varieties of corn, I look at everything in terms of Canadian heat units (CHUs). A 2,350 or 2,400 CHU corn is equivalent to what we would classify as a seventy-eight- or eighty-day corn. The Pioneer 3921 corn that I successfully grew for so many years was an eighty-five-day corn, for example.

Climate and microclimate are so variable from one region to the next here in New England. Soil type (loam versus clay), exposure, and elevation all play a part in how much heat is accumulated in a particular location. I am definitely on the low end of the heat-unit scale in my little region of Vermont's Northeast Kingdom. However, I can still find a warmer microclimate ten miles to the north and eight hundred feet lower in elevation. I might only be able to mature a seventy-five-day (twenty-two hundred CHU) corn on my home farm hilltop, while I can grow an eighty-five-day (twenty-six hundred CHU) corn a short distance away down in the valley. If I travel twenty miles to the west over the Green Mountains and into the Champlain Valley, the heat-unit factor increases substantially. Fifty miles from here in St. Albans Bay, it is possible to ripen ninety-five-day (thirty-two hundred CHU) corn. In other words, take the time to find out the heat potential and growing season of the locale where you plan to plant your corn. It's as easy as talking to your farmer neighbors. They will tell you what corn works best for them. If your neighbor is growing silage corn, remember that you will need a shorter-season variety if you want to ripen it for grain—silage corn only needs to be half ripe for harvest. Climate maps are available from NOAA and the extension services of the various state land grant colleges as well. When selecting a variety of corn to grow for grain, it is always best to err on the side of caution. Mother Nature has no mercy when an early freeze descends upon the Northeast in late September. A longer-season corn variety may give you a lot more yield, but if it gets killed by frost before it ripens, all that potential harvest is lost, and a shorter-season variety just might have made it.

When to Plant

The trend in corn growing these days has been to plant earlier and earlier, but the hard-and-fast rule has always been to wait until the soil temperature reaches fifty degrees before planting corn. Large commercial operators have ignored this rule for years because they use seed that is treated with fungicides and other chemicals to prevent it from rotting in the ground (as I did, many years ago). When you've got a very limited time to plant a very large crop of corn, time is of the essence. Most commercial corn growers are trying to maximize yield, so they plant the absolutely longest-season variety possible, which means that the crop has to be planted very early so that it can germinate and get growing. It's a bit of a race against time when you need to accumulate every little ray of sunshine and every heat unit to make sure your corn reaches maturity at the other end of the growing season. There is a lot of corn that gets planted in forty-five-degree soil, and sometimes it can take more than two weeks for this early-planted corn to emerge. If you're protected by chemicals on your seed, you'll be all right, but when you grow organic corn, you have to work with the cycles of nature. You certainly aren't going to use treated seed, so you have to wait for the soil to warm up. It's better to wait until the ground reaches fifty-five or sixty degrees: If you plant corn and it is up and growing in three to five days, your crop will definitely get the jump on the weeds. If your seed has been sitting there in the soil for a week and a half or two, on the other hand, the weeds will definitely have had a chance to establish themselves before the corn comes up. Planting your corn in extra-warm soil might mean that you wait until the third week of May. It also means that you might select a variety of corn that has a shorter growing season than what your conventional neighbors planted two weeks earlier in cold ground. The biggest concern that any grower of corn, conventional or organic, has is making sure the corn reaches maturity in the fall. Therefore you must strike some sort of balance when it comes to planting dates. In a normal year, I like to plant my corn between the fifteenth and the twentieth of May. This is the middle ground. Fifty-degree soil is fine for untreated seed, but it should be relatively dry, as corn grows much better on lighter-textured loam soils than it does on the heavier, wetter clays. This is not to say that you can't grow corn on clay, but for the sake of the soil and the seed, stay off clay soil until it is dry. You will be rewarded for your patience. Suffice it to say that once the ground has dried out and been prepared with tillage machinery, it's time to plant corn if the soil is over fifty degrees. Farming is indeed a gamble, however, and you never know what Mother Nature will hand you next week. If you're somewhat prudent in your actions and management decisions, you stand a pretty good chance of harvesting a decent crop of corn in four or five months.

Methods of Planting

Once you've selected a hybrid or OP variety that will work for you, it's time to plant. If you're doing a small-scale plot, hand-planting is the way to go. Simply bolt a long handle to a crosspiece that is equipped with two attached wooden furrow makers. After the seedbed has been prepared, pull this homemade tool through the plot to mark the first two rows. Once you've made the first pass, double back with the row marker to deepen and better mark the furrows. To mark row three, jump over one row with the handy little marker so that one furrow maker follows the existing row while the other digs a new furrow thirty to forty inches away. Make sure your little trenches are two inches deep, and you can dress up your furrow marks with a hoe if necessary. Once you are satisfied with the preliminary row preparation, drop a seed into the furrow every six to eight inches. Pull the soil back over the seed so the corn is covered by about two inches of dirt. Tamp the ground with the flat back of the hoe to ensure good soil-to-seed contact, and planting is finished. This job can be much more easily accomplished, however, with a push-type garden seeder like a Planet Jr. or an Earthway. Choose the proper seed plate and adjust the spacing to drop a seed every six to eight inches. Most hand-push seeders have a row marker for you to follow when you plant the next row. Modern commercial corn is usually planted in thirty-inch rows, although forty-inch rows were more common in the days of horsepower. Row width is certainly something to consider when it's time to cultivate your corn. Wide is better than narrow if you are going through with a rototiller or some other hand-operated cultivating device. Hand-seeding of corn is certainly slow, but it is reliable—if you are dropping the seed into a furrow and covering it with a hoe, you know the seed is actually being planted.

Using a Corn Planter

Most of us will be using some kind of a corn planter. Entire volumes could be written about the subject, but I will try to tell you as much as I can in a few paragraphs. Corn planters "drill" seed into the ground in much the same manner as the grain drill discussed at length earlier in this book. The main difference between these two pieces of equipment is that the grain drill drops large amounts of seed into the earth in rows that are only six or seven inches apart, but corn seed must be much more carefully metered in rows that are usually thirty inches apart. The older style of corn planters had shoe-style furrow openers. These blunt, metal bullet-like pieces of steel slid through the soil, parting it just enough for the seed to fall from above into the furrow below. Closing and packing wheels followed right behind the opener to finish the planting job. My first corn planter was a two-row pull-type International Harvester 56 with shoe-type openers. Planting was sometimes difficult because little pieces of loose sod would constantly hang up on the openers, but machinery manufacturers outfitted the next generation of corn planters with double disc openers like those used on grain drills. My corn-planting situation was much improved in 1994 when I stepped up to a later-model, six-row IH 56 planter with double disc openers, each row unit outfitted with an individual seed hopper box.

Accurate seed metering on most pre-1980 corn planters was accomplished by means of a revolving seed plate at the bottom of each hopper. Seed corn companies sized their seed as either rounds or flats, and a tag on the bag would usually delineate the exact size of the seed as small, medium, or large, along with round or flat. Some companies like Pioneer used numbers to further identify seed size. An "F-12" might be a medium flat, and an "R-16" would be a large round. Once you knew the specified seed size, you consulted a chart on the back of the seed bag to find out what plate to use in your planter. There are hundreds of different seed plates for corn planters; each equipment manufacturer designed seed plates for their own line of corn planters. Before 1970, seed plates were made of cast iron, but as corn planter technology began moving away from plate planters, hard plastic seed plates came into fashion.

Matching the seed to the plate was and still is very important if you use a plate-style corn planter. Seed spacing is controlled by how fast the plate revolves in the bottom of the seed hopper. Corn seeds fall into the little cells in the spinning plate and drop through a spring-loaded little opening that lets them fall one by one to the furrow below. The faster the plate spins, the thicker the seed gets planted. Corn planters are equipped with a ground-driven chain to power these seed plates, and there are several sets of drive and driven sprockets from which to choose the speed of seed plate revolution. Controlling how thick or thin the corn is planted is akin to shifting a ten-speed bicycle. The best piece of advice I can give about setting up a corn planter for proper plant spacing is to make sure that you have an operator's manual for your planter. Study it thoroughly and ponder everything in advance of when you will actually be using your planter. A knowledgeable neighbor who is willing to come over and hold your hand isn't a bad thing, either. Once you've chosen a seed plate and a seed spacing setting, put some seed in the hoppers and try the planter out in your driveway, in your farm yard, or on some other impervious surface. Most hybrid corn is usually planted at a population of around thirty-two thousand plants to the acre. Set your planter down so it just skims the surface and let it plant for twenty-five feet or so. Then take a tape measure and mark off seventeen and a half feet. This represents one one-thousandth of an acre if you are seeding in thirty-inch rows. Count the seeds in the seventeen and a half feet to see how close to thirty-two you can come. Keep experimenting and adjusting the planter until you are satisfied with the results.

My Transition to the Plateless Planter

I have to admit that for the last twenty years, I've struggled with plate planters. Perfect and precise seed placement has always been just out of reach for me. In years past, I've looked at other people's cornfields and noted that every plant was exactly the same distance apart. My cornfields seemed a bit more random and haphazard with skips and double drops, some years being better than others. The reason for this is that corn planting has gone the way of the plateless planter. Both John Deere and International Harvester introduced plateless corn planters in the 1970s; by the '80s, the John Deere 7000 plateless planter and the IH 400 Cyclo air planter were standards in the industry. Over time precision seed sizing became less and less important to the seed corn industry, because fewer and fewer plate planters were being used, but my tendency to look to the agricultural past for inspiration led me to stick with seed plate technology. Plate planters were also quite a bit cheaper to purchase. I began to grow a variety of row crops like sweet corn, soybeans, dry beans, and sunflowers. Since everything required a particular seed plate and I had a six-row planter, I soon ended up with hundreds of these round plastic discs in my collection. I must have been one of Lincoln Ag's best customers, because at six dollars apiece, I soon had several thousand dollars invested in seed plates. With this much money tied up in plastic, there was no turning back.

In 1999, I was ready to trade up to a new planter. This would have been the time to switch to a plateless model, but instead I looked all over the country for a "better" new style of plate planter. I ended up buying an IH 800 plate planter from a farmer in Indiana for five thousand dollars; he was switching to a plateless planter. I was too in love with the old to embrace the new. As time went on, I began to consider myself an expert in plate planting, and once I began producing my own Early Riser corn seed, I learned how to sort and size my own seed so that I could plant it with my own planter. I purchased an old seed-corn-sizing machine and a plate stand from Illinois for two hundred dollars. The seed sizer consisted of two tapered revolving cylinders on a horizontal axis. The top cylinder had a series of different-sized slots while the bottom cylinder was outfitted with round holes of varying sizes. Once corn seed was poured into a top-mounted hopper, it made its way through the two cylinders dropping down various outlet chutes as small, medium, and large rounds or flats. The next step was to take all of the sorted seed and run it through the International Harvester test plate stand. This little collector's item was something that would have been found in an IH dealer's shop in the 1950s or '60s. It was basically a frame-mounted plate planter seed hopper driven by a tiny electric motor. You put a particular seed plate in the hopper and then poured in some corn to try it out. How the corn seed dropped onto a little flat belt underneath the machine would indicate if you'd chosen the right plate for the seed.

I considered myself rather proficient at using all of this older outdated technology, but my stands were never as perfectly seeded as those of my neighbors who were using more modern technology. In 2009, I broke down and spent six thousand dollars on an older John Deere 7000 plateless corn planter. It took me a while to learn how to use this new planter, and there were a few glitches in the beginning, but my stands are now as perfectly spaced as anyone's. I don't regret my twenty-year love affair with plate planter technology. This is no different than any other agricultural adventure that I've had in the past thirty-five years—since I didn't grow up planting corn in the '50s and '60s, I had to go and try it the old-fashioned way first. My advice to anyone who wants to plant some corn, however, would be to find an inexpensive plateless planter. As the years have passed, prices have dropped, and unless you're an agricultural history buff like me, there is no need to torture yourself with plate planting.

Using Your Planter in the Field

Let's assume that you've got a corn planter and it is all ready to go to the field. All the drive chains are tight and properly adjusted, the machine has been greased, and the tires have been inflated to the proper pressure. You've lined up the plant population sprockets for the desired amount of seed per acre. If this is your first time planting corn, it might be a good idea to enlist the help of an experienced farmer friend to teach you the basics of corn planter operation. One very important thing to remember is that eventually you will be following your corn planter with a row-crop cultivator. When planting a field to a row crop, it is important to keep future trips through the field in mind. Proper layout will simplify future cultivating and harvesting trips through the field. Once you're in the corn business, you will carefully inspect everyone else's fields as you travel through the countryside, and you'll probably notice that most conventional corn silage fields are planted around the perimeter in a circular manner. The common practice is to plant around and around the field at least three or four times; after this, the tractor and corn planter are driven back and forth from one side of the field to the other. The initial three or four passes around the field give the corn silage grower plenty of room to turn equipment around at the end of the rows. This is especially necessary when it comes time to chop the corn for silage in September, as the forage harvester follows the path of the planter around and around the field. Once the corn has been harvested around the perimeter, there is adequate space to negotiate the train of tractor, forage harvester, and silage wagon through the rest of the field.

Indeed, organic corn requires a little more forethought when it comes to field layout. You will need to leave enough space to turn your cultivator around. Some organic farmers who don't want to waste any precious space will plant corn in the headlands; they figure, a few cornstalks knocked over by a turning cultivator is a small trade-off for the gained space at the end of a field. Another option is to plant the headlands to a quick-growing annual crop like oats, as having something growing in the headlands will keep the weeds from growing. Most corn planters four rows or wider have disc markers to lay out a line in the field delineating the next pass, and as the planter is pulled through the freshly tilled field, a metal arm with a disc attached to its end hinges down from either side of the machine. When you get to the end of the pass, the planter is turned around 180 degrees to head back across the field. The little line left in the soil by the marking disc provides a reference point to follow and make sure that the first row of the next pass will be exactly thirty inches from the last row of the last pass. It is very important to adjust the length of the marker arm to ensure the proper spacing between planter passes, so carry a tape measure in your tool box and measure the distance between planter passes. Shorten or lengthen the corn planter marker arms to achieve that thirty-inch row spacing, which will make cultivating and harvesting totally foolproof later in the season. The big guys with their sixteen- and twenty-four-row units all use GPS for this type of situation, but I still happily rely on finely tuned steel for my brand of precision planting.

Planting Seed with a Planter

Seed depth is just as important as plant population or proper row spacing. Corn is generally planted an inch and a half to two inches deep in our part of the country, and I always set my planter for two inches. Gauge wheels on each individual row unit can be raised or lowered to regulate how far into the earth a corn seed is placed, but once you are out there putting seed in the ground, the whole process can become an act of faith. Unlike a grain drill where you can watch seeds dropping from the tractor seat, seed drop on all makes of corn planters is hidden from view. This can be a little nerve racking, but many planters out there are equipped with seed monitors. These amazing units consist of an electric eye in each seed delivery tube and a series of electric wires that carry signals back to an electronic box near the tractor seat. If the flow of seed in front of the electric eye is interrupted for some reason, lights will flash and an alarm will sound. Most seed monitors have separate circuits and indicators for each individual row of the corn planter; if your machine plugs, the monitor will tell you what row is giving you trouble. These systems run off twelve-volt electricity from the tractor's battery and charging system. A seed monitor is only as good as its wires and electrical grounds. We all know how difficult it is to keep the wiring up to date and operational on any tractor, and the same holds true for the wiring between the control box and the row units of a seed monitor. Nevertheless, if your planter is equipped with a monitor, it is well worth maintaining it. There are enterprises out there that are devoted to repairing agricultural electronic equipment, but if you need to send your monitor away for repair, make sure you do it in the dead of winter instead of waiting until the last minute when you need it to be working. But it's not the end of the world if you don't have a monitor.

After you make your first pass with the corn planter, get off the tractor and dig around in the row to see how deep the seed has been planted. You can check to see if your planter is dropping seed from each of its row units when the machine is lifted out of the ground at the end of a seeding pass. If you are still moving forward while you hydraulically raise the planter, a few seeds will fall out on the ground. I always check the ground when I come to the end of a pass and turn the planter around to head back across the field. The biggest cause for seeding failure in corn planting is a row unit plugged with dirt, and if the disc openers are worn or the bearings that carry them are loose, chances of plugging the planter are increased. Planting in soil that is wet and sticky can also increase the likelihood of a row unit that gets packed with soil. Since row opener double discs touch each other in the front and are spread open in the back, the slightest backward movement with the planter can pack soil in the opening. This is why you want to be moving forward with the planter as you drop it in the ground to begin a journey across the field. If you lower the planter down while the tractor is sitting still, it might move backward ever so slightly and plug a row or two.

All of this corn planter information may seem a little daunting and overblown, but once you are out there on a beautiful May day, everything will seem like it was meant to be. Hopefully, you can learn some lessons from my mistakes. The answer to avoiding all these perils and pitfalls is to take the time in the quiet winter months to maintain your corn planter and keep it in top working condition.

Using a Fertilizer Hopper

Most corn planters have fertilizer hoppers—large rectangular fiberglass boxes mounted to the frame in front of the machine. A small cross auger that runs across the bottom of these hoppers pulls the fertilizer toward special outlet holes, and flexible rubber drop tubes are attached to these openings in the bottom of the fertilizer hoppers. Fertilizer is delivered by gravity down these tubes to another set of double disc openers that are located three to four inches to the side of the seed disc openers of each row. (Corn planters are set up this way to prevent chemical fertilizers from burning the roots of young plants.) The speed of fertilizer delivery augers is regulated in exactly the same manner as the seed drop mechanism on a corn planter. There is a choice of sprockets for the drive chain to transmit power to the fertilizer augers; the faster everything turns, the more fertilizer falls down the tubes between the openers. We certainly are not going to use the same sort of starter fertilizers that are used in the growing of conventional corn, but we do have some options as organic growers, and we might not even need any fertilizer if we are planting corn on a well-manured piece of plowed-down sod.

Most commercial corn starters are well supplied with soluble phosphorous to help young plants with root development, but phosphorous uptake by plants is very dependent on warm soil temperatures. Conventional logic has it that extra phosphorous is needed early in the season to get corn plants off to a great start because the soil is too cold early in the season to naturally supply this element. There is a bit of truth to this. As organic growers, however, we have to find alternative sources of phosphorous that are different from the conventional fare of super phosphate, triple super phosphate, MAP (monoammonium phosphate), and DAP (diammonium phosphate). Many pelletized organic fertilizers have come to the market in the last little while. The most common of these materials are the pelletized chicken manure composts that come from large poultry concerns like Perdue from the Eastern Shore of Maryland and Kreher's in western New York. There are also various blends of blood, bone-, and feather meals that all deliver nitrogen and phosphorous in a form that is acceptable to the national organic standards. These commercial organic fertilizers cost between three and five hundred dollars a ton, and a normal application rate is three hundred to five hundred pounds to the acre. Applying fertilizer to a crop through the planter boxes is costly, but it can provide piece of mind. There is already so much invested in putting in a crop of corn, and many people figure that dropping a moderate amount of organic fertilizer right next to the row is a cheap guarantee that the crop will have what it needs right there close to the seed. I've done it both ways and have had success with and without additional fertilizer. If your reserve fertility is marginal, an additional fertilizer contribution to the corn crop might be good insurance.

Emergence, Early Growth, and Primary Weed Control

Once a dormant corn seed is planted, it begins to absorb water from the surrounding environment. Germination begins when the kernel has absorbed 30 percent of its weight in moisture and 110 growing degree units have accumulated since planting. Corn planted in mid- to late May under just the right conditions can pop up in as little as three to five days. In much the same manner as wheat or barley, the radicle or seed root is the first thing to emerge through the seed coat.

The spike or coleoptile appears next, followed by several seminal roots, and the initial root system serves to anchor the seed as well as provide it with the necessary moisture for the growth process to continue. At this point, the germinating seed is living on the food reserves provided by its own endosperm. Eventually, these reserves will be depleted and the young root system will assume the job of absorbing nutrients from the surrounding soil.

The first node or point of attachment between the coleoptile crown and the kernel is called the mesocotyl. The coleoptile and the mesocotyl both elongate during emergence. Once the coleoptile pushes through the soil and senses sunlight, it will rupture and allow the first two true leaves of the young plant to unfold. Meanwhile, the actual growing point of the plant remains in the coleoptile node, at least one to one and a half inches below the surface of the soil. The fact that the plant's growing tip is still well underground for the first three to four weeks is the main reason why a late-spring

A recently emerged corn plant. PHOTO COURTESY OF SID BOSWORTH

frost is usually only a minor concern to a recently emerged corn crop. Exposed leaves may get burned by below-freezing temperatures, but the growing point will live on to push up another set of first leaves. Waiting for planted corn to appear in neat little rows is an act of faith, and I always breathe a sigh of relief upon seeing those first little plants glowing in the late-afternoon sunlight.

Pre-Emergence Cultivation

When I first began growing corn on any kind of scale, I watched little tiny weeds sprout and begin to grow while I was waiting for the corn to come up. The weeds had a good head start on the corn crop. I remember thinking to myself that this had the potential to be a big problem, and my hunch was correct. I was able to cultivate out the weeds between the rows of corn. The weeds that were in the actual rows between the plants were impossible to eradicate mechanically. I came to realize that I should have been doing something about those little weeds while I waited for the corn to germinate. I heard several stories from older farmers about how they used to pull a spike-tooth drag harrow through their corn before it came up. One old guy whom I met at an Acres USA conference told me that "blind harrowing" had been standard fare on Illinois farms before the advent of herbicides. Earl Spencer, an older dairy farmer from the Mohawk Valley of New York, called this practice "dust mulching" his corn. He liked to pull the spike drag through his rows of corn just as the little plants were pushing their way up through the soil. Earl felt that maintaining a dry, dusty soil surface would keep weeds from sprouting in his corn. The more I thought about it, the more this practice made sense to me. The most aggressive form of cultivation is indeed possible at this stage because there are no plants yet to damage. In 1991, I dug my old spike tooth out from the back of the machine shed and put it to work in my just-planted cornfield, and the results were phenomenal. The corn emerged from the soil free of weeds.

As seems to be the case in most of my crop misadventures, beginner's luck must have been on my side. I have since found out that there is a whole lot more to pre-emergence cultivation than madly running around the field with a spike drag. The observant crop farmer actually takes the time to monitor the emergence progress of a recently planted corn crop, which means digging around in the dirt every day to see how close to the surface the corn sprout is. When the little coleoptile has only half an inch to go, it's time to hit the field fast and hard with the steel. Over the years, I have modified this approach to suit my needs. If the weeds are coming fast and furious, I will do a blind harrowing several days after the corn has been planted. Sometimes, I will come back a second time. We definitely want to unearth these little weed sprouts when they are still in the infantile white hair stage, and a hot sunny day is best for this because these little sprouts that have been disturbed will bake in the sun and die. So if you have only one nice day before a long stretch of wet weather is due, it would be best to get out there and scuffle up the soil in that recently planted cornfield.

Weed Control

I'm not sure if this is the case for other organic crop farmers, but I have the tendency to think that some newly acquired piece of machinery is going to solve all of my problems. ("If I just had one of those, I wouldn't have to worry about weeds anymore.") How many times have I seen this happen? It's amazing the desire I can have for a hunk of iron. This has been particularly true when it comes to cultivation and weed-eradication equipment. I kept seeing these ads in *New Farm* magazine for the McFarlane Flexible Tine Tooth Harrow, basically a fancy hydraulically controlled, frame-mounted version of the old spike drag. The glossy ads were filled with testimonials of how these machines eliminated weeds in recently seeded fields of corn, and so I saved my pennies and paid close to two grand for a McFarlane Harrow in 1993. This piece of equipment did save time, because it folded out to twenty feet in width. It was also quite easy to maneuver around a field, because it could be raised up at the end of a pass. I don't regret buying the thing, but I have to admit that it was no more effective at weed elimination than the old twenty-five-dollar spike drag. One very important lesson I have learned is that corn that has not yet broken through the soil can withstand some pretty intensive soil disturbance. There have been several instances when I have arrived at a field to give it a little pre-emergence scratching only to find that I could see the little green rows of corn just beginning to poke through. I remembered Earl Spencer's dust mulch and went right ahead and hit it with my McFarlane Harrow. I worried and anguished over my decision, but in the end the corn survived and in-row weeds were not a problem.

The Rotary Hoe

Long before I discovered how simple and easy it was to pull a spike-tooth harrow through my cornfield, I thought the rotary hoe was the tool for the job. Twenty years ago, Rodale's magazine *New Farm* was one of the few sources for information about tending organic grain crops. The magazine was filled with articles featuring midwestern grain farmers who had made the switch from conventional to organic and sustainable methods. It seemed that just about every one of these converts sang the praises of their rotary hoe, which most used for both pre- and post-emergence cultivation. Of course, Jack Lazor, the little wannabe grain farmer from northern Vermont, wanted to be just like these guys, so, in the spring of 1990 I bought a used fifteen-foot Yetter rotary hoe for fifteen hundred dollars from Green Mountain Tractor in Middlebury, Vermont. It consists of a four-inch-square tool bar that attaches to the three-point hitch of a tractor. A series of spring-loaded cast-steel arms are mounted every six inches running the entire length of the tool bar, and these little arms curve downward toward the ground. The heart of the rotary hoe is an eighteen-inch-diameter "star wheel" that is mounted to the bottom of each arm by means of a stub shaft and bearing. A rotary hoe wheel is quite slim in profile and has a series of little protruding prongs that radiate from its center 360 degrees around its entire circumference. (Picture the face of a clock with a hand at every five-minute mark.) A specially shaped piece of metal called a spoon is attached to the tip of each of the radiating arms of a rotary hoe wheel. This is elliptically shaped, indented, and slightly pointed on the end in much the same manner as the utensil you would use to eat breakfast cereal. Once the tool bar of a rotary hoe is lowered on the three-point hitch of a tractor, these spoons come in contact with the soil surface. As the machine is driven forward, the rotary hoe wheels revolve along the top of the ground, and a light cultivation is accomplished as the spoons dig into and out of the soil, penetrating somewhere between half to three-quarters of an inch. Speed is quite important in rotary hoeing; the faster you travel, the better weeding job you do.

When I first tried the rotary hoe in 1990, I was told by a few experienced farmers to drive through my cornfield at just under road speed. "As long as you can stay on the seat of the tractor, you'll be fine," one midwestern farmer told me. My Yetter rotary hoe has a curved wire shield attached to the tool bar just behind the tractor. Once I started tearing through my cornfield, I found out why it was mounted there. Dirt, clods of sod, crop residue, and everything else went flying up in the air as I drove through my field at breakneck speed. There was a literal cloud of brown around the tractor. The shield was there to protect the operator. The rotary hoe was supposed to be the ticket to weed-free corn by knocking out the little sprouts before and after the crop came up. It seemed hard to believe that a piece of steel like this could be selective driving through four-inch-high corn. Much to my surprise, though, the spring-loaded hoe wheels rolled over the young corn plants with only minimal damage; a plant was plucked out here and there, but that seemed of minor consequence. I did do some rather careful inspection of the soil surface after every episode of rotary hoeing that first year. It seemed that the ground had received a good light cultivation and that many white-hair-stage weeds had been unearthed. I did notice, however, that not all of the ground had been worked by the rotary hoe. It stood to reason that as the little spoons revolved around their axis, some of the soil surface was left undisturbed. This was especially true in my glacial till ground, which was loaded with tiny rocks; small stones had dulled the spoons and prevented them from achieving 100 percent soil penetration. After a couple of years of trial and error, I came to use the rotary hoe less and less as I gravitated more to spike and scratch tillage in early planted corn. The rotary hoe is the only pre-emergence and early-cultivation tool that will perform adequately in wet soil conditions, and row-crop cultivation can be quite frustrating during a wet spell because adequate sunshine is necessary to kill the weeds that have been turned up by our iron.

The Growth Process

About a week after a corn plant has unfurled its first two leaves, the growth process begins. The radicle and seminal roots will have begun to develop branches and root hairs by this time, and the food supply from the original kernel has been exhausted as nutrition begins coming from the roots and leaves of the young corn plant. Top growth seems quite slow at this early stage; it's hard to believe that this tiny little plant will tower over your head in just a few short months. In fact, at this very early stage in the life of a corn plant, more is happening beneath the soil surface than above it. In the best of growing conditions, one leaf will emerge from the whorl of the plant every three days. Meanwhile, belowground, the nodal roots are beginning to emerge from the crown, which is just under the soil surface. This will become the plant's permanent and primary root system. The radicle and seminal roots have gotten the infantile plant off to a great start, but they are no longer very important after a week to ten days of growth. This process is well on its way after the emergence of the fourth leaf, and the growing point of the plant remains below the surface of the ground. The tassel is already initiated in the tip of the stem; future leaves and ear shoots are already genetically determined in the tiny stalk of the plant as well. (Ideally, the corn plant will have six leaves about three weeks after the young seedling has emerged.) Tillers are beginning to develop and grow from the belowground nodes, the nodal roots are now feeding the plant, and the growing point is at the surface of the soil. As these first three or four weeks pass by, the corn plant is anchoring and establishing itself. And, as I've described, top growth is not rapid as the plant's root system establishes itself. This all begins to change once the plant gets to the eight-leaf stage. At this point the corn plant is getting ready for takeoff. Rapid root growth, leaf elongation, and stem enlargement will soon follow, and determination of how many rows of kernels a plant will have on its ear happens at this time. The growing point is now two to three inches above the soil surface. Internally, ear shoots and the future tassel of the plant are beginning to develop. By this time the little two-leaf baby plant is now eight to ten inches tall, and a month or a little less has passed since you first noticed the appearance of the rows of tiny corn spikes out in the field.

Post-Emergence Weed Control

It is essential to lay down the foundation for weed control during these first three or four weeks in the life of a corn crop. The plants may be growing slowly, but the weeds certainly are not. Whatever piece of equipment you are using, it must be pulled parallel to the direction the corn was planted. The tires on the tractor should be lined up so they are centered between the rows of corn. If you are planting corn in thirty-inch rows, measure the distance from the center of the tractor's drawbar to the center of the rear tire treads. The distance should be thirty inches. The same holds true for the tractor's front axle and tires. If you are using a tricycle-style narrow-front-end tractor, make sure the two tires line up as close to the center of the row as possible.

The Tine Weeder

The same tine weeder that we discussed for use in cereal grains will work quite well in young corn. If you have been fortunate enough to start out with rows of tiny corn plants in a clean seedbed, tine weeding during early growth will help you keep the weeds at bay, especially between the plants in the rows. The long spring-steel tines will gently stir the soil around young corn plants as they pass over the top, but it's best to wait until the corn has two good leaves before attempting to tine weed it. Whether you use an inexpensive Kovar model or a more sophisticated German Einbock, adjust the angle and down pressure of the tines to suit the crop and the soil conditions. If you are running the tines pulled all the way forward in their most aggressive position, you may end up damaging a lot of corn plants. It's always best to stop, get off the tractor, and inspect your work—trial, error, observation, and adjustment will serve all of us well.

I have found that the weeder damages young corn far less than a rotary hoe, and I've successfully weeded corn with a Kovar up until that eight-leaf stage when the corn is close to ten inches tall. But I have one little word to the wise about any sort of early row-crop cultivation. You might think that that weeder or rotary hoe worked so well that you would want to turn right around and redo what you just cultivated, but that is the best way to injure and kill plants. A young corn plant will most likely tolerate one thrashing with a weeder or a rotary hoe, but not an immediate second. The first pass will loosen the roots of a four- to six-inch corn plant, but the second pass will definitely pull it out. Let your field rest and reestablish itself for a day or two before you return to try again. Work in full sunlight in the heat of the day if possible—you want any weeds that you unearth to bake and dry out in the sun. I wish that I had been more familiar with tine weeding when I first started growing corn in the early 1990s, but for some reason I thought this tool was only for use in cereal crops. After several decades of trying to grow organic corn, I have to say that tine weeding is first and foremost done for the control of weeds.

The Pencil-Point Weeder

The modern tine weeder was designed and patterned after the horse-drawn "pencil point" weeder from bygone days. These old units consisted of a minimal hardwood frame and a set of shafts for one horse, and had several rows of very unique spring-steel tines attached all the way across this frame. These pieces of metal started out as thin, flat one-inch-wide strips where they were bolted to the main frame. These flat springs ran parallel to the ground below before they turned ninety degrees downward; the last six to eight inches of this spring-steel bar was the actual pencil point because it transitioned to round steel about half an inch in diameter. These little round rods vibrated along the ground, pulling out small weeds between the corn plants in the row. If you are a small-time grower of corn or any other grain crop, the old pencil-point weeder will serve you well. You might see the remnants of one of these wonderful old horse-drawn tools on top of a stone wall in some back pasture, and if so, grab this hunk of old metal and restore the wooden frame if you can. Every once in a while, an old weeder will turn up at an auction sale as sheds are being emptied and sold off. (I bought a pencil-point weeder ten years ago at a local farm auction for fifteen dollars.) These versatile tools are so light that they can be pulled through a small garden patch of young corn by hand. We may think that we have all the answers and tools for weed control in these modern times, but previous generations who farmed with less petroleum and no access to herbicides knew a whole lot more about weed control than we can ever imagine.

The Row-Crop Cultivator

The row-crop cultivator is the next weapon in our arsenal of weed-control machines. Cultivators come in many styles and shapes, but they all do the same thing—pulling some sort of steel through the soil between the rows of corn plants to till the soil and cut off weeds at ground level. The most basic cultivator is the garden hoe. The simple slicing action of a hoe in the hands of a skilled operator will clean up a small plot of corn in a hurry, and since you are controlling this implement directly with your own body, it is possible to accurately and precisely remove weeds from your young crop. You can very carefully hoe right next to the row, slicing off the weeds without harming the corn plants.

There are also a variety of wheel hoes that can be used in a garden or small-plot situation. These devices consist of two handles attached to a lightweight frame and multifinger cultivator. You walk along pushing this device in the row and alongside the corn plants. This hand-cultivation machinery is very inexpensive and effective, and wheel hoes are certainly much easier on the body than the handheld model. For some reason, antique wheel and hand hoes seem to do a much better job than the modern tools that you can buy at a hardware store or garden center—perhaps because the old stuff was made out of better steel by people who actually depended on these tools for their food and survival.

Horse-Drawn Walking and Riding Cultivators

Horse-drawn walking and riding cultivators, which are also antiques, are equally effective at the job of weed removal, and there still seems to be a good supply of walking cultivators stored in old sheds and barns around the countryside. This classic old farm tool is cleverly constructed of five to seven spikes or shovels attached to a V-shaped metal frame and two hardwood steering handles. The bottom of the V faces forward and hinges to allow adjustment for row spacing. Corn was planted in hills on forty-inch centers in the days of horse-drawn cultivation, so if you choose to use this form of old technology for cultivating your corn crop, you might want to widen your row spacing to at least thirty-six inches. The standard thirty-inch row of today is much too narrow to accommodate a horse and a walking cultivator. We did a little of this kind of cultivation in our earlier days. We always had someone riding the horse. The skilled teamster of yesteryear guided horse and cultivator by draping the reins around his shoulders while guiding the two wooden handles with his hands. If you have a few of these old walking cultivators kicking around, you can also attach this tool behind a very small tractor. If you have enough people power and equipment, it is also possible to hitch two or more of these old cultivators to a tractor-mounted tool bar for a "multiple hitch."

The horse-drawn riding or "sulky" cultivator was a very sophisticated weed-control tool in its day; modern tractor cultivation machinery was patterned after the design of these old rigs. Riding cultivators had two steel wheels attached to a metal pipe frame that arched over and straddled a row of corn. A very long hardwood pole protruded from the front of the cultivator to connect the machine to a team of horses, and the teamster/operator sat on a seat directly above the straddled row of corn guiding the team. One horse walked to the left of the row while the other walked to the right. Cultivator shovels and shanks were attached to metal rigs in much the same manner, as would be found on a modern tractor-mounted tool-bar model. However, the horse-drawn machine offers more options and control because each rig on either side of the row can be manipulated by the feet of the person riding in the seat above. As the teamster guides the team and the cultivator down the row, the shovels and shanks can be steered closer to or farther away from the corn by means of foot stirrups. Older horse-drawn machinery was so well made that it can usually be restored to workable condition with a bit of effort. Start looking around as you travel around the countryside; I guarantee that you will eventually see all kinds of horse-drawn sulky cultivators out there. There are also many draft-horse enthusiasts who have large collections of these old machines. But sulky cultivators can also be pulled through a corn patch by a small tractor. In order to do this, the long wooden pole must be cut shorter and outfitted with a tractor hitch. This very quickly turns into a two-person job. One person drives the tractor while the other rides the cultivator and guides the rigs with their feet.

What Weeds Tell Us

Most problematic weeds are telling us a story about our soil. The two most common problems found on almost all crop farms are soil compaction and low available calcium. Compacted tight soils have no room for air, which is as important as anything for good crop growth, and water infiltration and good drainage are seriously impacted in a tight soil with poor crumb structure. This is the best reason I can think of for not tilling through a wet hole when preparing a field for a crop. Weeds like beggar-ticks and velvet leaf love to grow in compacted places, whereas weeds like hemp nettle and ragweed indicate low available calcium, which can be remedied by applying gypsum or other soluble sources of calcium like micronized limestone in liquid form through a crop sprayer. Lack of adequate soil biology coupled with an overabundance of soluble nitrates are both responsible for a plethora of common weed species like lamb's-quarters and redroot pigweed. Sometimes a sprayed-on application of molasses, liquid calcium, and soil inoculants will stop an invasive plant like quack grass dead in its tracks. Seeds inoculated with mycorrhizal fungi can also help a soil deficient in biology. We know that we have to do battle with weeds no matter what. A good program of balanced fertility will make it much easier for us as crop growers and for the plants themselves.

Tractor-Mounted Cultivators

The first tractor-mounted cultivators were patterned after the steerable horse-drawn rigs described several pages earlier. The three-point hitch had not yet been invented in the 1920s and '30s when tractor cultivation began to come into fashion, so agricultural engineers took the basic round tubular and square stock steel framework of the horse cultivator and attached it to the side rails of the tractor frame alongside the engine just behind the front wheels. These early mid-mount row-crop cultivators did two rows of corn and were raised and lowered by means of a hand-operated lever-type handle. Hand-operated lift handles were replaced by hydraulic cylinders when manufacturers began building tractors with hydraulics in the late 1930s and early 1940s, and solid pieces of round steel about two inches in diameter were attached to the cultivator frame by means of heavy-duty clamps. These so-called standards went from being round on top to being square on the bottom with bolt holes for attaching cultivator shovels or sweeps. Most of these units were designed in two pieces with a hinge and heavy-duty spring joining the round upper part to the square lower section. This sort of design allowed the cultivator sweep to be adjusted at various angles to the ground it was tilling. If you only wanted to gently slice off weeds and not throw much dirt, the sweep mounting was adjusted forward so the cultivator shovel ran almost parallel with the earth being tilled. If you wanted more of a plowing action and more dirt thrown up against corn plants in the row, the bottom of the standard was pulled farther back at a more perpendicular angle to the soil below. The spring served as a safety latch; if a cultivator shovel hit something solid like a ledge or a big rock, the safety spring would trip and allow the whole mechanism to break away backward. If your cultivator shank tripped, you had to reposition it forward with the help of a hammer and a wrench. Most of these early tractor-mounted cultivators had some sort of a rear tool bar with shanks or spring teeth to till out the wheel tracks of the tractor. These early units were a big improvement over their horse-drawn predecessors because the operator only had to steer with their hands. Foot stirrups and reins were traded in for a steering wheel, clutch, brake, and petroleum, and motorized cultivation allowed farmers to cover a lot more acres in a day.

My first exposure to this sort of cultivator was at John Ace's farm in Oregon, Wisconsin, in 1974. John kept his ancient Farmall F-20 with cultivator in a back shed for once a year use on his corn acreage. The tractor was vintage 1933, and the cultivator was permanently bolted to the tractor frame. The cultivator rigs on each side of the tractor could be raised individually and independently of each other by means of two long lever-type handles. Since I was so young and impressionable at the time, I pictured myself owning something similar to this when I would be growing corn on my future dream farm. Three years later in 1977, when the time did come, I stepped twenty years forward in time and bought a little John Deere 40 tricycle-style tractor complete with a mid-mount two-row cultivator. I set out to make the thing work on my one acre of flint corn that season. The whole unit was quite rusty and needed some adjustment, and I found out very quickly that you needed to carry a complete set of wrenches and a can of penetrating oil along on the tractor when it was time to go to the field. I was lucky enough to obtain an original copy of the owner's manual for the cultivator from the early 1950s, and the first thing I learned was that modern corn planted in thirty-inch rows was just a little too narrow for my old machine. I loosened up all the bolts and clamps and tried to slide everything in so that it would fit between my narrow rows. When I did this, the top of two of the round standards that held the shovels below hit the bottom of the tractor's oil pan when the rigs were raised up out of the soil. A quick perusal of the manual told me that the machine was actually designed to cultivate the thirty-six- to forty-inch rows of yesteryear. I made the thing work to the best of my ability, but the cultivating job was less than perfect because I had to remove several of the standards and shovels to squeeze between the thirty-inch rows of corn. I'm glad that my flint corn field was only an acre, and I did a lot of hand-hoeing.

Cultivating Corn

It doesn't matter what kind of weed-control implement you are using, you need to cut off the weeds and preserve the corn plants, and how the sweep or shovel contacts the soil and what happens as it is being drawn through the rows determine the effectiveness of the cultivation process. (I should mention here that row-crop cultivator and field cultivator sweeps look almost identical. They are both triangular in shape with wings that protrude from their sides. Field cultivator sweeps have a steeper stem angle of forty-seven degrees, since they are meant to move more dirt in the secondary tillage process, while row-crop cultivator sweeps have a shallower stem angle of fifty-four degrees because they are intended to slice through the soil profile on a much flatter plain.) We want to cut off the weeds just below the surface and throw a minimum of dirt up against the plants in the row. This is especially true on the first pass through the field. It doesn't take too much dirt to bury a five-inch-tall corn plant. Some cultivators have protective shields that run between the sweeps and the rows; older mid-mount models were equipped with long flat metal shields while the newer three-point hitch cultivators are outfitted with rolling shields right next to the sweeps. (Most farmers find shields a bit cumbersome and prone to hanging up with plant residue and field trash.)

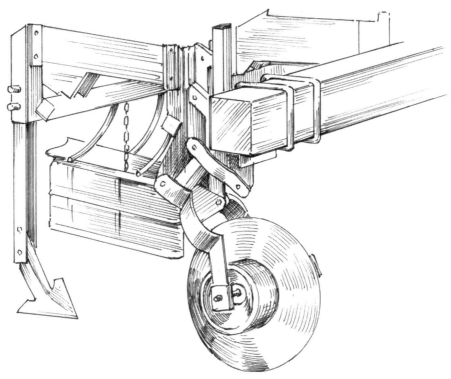

Cultivator gauge wheels.

Six-inch sweeps are the most common size, and they come in right-hand and left-hand configurations. The wing closest to the row is lopped off to keep dirt from damaging the young corn plants. It's a common practice to move these one-sided sweeps as close to the row as you dare for the first cultivation. Four inches is about as close as I like to get, as young corn plants at this stage have not yet begun to send lateral roots out into the row. A close and somewhat deep cultivation will cause minimal harm to a corn plant at this point. Ground speed of the tractor and cultivator needs to be rather slow and deliberate at this early time in the life of a corn plant. As the plant grows taller and bushier, ground speed can be increased and the sweeps pulled back a few more inches from the row. You want to avoid root pruning at all costs, and the older solid round shank style of standard offers one more opportunity for fine-tuning and adjustment not found on newer S-tine and C-shank machines. Loosening the upper clamp and turning the round standard one way or another will permit the sweep down below to throw more or less dirt at the row of corn alongside. This feature is especially handy later in the season when it is time to "lay by" the corn crop. A simple little twist of the round standard will allow us to throw a good amount of soil up against the stalks of now very sturdy knee- or thigh-high corn.

The last major adjustment for consideration in the proper use of a row-crop cultivator is the depth of operation. You only want to slice off weeds as near the surface as possible, and for this reason the framework of the row-crop cultivator that carries the shovels and shanks was designed to float independently of the main frame. Just about all machines are set up as hinging four-sided parallelograms that are attached to a crossbar. The front of the parallelogram (also called a rig) is equipped with an adjustable-gauge wheel that will only allow it to operate at a specified depth.

As seen in the accompanying illustration, this twelve-inch-diameter gauge wheel is attached to a movable metal bar that can be moved up or down by means of a clamp. If the gauge wheel is lifted, the sweeps will dig deeper. If the gauge wheel is lowered, cultivation will be shallower. Once you get in the groove with your cultivator, you will become familiar with all of the possible adjustments. Depth, sweep spacing, cutting angle, and proximity to the row are all critical to doing the best job possible, and a set of the proper-sized wrenches, a bar, and a good heavy ball-peen hammer should occupy the toolbox on the tractor. Operating a row-crop cultivator can be a fun experience—it's basically a moving erector set designed to eliminate weeds from your cornfield.

Cultivating Corn at Butterworks Farm

It wasn't until I began growing larger acreages of corn in a more serious manner in 1990 that I really began to master the art of row-crop cultivation. I still had my little John Deere 40 two-row mid-mount at this time. The first change I made was to increase the row spacing from thirty to thirty-four inches, which helped tremendously. There was now enough room to comfortably position the three shanks and shovels that were intended to fit between each row of corn. I still struggled a bit with hitting immobile objects like big stones. In 1992, Earl Bessette of New Haven, Vermont, gave me his old John Deere cultivator, which was equipped with spring shanks instead of the solid round standards that I had. The spring teeth worked much better in my rocky upland soil, and they worked their way around the rocks and pulled out a lot more quack grass. Once again in my farming career, I experienced that dying-and-going-to-heaven feeling when I stepped up to a new and better older machine. Cultivating corn was no longer such a fight.

In 1994, when I made the quantum leap from ten acres of corn on the hilltop to thirty acres of corn down in the stone-free valley, I needed to find a newer and better machine. Since I was now putting corn in the ground with a six-row planter, I needed to find a six-row cultivator. I had vintage-1960s Farmall tractors and began to look for a mid-mount cultivator to fit my 756. I kept watching the auction bulletins in the *Country Folks* paper until I found one on the Pennsylvania–Maryland border not too far from York, Pennsylvania. I called my friend Billy Ware who lived near Hagerstown, Maryland, and he agreed to go to the sale and bid on the machine for me. Billy drove through an ice storm in early March and bought the thing for six hundred dollars. (It would have been cheaper, but there was a Mennonite fellow there who wanted it just as badly as I did.) A local trucker from my area brought the thing home as a back haul for another six hundred, and when the machine arrived at our farm, I was surprised to see that it was also a spring-shank model. Cultivating six rows took a little getting used to, but it sure beat doing only two rows at a time.

It was (and still is) imperative that you follow the planter with the cultivator. If you get one row off, you are sunk because it is quite difficult to leave an exact thirty-inch space between each pass of the corn planter. My Farmall mid-mount cultivator that had been made in the early 1960s was very compatible with thirty-inch rows. It used the same row-crop sweeps that were intended for the solid round shank type of cultivators. The beauty of the mid-mount cultivator was that all of the rows were out there in front of me. I didn't have to turn around and look behind me at the three-point hitch. If something clogged or got jammed, I could see it before too much damage had occurred. I found the best way to operate this fifteen-foot-wide monster was to watch one row just to the right of the tractor frame. The cultivator was equipped with two independent hydraulic cylinders so that each side could be raised and lowered individually.

Over the years, my six-row mid-mount began to show signs of wear and tear. Gauge wheel bearings began to fail, and I broke a few spring shanks. In

1999, another friend from central Illinois purchased a spare machine for me at an auction sale in his neighborhood. The two-hundred-dollar price tag was much better, but what I saved was eaten up by increased trucking costs. This new mid-mount was constructed of solid round shanks much like my original John Deere two row, and we ended up making a hybrid cultivator out of the two machines by putting solid shanks next to the rows and leaving a spring shank in the middle of the row. My Farmall mid-mount has served me well over the years, and we eventually mounted it on a Farmall 806 with a tricycle narrow front end. The narrow front works even better for row-crop cultivation because it allows the tractor to turn sharply in some very confined areas at the ends of the rows. A six-row mid-mount can be a real bear to drive on a public road—at fifteen feet in width, it hangs way out there. We have had several close calls on the way to our fields, and we were forced into a guardrail once by a speeding vehicle. I was glad we had an extra cultivator for replacement parts, but there is something magic about cultivating with an old relic like this. It transports you back to the 1960s when farming was a less toxic and more straightforward business.

The Three-Point Hitch Cultivator

The three-point hitch row-crop cultivator became the industry standard in the 1960s as tractors got bigger and hydraulic systems stronger. The base of these units consisted of a six-inch-by-eight-inch heavy steel box beam mounted to the rear hydraulic arms of a tractor. The individual cultivator row units—which were parallelograms with gauge wheels—were then mounted onto this tool bar. Shovels, shanks, and sweeps were fastened to each row rig by means of the same types of clamping devices used on the older mid-mount models. A three-point hitch cultivator had the exact same configuration of three C shanks per row as its predecessor, the only difference being that the entire unit sat behind you instead of up front where you could see it from the tractor seat. I was told that the first rule of operation for this type of machine was to not look back very often—a quick turn of the head and you would be off the row, wiping out corn plants. The best advice I received was to center the hood ornament of my tractor on the row immediately in front of me and have faith that everything back there was in proper adjustment. Some farmers install large rearview truck mirrors on either side of the tractor for a better view of what is behind. This innovative little feature allows you to look forward and still be aware of how the rear-mounted implement is performing. Side sway can also be a major problem with three-point hitch rear-mounted row-crop cultivators. If you are working across a slope, gravity can allow the cultivator frame to "walk" sideways down the hill just enough to begin wiping out your crop.

It's best to check the tractor's three-point hitch linkage for wear and slop well before the cultivation season begins. Most older tractors will have worn three-point hitch arms. Some makes like Ford were equipped with an extra set of stabilizer arms (called stay bars), which added rigidity to the three-point hitch system. Most of the other brands like International Harvester and John Deere have special adjustable stabilizer chains that connect the three-point hitch arms to the rear casting of the tractor. These chains can be tightened to stiffen up the rear hydraulic lift system of these tractors. All of this steel will wear over the years and eventually will need to be replaced, which will happen quite often when you buy a beautiful classic old Farmall or John Deere to use as your cultivating tractor.

It's very easy to determine if your three-point hitch needs work. Simply mount the cultivator on the back of the tractor and lift up with the three-point hitch. Stand out at the end of the cultivator main frame and attempt to wiggle it back and forth. If the whole unit moves more than an inch or two, you've got work to do. First, attempt to tighten the turnbuckle on the stabilizer chains. If this doesn't solve the problem, the stabilizer mechanism will need replacement. This shouldn't break the bank; there is quite a bit of aftermarket material available for making this sort of repair. Most three-point hitch tool-bar cultivators are also equipped with large rolling discs called coulters. These sixteen-inch-diameter coulters, positioned at either end of the tool bar, keep the implement from jumping from side to side once it is working in the field.

Three-point-hitch-style cultivators certainly have many advantages. They are quick and easy to attach to and detach from a tractor. Mid-mounts have their advantages, too. Everything is out there in front of you where you can see it, and the machine is solidly attached to the side rails of the tractor frame, so side sway is really not a problem. This is certainly not the case with rear-mounted units; if the front wheels of the tractor are turned the slightest bit, the movement is amplified many times over at the three-point hitch. A one-inch front movement could very easily translate into five inches at the cultivator following the tractor. This might be just enough to cause "cultivator blight" and wipe out some corn. Although this is not as much of a problem with older-style mid-mount machines, the ease of operation and flexibility provided by the newer rear-mount models is still preferred by many farmers because it takes several hours to attach the older unit to the front of the tractor. Once a mid-mount is installed, it's there for the season on a dedicated tractor.

Three-point hitch cultivators come in as many configurations as their older mid-mount predecessors. The very earliest models from late 1950s were built with solid round standards and adjustable sweep mountings complete with breakaway spring trips. This older style of cultivator iron had several advantages over the newer spring shank system that came into use in the 1960s. In addition to being able to change the cutting angle of the sweep in relation to the soil, the round standards that provided attachment to the main frame could be turned within their mounting clamps. A bit of a turn one way or the other would allow the sweeps below to throw dirt into or away from the row of corn. These older machines may have been archaic and clunky, but the possibilities for adjustment were enormous. This is where the true art of row-crop cultivation manifests itself; ten minutes with a few wrenches and a ball-peen hammer can make the difference between a mediocre and an excellent cultivation. International Harvester rear-mounted cultivators from this era were attached to Farmalls by means of a two-point or "Fast" hitch; IH didn't have a three-point hitch system until the early 1960s. The Farmall Fast hitch was solidly attached to the tractor so side sway was not a problem. Every once in a while you will see Fast hitch implements in piles of old agricultural salvage iron. Old cultivator iron is a valuable asset to those of us trying to keep our row-crop machines in working order, and extra gauge wheels, shovels, shanks, springs, and round standards come in handy when we're out doing our row-crop cultivation. It's inevitable that something will need to be replaced or repaired during this process, because as you are making your way through your many rows of corn, some part of your cultivator is going to get loose and fall off out there where you won't be able to find it easily. A shank or a standard will drop out of a clamp, or a gauge wheel will fall off. This is why it's a good idea to have plenty of extra parts on hand. There is no better way to have a ready stock of needed items than by hoarding old cultivator iron. Most of this old stuff is fairly universal across the different brands of farm machinery. You will be dealing with square stock or round standards, and a John Deere part will most times fit into an International cultivator. Aftermarket companies like Shoup and Sloan Ag Express, both located in Illinois, are also good sources for wear parts like gauge wheel tires and bearings as well as cultivator sweeps. Daily maintenance on the row-crop cultivator is a must, if you don't want things dropping off out there in the field, and it is imperative that bolts are tightened often. Any grease fittings should be lubricated often as well. Row-crop cultivation is certainly a challenge, but once you've got some knowledge and experience, it's fun to keep your implements in perfect working order to do the very best job possible.

The C-Shank Cultivator

Most of the three-point hitch cultivators available to us as modern-day organic row-crop farmers are of the spring-shank type. These implements are designed to cover two, four, six, or eight rows and are equipped with either C shanks or Danish tines. The C shank replaced the older solid round standard with curved shanks, very similar to what is used on the modern field cultivator. This type of unit employs three of these curved spring-steel shanks per row. The middle shank sports a six- to eight-inch-wide sweep to cut weeds from the center of the row, and the two outside shanks are outfitted with half sweeps so as not to root-prune the row or throw too much dirt on it. Generally these three shanks are positioned far enough apart fore and aft to permit the flow of crop residue through the cultivator. C-shank-style cultivators throw a fair amount of soil, so crop shields and reduced ground speeds are necessary to protect young plants on the first pass through a field. When you have a field that is solidly grassed in or has well-rooted annual weeds, the C-shank style of cultivator is the most aggressive and does the best job.

Since the 1970s, Danish tine cultivators have become the most popular weed-removal tool. The Danish or S tine is constructed of lighter-weight spring steel and has a unique double-curved shape. These tines vibrate and stir as they are drawn through the soil, but they don't throw a lot of soil. Five S tines per row is standard on a row-crop cultivator designed for thirty-inch rows, and there is an assortment of special shovels and sweeps for attachment to the standard Danish tines. These tines are also available in different height configurations depending on how much room is needed for the passage of soil and weeds through the unit. The most common Danish tine measures eighteen inches from ground level to the top of the S curve. This style of cultivator can be used much earlier in the growing season in the battle against weeds because it disturbs the soil very little. I recently purchased a six-row Danish tine cultivator from a neighbor and have been delighted with its early-season performance in tender crops like soybeans and dry beans. I've been able to increase ground speed tremendously, which shortens my workday in the field.

Rapid Growth and Lay-By Cultivation

Once the corn plant has eight to ten leaves and has achieved knee-high stature, it is ready to enter a period of extremely rapid growth. New leaves are unfurling at the rate of one every two to three days, and brace roots begin to emerge from the first aboveground node. These are the future shoulders of the corn plant's roof system. By this time, two passes through the field have been completed, and it's likely you're quickly running out of time to cultivate because plants are growing taller by the minute and the root systems are growing laterally out into the rows. It's now time for the final or lay-by cultivation. I've always looked at this job as putting the corn to bed just before the leaves of the plants are getting ready to form a canopy over the row; once a canopy is formed, shade will take care of weed control for the rest of the season. When you've reached this point in the growing season, time is of the essence, because if you wait a few days, you might not be able to get into the field.

Heavier-duty C-shank-style cultivators work best for this job: You want to throw as much dirt as possible up against the stalks of the corn plants. If you are working with a three-shank system, you'll want to remove the outer two shanks to protect plant roots; one shank with a fairly wide eight- to twelve-inch sweep right in the middle of the row will do the trick. The job won't take long because the tractor should be operated at as fast a ground speed as is possible, which will throw even more dirt up against the newly developing brace roots. Choose a really hot day when corn plants tend to be the most supple; you'll be able to make your way through waist-high corn without causing any major damage to the crop, and the plants bend over easily and spring right back up if the sun is shining. Of course, cultivators with higher tool bars do the best work in these sorts of circumstances. The older mid-mount style of machine has a tool bar that is considerably higher than some of the three-point hitch rear-mounted rigs. Once the lay-by cultivation is completed, it will be time to put your row-crop cultivator back into storage until next year. The corn is on its own after this.

Side Dressing

Crop fertility needs are at the absolutely highest point when corn enters its late vegetative stage, and if you need to add any extra fertility to the crop, this is the ideal time for it. This practice of giving a row crop a little midseason boost is called side dressing. Since most New England farmers who are growing corn conventionally don't have the proper equipment, side dressing is accomplished by driving a three-point hitch cone-style broadcaster through the field to apply chemical nitrogen fertilizers like urea. You can always tell when this practice takes place because the corn leaves will turn white for a few days thanks to nitrogen burn. Many row-crop cultivators, however, are equipped with side-dressing applicator units that drop fertilizer down a tube to ground level, where it is worked into the soil by shovels and sweeps. This protects the foliage of the corn plants and prevents the urea from volatizing into the air as ammonia. Since nitrogen fertilizers are so expensive, it is highly recommended that a Pre Sidedress Nitrogen (PSN) test is performed before undertaking this practice, as there just might be enough nitrogen there to finish the season without the added expense and labor of side dressing.

There are a few side-dressing options available to organic farmers. Your crop may be in dire need of nitrogen because of weed competition or from leaching due to an extended period of very wet weather. If this is the case, an application of Chilean nitrate may be in order. This is almost identical to urea, but comes from the ground in the Atacama Desert of Chile. It is about 10 percent nitrogen and will provide a fine boost to an ailing corn crop. Some people call it "organic urea" because it can turn into a bit of a crutch. Organic standards prohibit the addition of any more than one-third the total nitrogen needs of a crop in the form of a soluble salt fertilizer, however, and since an average crop of corn requires somewhere around one hundred pounds of nitrogen, the maximum application of Chilean nitrate is limited to thirty pounds of elemental N or three hundred pounds of the actual material. It's worth noting that proposals have been put forth before the USDA National Organic Program (NOP) to prohibit the use of Chilean nitrate in future years; it is not allowed in Canada or Europe. Some other nitrogenous materials that can be used to side-dress corn are peanut, feather, and soy meals, and commercial pelletized chicken manure compost is also allowed. These materials take longer to break down from their organic state and release nitrogen to a crop, but they do work, although a bit more patience and forethought is required. Whatever you use to supplement the nitrogen needs of a corn crop, it is far better to actually drop the material directly to the base of the plants through a tube and work it into the soil. Row-crop cultivator side-dressing units like the ones manufactured by Cole are the best way to go.

Late Vegetative and Reproductive Stages of the Corn Plant

The last few weeks of vegetative growth occur once a corn plant has reached knee height. If you haven't had a chance to go through one last time with the cultivator, now is the time, because in a matter of days the plants could be too tall—growth becomes rocket-ship-like at this point. Plant roots have spread out in all directions to garner as much water and fertility as possible, while new leaves unfurl at a rate of one every two to three days. During this period, the tassel is developing down in the whorl at the same time as multiple ear shoots are emerging between the leaves and the stalk lower down on the plant. Stalk nodes are elongating as well, helping corn plants reach for the sky. The number of rows of kernels per ear has already been genetically determined by this stage of growth. Within two to three weeks' time, a corn plant can progress from knee height to well over your head in stature.

Pollination

Plant growth peaks a week before pollination. By this time, a new leaf is appearing every day or two, and upper ear shoot development has surpassed that of the lower ear shoots. Once the tassel has emerged from the whorl at the top of the corn plant, vegetative plant growth comes to a halt. This is as big as the plant and its root system are going to get. Two to three days after the tassel pops out, the first silk makes its appearance at the end of the uppermost ear shoot, and the reproductive phase of the corn plant has begun. Pollen shed begins two to four days after tassel emergence and continues for the next five to eight days. Corn silk is an amazing part of plant physiology. The first silks to appear are connected to the butt of the ear; later-appearing silks are connected to the middle and tip. Silk can grow from one to one and a half inches per day. Each strand of corn silk is connected to the future site of a corn kernel on the cob. Pollen shed from the tassel to the silk below is an intermittent process, usually happening only in the midmorning before noon and only when the tassel is dry. Once a grain of pollen has landed on the silk, it grows its way down the silk to the ovule where fertilization takes place, and each ovule becomes a future kernel of corn. This process takes twenty-four hours. Within two to three days, every silk on a developing ear has been exposed and pollinated. An average ear of corn will have 650 to 1,000 future kernels on an even number of rows. Corn has an amazing potential to remain diverse and strong because 95 percent of corn silks are fertilized with pollen from a neighboring plant; inbreeding suppression is naturally avoided in the reproductive process. Good weather is essential at time of pollination, as cloudy and wet weather can affect the process. Often, an unfilled ear tip can be traced back to undesirable weather at pollination time. Overly hot and dry weather can be just as destructive to kernel development and eventual yield—yet more proof that we are so dependent on what Mother Nature has in store for us when we are growing grain crops.

The Blister Stage

After pollination, the silks begin to turn brown and the cob continues to elongate. The grain will now form and begin a very rapid increase in volume and weight. Ten to fourteen days after silking, tiny little white kernels will form on the cob. These kernels are filled with a clear liquid and resemble tiny little blisters; hence this is called the blister stage. The little white blister has all inner trappings of a mature corn seed, and the radicle, coleoptile, and first embryo leaf are all present in its genetics. A tiny embryo has also begun to form. Total ear moisture at this point is about 85 percent. Dry matter will accumulate rapidly from this point onward.

The Milk Stage

By the end of the third week after pollination, the exterior of the kernels begins to turn from white to yellow. The clear inner fluid turns milky white in color as the kernel takes on weight and bulk. This is called the milk stage because the creamy interior of the kernel looks just like milk. This is the point when you would harvest an ear of corn for eating off the cob, as the kernels are still quite juicy and sweet at this stage of development. Dry matter has increased a little from the blister stage because total kernel moisture has dropped to 80 percent, and the inner embryo of the seed is also developing quite rapidly. A quick peek at the ear will tell you when a corn plant gets to the milk stage; The ear begins to acquire length and girth as it hangs away from the stalk a bit, and the silks shrivel and turn dark brown. We've been waiting for this all season—the sight of a fat ear in the milk stage gives us hope that a corn crop just might materialize.

The Dough Stage

Starch accumulation continues into the fourth week of development when the milky inner fluid begins to thicken and take on the consistency of paste. This period when the endosperm thickens and whole kernel moisture drops to 70 percent is known as the dough stage. By this time, the corn kernel has really begun to put on some weight, and moisture is now at 70 percent. This is the halfway point on the way to grain: Half the mature dry weight has been accumulated up to this point. The embryo or germ of the corn kernel is also quite well developed and noticeable during the dough stage. Starch is building up very quickly, and if the corn kernels are borne by a red cob, this is the time when a pink color manifests itself in the developing cob.

The Dent Stage

Sometimes it seems as if the dough stage lasts forever, and fall is approaching when you want the corn to mature before frost. Finally you go to the field and pull back the husk to find that the corn has finally reached the dent stage. Kernels have now begun to dry as their moisture has dropped below the 70 percent level. The corn kernel begins its final drying process by forming a layer of hard white starch at its very top. At the same time, a little depression or "dent" manifests itself on the upper surface of the kernel. (This process of the kernel's internal starch shrinking just enough to allow its crown to pucker ever so slightly was described earlier in this chapter.) The dent stage is the grand finale for the corn plant. Things are winding down to final physiological maturity at this point. Over the next few weeks, the hard starch layer at the top of the kernel begins to descend and make its way to the base, where it attaches to the cob. If you break a cob of corn in half and look toward the tip or upper end, you will see a distinct line on the side of the kernels. This is called the milk line. Everything above this line has become hard starch while everything below is still soft and pasty. As the milk line descends from the top of the kernel to the bottom, the plant is metabolizing the doughy, soft starch into finished mature grain-quality starch that will feed humans and livestock.

It can take two to three weeks for the milk line to reach the bottom of the corn kernel. Meanwhile, dry matter and mass are being accumulated in the kernel the whole time as the plant pulls the last of its life from the stalk and leaves to contribute to its finished fruit. Kernel moisture will drop from 55 percent to 40 percent during the final ripening process. If you plan on cutting your corn for silage, it is best to harvest the whole plant when the milk line is halfway down the kernel, because at this point the kernels are still soft and whole plant moisture is at the 65 percent, which makes for ideal fermentation conditions. If grain is the end game, you have to wait until the milk line gets all the way to the bottom of the kernel. This is where things begin to get a bit dicey, especially if you live in an area that is prone to an early- or mid-September hard frost. If the corn plant gets killed by freezing temperatures, there will be no remaining life force left within to metabolize soft starch to hard starch in the kernel. Kernels of half-ripe corn on a dead brown plant will continue the ripening process by shrinking instead of "growing" to maturity. When this happens, the final density and test weight of the grain is reduced. Corn that should be fifty-six pounds to a bushel might only weigh forty pounds. Corn that has been frosted early will also dry more slowly than corn that has ripened in a normal manner. Sometimes a light frost will burn the top leaves of a cornfield but leave the stalks and lower leaves unscathed. If this should happen and the milk line is three-quarters of the way down the kernel, ripening will proceed, albeit at a slightly reduced rate.

From my own personal experience as a corn grower in a rather short-season location, I can tell you that the dent stage during the month of September is fraught with worry and anticipation for me. Corn silage growers needn't be as concerned, because a half-mature corn plant will make high-quality silage, but grain growers need another week to a week and a half of good weather and above-freezing nights to ensure a good crop of high-quality grain corn. You end up paying very close attention to long- and short-term weather forecasts during this time. Choosing a good location with a warm microclimate can make a big difference in getting a corn crop ripe. I have found that river bottom fields that fill up with fog every evening in September are a good choice for planting a crop of grain corn—the fog keeps the frost away. We certainly hope for good weather and success after a long season of nurturing our corn.

Black Layer

Your corn has now reached physiological maturity, and you have made it through the three-week period waiting for the hard starch layer to advance completely to the cob. The growing process is now complete. The corn kernels have reached their maximum dry weight and are no longer receiving any photosynthate from the rest of the plant. Most of the green color has disappeared from the leaves of the plant by this time. A little brown and then a black dot will appear at the very tip of the corn kernel where it attaches to the cob. This so-called black layer is caused by the collapse and compression of cells at the bottom of the kernel. The grain is basically sealing itself off from any more transfer of nutrients and growing forces from the cob and the remainder of the plant. Kernel moisture has now fallen to the 30 to 35 percent level. Now you can breathe a sigh of relief, as your corn has matured for another year.

A Note on the Full-Season Nature of Corn

Corn is definitely a full-season crop, and every day during the growing season matters. A slow start because of cool and extra-wet weather in May or June can delay ripening later in August and September, and early-season vegetative growth is totally dependent on the accumulation of heat units. The better the weather in terms of heat, humidity, and adequate (but not excessive) rainfall, the sooner the corn will silk, tassel, and begin the reproductive process. Once pollination occurs, corn development becomes more of a timetable process; it generally takes fifty-five to sixty-five days to go from pollination to black layer. If September is warm and sunny, the milk line will move down corn kernels more quickly. A damp, cool September will have just the opposite effect, but one consolation to cloudy and rainy weather in September is that it generally won't freeze if it is raining all the time. It's those extra-clear starry evenings after a long, hot sunny day that will bring frost in September. It's often been said that farming is not for the faint of heart. We take what the weather gives us. We can manage our risk, however. Corn varieties range from extra short to medium to long season. A shorter-season corn variety will have a shorter vegetative period, which means that plant stature and size will be much less than those of a longer-season variety because the growth period will be shorter. We must remember that there is no free lunch; we can only grow a plant with the heat and climatic conditions that Mother Nature provides for us. I usually choose a corn variety in the shorter-season range of seventy-five to eighty days. I like to see tassels and silks emerging during the third or fourth week of July. Assuming that it takes two months to get to ripe grain from time of pollination, this will put me at black layer late in September before the first killing frost. I could select a longer-season variety and get more yield, but my odds of success will be reduced. If it is an extra-long hot season, I will have success, but it's a gamble enough to grow grain corn in my climate in a normal growing season. I choose caution and moderate yield over the bin-busting sweepstakes of longer-season corn.

Dry-Down and Harvest

Even though your corn is physiologically mature, it is still not ready to harvest. Kernels at 35 percent moisture must dry down to 15.5 percent to store safely without spoiling, and so begins the long process of waiting for the ear and the whole plant to lose moisture. By this time, we are already well into the month of October when the growing season is coming to an end, and cold, damp weather is imminent with the approach of winter. Corn kernels will lose one point of moisture per day if the weather is sunny, warm, and breezy, which is why those crystal-clear autumn days with glowing fall foliage are so revered by those of us who are raising corn for grain. If by now the frost hasn't turned the cornstalks and leaves brown, physiological maturity will, and once you are familiar with the ripening and finishing process, plant and husk color is as good an indicator as any of the stage of maturity. Once an ear of corn begins to dent, the green color in the husk starts to fade. By the time the ear gets to black layer, the husk covering has turned a yellow-brown color. Plant leaves follow this same progression of color loss.

It's always best to achieve physiological maturity in mid- to late September instead of October, and this will happen in a year that is blessed with above-average heat-unit accumulation and good weather. (Corn that is dead ripe in September has that much more time to dry down before the arrival of poorer weather.) There are numerous other factors that affect the rate of dry-down in grain corn. Varieties of corn with harder-textured starch like the flints and the flint/dent crosses dry more slowly because moisture escapes less quickly from hard glossy kernels, whereas lighter-colored flour-like varieties with softer endosperm lose moisture much more rapidly. The tightness of the husk around the drying ear is also a factor in dry-down; some varieties have a very loose-fitting husk that permits better drying, and yet these are also the same corn varieties that are susceptible to bird damage because ear tips are easily exposed. The type and size of the ear shank—which holds the ear to the stalk—found on a plant can also play a part in this final journey to harvest. A pencil-thin shank will allow an ear of corn to tip well away from the main stalk, allowing more air to circulate around it. Eventually, the shank will let go and the ear will tumble downward so that it is pointing straight toward the ground as it hangs from the stalk. This is the best scenario possible, because the drying ear will shed rain and snow during spells of inclement weather. Last but not least is the general stature and standability of the corn plant itself. Not all varieties are created equal in this department. It can be a major disappointment to have a field of ripening corn begin to lodge and fall over because of strong fall and early-winter winds, and once ears hit the ground they are fair game for roving rodents and mold spores. Hybrid corn usually has much stronger stalks than the older open-pollinated varieties. Be prepared to be patient during the final drying process: Corn takes much more time than any other grain crop. We plant seeds in early to mid-May, but we may not harvest ripe grain until five months later in mid-November.

When to Harvest Corn

As the month of October progresses, it is time to monitor the moisture of your drying corn. You should frequent the field every few days to peel back some husks and inspect your crop. When you are out there on a beautiful fall afternoon listening to the dry leaves of the corn plants rustle in the wind, it sure seems like the stuff ought to be drying down. Pick an ear or two for samples and then hand-shell them to get enough kernels to dump into the moisture tester for a quick analysis. I usually carry my battery-operated Dole moisture tester right along in my truck, and I set it on the tailgate and perform a quick test. If black layer wasn't reached until the end of the first week of October, you might find the moisture in the high 20s; if so, you've got a long way to go to get down to fifteen percent. On the other hand, if kernel moisture has dropped to the low 20s, you might think about getting your harvesting and final drying equipment ready. A moisture level in the high teens might be even better, but it might be November already and you would be very quickly running out of time before the onset of serious snow. Moderate amounts of snow during corn harvest aren't a major concern, because ears are borne high enough on the stalks to keep them above the white stuff, but a November blizzard can be debilitating. Once again you're back in gambling mode—the longer you wait into November (and in some cases early December), the drier the corn will be in the field. Cold clear nights when temperatures drop into the teens and single digits will freeze dry corn almost better than those sunny dry days earlier in the fall.

There are many other factors to consider when you drag your combine or corn picker out of the shed to begin the harvest. How many acres are you planning to harvest, and what sort of final drying facilities do you have at your disposal? If you have only a very small amount of corn to harvest, you can afford to wait until the last minute to take advantage of field-drying. Larger acreages necessitate starting a little earlier at higher kernel moisture than would be ideal sometimes. I think back to many a cold damp and gray November when the weather was switching back and forth from cold rain to wet snow, and we would bring heavy gravity wagons of shelled corn up icy hills to our hilltop farm for final drying and storage. I would ask myself, *Am I having fun yet? Wouldn't it be better to buy this stuff instead of going through all of this hassle?* But then the sun returns the next day as golden-yellow shelled corn pours from the grain dryer into the auger and up into the bin, and somehow all doubt disappears.

Methods for Harvesting on a Smaller Scale

There are many ways to harvest corn. Part of the beauty of corn as a grain crop is the fact that it can be easily hand-harvested without the mechanical fanfare that is necessary to deal with beans and cereal grains. Whole ears can simply be pulled from the stalk and deposited in a basket or wagon bed. Millions of acres of corn were harvested this way until the 1930s, when the mechanical corn picker arrived on the scene. The process of handpicking five or ten acres of ear corn seems impossibly daunting to all of us who were born into the machine age. We have handpicked an acre of flint corn here on our farm, and I can truly say that it was time consuming and hard work; it took a group of three or four people many days. An eighth- or quarter-acre plot of corn is a different story, however. There are certainly many advantages to this sort of hand-harvest. If there is someplace available to lay out the ears for final drying, you can begin much earlier when the corn is higher in moisture. A plastic greenhouse is ideal for this, and plant benches that are covered with quarter-inch wire mesh are a great place to lay out your small crop of hand-harvested ear corn. Turn on the greenhouse fan for a little extra air movement to speed up the drying process. Sample the kernels now and then for moisture level. When the moisture has dropped below 15 percent, it's time to think about storage. Corn can be left right on the cob and stored as ear corn in a wire cage or corncrib. You can also shell the ears with an old-fashioned hand-cranked or motor-driven corn sheller.

As seen in the accompanying photograph, the antique corn sheller consists of two cast-iron nubbed wheels that revolve inside a wooden box. Dry ears of corn pass between the cast-iron teeth, and shelled

Antique corn sheller. PHOTO COURTESY OF SID BOSWORTH

corn drops from the bottom of the box while the cob is ejected from the top. The corn kernels will still need to be cleaned with some sort of a blower or fan because they will be full of glume or "red dog"—the fluffy material that actually attaches the kernels to the cob. This whole process can be a lot of fun for anyone who wants to produce cornmeal for their own consumption. All you need is a hand hoe, some gumption, and a small plot of land.

The Corn Picker

Ear corn is the best harvest and storage choice for the small livestock producer with limited funds and infrastructure. One brand that comes to mind is the New Idea Model 323 picker-husker, which is a more modern version of the corn picker that revolutionized American agriculture in the 1930s. These machines were all-pervasive in the Corn Belt and elsewhere in the 1940s, '50s, and early '60s. They began to fall from favor in the '60s when combine harvesting of shelled corn became more prevalent. Nevertheless, corn pickers are still used by many smaller producers today. There are many of these machines still around in sheds and on fence rows. They sell for as low as several hundred dollars to as much as two thousand if you are bidding against an Amishman at an auction.

The corn picker is a marvel of agricultural engineering and ingenuity. Power is supplied to the machine by a PTO shaft from the tractor. Two galvanized "snouts" or row dividers straddle either side of the corn row. Internal gathering chains pull the stalks into the machine and down through two revolving rollers. These cast-iron stalk rolls revolve in opposite directions and serve the purpose of "snapping" the ears from the stalk as it passes through the rolls. Once the ears have been snapped, they are propelled upward toward the "husking bed," which is located toward the rear of the machine. The ears roll around on these rubber rolls, whereupon the husks are removed. Whole ears of corn and chipped shelled corn tumble into a rear elevator and are delivered vertically to a grain wagon pulled along behind. Corn pickers were a miracle in their day, but they were also responsible for maiming many a farmer; since the machine was designed to grab and pull in cornstalks, it also grabbed fingers, hands, and arms of many an unsuspecting farmer who was trying to unplug it while it was running. If you are using a corn picker to harvest your corn, *be careful!* Always turn off the PTO before you dismount from the tractor to make an adjustment or unplug something that is clogged or jammed. Stay well clear of a running machine, and don't wear any loose-fitting clothing that could be grabbed by a chain, belt, or gear.

A bit of practice and in-field experience is the best way to learn how to operate a corn picker. However, some basic information and a discussion of harvesting techniques will make it easier to get started. There are numerous brands of pull-type machines available to choose from in one-, two-, and three-row configurations. Just about every major farm equipment manufacturer had a corn picker line in the golden age of ear corn. John Deere made the Model 400 with the ability to pick as many as three rows at once; J. I. Case, International Harvester, and Oliver all provided machines to the market. Tractor-mounted two-row pickers were also quite common during this era. The snouts of these units were bolted to either side of a tricycle tractor, with the husking bed and ear corn elevator behind the seat. Tractor-mounted pickers turned this piece of equipment into a self-propelled unit. New Idea came out with its "Uni-System" self-propelled corn picker in the 1960s. The Uni, with its enclosed cab and ability to pick up to six rows, represented the ultimate in high-volume, efficient corn picking in its day, and I purchased a Uni in Ontario several years ago for two thousand dollars.

When you talk to old-timers who picked corn in their younger days, you will hear many different opinions as to which are the best machines. Whatever you buy, make sure to get an operator's manual and a parts book along with the machine. Parts availability can be a problem for these older machines, and you might find yourself frequenting farm machinery junkyards or Internet sites to find the repair and replacement parts you need. Having a second "parts machine" around is probably a very good idea. Agco dealers are able to obtain most parts for New Idea models; John Deere parts are also pretty readily available. If you end up finding a multiple-row machine, check the row width before sealing the deal. These machines were made in the era of thirty-six- and forty-inch row corn, but you'll want to buy a machine that matches your row width. Simply measure center-to-center in the head to determine if it is a wide- or narrow-row picker. The two row New Idea 323 was the narrow-row machine while the New Idea 325 was designed for wider rows.

Maintaining Your Corn Picker

The potential for wear on a corn picker is quite high since there are so many moving parts. You don't want to spend a fortune replacing chains, sprockets, and rolls before you even get to the field. The first place to inspect is the pickup head in the very front of the machine. The floating gathering points on either side of the head should be free moving and flexible. These little galvanized pieces of metal are intended to float along the ground and pick up downed and lodged stalks, but they have a tendency to run into rocks and other obstructions, which dents and jams them. I hit a stone with my New Idea one-row picker once. It permanently disabled one side of the head, which has compromised its operation ever since. Check the gathering chains and snapping rolls inside the head, as chains can stretch and wear. If the stalk rolls are smooth, you know that the machine has seen a lot of corn, and if you have the opportunity to hitch the picker to a tractor and run it, you'll know a lot more about how the machine actually operates. Corn pickers are loaded with slip clutches to protect the machine when it plugs. If you turn on the PTO and hear the characteristic chatter of a slip clutch, there is a shaft somewhere that is frozen and not turning. This is often the case when an abandoned corn picker spends years out in the elements. Sometimes it's as simple as squirting some penetrating oil into a bearing or bushing and encouraging a shaft to turn with a pipe wrench. You'll want to take a good look at the husking portion of the machine as well, because the rubber spider wheels that move the ears over the husking rolls can also wear. Make sure the belt-driven cleaning fan that blows away corn husks is operational. Last but not least are the shell corn saver and ear elevator. There are numerous drive chains, sprockets, and bearings to transfer power to various parts of the machine.

Corn pickers are few and far between these days, but once in a while you will see one in an auction flyer or advertised in a farm paper. There are a few of these machines that have been shed-kept and are in mint condition; these will be the more expensive options that sell for several thousand dollars. More often than not, you will find one in less-than-ideal condition that has been kept outdoors in "the Lord's garage." If the price is reasonable, you may just have to take what you can find and make the necessary repairs. That fine line between junk and serviceability is in a different place for all of us depending on our mechanical skills and creativity. Personally, I prefer to buy a machine that is field-ready.

Harvesting with the Picker

When you are satisfied that your corn is dry enough to harvest, it's time to head to the field with the picker. You don't need a huge powerful tractor for the job. We used a Farmall Super M to pull our corn picker and trailing gravity wagon around the field in the early 1990s. Traction is more of an issue than horsepower, especially when the gravity box that follows the corn picker is fully laden with a big load of ear corn. When it's time to pick corn, you simply drive up to the first row on the outside of the field. The head of the machine will be sticking out from the right side behind the tractor. Center the head on the row, turn on the PTO, and go. If you didn't leave a little space along the outside edge of the field for the tractor, you might have to run over the first row of corn with the tractor while the corn picker harvests the second row in. If the tractor wheel misses the first row and the plants only get bent over, you can turn the machine around and attack the row from the other direction once you have picked the first five or six rows. This is very similar to mowing the back swath around the edge of a hay field. It is also advisable to narrow the tread width of the tractor pulling the picker so that the right-hand tractor tire doesn't track too close to the row being picked. If you pick corn with the same tractor you used to cultivate, you will be just fine: The wheels are already positioned to center on thirty-inch rows.

There is nothing more frustrating than running over the stalks of corn you are trying to harvest. The snapping head of the corn picker should be run as high off the ground as possible. Head height is controlled from the tractor seat by means of a remote hydraulic cylinder. Lift the head up so that it engages the standing corn just below the ear. This will simplify the picking process because it will lessen the amount of stalk and leaves that must pass through the inner workings of the machine. Raise and lower the snapping head to suit the field conditions. The only time you need to drop those gathering points right down to the ground is when there is lodged corn that needs to be picked up. Out behind the picker, make sure that the trailing gravity box is properly lined up with the ear corn delivery elevator; the elevator should deposit the ears well into the middle of the wagon. If the ear corn is falling just behind the front wall of the gravity box, you may lose ears on the ground as you bump across rough ground. If this is the best you can do with the machinery you have, you will have to stop often to shovel ear corn backward in the wagon. Also take care when you come to the end of a row and it is time to turn your whole wagon train back around to return on the next row. It won't take long to realize that as you turn the tractor and picker to the right, the ear corn elevator will move over the side of the wagon as you make the corner, and all of a sudden ear corn will be dropping on the ground instead of in the wagon. This situation can be remedied several ways. When you come to the end of the row, stop for a minute or so to let the corn picker digest its corn and dump it into the wagon. When the stream of ears going up the elevator slows down to a trickle, shut off the PTO and make the corner, but don't shut off the machine when it is full of corn. Most picker elevators have an on-and-off clutch that's controlled from the tractor seat by means of a rope. These clutches can give you fits at times, but if you are able to control your elevator reliably from the tractor seat, it is still a good idea to pause for an instant to let it empty before you shut it off. This will keep your ear corn elevator from overloading. Operating a corn picker is as much an auditory task as it is physical and visual. Keep your ears peeled for the characteristic chatter and rattle of a slip clutch. Little things like too many ears in the bottom of the elevator or too many husks underneath the husking rolls can cause a plug-up. Damp conditions can also make corn picking more difficult. But there is no nicer culmination to a long season of nurturing a corn crop than watching bright yellow ears traveling up a corn picker elevator on a breezy sunny late-fall day in a yellow-brown field of rustling corn.

Corncribs for Larger Harvests

Although its heyday has passed, ear corn has many advantages for those of us trying to farm on limited budgets. Picking and storing corn on the cob is much less energy-intensive than combining shelled corn because there are no drying costs—corn picking can begin when the corn moisture percentage is still in the high 20s. However, things will go much better if you wait until your crop has dried a bit more. Wet corn doesn't husk as well or flow out of wagons and up elevators as easily. You will also need a place to store your ear corn—a corncrib—so that it can finish drying and be protected from the elements. Corncribs come in all sizes and shapes. The first storage cribs used in the nineteenth century were constructed of wooden slats. The archetypal corncrib was narrower at the bottom, with its walls sloping outward. In New England, these buildings were elevated on large rocks to provide ventilation from underneath; a flat rock underneath each corner of the building served as a rodent guard to keep rats and mice away from the corn. This same style of slatted corncrib was also hung from the second- or third-story wall of many an old post-and-beam nineteenth-century barn.

The more modern round wire cage style of corncrib is pretty universal in most parts of the Midwest where ear corn is still being grown, and is a good choice for corn growers with large acreage. These cribs abound in Amish settlements in parts of Ohio and Indiana. Round wire cribs have a galvanized-metal roof and are generally twelve to sixteen feet in diameter. Most of these units are outfitted with a four- or five-foot-diameter ventilation shaft in the center—it's basically a cylinder constructed of large wire mesh. A revolving air scoop that turns into the wind like a weather vane is attached to the peak of the roof to channel air down into the middle of the corn. This style of crib is like a doughnut: Air can get to the drying ear corn from the center as well as from the outside edge. When and if you can locate one on an old abandoned farmstead, round wire corncribs can be purchased quite reasonably. Just dismantle it or put it on a trailer in its entirety, bring it home, and set it on a concrete foundation. But really, this style of corncrib works much better in the Midwest where it's possible to ripen corn to lower moisture levels because of better growing conditions. Even though the drying corn is exposed to air from the inside and out, it is still six feet thick—which makes it difficult to quickly dry wet ear corn in our damper and cooler New England fall weather.

Corncribs for the Northeast

Long, narrow corncribs constructed of poles and wire mesh are much more common here in the Northeast. I was blown away by the sight of half-mile-long cribs when I first started driving around the wide-open expanses of southern Québec thirty years ago. These corncribs were no more than four feet wide, which allowed for simple air-drying because the wind could easily blow through the narrow width. These structures were twenty feet or more in height and oftentimes didn't even have a roof. I found out many years later that farmers would come back to these corncribs in the late winter or early spring with a portable shelling unit. The shelled corn from this cribbed corn would bring the higher price for number one corn because it had been slowly air-dried, whereas corn that had been combined and propane-dried the previous fall would only bring a lower number two price because of lower test weights and kernel damage. As I mentioned earlier in this narrative, I constructed my first corncrib from cedar poles and two-inch mesh wire in 1990, and we loaded some pretty wet corn into that crib in a howling snowstorm in early November of that year. Everything worked, but we still had a small amount of mold in our cribbed ear corn the next spring. My inexperience and mistakes taught me quite a bit about cribbing corn on a wintry hilltop in northern Vermont. First of all, it's a lot colder and wetter here than in the lower elevations near the St. Lawrence River. If I had it to do over again, I would build my rectangular wire corncrib three feet wide instead of four. This would allow the air to travel through the wet ear corn more easily. I also found out that year that it is imperative to keep all of the loose chipped shell corn out of the crib. Corn kernels that chip off ears in the picking process end up in the wagon as well, and if this loose corn travels up the elevator and into the crib, it can block the passage of air through the ears. To remedy this situation, many solid-bottomed ear corn elevators have small holes or a screen in the bottom that permits loose corn to fall through. Once I figured this out, corn storage and crib drying got much better for me. One other little caveat is to build your crib strong enough to withstand the outward pushing forces of fifteen to twenty feet of heavy ear corn. We had such a bumper crop of corn in 1999 that we ended up filling our corncrib as well as several outdoor steel silos with shelled corn. The corn was of such high quality and density that it started to push the crib apart at the top, and I feared that the sixteen-foot cedar poles were going to break from the force. We ended up running steel cables between the poles to hold everything together that year. If you are interested in building this style of corncrib, you might do well to invest the time to travel and inspect how other people have built these structures. There are many empty corncribs in the Montérégie region of Québec just north of the Vermont border, and many are being used to store firewood these days. These structures are built on cement foundations of dimensional lumber. Ear corn is the cheapest low-tech way to store your harvest of grain. It is a relatively stable commodity with relatively simple storage requirements—good aeration and minimal protection from the elements.

The Drawbacks of Crib Storage

Although it is nutritionally superior, ear corn is bulky stuff. When you are grinding it with a hammermill or PTO-powered grinder-mixer, every last bit must be physically moved with a scoop shovel. Shelled corn flows either by gravity from a bin above or through an unloading auger, but ears of grain corn will tumble out of a crib and into the feed hopper of a grinder to a certain extent, and sooner or later you must climb into the crib and begin pushing things around with a shovel. Some of the more ingenious Québec farmers installed drag line paddle chains in the bottom of their quarter-mile-long corncribs. Leave it to an inventive farmer to find a better way. However, even these clever folks have abandoned their corn pickers and cribs in favor of modern shell corn combine harvesting.

Another factor to consider about corncrib grain storage is that you will have the potential to feed all of your local wildlife, from rodents like rats and mice to raccoons and birds. Pigeons, starlings, and mourning doves especially will really appreciate the giant bird feeder that you have provided them. I used to love viewing the glow of the bright yellow corn in my crib in the late-afternoon sunlight. Then one day I looked out and found the entire side of the corncrib covered with birds stocking up on some late-day carbohydrates. After a month of this, the beautiful late-day yellow color morphed into the brown and red hues of bare cobs whose grain had been removed by my feathered friends. I told myself that I needed to share the harvest and that it was only one little layer of corn that had been lost to the birds. However, I soon found out that the interior of the corncrib was at risk from the onslaught of other forms of wildlife. It was an ideal home for other little creatures: Here was shelter and food all in one place for every mouse, rat, and raccoon from miles around. It stands to reason that my corncrib could double as an apartment house for these animals. The invasion of the rodents wasn't such an awful thing in and of itself; these little creatures didn't make that big of a dent in my supply of ear corn. The smell and scent of the urine and feces left behind by rats and mice, however, was pretty hard to stomach. Almost everyone who has grown and stored any appreciable amount of grain knows the special ammonia-like smell of rodent infestation. The odor lingers and contaminates the surrounding grain. This was one of the many factors that helped me decide to switch from cribbed ear corn to bin-stored shelled corn in 1994.

The Rise of Combines for Corn Harvest

The switch from ear corn harvest to the modern grain combine is representative of the general trend in agriculture toward higher efficiencies gained through the increased use of petroleum-based energy resources. Ear corn finishes drying in the crib aided by the power of the sun and the wind, whereas shell corn, in most cases, needs to be artificially dried with propane. Circumstances are a bit different in the Corn Belt, where additional drying isn't necessary in a normal growing season. On a mid-September visit to Iowa several years ago, I noticed that the corn had already dried up and turned completely brown; physiological maturity had most likely occurred the first week of September. Two or three more weeks of field-drying in the warm Iowa sunshine would put this corn at the 15.5 moisture level during the first week of October. But here in New England, we would be lucky to find our corn beginning to dent in the first week of September and black layered by the first week of October. This explains why the first corn head attachments for self-propelled grain combines were developed in the Midwest during the mid-1950s.

A quick look back at sales literature and extension bulletins from this era shows most farm machinery manufacturers releasing two- and three-row corn heads for their combines. Most of these units were built for wider thirty-six-inch rows planted with two-, four-, and six-row corn planters. Combine corn heads were basically identical to the snapping head portion of the pull-type corn picker. Metal snouts with gathering points guided the ripe corn plants into gathering chains; once inside the unit, the ripe ears of corn were pulled from the stalk by the action of rotating stalk rolls against stripper plates below. After being snapped from the stalk, ears of corn traveled up the feeder house and into the threshing cylinder of the combine. It turned out that corn shelling was a relatively easy job for the modern grain combine. The concave was lowered to allow for the passage of ears between it and the cylinder, and cylinder speed was reduced to five hundred RPM. One of the reasons why corn was so easy to harvest with a combine was that only the ears and small amounts of husks were entering the machine. The cleaning functions of the combine were not being taxed because most of the corn plant remained in the field. In really good picking conditions, even the husk remained on the stalk. This was entirely different from combining wheat or oats, where the whole plant was being cut and processed by the machine.

Indeed, American agriculture has always been quick to adopt technological change and increase productivity. Think back to the 1930s when hybrid corn was almost universally adopted in a five year period. The late 1950s and early '60s were a time of tremendous change in the Corn Belt. Crop yields increased dramatically as fertilizers and herbicides became more commonplace, and plant populations were inching upward all the time as corn growers began to seed thirty thousand plants per acre instead of the twenty thousand that had been the norm ten years earlier. By 1960, larger, more powerful tractors became available to the American farmer. Along with these came wider planters, narrower row spacing, and combines with increased harvesting capacity. The two- and three-row-wide corn heads developed in the '50s were replaced by four-, six-, and eight-row heads a decade later. The agricultural landscape underwent dramatic change during this same period. Mixed farming with its emphasis on feeding farm-harvested grains to livestock was being replaced by the cash grain system. Small dairy herds as well as beef and pig operations began to disappear. The archetypal two-sided drive-through wooden-slatted corncrib so commonplace on midwestern farms soon became a storage shed because ear corn was no longer being harvested.

Round galvanized-steel grain silos started popping up everywhere to store the harvest of shell corn that was now being sold off the farm. A watershed moment was taking place at this time in American agriculture. Small decentralized local and regional food production was being replaced by the larger, more centralized system of industrial agriculture that we know today. There were certainly many factors that contributed to these changes. At the top of this list was the rather cheap cost of petroleum energy. As our productivity increased by leaps and bounds, so did our use of fossil fuels for chemicals, crop production, grain drying, and transportation. None of this would have been possible without the development of corn combine harvesting.

A Middle Ground: The Picker-Sheller

There is a middle ground of harvesting technology that exists between the corn picker and the modern combine. The New Idea Farm Machinery Company of Coldwater, Ohio, also manufactured a piece of equipment called the picker-sheller, and most of these machines were configured with two rows. The picker-sheller was basically a standard corn picker with a corn-shelling attachment added onto the back. The shelled corn travels up an auger and is delivered to a trailing wagon. These units are easy to spot in a field or a used-machinery yard because they have an auger tube instead of an ear elevator behind the machine. New Idea also developed an advanced cage-type sheller, which made the company a pioneer in corn shelling technology. This unit consisted of a cast-iron cage about a foot in diameter; a spiral metal rotor much like the blades of an old-fashioned push lawn mower revolved inside this metal cage. As the ears of corn traveled through this unit, friction between the spiral flighting and the outside cage removed kernels from the cobs. Corn kernels would then drop through large holes in the cage; after a little sifting and fan cleaning, the corn would make its way up the rear delivery auger and into the gravity wagon behind.

Picker-shellers were quite common on dairy farms in the 1970s during the high-moisture corn craze. If you want to shell corn and don't want to go the grain combine route, one of these older machines might be just what you need. They aren't quite as popular as regular picker-huskers, which makes them inexpensive and affordable for someone just starting out in the shell corn business. Some inventive farmers in Québec have detached the shelling units from these old machines to use as high-capacity freestanding corn shellers on their operations. When compared with combine harvesters, picker-shellers are somewhat crude and outdated, but if the price is right this may be a viable alternative for a beginning grain grower.

The Importance of Dry Conditions

There are many similarities between combining shell corn and picking ear corn. Crop condition and harvest moisture are important in both systems. Like a corn picker, a combine will do a better job under drier conditions. Ears will snap better at 20 percent moisture than 30. This also holds true for the shelling that takes place inside the combine. Wetter corn will be more difficult to shell. This is evident when you walk behind a combine and inspect the cobs that are coming over the chaffer and out the back of the machine. If you see kernels of corn still attached to the cobs, you are losing grain. Corn at 30 percent moisture is much more difficult to shell. If you are losing grain, it will be necessary to more aggressively thresh the corn by raising the concave closer to the cylinder. Cylinder speed might have to be increased as well. But increasing the aggressiveness of threshing comes along with its own set of potential perils and pitfalls. High-moisture corn is soft and easily damaged; corn combined at 28 to 30 percent moisture produces lots of "fines." These are tiny little particles of cornmeal that get rubbed off the shell corn during the threshing process. Fines represent lost and damaged corn. In addition, corn that contains a high proportion of fines will not dry or store as well, because the passage of air from dryer or bin fans through the sample is impeded by this extra material. This is the primary reason why corn is put through a rotary screen cleaner before it is dried and also before it goes into the bin for final storage. Corn combined at higher moisture turns into one giant headache in the fall. Get ready to use an inordinate amount of propane to fire your grain dryer because extra-wet corn takes a lot longer to dry. (We've had to keep batches of 28 percent corn in the grain dryer for close to eight hours.) Meanwhile, you are watching the needle on the gauge of your five-hundred-gallon gas tank head toward empty. This gets expensive and discouraging when propane sells for well over three dollars a gallon. When the corn finally does reach 15 percent, it has shrunk by more than a quarter in volume, and grain quality and test weight have been lost in the prolonged drying process. Corn kernels will no longer be bright yellow in color; they may contain cracks and other deformities that will affect their keeping quality. This wet corn scenario is generally the by-product of an early frost, an extremely wet fall, or both. Molds and mycotoxins can also invade wet corn, causing potential health problems in livestock that might consume it.

All of this information on the perils and pitfalls of wet corn at harvest is not intended to discourage you from growing corn for grain here in our short-season climate; I only want to caution you to avoid some of the mistakes that I have made over the years. If your crop gets stopped in its tracks by a hard frost before it is close to physiological maturity, you might want to consider options besides grain before proceeding with harvest. Corn silage might be the best thing to do with your damaged crop. If you choose this route, chop the corn sooner rather than later to preserve it at its highest quality when there are higher levels of fermentable sugars and starches still remaining in the whole plant. Another option in this situation would be to harvest the crop as high moisture ground ear or shell corn. The corn is combined or picked at the 28 to 30 percent moisture level and then left to ferment in a sealed storage container like an upright silo or an airtight plastic "ag bag." Cows love high-moisture corn, but it should be fed in moderation because it is an extremely "hot" feed, high in soluble energy; too much high-moisture corn is a ticket to acidosis, also known as sour stomach. Let's hope that your corn harvest will go smoothly for you and that you won't have to worry about any of these less-than-perfect outcomes.

Using a Combine to Harvest Corn

Corn combining is actually much simpler and more straightforward than harvesting cereal grains. The first thing you will need to do is change heads. First, detach the grain head and back the combine away from it. Before you hook up the corn head, however, several small changes must be made to the feeder house—the inlet conveyor that will deliver the corn to the threshing cylinder. Get out your operator's manual and follow along. The drum that drives the "raddle" or slatted feed chain must be lifted up to a higher position at the front of the feeder house to provide more space for the snapped ears to travel underneath it. Ear corn definitely needs more room than cereal grains, and this is a relatively easy adjustment to make. You simply loosen a few lock bolts and turn a cam that raises the front feeder house drum. Retighten the lock bolts, and the job is done. This feeder house drum is driven from the main machine by a chain and sprockets. Corn heads need to turn much faster than grain heads. To increase the RPMs it will be necessary to change the feeder drive chain over to a smaller sprocket on the drum drive shaft. This type of drive system generally uses a double sprocket, which consists of a large and a small gear side by side as part of the same metal casting. This gear needs to be loosened and moved over on the shaft so the smaller sprocket side lines up with the drive chain. Sometimes this gear can freeze to the shaft and needs to be persuaded with heat from a torch and a big sledgehammer. Once these two adjustments have been made, it's time to drive up to the corn head and attach it to the combine. You might have some extra hydraulic hoses that were necessary for the grain head, but not for the corn head. Couple these hoses back together so hydraulic oil from the main combine will be able to recirculate back through the system. If you have a spare pressure line that was used to drive the pickup reel of the grain head, the oil will need to get back to the main hydraulic pump. I remember all the troubles I had when I hooked the corn head up to my old John Deere 3300 combine for the first time in 1994. Every time I engaged the corn head, the combine motor would lug down and just about stall the combine. I was "deadheading" the hydraulic pump, but as soon as I hooked the male and female portions of the two hydraulic quick disconnects together, everything worked just fine. One other thing to check is the oil level in the gear boxes that run the stalk snapping rolls. Each row of the head has its own individual gear case. Low oil is usually not a problem, but a general gear case inspection should be done to ensure proper lubrication and to extend the life of the corn head drive chain. Attaching the corn head to the combine is probably the most time-consuming part of getting ready to combine corn, but everything after this is easy.

Tips on Operating Your Combine

Instructions for proper combine adjustment vary slightly from machine to machine. As always, you will definitely want to consult your owner's manual for the finer details, but there are a few basic principles that hold true for all machines. First, slow the cylinder speed down to around 500 RPM; corn needn't be threshed too aggressively. Drop the concave down low enough to completely remove all the kernels from the cobs. This is an important adjustment; if the concave is set too close to the cylinder, it will break cobs as well as shell them. Broken cobs are to be avoided, because a combine's cleaning system will have difficulty separating little pieces of cob from the grain sample. As with all crops harvested with a grain combine, a periodic check out in back will keep you abreast of how good a job the machine is doing in the field—you want to see completely shelled whole cobs shaking out over the chaffer at the rear end of the combine. Once you are in the cornfield, line up the points of the corn head with the middle of the rows and go. Remember to follow the planter just as you did when cultivating. The threshing capacity of your combine and number of rows on your planter will be considerations when you figure out how large a corn head to put on your machine. For example, a four-row corn head works well when following a four-row planter, and a three-row corn head works quite well with a six-row planter because two passes with the combine will match the planter. We use a four-row corn head with a six-row planter. When doing this, it is very important to be as precise with your planter as possible. Strive for that thirty-inch spacing between each pass with the corn planter so odd rows will line up with the corn head. This whole business is a matter of geometry. Once you are out there in the field traveling up the rows, the whole process will come to light. Lift the corn head with the combine hydraulics as high as you can to barely catch all the ears in the field. Ideally you would want the head to grab stalks of corn just below the ear. When the corn head is operated in this sort of raised position, it is running nearly level. Ear loss will be minimized because snapped cobs will not have to climb up a grade to enter the feeder house. If the crop is lodged in spots, lower the head and let the points pick up the downed corn. Each row of a corn head will have two little rubber flaps called ear savers on either side of the gathering chains and snapping rolls. The purpose of these little pieces of flexible rubber is to save flying ears from ricocheting out the front of the corn head. Ear savers seem to tear and fall off quite easily. Keep a few spares in the cab of your combine for replacement out in the field. Last but not least, match your ground speed to field conditions, crop yield, and combine capacity. Believe it or not, there is such a thing as going too slow when combining a crop of corn. With all of these hints under your belt, it's time to go out there and give it a shot. You can't beat learning by doing.

Drying and Storage

I have mentioned several times how much pleasure I derive from watching golden kernels of shell corn streaming from the combine unloading auger into a grain wagon on a sunny late-October afternoon. Part of this magic comes from the fact that corn is the best yielder of all the grains. It's absolutely amazing how quickly the combine's grain tank will fill up in a good stand of corn. My John Deere 6620 with its four-row corn head and 166-bushel grain tank can barely get down and back twice in a quarter-mile-long field before the bin is overflowing. Compare this with a field of oats or barley and you can see why corn is king. Cereal grain yields usually average between one and two tons to the acre, but grain corn grown under ideal conditions in an area with plenty of heat and a long growing season can top five tons to the acre. In my more marginal area of northeastern Vermont, I am overjoyed with corn yields between three and four tons to the acre, which is still twice the yield I would get from a cereal grain.

Using the Grain Dryer for Corn

Although we have already discussed the principles of grain drying and the operation of a dryer, there are a few more things to know that are particular to drying corn. For one, if the corn is quite wet and full of fines, it should be rotary cleaned on its way into the dryer. Avoid operating your grain dryer in the rain, especially if it is an open-topped circular batch model. Perform all necessary maintenance and repairs on your dryer well in advance of the corn-harvesting season—it's not a bad idea to have a certified propane service person check over everything. Since gas requirements are so high, grain dryers burn propane in liquid form. This is the only way that enough fuel can be delivered to a one-million-plus BTU burner. Gas in its liquid form is piped from the bottom of a large propane storage tank through a special flexible high-pressure hose. Once it enters the dryer, the liquid propane is vaporized by residual heat from the burner. (Safety is the number one concern here, because liquid propane is very volatile; under extreme pressure, a leak could turn into a major disaster.) A solution of bubbly dish detergent painted with a brush onto all gas connections will indicate the presence of a leak.

Grain drying is serious business. There are a multitude of systems on a dryer for regulating propane delivery and grain temperature. Electric solenoid valves open and close to regulate the flow of gas to the burner. Signals are sent to these gas valves from a series of thermostats that set the temperature of the main heating plenum and of the grain itself. Plenum temperature for shelled corn should be set between 200 and 230 degrees. Grain temperature for drying corn should be set between 130 and 140 degrees. Probes and high limit switches will shut off the flow of gas if temperatures climb above preset values. The most important safety feature on a grain dryer is the sail switch. This little device sets between the main fan and the gas burner and is powered by the airflow from the fan. If the fan stops for any reason, the sail switch will shut off the flow of vaporized gas to the burner. If the supply of gas was able to continue to the burner without the airflow to move the heat into the plenum, the corn within the dryer would soon catch on fire.

Most dryers are equipped with an ignition button, spark plug, and coil, and twelve-volt electric power is provided from the tractor. When you want to start the dryer, engage the PTO lever from the tractor and bring the RPMs up to 540. Once the fan is running and grain is circulating, slowly open the valve at the bottom of the supply tank. If you open the main valve too quickly, the liquid propane can slug and shut itself off for safety reasons. Push the ignition button to light the gas burner. You will actually hear the click of the coil sending voltage to the spark plug. The roar of the burner should become audible immediately. Hold the button in for a minute or two to give the burner time to warm up and vaporize the liquid propane. Grain dryers are also equipped with a spark plug ignition system to light the gas burner before adequate vapor is available. Once everything has warmed up, you can let go of the ignition button because the liquid will vaporize naturally from the heat of the burner. If the burner fails to light, start by checking the sail switch. It may have stiffened since the last drying season. The spark plug and coil could also be faulty. This has only been a little overview of the basic electrical and propane workings of an old fashioned batch dryer. Review the grain dryer section in chapter 7. There is so much more to know. Do everything you can to familiarize yourself with your grain dryer. A good place to start is the operator's manual.

Larger commercial operations store recently harvested corn in a dedicated hopper bottom bin called a wet tank. If you are driving past a large-scale grain system somewhere in the farm country of Québec or New York State, you will usually see the wet tank perched on stilt-like legs over the dryer nestled between a collection of grain silos and elevator legs. With this sort of setup, the combine doesn't have to stop and wait for the dryer. But gravity wagons and truck bodies work almost as well as a wet tank and are a whole lot cheaper. We can only dry one batch of corn at a time with our old cast-off (but reasonably priced) grain dryer, and a reliable grain auger of moderate length is all we will need to load recently harvested corn into the dryer. Longer, larger-diameter PTO-driven augers work best for this job, as wet corn is heavy; the task of moving it up the slope and into the dryer demands a bit of power. The PTO drive will ensure that there is adequate torque. The longer length of the auger will allow the gravity box of wet corn to be unloaded farther away from the dryer. This gives you plenty of room to maneuver at a safe distance from the grain dryer and the tractor that is powering it. A longer auger will also run at a much shallower incline, which makes it easier to elevate the corn into the dryer. We use a fifty-three-foot PTO-powered auger ten inches in diameter here on our farm, and have had problems when we tried to use a shorter seven-inch-diameter electric auger for this job. The heavy wet corn climbing up the steep angle kept overheating the motor and tripping the overload bottom. We found that this shorter auger worked much better to move dry corn at the other end of the process.

Be careful not to overfill the grain dryer; the temptation is to squeeze as much corn as possible into the unit. The thought is that corn shrinks when it dries, so why not mound the corn right up to the top ring? The truth of the matter is that corn expands for the first half hour as it is heated by the forced hot air coming from the plenum of the dryer, so you will need to leave three or four inches of room between the top of the corn and the top ring of the dryer. Watching corn spill over the top lip of a grain dryer is frustrating and preventable. The dryer should also be level and well supported by its adjustable bottom legs. There is a lot of weight being carried rather high in the air here. Most round batch dryers are at least twelve feet tall, and something this top-heavy needs the support of a good foundation underneath it. Now that you have taken care of all the physical logistics of the grain dryer, it is time to stand back and watch the grain heat up as it circulates.

Monitoring the Dryer

Drying corn in the late fall is always quite a spectacle. Noise levels are quite high because the tractor powering the grain dryer is roaring at full throttle. The weather has cooled off into the thirties and forties. There might be snow flurries in the air and a dusting of snow on the ground. But once the wet grain heats up to one hundred degrees or better, steam begins to billow from the top of the circulating corn and the sides of the dryer. If you climb the ladder up the side of the grain dryer to get a grain sample, it's like being in a sauna. All of this steam escaping to the atmosphere is a good thing because you know that water is being driven from the drying corn. It's a good idea to be present during this process. A quick trip to the house for a bite to eat might be all right, but you will want to be there monitoring the situation most of the time. Keep your eye on the tractor's fuel supply, and fill the tank before you start the drying process. If the tractor runs out of fuel and the sail switch doesn't shut off the gas supply to the dryer, there could be trouble. Corn stoves have become popular over the last little while because shell corn is such a great source of heat from its combustion, which means a dryer full of half-dried corn would make ideal fuel for a rip-roaring inferno. The propane tank isn't far away, either. The seriousness of this situation and the imminent dangers involved cannot be overstated. Years ago in the 1990s we had a dryer fire when a tractor ran out of fuel. I walked around the corner of a nearby building and smelled the popcorn-like odor of burning corn. Luckily, the fire had only been burning for five minutes or so, and we quickly shut off the propane and put the fire out. Memories of this lesson still make me tremble. Grain drying is a totally unregulated activity in a world that seems to have a law about everything. It is possible to set up and operate a propane-fired grain dryer with virtually no knowledge and experience. If you are a novice and fairly new to this, I would recommend seeking help and advice from someone who is very familiar with propane, burners, and the operation of grain-drying equipment. This sort of help or consultation may be expensive, but the safety of you and everyone around you as well as your property is more important than anything else.

The amount of time a batch of corn spends in the dryer is totally dependent on its harvest moisture. Corn that is above 25 percent moisture is the most time (and fuel) consuming. We have kept 28 percent moisture corn in the dryer for as long as eight hours. Check the total efficiency of your dryer's burner well before corn harvest. A batch of drying corn will usually only steam for the first hour or two. When the steaming ceases, it is usually a good indication that the corn has fallen below the 20 percent level; the lion's share of drying has been accomplished by this point. This is where it pays to monitor the moisture frequently. Most grain dryers have a smaller sample spout conveniently located on the bottom of the unit; otherwise you have to climb the ladder up twelve feet and scoop a sample from the top. (See chapter 8 for more on the relationship between grain temperature and moisture content.)

It is worth mentioning again at this point that corn can also be overdried. If the corn stays in the grain dryer for too long, it will shrivel a little and lose some of its bright yellow color. Corn that goes into the dryer at 28 percent and comes out at 13 or 14 percent will always have a dull yellow color. Nineteen percent corn will always retain its shiny yellow-orange hue because it will only need an hour or less drying time. Basically, prolonged heating of corn is not a good thing, as precious enzymes and other growth factors are lost in the process. This is why we are constantly hoping for that exceptional corn year where we can get the final harvest moisture of our corn in the field down to 19 or 20 percent.

Shutting Down and Unloading the Dryer

Once you have achieved your desired final moisture level in the drying process, it is time to shut the gas off. Simply walk over to the big propane tank, squat down, and close the valve that is located underneath. The gas burner in the dryer will burn for several more minutes as the last of the propane in the system is utilized. You have created a much safer environment around the dryer by purging all the volatile gas from the system. This is a fun moment on a cold clear October or December night. You can stand in front of the dryer (with ear protection against the howling noise of fan and tractor engine) and watch the blue flame of the big burner gradually flicker away to nothing. At this point, the dried grain needs to cool down to the ambient outdoor temperature, but first let the dryer continue to circulate and the fan run for at least another half hour to forty-five minutes, as you definitely don't want to put hot grain on top of cold stuff in the bin. It's now time to unload. Most round batch dryers are top unloading: The circulating corn that is being augered up the center tube of the dryer is intercepted and directed down an unloading shoot into a waiting truck or gravity wagon that is parked alongside, which is accomplished by turning the unloading shoot ninety degrees from the back to the side of the dryer. A specially placed unloading hole at the top of the center auger is exposed, letting the dry corn make its way down the unloading shoot by gravity.

Unloading is actually a quick and easy process; we can unload ten tons of corn from our 575-bushel dryer in ten minutes or less. Flowing grain always seems to provide me with lots of viewing pleasure. We usually unload our dried corn into a truck body on an old 1954 army truck. I love to perch myself up there twelve feet off the ground and guide the unloading chute back and forth to evenly distribute the flowing corn into the truck body. When the load gets near the top, I have to jump in and shovel grain around a bit to keep it from running over the sideboards. If the load is extra large and there isn't enough room in the truck, I have to quickly climb down a ladder and shut off the tractor PTO to stop the unloading process. My shoes and pant cuffs are full of shell corn at this point. This whole process began back in May and now it's November—what a great feeling this is after a half year's worth of work and anticipation.

Processing Corn for Animal Consumption

More corn is fed to livestock in North America than any other grain. Corn and its by-products—including gluten feed, hominy, and distiller's grains—dominate the makeup of the grain rations fed to chickens, hogs, and cattle in modern commercial agriculture. I must admit that I had a bit of an anti-corn bias during the first decade or so of this farming adventure that I have been pursuing for the past thirty-five years. However, this all changed when I began buying in some organic corn from New York State in 1986. Once I began combining ground shell corn and a small amount of roasted soybeans with my homegrown barley and oats, my dairy cows began doing much better. Milk production climbed, and overall herd health improved. My personal experience demonstrates why corn is such a good livestock feed when it is mixed with other common grains. Aside from being the highest yielder, corn provides large amounts of starch and concentrated energy to livestock.

Indeed, no other domesticated plant can convert as much of the sun's energy into grain as corn. The speed at which the energy in the starch portion of feed grains is metabolized by livestock is a very important factor in animal nutrition. This solubility of energy varies from grain to grain. When consumed by ruminants (dairy and beef cattle), oats, wheat, and barley release carbohydrate energy immediately and completely into the first of the four stomachs. There is very little energy left for utilization farther down the line in the other three stomachs and the small intestine. Both protein and energy are highly soluble in the common cereals. This seems a little hard to believe, because grains like barley and wheat are much harder-textured than corn; you would think that grains with hard starch would stick around awhile in the rumen. However, just the opposite is true. Corn, with its soft-textured endosperm, has loads of "bypass" energy. When included in a dairy ration, corn is still gradually releasing small amounts of energy in the third and fourth stomach as well as the small intestine of a cow. This explains why corn is such a great grain to combine with barley when formulating and feeding a grain ration to a dairy cow. Digestion is more balanced: The barley's energy is consumed early in the process while the corn's energy is being utilized more evenly throughout its entire duration. It took me many years of reading articles and talking to feed experts to find out this information. In the short term, I realized quite quickly that feeding corn along with my oats, barley, and wheat was a major benefit to the production, health, and well-being of our dairy herd.

Combining Corn with Other Grains for Livestock Rations

Whereas corn is king of starch, carbohydrates, and energy, it is relatively low in protein. Most modern hybrids are lucky to break the 9 percent level when it comes to protein, and protein levels of 8 to 8.5 percent are standard in the modern high-yielding cultivars. Wheat will usually test close to 14 percent protein, and barley generally comes in at 13 percent or better. This fact of life simply gives us one more reason to combine corn with other grains when formulating a ration. (Specific ration formulation will be discussed in a later chapter.) If the protein requirements of the grain mix are high, additional high-protein concentrates must be added to corn. A small amount of ground roasted soybeans or common soybean oil meal with protein levels over 40 percent will balance quite nicely with ground shell corn. Baby chicks, turkey poults, and growing calves require significantly higher levels of protein, and the percentage of corn in a ration can be increased once the finishing stage of growth is reached. Often pigs and beef cattle are finished and fattened on corn alone. Ear corn is ideal for this sort of situation. It can be ground for beef cattle or fed directly on the cob to pigs.

There are many different schools of thought when it comes to the actual grinding of corn for livestock feed. When I first began using hammermills and grinder-mixers in the early 1980s, coarsely ground corn was the norm. We always changed to a larger five-eighth-inch-hole screen in the hammer mill when we switched over from barley to ear corn. There were lots of small chunks of cob in the cow feed, and these bits of cob were always left behind in the manger because the animals didn't really care for them. As time progressed, I began to experiment with the grind, and I started using the same small-hole screen for corn as I did for barley. The ration became finer, and the cows ate it better. At the time, I was concerned that too finely ground corn would "paste up" the rumens of our cows, but this didn't seem to be the case. However, a bit of caution is in order at this point in the discussion. When dealing with ruminants like dairy and beef cows, practice moderation when it comes to feeding corn and grain. The rumen of a cow is basically a large fermentation vat filled with microbes that help digest forages—primarily grass and legumes. If corn is overfed, the rumen can become acid and sour. All of a sudden the bugs down there are impeded by the lower pH. This condition is called acidosis and is quite commonplace when lots of grain is fed to dairy animals. Acidosis makes cows sick—they go off feed and can get lame. This can be easily avoided by concentrating on a high-forage, low-grain diet for your dairy cows.

Pigs and chickens are monogastric, and a different story. Most chicken rations contain at least half corn. The chicken's crop, located in the neck region, provides a secondary feed-grinding mechanism to the species. This is why chickens love scratch feed of cracked corn and wheat; the little stones in there roll around with the grain to grind it for better utilization by the chicken. Pigs aren't very fussy about how they get their corn, either. Many people like to mix ground corn into warm water or milk to create a mash or "slop" for their pigs. Once a hog is on its feet and growing, whole ear corn is a great supplemental feed. This feed regimen is particularly practical for the low-tech homesteader sort of individual who wants to raise small amounts of ear corn by hand without all the petroleum and harvesting machinery upon which the rest of us are so dependent. If you've got pigs, you don't have to wait until your corn has made black layer to feed it. As the ears are ripening throughout the month of September, you can handpick enough ear corn every day to feed your pigs. This softer immature corn is easy to chew and will fatten up your animals cheaply and easily.

Processing Corn for Human Consumption

It generally takes several months for fall-cribbed corn to drop to a moisture level where it can be ground into livestock feeds that will not spoil in the bag or the bin. Cob corn provides many advantages to the beef or dairy producer. The general practice is to grind the whole ear, cob and all, in a hammermill. The cob is just as valuable as the shelled corn because it contains very digestible fiber and beneficial pectate energy to cattle. Ground ear corn is considered by many animal nutritionists to be equal in value to ground shelled corn on a pound-for-pound basis. If you are using ear corn in your rations, you won't need as many acres of corn because you get almost double the total yield when the cobs come along with the shelled corn. The inclusion of the cob in the ration lessens the possibility of acidosis or sour stomach disease in cattle that comes along with high levels of corn grain consumption. Fiber in the cob acts as a rumen buffer, which helps to keep your cattle on track and healthy.

Even more rewarding than feeding homegrown corn to your animals is providing corn to feed yourself, your family, and your local community. Corn for human consumption must be of the most excellent quality. More so than any of the commonly consumed grains, corn lends itself ideally to hand-scale production. A gardener or homesteader can cultivate a tiny plot of human-consumption grain corn in a backyard with hand tools. An eighth- or quarter-acre corn patch can easily provide a family with a year's worth of cornmeal and all of the concurrent goodies made with it. It's also much easier to protect a plot of this size from raccoons with electric wire. When planning a homegrown cornmeal patch, carefully consider available varieties and what your ultimate use for the corn will be. If soft-textured corn breads are your favorite thing, a flour variety like Mandan Bride might work best. If you're more partial to polenta and steamed recipes like johnnycake, a flint variety like Roy's Calais, Garland, or Longfellow might be a better choice. Open-pollinated varieties are generally better suited for human consumption than the modern hybrids. Modern corn is bred for bushels and bins, not for flavor and taste, and if you're concerned about traces of GMO contamination in your food, it's best to steer clear of these modern hybrid corn varieties. (Genetically modified corn is so all-pervasive these days, however, that even the conventional non-GMO hybrid varieties may inadvertently contain GMO germplasm.) Test weight is also very important when you're setting out to grow corn for meal. Denser kernels with bushel weights in excess of fifty-six pounds will make for the highest-quality flours and meals, because there is more "meat" to start with when grinding. Early harvest is certainly much easier to accomplish on a garden plot scale, and test weight and grain quality of timely harvested field corn is best preserved by leaving the kernels on the cob and using very low temperatures for drying the ears. Drying temperatures should not exceed ninety to one hundred degrees. Ripe ears of corn can even be laid out on greenhouse plant tables for sun-drying or dried in a makeshift bin with forced lightly heated air. Once kernel moisture has dropped below 15 percent, the corn can be shelled with a hand-cranked or motor-driven corn sheller. The extra care taken by harvesting ear corn and slowly drying it on the cob represents the difference between ordinarily processed commodity corn and the artisanal fare we want to consume ourselves or sell to others for premium prices.

Storing Corn for Human Use

Once you've shelled your precious corn, it's time to finish the process and store it for use or sale throughout the upcoming year. Cleaning is the first order of business. The first step in the process is to run the shell corn through a fanning mill or an air screen cleaner. Consult your manual for the proper screen sizes. Corn usually takes a larger number thirty screen for the top and a medium-sized number fourteen or fifteen for the bottom. Corn is heavy, so you can use the maximum amount of air blast to blow off as much of the red fuzzy material—called red dog or glume—as possible. Corn that comes through a sheller is always full of this connective tissue that joins kernels to the cob. The ideal end product after corn has been through the cleaner is one free of chaff, smaller kernels, and cracked corn. If you are still not satisfied with your grain sample, pass it over the gravity table to separate out kernels with lower density. Depending on the amount of your final harvest, you can choose to store it in bags in a rodent-proof vessel like a commercial shipping container or in bulk in an aerated bin. As with any stored harvest, it's important to monitor your human-consumption corn often for problems during storage. Hopefully, if small amounts of corn are being processed for people food every week or two, changes in the stored grain will be noticed and dealt with promptly. If you double or triple the amount of care you take for your special crop, everything will be just fine.

Corn Products for Good Eating

The potential for processing homegrown corn into products is many-fold. Cornmeal is the first thing to come to mind. Remember that most of the commercial cornmeal produced by larger industrial mills is degermed to extend shelf life: The germ of the kernel contains the oil fraction, and rancidity is avoided by removing the germ from the kernel before grinding. Specialty processors will then press the germ for corn oil, but this would be difficult to do on a farmstead scale. Therefore, you grind the corn up germ and all. You'll still get a month or two of shelf life, which is all you need—you are not big industry. You can keep grinding fresh cornmeal, refrigerating it if you have to. A table model hand-cranked Corona-style mill will suffice for household use. These units are generally constructed with steel grinding buhrs. If you need to scale up to a slightly higher output of cornmeal, a Meadows vertical stone mill will grind corn into meal quite nicely. Coarseness and fineness of grind is controlled by adjusting the distance between the two stones. Hank Duncan, the itinerant Meadows Mill expert from North Carolina, has given me lots of advice about grinding corn in a Meadows stone mill, and recommends having a mill dedicated just to corn. Corn and wheat grind very differently; corn is softer and much larger in size. It is also much more abrasive and will wear your stones more quickly than wheat. After ten years or more of switching back and forth between wheat and corn in my Meadows Mill, I finally bought a second mill to remedy this problem. Once you're set up and properly adjusted, grinding corn into meal is a breeze, and I have found that people generally want a coarse meal as an alternative to the fine powder available at the store. You can't beat fresh-ground cornmeal that still contains all the germ and oil for taste and health.

Good corn recipes can be found anywhere people like to cook and bake. Methods of corn preparation will vary from region to region. Grits and hominy grits are made with softer-textured dent corn and seem to dominate in the South. Flint corn has been popular in New England since early settlers began making johnnycake in the eighteenth century. The rising popularity of Mexican food and the increasing Hispanic influence in our culture has promoted the use of masa harina, a corn flour, for tortilla making. Hominy grits and masa harina are prepared in a very similar fashion. The outer skin of the corn is removed by soaking kernels in an extremely alkaline solution of water mixed with either baking soda or wood ashes. The hard outer skin of the corn kernel is then separated from the soft inner corn flour. If grits are being made, the wet corn flour is dried and reground for cereal. A good bowl of southern grits is light and fluffy with a texture similar to that of cream of wheat. Masa can either be dried for later use or immediately pressed into tortillas, which are traditionally cooked on a hot stone griddle. This sort of Mexican-style corn flour can be made at home using your own corn. A flour variety of corn will work best, but dents and flints will also work. The process is long and drawn out, but doable on a home scale. The creative possibilities for baking and cooking with your own homegrown corn are endless; combine your own will and creative genius with the raw material and some guidance from others. Then get ready to eat some really good local food.

Corn Breeding and Seed Saving

The beauty of open-pollinated corn is that you can produce your own seed from what you grow. This is a basic underutilized right given to all of us who grow gardens and crops. Seed saving takes knowledge and resolve combined with a small amount of skill and some hard work. If you want to save your own corn seed, start by finding an open-pollinated variety that you like and that matures well in your region. This can be a tall order—there just isn't a whole lot of OP germplasm around to choose from. It took me more than a decade of experimenting and trialing corn varieties before I so luckily stumbled on Early Riser, but this process of searching for the right variety can be simplified and sped up by involving yourself in OP farmer networks. Frank Kutka is quite accessible either at SARE at North Dakota State University in Dickinson or through the Northern Plains Sustainable Agriculture Society. His *Corn Culture* journal is available online as well. Walter Goldstein at the Michael Fields Institute in East Troy, Wisconsin, is also a very helpful source of knowledge and experience in the world of open-pollinated corn. Joining the Seed Savers Exchange in Decorah, Iowa, and perusing their catalog will also provide you with hundreds of contacts and potential varieties. There is some help available from extension here in New England as well as from the Northern Grain Growers Association. OP corn people stick together and help each other out, as we are such a minority in the larger agricultural community of commercial hybrid seed corn users that we have to band together to further the cause of seed independence in an increasingly corporate-dominated world.

Let's assume that you've found an OP corn that performs reasonably well for you. It might have a few problems, like less-than-ideal standability or reduced yield, but it does ripen and dry down where you live. At least you have a place to begin, and you can select for improved traits. Saving corn for seed differs very little from producing corn for human consumption. Quality is paramount.

Shelling Corn for Seed

Once your corn is dried (as I've outlined previously in this chapter), the next step after drying the ears is shelling. Once again, you need to take the utmost care to not damage the corn. High-speed, high-powered shellers like a New Idea or Haban rotary model are much too rough for this job. The old-fashioned antique hand-cranked one- or two-hole shellers from the days of yesteryear work best, and the big interior cast-iron shelling wheel with its little nubs seems to do the gentlest job rubbing the kernels of seed corn from the cob. We have modified our old corn sheller with an electric motor instead of a hand crank. We set it up in the shop right in front of the homemade dryer full of ear corn. One person up in the dryer drops corn down an old galvanized silo filling pipe into a plastic fifty-five-gallon drum that stands right in front of the sheller. A second person on the floor below inspects each ear before tossing it into the slowly whirring corn sheller. Some ears don't make the grade because of defects like small size or immature grain. This is a slow peaceful process complete with the very distinctive sound of corn kernels rattling off the cob as it spins through the cast-iron inner works of the corn sheller. Cobs are ejected from the other end of the sheller while the shelled corn drops out the bottom into a box on the floor. The operator periodically hand-scoops corn from the box below into grain bags to await further cleaning. If you've got any significant volume of seed corn to shell, be prepared to invest many days of your time performing this noble chore. It takes us well over a week to shell out our dryer full of corn. When the job is finished, we have transformed a volume of ear corn that measures six feet in diameter by six feet in height to about five thousand pounds of good seed corn. The corn cobs are bagged up and given to a local freezer locker for meat smoking. It's always a relief when that last ear of corn goes through the sheller.

Cleaning Your Seed

The seed corn must now be cleaned just as it would if it were being used for human consumption. The first cleaning step will be the fanning mill. You might change the bottom screen to one with holes a bit larger to clean out extra shriveled and shrunken kernels, and a pass over the gravity table will also help to make sure you'll be planting only the very densest kernels for seed. At this point the seed is pretty well cleaned, and it's time to put the finishing touches on everything. Seed can now be graded and sized for use in a plate planter, and there are a number of simple seed graders out there that will sort corn kernels into small, medium, and large rounds and flats. Then, bags are marked and filled with corn seed, and everything can be put away in dry shipping container storage to await spring planting—only five or six months away. The process of growing a crop of corn is now complete.

Testing the Grain

You can perform a simple test for germination by putting a hundred representative seeds between two wet paper towels. A certified seed laboratory, however, can provide more and better information about your seed. They can also do a PCR test to detect the presence of the most minuscule amount of GMO germplasm in your seed. The PCR test costs well over two hundred dollars, but it is well worth it for the peace of mind it provides when you know that Monsanto and Dupont haven't invaded your crop. Seed labs can do cold and warm germ testing of corn seed to simulate the cool soil temperatures of early spring or the warm soil of late May, which is far more accurate than counting the number of sprouts between the two paper towels up on the warm shelf. There is also something called a sand test where seeds are germinated in a sterile environment of beach sand. This sort of test will tell you if there are any diseases or fungus present in the seed. I always hope for a germination rate of 90 percent or better; the best I've ever done is 98 or 99 percent. A poor growing year with low heat units and an early frost might only produce seed with 80 or 85 percent germ. I either plant low-germ corn at a higher population or hope that I saved some of the good stuff from the year before. I believe it is always a good idea to keep a stock of older high-quality seed around as an insurance policy against the vagaries of nature.

Corn Breeding

Once you've mastered the art of seed saving, you can move on to corn breeding and variety selection. It's actually wise to improve your corn by selecting it for better genetic traits like higher yield, better standability, and earlier maturity. Farmers were the first corn breeders long before the introduction of hybrids and university breeding programs—they simply walked their fields and chose the best ears from the best plants for seed. You can do the same sort of improvement work with a little help and guidance from those who know more about this. I have taken much inspiration from Frank Kutka and his teacher, Cornell plant breeder Margaret Smith. The biggest take-home lesson from these professional plant breeders is to understand the difference between phenotype and genotype and put this knowledge to work when you select ears of corn to save for seed. *Phenotype* refers to a plant's response to its growing environment. A lone plant grown at the end of a row in full sun with plenty of organic fertility might produce a tremendous-looking ear, but this characteristic will not necessarily transmit to the next generation. *Genotype*, on the other hand, refers to a plant's ability to pass desirable genetic traits on to subsequent generations. A healthy strong-stalked plant that yields moderately well and withstands the pressures of wind, high population, and European corn borer will be a much better candidate for selection than the plant grown under ideal conditions. This phenotype/genotype consideration explains why it is better to walk the field choosing seed ears from superior plants than it is to pick out the biggest, girthiest ears from the grain wagon. You need to actually get to know your plants out in the field as they are growing and maturing—not all plants are created equal. This is especially true when growing open-pollinated corn with lots more genetic diversity than the hybrids. You need to look beyond ear size to the actual plant itself. Good husk coverage will prevent bird damage. A thin shank below the ear will permit the corn to drop down and shed water during the fall rains. Last but not least is the general strength of the stalk during the ripening and dry-down period. Plant breeders like to put more pressure on plants than we do as farmers.

When Margaret Smith came to our farm for a corn-breeding field day several years ago, she walked through the field kicking stalks to test their strength. This was something I was afraid to do because I wanted every ear of corn to be harvested and make its way to the grain bin. But breeders aren't farmers and farmers aren't full-time breeders, and we will all have to find that happy medium between the two jobs of harvesting a crop and choosing plants and ears for seed selection.

In the short seven years that I have been growing Early Riser corn, I have been able to improve the standability of the variety considerably. I made the most progress on this front the second year I grew it. A heavy, wet snowstorm came along on about the tenth of October. There was a wicked wind as well, and it knocked over a very high percentage of my Early Riser. So I walked through the field with a grain bag and picked ears only from plants that were still standing. To my delight, the corn was much stronger the next season. The same principles hold true for other traits in addition to standability. Rigorous selection performed right in the field is the best way to go. It's difficult to muster up the same devotion as a professional plant breeder when you are trying to make a living at farming at the same time, but it is my hope that the few remaining university plant breeders will continue to work closely with the organic community to provide corn varieties that work well for us. Participatory plant breeding is the wave of the future: We can maintain our independence from the corporate agricultural sector by producing our own seed whenever possible. If we can add elementary plant breeding and selection to our roster, we will be a lot closer to returning to Thomas Jefferson's yeoman farmer upon whom the backbone of America was developed.

CHAPTER TEN
Wheat and Its Relatives

If corn is king of the grains here in North America, then wheat is queen. And while wheat ranks third behind rice and corn for total volume of grain produced, more acreage is devoted to growing wheat worldwide than any other grain crop. Wheat produces more usable protein for human consumption than corn and rice, and it also has the distinct advantage of being able to thrive in more challenging climactic environments. Corn and rice both require wetter, more temperate growing conditions, whereas wheat does quite well in more arid high-elevation regions with lots of cold wind. Surprisingly enough, this staff-of-life grain performs quite well here in northern New England. We have plenty of cool damp weather to help seeds get up and growing in the very early spring; in addition, we generally have a hot dry spell in August as harvesttime approaches. I found this out by accident in 1977 when I grew my first wheat crop here on our farm. We really surprised ourselves with the high-quality and good baking qualities of the wheat that we produced. Three and a half decades and many crop years later, I can still say with confidence and actual experience that wheat is well worth growing in the Northeast. Spring wheats seem to do the best in the cooler, shorter growing seasons of the mountains where I live, whereas winter wheat, which is sown in mid-September and overwintered, is the crop of choice for warmer, lower-elevation regions like the Champlain Valley and southern New England. A long succession of successes and failures over the years has taught me a few things about growing this "queen of grains," and you will certainly have to make many of your own mistakes as you learn to grow this staff of life here in our region.

Wheat: A Brief History

Much like maize, wheat descended from wild grasses, changing and evolving over many millennia. However, whereas maize originated here in North America, early wheat can be traced to the region of Southwest Asia that encompasses modern-day Syria, Jordan, and southeastern Turkey. Archaeological specimens of wild emmer wheat dating back to 9600 BC have been found at Iraq ed-Dubb in northern Jordan, and wild einkorn seems to have been growing in the Karacadag Mountains of southeastern Turkey during the same period. Carbon dating of early wheat from Abu Hereya on the Euphrates River in what is now Syria even traces these ancient grains back to 7800 BC. These predecessors of modern wheat were developed as the hunter-gatherers of the time cultivated and repeatedly sowed the grains of wild grasses. The spikelets on the heads of these primitive grasses contained the small seeds that were to become the staff of life for mankind. Genetic progress occurred as mutant strains with larger kernels were selected and replanted. In fact, the seed-containing spikelets of these ancient grasses did not adhere well to the *rachis* or head stem of these plants, which meant that a lot of this early grain fell on the ground instead of into the sack of some post-Neolithic farmer. This supposed disadvantage was actually an accidental boon to the development of modern wheat because only plants that had better seed retention were replanted, and seed shattering soon became much less of a problem in the early development of wheat. In his book *Guns, Germs, and Steel*, Jared Diamond claims that wheat was partially responsible for the rise of the urban city-state because it could be mass-produced and stored as a food source that would last the entire year. Wheat culture made its way to Greece, Cyprus, and India by 6000 BC. Shortly thereafter, the crop came to Egypt, and the Egyptians are credited with the development of ovens and bread baking. Germany and Spain followed suit by 5000 BC. Finally, by 3000 BC, wheat had become a staple in England and Scandinavia. The wheat that we take for granted in these modern times has been nearly ten thousand years in the making.

Einkorn

The ancient relatives of wheat emmer and einkorn once fed the entire population of the Near East. These two grains began as wild grasses and were eventually domesticated. The major difference between these ancient grains and wheat is that they remain encased in the glume or hull after being threshed; further pounding or processing is required to remove the grain kernel from the spikelet. Varieties of einkorn predate emmer by several thousand years; einkorn is and was especially well suited to cool growing environments and poorer soils with marginal fertility. Each spikelet contains one kernel and is topped off by a wispy little beard called an awn. The crop proliferated all the way from the Mideast to southwestern Europe. It is nutritionally superior to modern hard red wheats: Levels of protein, crude fat, potassium, and beta-carotene are significantly higher. Modern commercial cultivation of einkorn is nonexistent, but there are several accessions of this ancient grain still available to plant breeders and other interested parties in these modern times, although it's worth nothing that einkorn is much better suited to the semi-arid climate of North Dakota and Montana than the wetter, more humid weather of the Northeast where it tends to lodge easily and drop its kernels. This grain took a backseat to emmer thousands of years ago, although Eli Ragosa of the Northeast Heritage Grain Conservancy in Colrain, Massachusetts, has worked long and hard to bring back einkorn culture from the verge of extinction.

Emmer

Emmer seems to have a whole lot more potential as an edible and easily cultivated grain than its predecessor einkorn, which was as much the case three thousand years ago as it is today. Emmer is also predominantly awned with long, narrow spikelets. Each individual spikelet contains two kernels, which gives emmer a distinct yield advantage over einkorn. As post-Neolithic civilizations developed and advanced their agriculture, emmer soon became the grain of choice, although both of these early grains were consumed as porridge before bread baking was developed. But unlike einkorn, emmer has seen its popularity persist over millennia.

It also became a staple in and around the Volga River delta in the Crimean region north of the Black Sea in Eastern Europe. Welcomed by Catherine the Great in the eighteenth century, large numbers of ethnic Germans fled persecution in their native land to settle and farm in this breadbasket region of Russia. This same group of "Germans from Russia" immigrated to North America en masse between 1870 and 1917. They chose the high plains of North Dakota, Manitoba, and Saskatchewan because they wanted to live and farm in an environment similar to the land they'd come from. Entire areas of North Dakota became peppered with German settlements. These settlers from the black soil grain belt of Eastern Europe brought seeds of their own grain varieties with them to plant and propagate in their new homeland. Along with the numerous land races of wheat came the seeds of many different strains of emmer, which did exceptionally well as a feed grain for livestock under the harsh and unforgiving growing conditions of the northern plains. Historical narrative has it that when crops of oats or barley failed because of drought, hail, or early frost, emmer could always be counted on to provide nourishment to farm livestock. Emmer did so well in this region that the USDA began to promote its culture by importing and releasing even more strains from Ukraine; a 1906 USDA publication encouraged farmers to cultivate this crop for livestock feed. Meanwhile, back in the Mediterranean region, emmer also found a home as a human staple on the Italian peninsula. Known also as farro, this ancient grain was dehulled and used whole in all sorts of dishes. Farro also makes incredibly good pasta and has long been used for that purpose by the Italians. Emmer's greatest asset is that no one has tampered with its genetics for the past ten thousand years. And while this ancient grain still contains gluten, for some unknown reason many celiacs and others with allergies and intolerances to wheat can consume and digest emmer grain. We are so fortunate that emmer has persisted over the centuries and has not joined the ranks of extinct plants. Recently, emmer has been rediscovered as a staple here in North America.

Modern Emmer in North Dakota

North Dakota has been the center of emmer culture for the past one hundred years, and Steve Zwinger, a farmer and researcher at the North Dakota State University Carrington Experiment Station, deserves the most credit for bringing this ancient grain back into modern-day agriculture. Steve has been increasing seed as well as planting and studying plots of emmer for the past ten years, and has been my go-to person for innovative grain research for quite some time now. Steve claims that emmer really shines in a difficult year when moisture for crop growth is in short supply. He has been able to lay his hands on many of the old land race varieties of emmer that came from older German farmers who saved their own seeds over the years. McIntosh County in south-central North Dakota has been a great source for seed stock because it was at the center of emmer production until the 1960s. Many towns in this part of the state were German speaking until a few decades ago, and Steve tells me that on-farm seed saving of emmer was quite popular right up until this time; every old German farmer who cultivated the crop to feed his beef cows had his own particular land race selected and developed for his farm.

The 1960s were a watershed time between an older, more traditional farming culture and the newer high-yield model of agriculture that is today's norm. As a result, modern high-yielding varieties of wheat and barley supplanted emmer culture on the high plains. But Steve Zwinger has managed to ferret out what was left of this old grain. In many cases he has had to act like a detective and an archaeologist, searching old farm granaries for leftover emmer seed that hasn't been planted for several decades. Gil Stolmac, a retired plant breeder from Montana State University, has also contributed to the advancement and proliferation of emmer as a viable grain in this region. Over a twenty-five-year period, he has selected and crossed many of the old farmers' varieties to come up with a modern emmer called Lucille. Steve Zwinger's emmer project has distributed emmer stock seed to a number of North Dakota organic grain producers for trial and commercial viability analysis. My friend Blaine Schmaltz of Rugby, North Dakota, has been the largest emmer producer, with over one hundred acres in cultivation for several years in a row. Blaine reports that the crop has done quite well for him; yields have been around twelve hundred pounds to the acre on marginal, poor-fertility land. However, if the crop is seeded on better land as part of a good rotation, yields rise to twenty-eight hundred to three thousand pounds to the acre, which is comparable with wheat and barley. The important difference and advantage to emmer is that it is a scavenger crop and it requires few if any inputs. Although Blaine is sold on emmer, he does express some reservations about growing it. For one thing, he has several hundred tons of the stuff sitting in a grain silo waiting for a home, as established commercial outlets for the grain are nonexistent. The fact that emmer needs to be dehulled for human consumption is a major sticking point; this specialized infrastructure does not yet exist even in an important grain-producing state like North Dakota.

Growing Emmer in the Northeast

Emmer offers us many possibilities and opportunities here in the Northeast. First and foremost, we are in a region where it is in demand; Italian farro and emmer pasta have become quite popular in specialty markets. Second, emmer grows well when planted on fertile soils in our climate. In 2007, Steve Zwinger sent me twenty pounds of emmer seed to plant as part of an heirloom wheat trial. I seeded a small plot of the grain with my grain drill during the first week of May. The emmer had so much vigor that it almost literally jumped out of the ground five days after seeding. The young emmer plants tillered profusely, providing many additional shoots for each seed planted, and the resulting rapid establishment of a canopy and complete ground cover gave this crop superiority over competing weeds. Like other heirloom grains, the emmer grew quite tall throughout the season that year. Lodging was a bit of a problem—wind and rain from July thunderstorms pushed down the stalks of this ancient grain—but a timely harvest with the combine in early August rescued the emmer before it went totally flat. When all was said and done, the twenty pounds of emmer seed planted on less than one-fifth of an acre yielded a bit more than 150 pounds of grain. This provided enough seed to plant more than an acre of emmer the next season, and at the end of the 2008 crop year I had increased my supply of emmer to fifteen hundred pounds. We were able to dehull some of this emmer for use as entrées at several specialty-food localvore dinners. Elizabeth Dyck, a research scientist and grains specialist from New York State, has also become quite interested in emmer as a potential grain crop for this part of the country. As part of her Organic Growers Research and Information Network (OGRIN), she has collaborated extensively with Steve Zwinger to test and trial emmer at numerous locations throughout the Northeast. Troy Oechsner, a farmer from the Finger Lakes region of New York, has become a major grower and market supplier of emmer. Most of the crop has gone to chefs in New York City supplied by Cayuga Pure Organics of Brooktondale, New York.

Agronomic Considerations for Growing Wheat

Wheat is a heavy feeder, pure and simple. Excellent soil fertility is required to grow a wheat crop that yields well and has adequate protein levels; soil nitrogen must be plentiful to ensure success when growing the queen of grains. Fertility recommendations from most seed companies as well as land grant universities equate specific amounts of soil nitrate nitrogen (NO_3) with a certain number of bushels of wheat expected per acre and desired levels of protein. Fifty-bushel wheat (about a ton and a half to the acre) needs seventy-five pounds of N, and according to the experts high-protein wheat requires even more nitrogen inputs—twenty to thirty pounds of additional N is required for each point of protein over the 12 percent level. As you might imagine, growers of conventional wheat dump tons and tons of synthetic nitrogen on their crops in the form of anhydrous ammonia, urea, and ammonium nitrate. All of these human-made substances are prohibited in organic production, so we must find another way to feed our wheat crops.

Medium-red clover, the prince of all cover crops. PHOTO COURTESY OF SID BOSWORTH

Nitrogen-fixing nodules attached to a plant's root system. PHOTO COURTESY OF SID BOSWORTH

You'll need a whole lot more than a bag of nitrogen fertilizer to grow a great crop of wheat. Organic farming concentrates on feeding the soil first. If the earth is teeming with microbial life and diverse biology, nitrogen and other essential minerals will be gradually released to the plants by microbial activity. Thirty pounds of nitrate nitrogen will be mineralized and released for every 1 percent of organic matter in a particular soil. Average soil organic matter levels run about 1.5 percent in North America. In an organic system, we'd like to see these levels in excess of 3 and 4 percent. A fall plow-down of a nice leguminous hay field with plenty of alfalfa and clover will go a long way toward supplying a successive crop of wheat with essential levels of nitrogen. A look back at chapter 2 on soil fertility is highly recommended at this point, as a balanced soil with the proper levels of calcium, magnesium, phosphorous, sulfur, and boron will provide ideal conditions for soil biology to feed a wheat crop. Proper soil aggregation (also known as tilth) goes hand in hand with diverse biology and adequate mineral levels; a friable soil with plenty of pore space will drain better and contain plenty of air, which the microbial life needs to do its magic. Test your soil well in advance of planting a crop of wheat. A complete soil test that measures cation exchange capacity, base saturation, as well as the major and minor elements will provide you with all of the information necessary to fine-tune your little piece of earth to grow the highest-quality wheat possible.

Soil Fertility for Wheat Production

Now that we know that an abundant crop of high quality wheat requires excellent soil fertility, let's figure out how to make it happen. First, you will need at least seventy pounds of available nitrogen to grow the crop. There are numerous ways to approach this situation. Certainly the best-case scenario would be to supply the nitrogen needs of the wheat directly from the soil in which it is growing. A healthy soil with organic matter in excess of 5 percent coupled with a plow-down of alfalfa or sweet clover will grow an incredible crop of wheat. But many of us who are still working on building up the native fertility of our farms don't have the luxury of a perfect soil. We must apply compost, manure, and fertilizer to feed the soil and the wheat, but it's important to know our options and the potential results of the fertility input choices we make. Raw manure should be avoided at all costs because of its high salt index and its propensity to encourage the growth of weeds. If you are fall-plowing for wheat the next spring, you might get away with a light application of some liquid manure on the land before plowing—the fact that the manure will be plowed down with the sod and the six-month delay between manure application and spring planting will give the applied slurry time to mellow out and break down in the soil. Well-rotted manure is a much better choice. There are numerous piles of abandoned manure next to old barns all over New England, and in many cases these "manure mines" have been sitting there for years. They are usually grassed over and difficult to see, but once they are broken open with an excavator or bucket loader, these treasures will yield tons and tons of "black gold" that's every bit as good as compost. In many cases an old "shit pile" can be purchased for a couple of hundred dollars. The biggest expense will be hiring someone with heavy machinery to move this carbon sink back to your fields.

This material can be liberally applied to your wheat field just prior to planting. It doesn't matter if you are planting winter wheat in September or spring wheat in April; lay the stuff on thick and work it in with a harrow or field cultivator. There is no danger of burning the crop with excessive nitrogen from this material. Purchased or farm-made compost can be used in much the same fashion. But since well-made compost is precious and expensive, you might want to cut back on the application rate just a bit to make sure that you can evenly cover the entire area to be planted. Remember that composted manure will usually shrink by at least two-thirds. Good compost is dense and packed with biologically active microbes as well as slowly released nutrients. Whereas ten to fifteen tons to the acre of well-rotted manure might be a good application rate, three to five tons of excellent-quality compost will have the same beneficial effect on a newly seeded piece of wheat.

There are numerous ways to apply these organic materials to your fields. The common manure spreader is probably the best tool for the job because most of us either own one or have access to one from a neighbor. You'll need to perform some sort of calibration for the manure spreader to ensure that you are applying the desired amount of manure or compost to the field. This is easily accomplished on a trial run with the loaded manure spreader. Simply spread the material over a measured sheet of plastic out in the wheat field. Weigh the plastic along with the applied material on top of it. Once you know the weight of the compost or manure and the square footage of the plastic, you can easily figure out your application rate per acre. (An acre has 43,560 square feet, so all you need to do mathematically is determine the proportion of your plastic sheet to that of an acre.) If you don't want to fool around with all this weighing, measuring, and figuring, you can wing it. Five tons to the acre is an extremely light coat of material. For light spreading, you will need to set your manure spreader on the slowest apron speed possible and drive like hell. The high ground speed will help you stretch out your pile over more square feet of field.

Other Fertilizers to Consider

Organic inputs like well rotted manure and compost are not always plentiful and available in quantities large enough to liberally cover a new wheat seeding. If this is the case, you have a choice of several commercially available more concentrated options for applying nitrogen to a wheat crop. Composted chicken manure is now widely available and has become quite popular over the last few years. Many large commercial poultry operations are processing hen and broiler litter with forced air to make a fertilizer product with a 4–3–2 analysis (4 percent nitrogen, 3 percent phosphorous, and 2 percent potassium). This material is pelletized to permit it to evenly flow through planters and other fertilizer distribution equipment. One thousand pounds to the acre of Cheep Cheep or any other brand of pelletized chicken compost will provide a wheat crop with forty pounds of slowly released N. When this amount of nitrogen is added to what is supplied by the decomposition of soil organic matter, the wheat crop will usually have its seventy-pound nitrogen requirement met. At over three hundred dollars per ton, I have found this commercial product to be a convenient but fairly expensive fertility source for wheat. The pelletized nature of this material is quite advantageous because it can be blended with other commercial amendments and evenly applied to the land at fairly minimal rates. I simply call up my fertilizer dealer and he brings me a bulk spreader loaded with four tons of the stuff. Sometimes I buy a blend of chicken compost and other organic fertilizers like gypsum and sul-po-mag.

There are several caveats to be considered when using commercial pelletized chicken compost, however. Most of this product comes from industrial-scale layer and broiler houses located on the Delmarva Peninsula. These operations are owned by people and corporations like Frank Perdue, and the chickens on these giant factory farms are consuming a diet of chemically produced, genetically modified corn and soybeans. If you don't want the Roundup Ready gene in your wheat field, be aware that it will come to you via this product. You might also wish to consider the inherent lack of carbonaceous material in this product. These large chicken farms use very little bedding, so the carbon-to-nitrogen ratio of the manure is not ideal, which is one of the reasons that conventional chicken manure has such an odor of ammonia. The tendency of commercial chicken manure is to overheat way past the 160-degree level considered acceptable for the production of good stable compost. To counter the overheating factor, the commercial chicken manure has vast amounts of forced air blown through it. This keeps the temperatures low and preserves the ammonia nitrogen from volatilization. The end result of this process is a pretty hot little fertilizer that we can call compost. Cheep Cheep and Perdue pelletized chicken compost still have the distinct nitrogenous odor of chicken manure, but at the same time, these products can supply a hungry wheat crop with nitrogen. You have to make your choices here. Your own farm- or garden-produced compost is certainly more benign and best, but sometimes there just isn't enough of this precious resource to go around.

Other organic nitrogen sources do exist. Just about any protein source will mineralize and release nitrogen to a crop. Human hair swept from the floor of a barbershop or a beauty salon would be a good source of slow-release nitrogen for a garden plot of wheat. Feather meal is 13 percent nitrogen, and like hair it's very slowly broken down and released to plants. Oilseed cake, which is a by-product of oil extraction from high-protein seeds, is a very good source of easily mineralized nitrogen for a wheat crop. Until relatively recently, castor bean pomace (7 percent N) from castor oil pressing was a very popular nitrogen source on golf courses and in the turf industry. Soybean oil meal can be a very practical nitrogen source because of its widespread availability and moderate cost. This animal feed additive stacks up with a fertilizer analysis of 7–2–1. Peanut meal (8–1–2), which I'll discuss in a moment, and cottonseed meal (6–2–1) have both been used as organic fertilizers for higher-value vegetable crops. You can also use animal by-products like fish meal and blood meal to fertilize a wheat crop, but these substances are much more concentrated and expensive. Blood meal is quite potent, with a nitrogen level between 12 and 13 percent, and is broken down quickly by soil microorganisms and speedily released to young wheat plants. Fish meal analyzes as a 10–2–2 fertilizer and is almost as readily broken down as blood meal, but not all fish meal is approved for organic use because it contains chemical antioxidant additives to protect its oil fraction from rancidity. Crab meal has about half the nitrogen of fish meal at 5 percent, but since crabs are processed with the shell, this waste material contains chitin, which has its benefits and drawbacks as a soil amendment. Chitin preys on soil nematodes, so you might want to think twice before applying crab meal—it destroys good as well as bad nematodes.

There are numerous other sources of relatively expensive organic fertilizers available to us. Peanut meal (8–1–2), a by-product of the peanut oil industry, has become quite popular with vegetable growers here in the Northeast. North Country Organics in Bradford, Vermont, has been able to procure large quantities of this material and distribute it to growers for a reasonable price. Paul Sachs, the owner of North Country Organics, has very cleverly blended many of these plant and animal materials into a packaged starter fertilizer called Pro Gro (5–4–3), which will help you build better soil while it feeds your crop of wheat. All of the previously described materials are pricey and should only be considered as a last resort for fertility supplementation. These oil seed meals and animal by-products must be processed and transported many miles before they can be applied to your field or patch of wheat. High-priced nitrogen applications might fit the bill for a small garden plot of wheat, but they're really not cost-effective for larger field-scale crops. Farm-produced manures combined and composted with local carbon sources like straw and wood products are ultimately the best fertility source when you scale up.

Chilean Nitrate

Inorganic nitrogen fertilization, urea, is the standard in modern nonorganic conventional cash crop farming. As I have said many times before in this narrative, organic farmers don't do it this way in an organic system for a number of reasons, one being that we want to feed the crop from the soil, not a fertilizer bag. There is, however, one exception to this rule. Organic certification allows the use of a soluble substance called Chilean nitrate to supply 30 percent of a crop's nitrogen requirements. Chilean nitrate, chemically known as sodium nitrate or nitrate of soda, is a naturally mined salt that comes from the Atacama Desert in northern Chile, and contains 16 percent readily available nitrogen. With its characteristic little round white pellets, Chilean nitrate looks exactly like its manufactured cousin, urea, and has been called "organic urea" by many of its detractors. A little dab of Chilean nitrate will definitely elicit an amazing response in a wheat crop; a 150-pound application will provide the wheat with twenty-four extra pounds of N, which is about a third the total requirements for a fifty-bushel harvest. Most of us choose to live without Chilean nitrate in our cropping systems, because at twelve hundred dollars per ton, it is cost-prohibitive. Also, because sodium nitrate is so salty and soluble, the USDA National Organic Program is considering removing it from its list of approved fertilizing materials. It is, however, one of the tools in your toolbox that can get you out of a jam quickly. If you've had a cold wet spring with lots of soil leaching and subsequent denitrification, that little bit of added Chilean nitrate could make the difference between the success and failure of a field of wheat.

The timing of nitrogen applications to a wheat crop is critical. Split applications seem to make the best use of this expensive practice of supplemental fertilization. Agronomist Heather Darby of the University of Vermont has studied different approaches to nitrogen fertilization in her wheat trial plots. She has achieved the best results in terms of yield and protein levels by applying half the nitrogen at planting time and the other half at the tillering stage. Winter wheat gets its partial dose at planting time in mid-September, while spring wheat fertilization takes place the following April. Heather has found this to be a real yield booster. The second application when plants are developing tiller shoots will help to boost the overall protein of a wheat crop. UVM crops and soils technician Susan Monahan studied nitrogen fertilization of wheat extensively when she was doing her master's study at the university; she found that the quality of the inherent soil fertility of each particular wheat field determined how much benefit could be expected from supplemental nitrogen fertilization. Wheat grown on lower-fertility sandy soils might demonstrate a yield increase from a nitrogen application, but not a protein increase. Naturally fertile fields with plenty of organic matter and balanced minerals did experience increases in wheat protein with the addition of more nitrogen. Before we beat your wheat to death, suffice it to say wheat needs more nitrogen to produce a respectable crop than any of the other common cereals. It's best to derive as much of this mineral as possible from a balanced, healthy soil. It goes without saying that the finest soil will produce large amounts of high-quality wheat.

Planting Your Wheat

We discussed tillage and planting earlier in this book, and planting wheat is really no different from planting any other cereal grain. Grain drills and seeding strategies are also discussed in chapter 4, but I will review some of the finer points as they apply to establishing a good stand of wheat.

Planting Spring Wheat
Fall plowing is essential for success with any spring-planted cereal—this is especially true for wheat—and seeds should be planted as early as you can get machinery onto the land in the spring. Almost all aspects of soil preparation are best performed the

previous autumn. Some of our Canadian neighbors to the north have taken to no-till seeding wheat into frozen ground as early as March. They get the field ready to plant by doing secondary tillage and fertilization the preceding November. Once planted, wheat seeds lie dormant in the thawing earth until they germinate when soil temperatures climb above forty degrees. This type of no-till seeding requires a specialized grain drill seeder that is extra heavy and equipped with special cutting disc openers that will penetrate the frozen earth of late winter/early spring. But you needn't be that compulsive about the seeding process; if you get out there with your disc harrow and field cultivator to prepare the field in the first or second week of April, you will be just fine. (It's also well worth spending a couple of extra days to liberally apply as much compost as you can muster to your future wheat field.) You'll want to shoot for seeding your wheat at 150 pounds to the acre by the fifteenth of April, because you want that wheat up and growing as early in the growing season as possible. A well-established early-planted stand of spring wheat has the best chance of beating the weeds and giving you the best crop possible.

Planting Winter Wheat

Winter wheat is usually planted in our region sometime between the fifteenth and twenty-fifth of September. Good seedbed preparation and adequate soil fertility are equally important for a fall-planted crop, and if you begin working the ground several weeks earlier in late August, the soil will have enough time to mellow as you progress through all of the various tillage phases from plowing and harrowing to readying the final seedbed. September is also a great month for applying compost and other mineral inputs like rock dusts and mineral powders—the earth begins to draw in as the days shorten and the regular growing season winds down. Timing is important when seeding winter wheat, as winter survival of a fall-planted wheat crop is essential. An early-planted winter wheat crop, seeded in late August, might put on too much growth. This can actually weaken the plants and make it difficult to establish true winter dormancy under a heavy layer of snow. The same holds true for a crop of winter wheat seeded in mid- to late October; The young wheat plants could have difficulty surviving the onslaught of winter if they haven't put on enough growth. Ideally, you would want to see two- to four-inch seedlings of winter wheat headed into the depths of winter. Contrary to what you might think, a little crust of frozen ground around the young wheat shoots is just what you want as the new wheat seeding heads into late fall and early winter. If a heavy blanket of snow covers the ground before it freezes, the winter wheat will enter its dormancy period in cold but not frozen ground, which sets the wheat up for the development of snow mold. Snow mold likes to grow under a heavy layer of snow in temperatures between thirty-three and forty degrees, and consumes the tender green shoots of winter wheat seedlings. When the snow finally melts off the fields in late winter, you want to see nice green rows of winter wheat seedlings ready to emerge from dormancy. If the young plants are white in color and crushed down to the ground, there is a pretty good chance that snow mold has damaged the crop. Sometimes, however, the snow mold doesn't reach the crown of the wheat plant. If there is evidence of a little green spot at soil level, the crop will probably survive, but it will be stunted because it will be slow to rally in the spring.

If winter wheat will survive where you live, it's a great choice for many reasons. It will already be green and growing when the snow melts in the spring, and the fact that young winter wheat shoots are already three or more inches tall in early April gives this crop a distinct advantage over spring-planted wheat. Weed competition is virtually eliminated because winter wheat is quick to establish a canopy with its tremendous early growth. Harvest will come two to three weeks earlier than spring wheat. All of these factors will usually add up to larger yields—in many cases they are double those of spring wheat. Fifty-bushel yields (three thousand pounds to the acre) are not unusual here in the Northeast.

The Importance of High-Quality Seed

The importance of procuring and planting high-quality seed cannot be overstated. You only get one chance to put a crop in the ground, so you want to make the best of it. At the very least, verify that the wheat you are planting has a good germination rate. This information should appear on the seed tag, which is normally sewn to the top of the bag. Ninety percent or better will guarantee a well-established stand. Seed vitality is also very important. Plump red kernels will supply young wheat seedlings with a whole lot more carbohydrates than shriveled and shrunken seeds. Good seed should almost glow with well-being. Buy your seed from a reputable source. Certified seed is probably the best choice because it has been raised, inspected, and tested under rigorously controlled conditions. Farm-produced seed is your other option. At the very least, a germination test will tell you whether or not you have quality seed. If time and money permit, send your seed away to a seed lab for analysis. (Germination testing procedure is described in chapter 4.)

Types of Wheat

Hard Red and Hard White Spring Wheat

Hard red spring wheat has been called "the aristocrat of wheat." It has the highest protein content of any of the common wheats and thrives in cool growing seasons. Hard red spring is wheat extraordinaire. Protein levels range from 13 to 16 percent. It is known for its superior gluten strength and wonderful baking quality. Often, flour ground from hard red spring wheat is blended with lower-protein fare to improve water absorption and the handling and mixing qualities of bread dough. The North American center of hard red spring wheat culture is found in the Canadian prairie provinces and in a four-state area that encompasses Montana, North Dakota, Minnesota, and South Dakota. Surprisingly enough, this crop also thrives in the cooler mountainous regions of northern New England, but it might not be such a good choice for warmer areas with heavier clay soils that prevent early planting. I have had excellent success growing it here on my farm in the Green Mountains of Vermont's Northeast Kingdom. In a year with ideal growing conditions, we have been able to achieve wheat protein levels in excess of 15 percent. Many times, my early-April-planted hard red spring wheat crops will experience tremendous adversity after the seed goes in the ground—newly sprouted wheat generally has to deal with spring frosts and a snowstorm or two, but it is never worse for wear. By the first week of May the narrow rows of little green wheat shoots have gained in size and stature to become a thick and bushy field. Early planting is the key to success when cultivating this crop, and if all goes well and the remainder of the growing season cooperates, you will be harvesting bushels and bushels of golden glowing red wheat berries in the sun of early August. More than any other type of wheat, the kernels of hard red spring have that shining vitreous characteristic that so embodies quality and a job well done by grower and Mother Nature alike.

Hard white wheat spring wheat has recently become an alternative choice for natural foods consumers who mill whole grains at home. The only difference between hard white wheat and hard red spring is that the endosperm of the white variety is surrounded with a lighter-colored layer of bran. Supposedly, whole-grain flour ground from white wheat is much lighter in color and less bitter in flavor than its red wheat cousin. Companies like Wheat Montana market "Prairie Gold" brand white wheat flour to people who want milder-tasting whole wheat flour. The first commercial varieties of hard white spring wheat were released to farmers in North Dakota and Montana in 1990. Since then, this wheat has become a popular ingredient in the

Recently emerged cereal grain drilled in seven-inch rows. PHOTO COURTESY OF SID BOSWORTH

Asian noodle market. I planted some white Prairie Gold wheat berries several years ago to see how they would fare in northern New England. The wheat grew well, but it barely got above my knees in height. It seemed more like a dwarf variety that might be better suited to the drier climate of the northern high plains. Because of its short stature, the hard white spring wheat that I trialed did not compete very well with weeds. Personally, I prefer the stronger wheat flavor of hard red spring wheat. It's just fine with me if my loaf of bread has a brown hue instead of a white look about it. The jury is still out on the suitability of hard white spring wheat for our damper New England climate. We will certainly have a lot more information about this relatively new type of wheat once some serious replicated trials have been conducted by agronomists and researchers at some of our state agricultural institutions.

Hard Red Winter Wheat

Hard red winter wheat is grown on more acres in the United States than any other type of wheat. This fall-planted crop is primarily grown in the central and southern regions of the Great Plains, especially Kansas, Nebraska, Oklahoma, and the Texas Panhandle. Most of this wheat finds its way to Kansas City, where it is marketed for bread flour. The famous heirloom Turkey Red was the leading wheat variety on the southern plains from 1870 until the 1920s, when it was replaced by more modern higher-yielding varieties. Much like emmer and some of the old North Dakota varieties, Turkey Red was brought to this country from the Black Sea Crimean region by Germans fleeing persecution in tsarist Russia. In recent decades, the culture of hard red winter wheat has moved far beyond the southern plains: Wheat breeders have selected and developed varieties that will thrive in more northerly climates like North Dakota, Ontario, and Vermont. These wheats seem more similar to European winter varieties than to those hailing from Kansas. In Vermont's Champlain Valley, yields of hard red winter wheat can easily top three thousand pounds to the acre. Yields may be high, but for some unknown reason, the protein of these winter wheats rarely exceeds 13 percent. Nevertheless, some bakers prefer winter wheat flour to that made from hard red spring wheat. It's definitely a matter of taste and what you're accustomed to.

In Vermont, hard red winter wheat is more commonly cultivated on the west side of the Green Mountains where the elevations are lower and the climate is more benign. Winterkill and snow mold infections make it difficult for me to reliably cultivate winter wheat where I live in my high-elevation location. However, if I choose a sheltered valley field with well-drained soil and a tree belt for protection from the west wind, I can raise a respectable crop of hard red winter wheat. Winter wheat is probably the best choice for a beginning grain grower who lives in a Zone 4 or warmer climate. (I live in Zone 3.) Annual weeds like mustard and lamb's-quarters are not usually a problem in winter wheat because they will not survive the winter in a fall-planted crop. The wheat is sown in the Northeast during the third and fourth weeks of September, and emerges from the snow in March, strong, green, and ready to grow. Winter wheat is normally ready for harvest in July, about a month earlier than its spring-sown counterpart.

Soft White Wheat

Soft white wheat was the mainstay of winter wheat production in the Corn Belt and mid-Atlantic states for many years, and the central and western New York area is still well known for the quality of the soft white wheat produced there. Kernels from this type of wheat are distinguishable by their light creamy brown color. This is definitely not red bread wheat; soft wheat has much lower protein and gluten levels than the hard varieties. For this reason, it is primarily used to make pastry flour used in unleavened baked goods like cakes, cookies, and crackers. There is nothing more divine than a piecrust made with whole wheat pastry flour; the silky soft wheat starch and butter go hand in hand to create a genuinely pleasing experience for the taste buds.

Soft white winter wheat is planted in late September in the same manner as any other fall-planted cereal. Whereas hard winter varieties can be difficult to locate, there are numerous soft wheat varieties available to us from New York State. Frederick and Ticonderoga have been among the most popular varieties for the last ten years. I have been able to quite successfully raise soft white winter wheat by choosing a Canadian variety named Borden, which seems to thrive in my harsher climate. Borden was developed on Prince Edward Island and has been quite popular in the other Maritime Provinces of New Brunswick and Nova Scotia. I'm not sure if it's because this type of wheat is lower in protein and higher in starch, but soft wheat is usually a much higher-yielding crop than its hard-textured relative—soft wheat yields in excess of four thousand pounds to the acre are not uncommon. There is also ample straw to accompany the bushels of grain harvested from this type of wheat.

Soft white wheat culture does pose a few potential problems for the organic grain grower. For one thing, marketplace demand for pastry flour is only a fraction of that for bread flour. If you plan to grind and sell flour for a local market, you might have more pastry flour than you can sell unless you find a commercial outlet. An even more serious drawback is that soft white wheat is rather disease-prone. The fact that these wheat kernels have a very soft-textured outer layer and endosperm makes them very susceptible to fusarium head blight, a rather nasty cereal grain disease that will be discussed in more detail later in this chapter. Suffice it to say, once airborne fusarium mold spores enter a pollinating head of wheat, they are there to stay and wreak havoc with your wheat crop a month later at harvesttime. Fusarium infection leaves behind the vomatoxin DON (deoxynivalenol), which will make wheat unfit for human consumption if found in levels above one part per million (ppm). The silver lining in all of this is that due to its high starch and low protein levels, this type of wheat doubles as a good source of livestock feed. Cattle can consume soft wheat with DON levels up to ten ppm. So all is not lost in this situation, as you can still feed the grain to your stock if you cannot grind it into pastry flour. Usually, extended wet conditions at the time of pollination are the forebears of fusarium problems in a wheat crop. As with any type of crop farming, you have to remember that you're gambling. When things work out, you're overjoyed and thankful. When you've got a mold problem in your wheat, you have to make the best of it.

Soft Red Winter Wheat

On the larger scale of modern commercial agriculture, soft red winter wheat has supplanted most of the soft white varieties. A quick glance through farm seed catalogs from large commercial suppliers will reveal that most of the soft winter pastry wheat variety choices are now red. Wheat breeders have developed these soft red varieties to be even higher yielding and more disease-resistant. But as always seems to be the case, soft red winter wheat lacks the character and flavor of white wheat. Modern high-input/high-output farming is all about trade-offs. Commercial agriculture has exchanged sublime flavor and taste for extra-high yield and some disease resistance. End users like Nabisco and Keebler and seed companies like Pioneer have teamed up to create special varieties of soft red winter wheat for baking applications in their crackers and cookies. If you are a red wheat grower in western New York or eastern Ohio, you might even be required to plant a particular Pioneer red wheat variety if you want to sell your grain to one of these large multinational baking companies. This is not something that we as organic growers will ever do, but it is a good thing to be aware of where the larger commercial scene is heading. If you want to grow large quantities of pastry wheat for the commodity marketplace, choose a numbered red variety from a multinational seed company. If you believe in local production of pastry wheat with character and flavor, take your chances with soft white winter wheat.

Soft White Spring Wheat

One of the most recent developments in plant breeding is soft white spring-planted wheat. We have the Canadians to thank for this development. I had seen some spring-planted soft wheat growing in the Beauce region of Québec about twenty years ago, but didn't hear of it again until a couple of years ago. Semican, the large cereal seed supplier from Plessisville, Québec, began offering spring pastry wheat in their catalog that year. The variety was called Kaffé and it seemed to have a lot of potential. I planted a three-acre field of the stuff in mid-April 2010, and it grew vigorously into a thick stand of beautiful tall wheat. Yields were in the ton-and-a-half (fifty-bushel) range. The only drawback was that this soft spring wheat was just as prone to fusarium infection as its soft white winter wheat counterparts. The jury is still out on this subject, but I'm sure that the Canadian breeders are working on more fusarium-resistant varieties of this sort of wheat.

Durum Wheat

Last but not least is durum wheat, often known as macaroni wheat. Durum is in a class all by itself, as it is distinctively different from other members of the wheat genus. First and foremost, it has the genetic makeup of a tetraploid—which means that it has four sets of chromosomes for passing along traits to the next generation. (Hard wheats are hexaploids with six chromosomes.) Durum is a direct descendant of the ancient grain emmer. It is always awned or bearded, and kernels are large and have a distinct yellow endosperm, which gives pasta its unusual color. Its total grain protein is higher than that of any other wheat while desirable gluten is low. For this reason, durum is ideal for Middle Eastern flat breads and pasta. Durum's extra-high levels of protein give the kernels a unique translucency, which also manifests itself in the appearance of the pasta. Unfortunately, growing durum is out of the question here in our region. This plant thrives in arid conditions. Even in North Dakota where more than 75 percent of the nation's durum is grown, durum culture is only possible in the western half of the state where rainfall is considerably less. I have tried planting this crop several times and have failed miserably with each attempt. Heather Darby of UVM Extension has trialed numerous varieties of durum in her plots and has come to the same conclusion. Durum wheat doesn't like a whole lot of water.

Wheat Diseases

Entire agronomy textbooks are written about crops and their diseases. We'll have to spend a little time on this subject because disease in our wheat is something you cannot simply ignore. Twenty or thirty years ago, no one really worried much about these sorts of afflictions. You sowed seeds and harvested a crop ninety or one hundred days later, as wheat was the new kid in town all those years ago and its afflictions were minimal. But times and weather patterns have changed over the past couple of decades, and now we do get diseases in our wheat. It's important that you have the ability to identify them and at least a rudimentary understanding of their causes and symptoms. Grain growing in our New England climate has its challenges—this business is definitely not for the faint of heart.

Fusarium Head Blight

Fusarium head blight, also known as wheat scab, has more impact on wheat crops in New England than any of the other cereal diseases. We have touched on it briefly earlier in this narrative, but let's look at it in more detail. Fusarium (*Gibberella zeae*) is a very common mold that is present just about everywhere in the growing environment. This reddish mold attaches itself to old crop residues like last year's cornstalks, and it thrives and multiplies in damp and humid conditions. The florets of wheat flowers are particularly susceptible to the invasion of airborne fusarium spores at the time of pollination

Wheat head shedding pollen. PHOTO COURTESY OF SID BOSWORTH

TABLE 10-1

Diseases That May Impact Organic Wheat Production in the Northeastern United States

COMMON NAME OF DISEASE	CASUAL ORGANISM	DISSEMINATION	FAVORED BY	CONTROL MEASURES FOR ORGANIC PRODUCTION AND THEIR EFFECTIVENESS (1 = HIGH TO 3 = SLIGHT)
Eyespot foot rot	*Tapesia yallundae* (*Ramulispora herpotrichoides*)	Rain-splashed spores from wheat debris	Moderate temperatures, rainy during tillering	2+ years between wheat crops (1) Delayed fall planting (2)
Fusarium head blight (scab)	*Gibberella zeae* (*Fusarium graminearum*)	Airborne spores (regional) from corn and cereal debris	Moisture at crop flowering	Plant moderately resistant varieties (2) Stagger planting dates (3) Follow a noncereal crop (3)
Leaf rust	*Puccinia recondita* f. sp. *tritici*	Airborne spores (long distance)	Warm, humid; thunderstorms in June	Plant resistant varieties (1–2) Plant at earliest recommended date (2)
Loose smut	*Ustilago tritici*	In seed (embryo)	Noncertified, bin-run seed	Plant certified smut-free seed (1) OMRI-approved seed treatments (2)
Powdery mildew	*Blumeria* (*Erysiphe*) *graminis* f. sp. *tritici*	Airborne spores (regional)	Humid; moderate temperatures	Resistant varieties (2)
Seed decay/ seedling blight	Various fungi, oomycetes, and bacteria	In soil; seed	Cool and moist or very dry soils; poor-quality seed	Plant certified high-vigor seed (2) Partially resistant varieties
Septoris tritici blotch	*Mycosphaerella graminicola* (*Septoria tritici*)	Wheat debris; windborne	Splashing rain; extended leaf wetness	Avoid wheat after wheat (2) Partially resistant varieties (2–3)
Stagonospora nodorum blotch	*Phaeosphaeria nodorum* (*Stagonospora nodorum* = *Septoria nodorum*)	In seed; wheat debris; windborne	Splashing rain	Avoid wheat after wheat (2) Partially resistant varieties (2–3) OMRI-approved seed treatments (?)
Stinking smut (common bunt)	*Tilletia tritici* (*caries*), *Tilletia laevis* (*foetida*)	With seed (external); in soil	Noncertified, bin-run seed	Plant certified smut-free seed (1) OMRI-approved seed treatments (?)
Tan spot	*Pyrenophora tritici-repentis*	In seed; wheat debris; windborne spores	Humid; moderate temperatures; rainfall	Partially resistant varieties (2) Avoid wheat after wheat (2)
Wheat spindle streak mosaic	Wheat spindle streak mosaic bymovirus	By a soilborne protozoan	Cool spring temperatures (affects only winter wheat)	Plant resistant winter wheat varieties (1–2)
Yellow dwarf	Barley yellow dwarf luteovirus and cereal yellow dwarf polerovirus strains	By aphids (short and long distance)	Early-fall planting or later-spring planting; large aphid populations	Plant after Hessian-fly-free date in fall or at earliest recommended date in spring (2)

Compiled by Gary C. Bergstrom, Department of Plant Pathology and Plant-Microbe Biology, Cornell University, Updated February 2010

Bleached head of wheat indicating the presence of fusarium. PHOTO COURTESY OF SID BOSWORTH

or anthesis, and this is especially the case if you get a rainy damp spell of weather at the anthesis stage of a wheat crop. Your eyes and nose will tell you when your wheat is pollinating. A week or ten days after grain heads first appear, you will begin to smell the fruity scent of wheat pollen grains being released, and a closer inspection of the individual wheat heads will reveal little bits of male pollen attached to the florets—you'll want to hope and pray for dry weather conditions at this time.

It's relatively easy to tell if you've had a fusarium invasion in your wheat crop. As the crop ripens and progresses from the milk stage into soft dough, look for heads of wheat that are prematurely bleached light yellow in among the remaining green heads. These heads of wheat that appear to be ripe way too early are infected with the fusarium fungus. A small number of prematurely bleached wheat heads probably won't be a problem, but if you see a sea of them out there in your wheat field, there might be cause for concern. As stated earlier, the fusarium in and of itself isn't really the problem—it's the DON mycotoxin produced by this mold that is poisonous and toxic to humans and livestock. The presence of mycotoxins in excess of one part per million in a sample of wheat makes it unfit for human consumption. A little blotch of pink or red on the tips of harvested wheat kernels is a sure sign of fusarium, so if there is any doubt in your mind about fusarium in your harvested wheat, send a sample away for a DON test. The good news is that we can now send our samples to the University of Vermont Grain Testing Laboratory. The little red kernels of fusarium-infected wheat are generally lighter and bit more shrunken than the uninfected kernels, and are generally known as "tombstone" kernels because they have the appearance of a little red gravestone.

The silver lining in this dark cloud of fusarium and DON business is that these lighter kernels can often be separated from the good wheat with density sorting equipment like a gravity table. If your wheat has fusarium problems, pass it over the gravity table before sending it out for the mycotoxin test. You might be pleasantly surprised to find DON levels below one ppm. Another little-known fact about the presence of fusarium in wheat is that sometimes the mold can be present, but mycotoxis will not be produced. Either way, mycotoxin testing is highly recommended for any crop that is intended for human consumption. The last thing you want to do is make your local grain customers sick.

Loose Smut

Fusarium is an airborne disease problem, which means it isn't carried on the seed like so many of the other wheat diseases, such as loose smut. This particular wheat affliction (*Ustilago tritici* and *U. avenae*) is bad news. It all begins with seed whose embryonic tissues are infected by invisible dormant mycelium. Once planted, the wheat won't show any signs of problems until the heads emerge from the boot. Loose smut development is especially favored by warm, humid, and wet growing conditions. The mycelium develop along with the growing point of the plant. At flowering time, masses of black spores replace the floral part of the wheat spike; in a matter of days, the entire seed spike of a wheat plant can be consumed by these black spores. This is scary stuff. Once you've had an outbreak of loose smut, your soil, seed, and growing environment are polluted with these spores. The common practice in conventional agriculture has been to treat seeds with fungicides like captan and Formalin. However, when I began crossing the border into Québec to buy cereal grain seed over thirty years ago, all of the seed was treated with pink fungicides. When I asked for seed that was *non traité* (untreated), people looked at me as if I was crazy. But untreated seed is much more readily available in today's world where organic agriculture is playing an increasingly important part each year. This issue of loose smut and its potential to harm organically produced cereal crops drives home the point that top-quality clean seed is of supreme importance. Don't buy wheat seed out of someone else's bin without the assurance that it's free of this terrible disease. Purchase your wheat seed far enough in advance of the planting season to allow for testing at a reputable seed laboratory, and whenever possible, buy certified seed.

There are some organic options for dealing with seed infected with loose smut. For one, hot-water treatment prior to planting has been shown to kill the mycelium on the seed, but be cautious: Prolonged exposure to hot water can damage the germ and vitality of wheat seed. Matt Williams, an organic wheat grower and miller from Aroostook County, Maine, has told me that special hard-to-get varieties of organic grain can be saved from extinction by turning the seed over to a conventional neighbor for one year to grow it out with fungicide treatment. But let's hope that we can be vigilant enough so this doesn't have to happen. I've seen the consequences of planting loose-smut-contaminated seed in several of Heather Darby's wheat trial plots in Alburgh, Vermont. Heather planted a certified organic wheat variety from a seed house in Minnesota. Once the wheat began to flower, the black spores and fungal bodies of loose smut annihilated the entire crop in a matter of days. As we grow more and more wheat in our region, accidents like this are increasingly possible. Make sure to buy excellent-quality seed and make every effort to ensure that it is free of loose smut and other seedborne diseases. Knowledge of the perils and pitfalls of growing wheat is even more important than the actual physical experience of growing a crop.

Common Bunt and Common Take-All

There are numerous other seedborne diseases that are relatively inconsequential in our part of the country. The list is long and detailed and can be found along with colored pictures in any agronomy textbook. The fact that wheat is such a minor crop in the Northeast works in our favor. Vigilance is still the best policy, because lots of the seed we plant is imported from regions where it is the dominant crop. Common bunt (*Tilletia tritici*), also known as stinking smut, is a major problem in other wheat-growing areas. I saw the devastation left behind by common bunt on a recent trip to Denmark, where Anders Borgen, a renowned scientist and cereal breeder, showed us plots that had been totally consumed by this nasty affliction. Anders was working on developing resistant varieties, and had also had very good success with hot-water seed treatments.

There are also seed- and soilborne diseases that stunt plants much earlier in the growing cycle. Common take-all (*Gaeumannomyces graminis* var. *tritici*) causes the stems and roots of young wheat plants to rot and perish. Basal stem and leaf tissues as well as the roots turn black in the process. There are numerous other wheat diseases in this category of seed- and soilborne maladies. The majority of these problems are very rare in New England, where wheat is only sparsely grown. If you find abnormal symptoms in your wheat plants, it is best to call in the experts for advice. Every state has a number of trained agronomists in the public and private sector who can identify diseases and propose remedies for problems.

Rust

Leaf and stem diseases like rusts weren't much of a problem for wheat crops until the last few years. Now all sorts of rusts afflict the leaves of growing wheat plants, however. Rust outbreaks in wheat crops are caused by urediospores, which are carried in on the wind from distant locations. There is nothing like a mid-June thunderstorm to set up wheat plants for an invasion of wind-carried rust spores. Wet leaves and warm temperatures provide an ideal environment for this sort of situation. Once a urediospore lands on a leaf, it begins to grow and produce pustules that can eventually dominate a leaf or a plant tiller. The many different kinds of rust have different colors and shapes as they come to dominate the surface area of plant leaves. Among the most common is brown rust (*Puccinia recondita*), which shows up as circular or elliptical orange-brown spots on wheat leaves. Infection sites are primarily found on upper leaf surfaces and sheaths. Once this malady takes hold, a new crop of urediospores is produced every ten to fourteen days under ideal conditions. Yellow stripe rust (*P. striiformis*) is another very common form of rust. Once leaves are infected by the windblown urediospores, narrow yellow stripes form; these stripes will dominate the surface area of a leaf rather quickly.

Rust on wheat leaves inhibits photosynthesis, and if a plant can't photosynthesize, it can't make chlorophyll. Once plant growth shuts down or becomes severly limited, the wheat plant is forced to ripen prematurely. Yields and test weight are also reduced by rust. Early planting and choosing rust-resistant varieties are probably the two best stategies for combating wheat rust. The earlier in the season a wheat plant can get established and growing, the more time it will have to grow and photosynthesize before its growing efficiency is reduced by rust. I found this out the hard way in the 2011 growing season. We planted some heirloom wheat trials in early June because an extremely wet April and May would not allow us in the field any earlier. The wheat plots grew rapidly with great vigor until rust invaded in July. Within a matter of weeks, the wheat turned totally brown and pretty much shut down its growth cycle. Grain quality weight was severly impacted, which reduced yield and test weight; wheat that should have weighed sixty pounds to the bushel weighed a mere forty-two. Some of the wheat plots yielded even less seed than was actually planted. This was all because the wheat's growth cycle was shut down too early by the presence of rust on the leaves.

Certain wheat varieties are also much more prone to rust infection. My North Dakota friends tell me that rust is a very large problem out there on the high plains, as it blows in from Kansas and Oklaoma almost every summer. When I praise the spring wheat variety AC Barrie, which works so well for us here in this part of the country, they tell me that they had to give up on it many years ago because it was simply destoyed by rust. We have had several wet summers in a row over the past few years. Unfortunately, wheat rust has begun to be a problem here in the Northeast.

The Hessian Fly

Wheat seems to have very few insect enemies here in our region. History tells us, however, that the Hessian fly, also known as the barley midge, devastated wheat crops in New England between 1760 and 1840. This pesky little insect made its way from Asia to Europe and then on to North America in the straw bedding used by Hessian troops who fought for the British in the Revolutionary War. Hessian flies usually lay two broods of eggs in a season, one in the fall and one in the spring. The eggs are laid on the first nodal stems and the early leaves of young wheat plants. Once they've hatched, the maggots from these eggs suck the juices from the leaves and stems of the young wheat plants. If this activity doesn't directly kill wheat plants in the fall, it leaves them with weakened stems, causing premature lodging and preventing harvest the next season. Fly pupae overwinter in little "flaxseed" sacs to hatch more eggs and wreak havoc all over again the next spring. The Hessian fly seems to be more of a problem on winter wheat in areas south of my region—most of the literature on Hessian fly control comes from Pennsylvania, North Carolina, and Missouri. Agronomists in these regions recommend delaying winter wheat planting until well after the fall brood of the midge has hatched out, which will prevent the eggs from being laid on the young wheat plants in the first place. Planting winter wheat varieties that are resistant to the Hessian fly is also recommended. Who knows if we will have to deal with the scourge of this little insect here in our region, but there is no doubt that growing conditions have changed over the past three decades. We are certainly dealing with other wheat problems that we would never have imagined all those years ago. Vigilance is the best approach in these matters. You should at least know what the barley midge looks like and what sorts of damage it could inflict upon the wheat you grow.

Prevention and Micronutrient Balance

This discussion of the potential of disease to damage our wheat crops forces the question: Just what is disease anyway? Divide the word into its two syllables, *dis* and *ease*, and think about it for a moment. Folly seems to strike at plants that are "out of ease," and fungal spores, insects, and bacterial infections will find a welcoming environment for growth and proliferation on wheat plants that are "out of sorts." It's the same for us humans: Healthy people with excellent immune systems don't get sick as often. In the same way, healthy plants grown on mineralized and humified soils will repel the ravages of disease. So the next question that we need to ask ourselves is: What sorts of agricultural practices keep our wheat crops healthy? The basics of soil fertility and crop rotations have already been covered. However, it's the nuances and finer points of soil and ecosystem health that will give our crops protection against the ravages of disease.

We begin by balancing the macro elements of calcium, magnesium, potassium, and phosphorous, because shortages or excesses of any of these elements can spell trouble for a crop of wheat. From here we move on to the micro elements, which play a very important part in plant metabolism and immune response. Copper, zinc, and manganese are the most commonly supplemented trace elements in agricultural crop production, and they can all be added to fertilizer blends in sulfate form. Boron, a common ingredient in laundry detergent, can and should be a standard part of any cereal grain fertility regimen. This trace element from Death Valley, California, allows calcium to move other minerals around in the soil system. Gary Zimmer, an organic soil consultant, describes calcium as "the trucker of all minerals" and boron as "the steering wheel." There are numerous other trace elements on the periodic chart that are necessary for proper plant metabolic function. Minute amounts of cobalt, selenium, and molybdenum all contribute to healthy wheat plants that will have increased immunity to disease. I have been applying copper, zinc, manganese, and boron blended with gypsum and sul-po-mag to my forage crops for years, and the results have been phenomenal.

Liquid seaweed and sea mineral concentrates sprays are the best way to apply the minor micronutrients to grain crops, as every minor element required for crop well-being is present in the ocean. This area of micronutrient fertilization of wheat is brand new to me this year. I bought a crop sprayer and have begun to apply micronized minerals to cereal crops through foliar feeding. The principle behind micronization is that minerals and humic substances are broken up into extremely small fragments that can be measured in microns. Once dissolved in solution and sprayed onto plants, these beneficial nutritive factors enter the plant directly through the leaf stomata. Foliar feeding is not a substitute for macro fertilization of a wheat crop with compost and rock powders, however. It does allow you to fine-tune the system by feeding your wheat plants directly with micronutrients that regulate plant growth and bring health to the system. Add to this the beneficial effects and buffering action of the carbon in micronized humates, and you have the beginnings of an elixir that will bring balance and well-being to your wheat field. Please be advised that this is not a guarantee that plant disease will not strike your wheat crop, but rather a best-foot-forward approach to growing plants that are nutrient-dense and vital. As far as I am concerned, this is the direction organic agriculture needs to be heading—it's about more than not using chemicals and pesticides. If we nourish life on all levels—soil, plant, and end user—we will certainly have more immunity to the vagaries of disease.

Modern and Heirloom Wheat Breed Varieties

We don't exactly live in a popular wheat-growing region here in the Northeast, but we know that wheat will grow well here because it is cultivated in nearby regions like the Canadian Maritimes, Québec, Ontario, and New York State. So where can we get seed and what varieties should we grow?

Sourcing Your Seed

Wheat seed is usually a special-ordered item at most New England farm supply stores. There are thousands of wheat varieties, old and new, to choose from. Those of us who live close to the Canadian border are fortunate because we can drive into Québec or New Brunswick to find top-performing varieties that will thrive on our side of the border. Up until fifteen years ago, seed importation was a relatively easy matter—you just declared your seed purchase and drove away from customs. But seed importation has become increasingly complicated and bureaucratic over the last decade and a half. All seed coming into the United States from Canada eventually had to be inspected by APHIS (Agricultural Plant Health and Inspection Service), which is part of the USDA, and you had to make an appointment with the inspection officer and meet them at the border at a designated time. The USDA person might take a representative sample from your seed shipment and send it off to a government laboratory for further quality analysis. I always thought that this was American chauvinism at work, because the Canadian seed standards are much more rigorous than ours. Thankfully, the whole process has actually gotten a bit simpler and more straightforward over the years. If you want to buy seed in Canada, all you need to do is get a Seed Quality Certificate from the supplier, which is a standard US Customs and Border Protection (CBP) form number 940. The seed that you are planning to import will have to be sent to a certified seed laboratory for a standard analysis, which is generally not a problem because most of the larger seed companies like Semican or Prograin have in-house seed labs. If you are purchasing seed from an individual or a smaller outfit, start the process a month before you actually want to bring the seed over the border. This will allow time enough for the whole process to run smoothly. Having the form number 940 in your hand when you drive up to the customs booth is as good as having a US passport. You will be welcomed back into the United States with open arms and sent on your way to go plant your field of wheat with good clean "approved" seed.

Canadian wheat varieties are also available from many US suppliers, like Seedway and Bourdeau Brothers. You might pay a little more because the seed must pass through one more pair of hands before it reaches your farm, but if you can get the variety you want at the time when you need it, it is definitely worth paying the extra money for this service. Seed dealers are in the business of importing and distributing seed—let them worry about the importation, paperwork, and transportation. There are numerous other ways of finding wheat seed sources. Tap into the existing network of wheat growers in our region, as there are more and more people growing wheat every year, and you might be able to piggyback your seed order with that of another grower. The Northern Grain Growers Association here in Vermont is a good resource for organic wheat seed sources. There are also a number of grain farmers in the group with seed cleaners who sell small amounts of wheat seed to individuals. If you are not looking for large amounts of wheat seed, High Mowing Seeds in Vermont and Johnny's Selected Seeds in Maine offer a limited number of varieties at home gardener prices. Albert Lea Seedhouse in Albert Lea, Minnesota, publishes a rather extensive catalog with numerous organic wheat offerings every year. Seed prices from this old-time midwestern company are quite reasonable, but shipping costs can outweigh the savings. Albert Lea Seed is an especially good source of organic seed if you are buying it by the pallet. Indeed, there are numerous commercial sources of wheat seed available to us from all over the United States and Canada.

Choosing the Right Variety

Is any one variety of wheat better than another? Every once in a while, I stumble onto a variety that seems particularly exceptional, although it can be hard to know if the successful crop I just harvested was due to the discovery of a great new variety or perfect weather. I grew a variety called Neepawa when I planted my first crop of wheat in 1977, which wasn't an intentional variety choice on my part, but was what the NOFA bulk seed order offered that year. There is nothing more sublime than beginner's luck. Looking back, I think the Neepawa wheat that I grew thirty-five years ago was as good as anything I've grown since. This variety of wheat had been developed and released by Agriculture Canada's Winnipeg Experiment Station in 1969. I liked its baking quality and its name, which means "land of plenty" in Cree. But it didn't take me long to find out about the world of wheat breeding and the introduction of new and so-called better varieties. Neepawa was only on the market for a few more years before it was discontinued. By 1980, I was planting Sinton because that was all I was able to procure from my Québec seed dealer. After a few more years, the wheat of choice became Columbus. A little peek into the world of wheat breeding clearly illustrated to me how similar all of these wheat varieties were to one another, and wheat breeders were constantly trying to backcross one of these varieties to another. About ten years ago, I settled on the variety AC Barrie. This beautiful wheat seemed to produce high-quality bread flour for me year after year. Research into the lineage and history of AC Barrie turns up some really interesting and telling details. *AC* stands for Agriculture Canada. This wheat was released from the Swift Current Saskatchewan Station in 1996, and was bred by Dr. R. M. DePaula and is a three-way cross between the modern variety BW 90 and the old standards, Neepawa and Columbus. After three decades of watching scores of new and improved wheat varieties come and go, we still have a lot of the same germplasm we were cultivating thirty-plus years ago. This whole business makes me think of the world of fashion, where what goes around seems to come around again.

Wheat breeding is quite limited outside of the main wheat-growing areas here in North America. Most of the work being done with hard red spring wheat varieties takes place in the Dakotas and the prairie provinces of Canada. The majority of the soft winter wheat breeding work being done in the country happens in the Palouse region of eastern Washingon, while hard red winter wheat breeding is concentrated in the southern plains states of Kansas and Oklahoma. Here in the Northeast, we are quite fortunate to have Dr. Mark Sorrells working and teaching in the Department of Plant Breeding and Genetics at Cornell in Ithaca, New York. Dr. Sorrells recently released a new variety of soft white winter wheat called Jensen. This particular variety suits our wet and humid climate because it is resistant to pre-harvest sprouting and fusarium head blight. Laval University in Québec City and the University of Guelph in Ontario both have wheat-breeding programs that can supply us with germplasm and varieties that are well adapted to our region. If you see the prefix *OAC* in front of a named variety, this is an indication that it was developed at the Ontario Agricultural College at Guelph. In many cases, eastern wheat breeders start with genetic material from the prairies or the Palouse and acclimatize these varieties to our wetter climate through selection and back crossing. There is certainly no shortage of wheat varieties out there for us to choose from. A little research and networking with existing growers will find you the seed you need to plant. You will also find that many of the wheat breeders at these various institutions are quite accessible for consultation and advice; these folks actually want to talk with the farmers who are out there in the field trying their varieties.

Hard Red Spring Wheat Varieties

Despite concerns about sourcing regionally adapted seed, most of the hard red spring wheat available to us here in the Northeast comes from Minnesota, North Dakota, Manitoba, and Saskatchewan. These varieties perform quite well for us for a number of reasons. Although we receive an average of forty-five inches of rain per year compared with North Dakota's twenty-two, we do have a few things in common. Length of growing season is about the same in both areas. Our winters are cold and lengthy, and spring planting comes at just about the same time in April in both regions. The summers of 2010 and 2011 were extremely wet in North Dakota, yet high-quality wheat continued to grow there. So we can discount the argument that it needs to be semiarid to grow excellent wheat. Prairie varieties like AC Barrie, Glenn, and Neepawa have really excelled here, and I think that our best strategy for developing locally adapted varieties of hard red spring wheat for northeasrtern growing environments is to save and replant the seed of these prairie cultivars. Several of us here in Vermont have been doing some wheat-breeding work under the supervision of Dr. Steve Jones of Washington State University. We have crossed some of these modern wheats with older heirloom varieties. After four generations of seed saving, selection, and increases, we are beginning to see the fruits of our labor. Regionally adapted wheat might just become a reality in this part of the country. The one word of caution that I have when it comes to wheat variety selection is stay away from the very modern dwarf varieties that are coming out of prairie wheat-breeding programs. Dwarf wheat is bred extra short in stature so it can be pumped with chemical nitrogen and not lodge. This kind of wheat is not suited to an organic program because it has minimal roots that are only there to suck up large doses of fertilizer. These sorts of plants don't know anything about scavenging nutrients from soils with a natural nitrogen cycle. Stick with the old standard conventional varieties of hard red spring wheat and give it a go.

Hard Red Winter Wheat Varieties

Selecting hard red winter wheat varieties is a bit more compicated, as varieties from Kansas and other southern plains states are not winter-hardy enough to survive in our part of the country. Fortunately, the production areas for hard red winter wheat have been moving northward during the last several decades. Breeders have developed winter bread wheat cultivars that thrive in frigid places like the Dakotas and Canada, and these are the varieties we want to plant. University of Vermont Extension agronomist Heather Darby has been conducting winter wheat trials at two different Champlain Valley locations for the past several years. In 2010, Heather and her Northwest Soils and Crops Team grew out twenty-six different winter wheat varieties in Alburgh, Vermont, and Willsboro, New York. The results of this study were quite telling. Heather tested varieties from all over North America for flowering dates, straw height, lodging, yield, disease, harvest moisture, and baking quality. Wheat varieties from Washington State were the poorest performers; they flowered later, yielded less, and were much more prone to diseases such as fusarium head blight. Canadian hard red winter wheats like AC Morley, Maxine, Harvard, and Redeemer were some of the top performers in both yield and grain quality. These wheat varieties are all produced and distributed by C and M Seeds of Palmerston, Ontario. Some of the other noteworthy wheat varieties in Heather's trials were Arapahoe from the Albert Lea Seedhouse in Minnesota and Warthog from Québec's Semican. Warthog has been my personal favorite for the last couple of seasons. Please note that this information is all applicable to the time of this writing: 2010 through 2012.

The one thing that is constant in this search for good varieties to plant here in the Northeast is change. A quick web search for any of these seed houses will turn up lists of new varieties—wheat breeding is alive and well, with lots of new varieties being released every year. Trying new unproven varieties is recommended, but be careful not to put all of your eggs in one basket. Before planting your entire hard red winter wheat crop to one particular unknown cultivar, try a small amount first. Stick with what you know works well for you and other growers in your geographic region. If the new variety shows lots of promise, increase your acreage next season. Caution, combined with good observational and trialing skills, will help you successfully navigate the morass of having hundreds of different varieties to choose from.

Soft Winter Wheat Varieties

Choosing a soft winter wheat is a bit more straightforward because there are fewer varieties to choose from. The Palouse region of eastern Washington doninates the North American soft white winter wheat market, however, with numerous varieties that simply don't thrive in our part of the country; midwestern Corn Belt varieties don't do well, either. But once we move eastward into New York State, we begin to find varieties of soft wheat that are more adapted to our northeastern growing conditions. The Finger Lakes and points farther west in New York have a long tradition of growing soft white winter wheat; varieties like Cayuga, Ticonderoga, and Caledonia have been around for decades and are pretty well adapted to eastern New York and New England. I remember seeding Ticonderoga wheat over thirty years ago. These older varieties were eventually supplanted by Richland, which remains one of the most popular soft white wheat cultivars in modern use. We are very fortunate to live in the outer reaches of the New York State wheat universe. Cornell continues to be a leader in developing new and improved wheat varieties with improved disease resistance and winter hardiness. For instance, Dr. Mark Sorrells's recently released Jensen resists pre-harvest sprouting and scab.

If you happen to live in one of the Northeast's colder subregions, these New York varieties might not be tough enough to survive extra-harsh winter conditions. This has certainly been the case for me over the years. In fact, I was ready to totally give up on winter pastry wheat because I had experienced so much winterkill in my crops. Then about ten years ago, my Canadian seed dealer, Michel Lefevbre of Compton, Québec, mentioned to me that he had just received some seed for a super-hardy soft white winter wheat from Prince Edward Island. The variety was called Borden, and I planted it here in Westfield, Vermont, with great success. It is not totally resistant to fusarium and other wheat maladies, but it does persist well through our tough winters. Borden has consistently performed well in Heather Darby's winter wheat trials, as well, and it was the top yielder with the longest straw in 2011. I've managed to keep my strain of Borden alive for ten years now by carefully saving seed after each season's harvest. New York State to the west and Canada to the north both offer us numerous possible varietal choices for soft winter wheat.

Heirloom Varieties

I would be totally remiss if I neglected to mention the possibilities of growing heirloom wheat varieties. There have been thousands of different wheats grown here in North America over the past two hundred years, many of them developed and selected by the very farmers cultivating them. Heirloom wheat has a lot to offer us, including sublime taste and the ability to outcompete weeds. These older varieties grow much taller with more leaf surface area than their modern counterparts, and their root systems are much better developed, too; heirloom wheat puts out numerous tiller shoots, so seeding rates can be reduced to 100 to 120 pounds to the acre. The Slow Food movement has been especially instrumental in bringing older wheat varieties like

Red Fife back to the forefront. Red Fife wheat is indeed the poster child of a very successful attempt at variety restoration and reintroduction. Folk history has it that in 1843 Peterborough, Ontario, farmer David Fife planted out some wheat seed he had received from a friend in Glasgow. Fife's entire planting was consumed by some sort of a blight or smut—except for two heads, which were unaffected. These heads were replanted over the next few years and seed selections were made. By 1860, Red Fife wheat was dominating the Canadian wheat market. It held this position of prominence until right around 1900. The one complaint that growers had with Red Fife was that its growing season was extra long, which made the crop susceptible to early prairie frosts. In 1904, Canadian "cerealist" Charles E. Saunders released the earlier-maturing Marquis variety, which was a cross between Red Fife and a short-season wheat called Hard Red Calcutta. By 1920, 90 percent of the wheat grown in Canada was Marquis. Another variety called Thatcher was introduced in 1937 and led the market into the 1950s. I have grown all of these varieties and more in my own heirloom wheat trials and seed increase efforts, and can tell you that these and other heirloom wheats grow vigorously and taste good. We actually have Marc l'Oiselle of Vonda, Saskatchewan, to thank for most of the Red Fife wheat grown in North America today. In 2003, Marc set out to increase the supply of Red Fife and market it to people who want to bake with older varieties of wheat. He has been quite successful in this project and has now moved on to include Marquis in his lineup of available heirloom wheats. I personally obtained my original supply of Red Fife wheat from Spearville Flour Mill in New Brunswick. This Saskatchewan wheat had been grown out in the Canadian Maritimes for several years before coming to Vermont.

Heirloom wheat is much lower yielding than the modern varieties; Red Fife yields can't even begin to compare with more modern hard red spring wheats like AC Barrie and Glenn. But there is no doubt that the heirlooms have superior taste and flavor, and Red Fife has tremendous protein and gluten, which makes it an excellent baker. The taller stature of heirloom wheats can also be a distinct advantage to those of us who are growing out small plots with antique grain-harvesting equipment. For example, Red Fife grows almost chest-high in my fields; if you're reaping wheat by hand with a grain sickle or by machine with a grain binder, this extra height makes the job considerably easier. The grain binder was perfected in the late nienteenth century in the days of Red Fife wheat, as the tall stems cut easily and travel flawlessly across the machine's platform canvas. Long stems also pack into tighter bundles that stand up much better out there in the field, whereas modern knee- and thigh-high wheat barely has enough stem to get a string around a bundle.

In addition to harvesting advantages, taller heirloom wheats provide more space for an underseeded hay crop of clover, alfalfa, and timothy. I have seen ready-to-harvest underseeded wheat crops with alfalfa growing at the same height as the heads of grain. This wheat is quite difficult to harvest because of the amount of green alfalfa that must travel with the wheat through the combine, but knee-high alfalfa isn't going to interfere with the harvest of chest-high Red Fife. If you find yourself in this sort of situation, you simply lift the combine platform up above the alfalfa and only harvest the top half of the wheat stem. If there is a need for more straw, just mow down the stubble and alfalfa after wheat harvest. Although heirloom wheat yields may trail behind those of modern varieties by as much as 40 percent, there are numerous other reasons to grow these old cultivars. For more information on specific varieties and growing characteristics of heirloom wheat, please see the Northern Grain Growers Association website. A feeling of reverence and respect for the past comes over me when I let a handful of beautiful Red Fife kernels slip through my fingers. This is the same wheat that was grown by farmers and eaten by the masses a century and a half ago.

How to Harvest High-Quality Wheat

Wheat flours are not all created equal. We all know that one flour will make a better loaf of bread or a nicer cake than another, and flour quality is totally dependent on the wheat from which it was ground. First and foremost, wheat quality is rather weather-dependent. We need enough moisture to germinate seeds and adequate fertility to nourish plants through their vegetative and reproductive stages, but too much water at any time during the growing season can lower the quality of a wheat crop. Wet and cloudy weather at pollination can infect our plants with fusarium, and an extended rainy spell at harvesttime can delay combining and encourage pre-harvest head sprouting. Wheat would rather have it dry. But despite the fact that we are gambling when we try to grow wheat here, there are some steps that we can take to reduce the odds and increase our chances for a better outcome at harvesttime. I still encourage you to try growing wheat here in our region even if it seems like success just might be too elusive. You can hedge your bets a little by adopting a few practices that will give you a bit of a leg up on this fickle climate.

What Defines High Quality

Bakers know more about wheat quality than we do as growers and farmers because they are the end users of what we harvest and store in the bin. I'm not a baker, but I have been around sourdough bread for most of my life; as a child, I consumed vast amounts of my father's sourdough rye. Since 1977, my wife and I have been eating our own sourdough breads made from our own flour. Until relatively recently, we had the rather simplistic notion that wheat grown in a dry year always made better bread than wheat grown in a rainy season under wet conditions. We really weren't sure why this was so, but the logic behind this conclusion was that since wheat thrived in semi-arid environments like the Great Plains, a dry summer simulated a prairie climate. It wasn't until six or seven years ago (around 2005), when a small group of Vermont farmers and researchers interested in grains began to meet, that I really began to understand the broader concepts of wheat quality. These grain meetings were led by UVM Extension agronomist Dr. Heather Darby and were the beginnings of what was later to become the Northern Grain Growers Association. We invited speakers from all over the country and Canada to share their grain knowledge with us. Washington State University wheat breeder Dr. Steve Jones and Milanaise Flour Mill owner Robert Beauchemin from Québec were particularly helpful in educating us about achieving consistent wheat quality. We were also joined by a number of local bakers who were very interested in using local grains in their breads. Jeffrey Hamelman from King Arthur Flour and Randy George from Red Hen Baking in Middlesex, Vermont, offered us their perspectives on how wheat quality related to turning out a great loaf of bread. While my knowledge of the baking craft remains pretty rudimentary, I have learned a few things over the last several years that help me produce wheat flour that bakes well. Good test weight is essential—lightweight wheat doesn't make good flour. Wheat with high protein is necessary to provide enough gluten that a loaf of bread doesn't end up like a brick. Last and most important of all is to minimize the amount of enzymatic activity in the wheat kernel. When ripe grain is exposed to extended wet weather, it actually begins to wake up and sprout right on the stalk, and wheat that has begun to sprout internally will have higher levels of the enzyme amylase. This process begins to consume the carbohydrates in the endosperm of the wheat kernel, which in turn makes low-quality flour with compromised baking quality. Accomplished bakers use a variety of terms like *crumb structure* and *dough hydration* to describe how a certain wheat flour will work for them. I will leave the detailed explanations of the finer points of baking to concentrate on what we can do in the field to produce the best-quality wheat possible in a climate that occasionally makes this task quite difficult.

High in Protein

We needn't spend too much time on the issue of protein because it has been discussed earlier in this chapter. Remember that seventy pounds of N that is required to harvest a fifty-bushel wheat crop? Excellent soil fertility and selecting the right variety are both necessary ingredients for high-protein wheat. Spring wheat will generally develop protein levels two or more points higher than its winter counterparts. Research done here in the Northeast by Susan Monahan and Heather Darby at UVM has determined that added nitrogen applied to wheat at the time of tillering will definitely produce a crop with higher protein. There is no substitute for excellent-quality biologically active soil with elevated organic matter levels. These sorts of soils release nitrogen steadily and gradually through the whole growing season as the microbes and other little critters down there live, die, and consume one another in an orgy of fungal and bacterial activity. The result is high-protein wheat grown here in the East. A number of growers in Vermont have even produced hard red spring wheat with 15 percent and above protein by judiciously using the manure and compost resources from their dairy farms.

Good Test Weight

Test weight is all about the density of the harvested grain crop. A bushel of good-quality wheat is supposed to weigh sixty pounds. We are a bit at the mercy of the weather in this department, as any prolonged spell of wet weather can cause a reduction in test weight. If the weather is wet at planting time and the ground is saturated, oxygen will be driven from the soil. An anaerobic soil slows down the growth process and will ultimately stunt a wheat crop, which will definitely lower the final test weight. Waterlogged soils can also inhibit grain fill during the seed-development process a few weeks before harvest, and once again test weight will be reduced. Photosynthesis is much less efficient when the clouds are blocking the sun, and the translocation of sugars from the leaves of the wheat plant and conversion to carbohydrates in the endosperm of the kernel are compromised under such weather conditions. The only option we have as northeastern wheat producers is to plant our crop as early in the spring as possible. Earlier-planted wheat always produces heavier kernels than later plantings because it puts on the majority of its growth on the spring side of the summer solstice when the days are increasing in length. Thus, these early plantings garner the most amount of sunlight possible. Better opportunities for photosynthesis translate into more carbohydrate production in the endosperm of the wheat kernel, and the end result is improved test weight.

This same principle of early planting also holds true for dodging the vagaries of mid-summer diseases like stem and leaf rust. Early-planted wheat will hopefully make most of its growth before wind-driven rust spores arrive in July. If leaf rust strikes before or during grain fill, you can pretty much count on drastically reduced harvest test weight, because the plants' photosynthesis process will be compromised too early. If grain fill is complete when the rust arrives, it is not as serious a problem: The plant has already begun the final dry-down process. Early-planted wheat will usually have completed its grain fill process by the middle of July, when rust spores begin to arrive on the wind from points south. Weed pressure is generally much lower in spring wheat planted in early to mid-April, and fewer weeds means less competition for nutrients—which ultimately translates into higher-quality grain with better test weight. We may not be able to control the weather, but we can certainly make every effort to plant our wheat early enough in the spring so that it can achieve its best potential in the upcoming growing season.

Avoiding Pre-Harvest Sprouting

The biggest revelation that I have had in learning about harvesting high-quality wheat is how important it is avoid pre-harvest internal sprouting in a crop of standing wheat awaiting the arrival of the combine. Once in a while in an extremely wet spell, I've seen green shoots emerging from wheat heads out in a field of grain, which is pre-harvest sprouting at its worst. A crop in this kind of condition is pretty well ruined. However, it's the internal sprouting that you can't see that you really have to worry about. If wheat is physiologically mature and exposed to too many days of continuous wet weather, the kernels begin to grow just as if they were planted in soil. The seed virtually "wakes up" as its internal clock and chemistry tell it to grow. The enzyme amylase will be produced within the kernel as this process begins. Enzymes consume starch and produce sugar, which means that the endosperm of the wheat kernel is being mined for starch to supply sugar to a future wheat plant. This is basically the same process that takes place when grain is malted, but you don't want to make sweet malt out of the carbohydrate portion of the wheat kernel up there on top of the plant out in the field. This endosperm starch would be put to much better use supporting the structure of a developing bread dough after it's been ground into flour. If there is too much sugar and not enough starch in the flour, the dough will become overly sticky, which usually results in a finished bread product with poor texture and crumb structure. Our group of fledgling wheat growers heard this story over and over again as we met and listened to the wheat experts who came to present at our early grain meetings.

Achieving a High Falling Number

The potential for enzymatic activity in wheat can be measured right after harvest by a falling number test. Five years ago, this was all completely new information to me. I saw my first falling number test in 2008 when I visited the Milanaise Flour Mill (Moulin des Soulange) in St. Polycarp, Québec. The test is relatively simple and easy to perform with the proper equipment: A seven-gram sample of finely ground flour is combined with twenty-five milliliters of distilled water in a laboratory glass tube that looks very similar to a graduated cylinder. A stirring device is placed in the tube, which is shaken and stirred vigorously to form a flour paste slurry. The glass tube is then placed in a boiling-water bath and stirred constantly until it boils and forms a thick paste of gelatinized starch. At this point, the glass tube is removed from the heat and the stirrer (which is much like a plunger) is lifted to the top of the tube. The plunger is released at the same time as a stopwatch is activated. The plunger makes its way slowly down through the gelatinous solution to the bottom of the tall glass tube; the number of seconds that it takes to "fall" from top to bottom is the falling number. The speed at which the plunger descends through the flour-and-distilled-water paste is determined by the viscosity of the solution. If the wheat being tested has suffered sprouting damage, it will stir up into a more sugary, much less viscous mixture through which the plunger drops rapidly. This will, in turn, produce a falling number on the lower end of the scale, indicating that the wheat sample has reduced baking potential. Falling numbers of 250 seconds and below tell us that there is high enzyme activity in our wheat sample and that we might want to consider using it as animal feed instead of people food. Falling numbers of three hundred seconds and higher are the sign of some pretty good-quality wheat. If there is only starch and no sugar in the flour-water solution, it will be extremely viscous and the plunger will take a lot longer to reach the bottom of the glass tube. Heather Darby was able to obtain funding for a Perten FN 1500 Falling Number Machine, which she installed in her new grain-testing laboratory in the Jeffords Science Building at the University of Vermont in 2010. Grain growers from all over the Northeast can now send in their wheat samples to UVM for a reasonably priced rapid-turnaround falling number test. Waiting for the falling number test results to come back on your wheat is probably just as bad as sitting there while your high school teacher passes out corrected test papers. Your hopes for a wheat crop that will bring top dollar as whole wheat flour will be either bolstered or dampened by the results of this test, which tells you how much sprouting took place inside those heads of wheat out there in the field.

So what can you do to harvest wheat with minimal sprouting damage and a high falling number? The answer is pretty straightforward. Don't leave your ripe wheat in the field any longer than you absolutely have to, and when your crop is ripe and physiologically mature, get the combine, threshing machine, or grain sickle rolling. In many cases, it might be necessary to harvest ripe wheat with a moisture content in excess of 20 percent, especially if your wheat is ripe and you see three days of rain coming. You will have much better grain if you harvest it before the rain and dry it in a bin with forced cold air through a drying floor. It doesn't matter what your scale is in this sort of situation; if you are harvesting a garden plot by hand with a sickle, the reaped wheat can be bundled and put under cover until the sun comes back out. If you don't have a bin with a forced-air fan and a drying floor, insert several screw-in aerators into your wagon or vessel full of wheat. If this isn't an option, spread your high-moisture wheat out thinly in a greenhouse with box fans running to move the air around. When it comes to saving your crop from wet weather, use all of your creativity and ingenuity. Where there is a will, there is a way. Wheat at 20 to 25 percent moisture will eventually dry into some pretty high-quality material if you are vigilant and undaunted. Every baker you meet will tell you that achieving a high falling number in a wheat crop is more important than attaining high protein levels—and they're all right.

Learning to Live with Uncertainty

We all want to produce wheat of the highest quality possible, and there are certainly many extra little steps that we can take to turn our desire for ultimate quality into reality. Unfortunately, the weather is still in control. We can jump through all sorts of hoops to ensure the harvest of high-quality wheat, but sometimes it just doesn't happen. The rain arrives too soon or leaf rust gets out of control too early in the season. But if you're committed to growing grains to feed yourselves and your community, you have to be willing to roll with the punches and give it another try next year. There's a certain amount of resignation that has to be part of the temperament of anyone who wants to defy the odds and raise wheat in this part of the country. Grain growers in North Dakota lose crops to hail every year, and we northeasterners have to be willing to live with excess rainfall. I find it rather interesting that twenty to thirty years ago, we never worried about grain quality, and we definitely didn't know anything about falling numbers. High-protein wheat never entered our thoughts. We simply ate what we grew and loved it. We didn't starve back then when wet weather made it difficult, and we won't starve now, either. If we're hungry enough, we can still make bread out of our so-called low-quality wheat. We need to remember why we are raising this crop today when the last wheat boom in New England was in 1917. We are twenty-first-century pioneers and we should know by now that the work we have chosen is by no means easy. Guarantees don't exist. So let's love our wheat no matter what the tests tell us about crop quality. We know what the ideal is, and we will rejoice when we achieve it in a banner year, but we will still find satisfaction even when our labors are fruitless.

Wheat Processing and Flour Production

The wheat harvest is complete and you've dried your crop to at least 13 percent moisture. So how do you go from kernel to loaf of bread? Cleaning is the first order of business. It's best to remove all weed seeds and other foreign material before the crop goes into long-term storage. (Now is when you get to use the bevy of grain-cleaning equipment described in chapter 8.) Moist annual weeds like ragweed that carry oil-based scents should be cleaned from the wheat within hours of harvest because of their propensity to flavor the grain. In most cases, a fanning mill air-screen-type cleaner will adequately sift and sort the wheat. Once in a while, it may be necessary to grade wheat berries by density over a gravity table. If you find hairy vetch seed in your harvested wheat, you may have to resort to a Carter disc cleaner to accomplish the removal process. Suffice it to say that you want impeccably clean, dry wheat without mycotoxins for milling into whole wheat flour for human consumption.

The Significance of Bran and Endosperm

Before you set about grinding your crop into flour, let's take a good look at that kernel of wheat you intend to pulverize. The wheat "berry" consists of three distinct parts. The entire kernel is covered with an outer layer of bran, which contains fiber and protein. Bran accounts for roughly 14 percent of the kernel weight and is loaded with the healthful B vitamins niacin, thiamin, and riboflavin, along with dietary fiber and iron. The germ, which is located at the base of the kernel, is the embryo or sprouting section of the seed. It makes up only 2.5 percent of the total weight of the wheat kernel, but is absolutely necessary for reproduction because this is where the genetic information for the future wheat plant is stored. This wheat germ portion contains 10 percent fat in the form of essential fatty acids. The remainder of the wheat kernel (83 percent) is endosperm, which is its starch component. The carbohydrate in the endosperm serves as food and nutrition for the young wheat seedling as it emerges from the germ. It is also the primary component of wheat flour. Commercial white flour is basically ground-up endosperm; the germ and the bran are removed in the milling process. Whole wheat flour, on the other hand, contains the germ and varying amounts of bran. Flour milling can be as straightforward as grinding up wheat kernels by hand with a mortar and pestle or as complex as sending wheat through multiple siftings and passes in a high-speed roller mill. Entire volumes have been written about this subject, as milling is both art and science. Kansas State University has a whole degree program on flour milling and grain processing. (It is also possible to attend a number of weeklong short courses at KSU specifically designed for laypeople interested in various aspects of turning wheat into flour.) Then there are the old-time artisans who have been operating water-powered gristmills all over the country for years. These individuals have been declining in numbers, and a lot of their knowledge is disappearing with them. It is my recommendation to seek these folks out and learn as much as you can from them. If you pack up a couple of sacks of wheat berries and bring them to someone with a flour grinder, you will leave the mill with knowledge as well as flour. Hopefully, the newfound interest in whole and local foods that is beginning to permeate our culture will encourage more learning and skill transfer from one generation to the next. I will do my best to lay out the fundamentals of flour milling here, but I encourage you to study with a master.

The Fundamentals of Flour Milling

Water-powered gristmills were once a common sight all over the eastern United States. Just about every community had a local mill located on a stream or small river. The Northeast was particularly well suited to water power because of the up-and-down nature of its topography. Millponds with their own small dams were constructed wherever water levels flowed downhill enough to harvest motive power from falling water with a waterwheel. Overshot and undershot waterwheels began to be replaced by hydraulic turbines after the Civil War. In the latter half of the nineteenth century, the James Leffel Company of Springfield, Ohio, revolu-

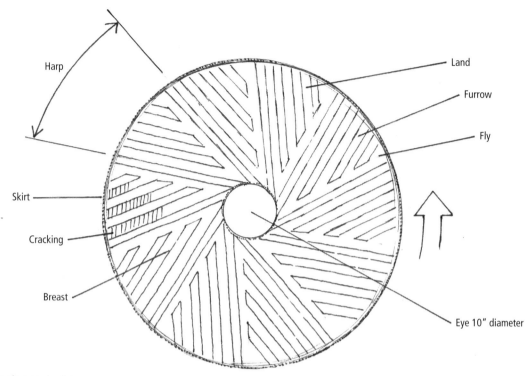

The layout of a typical millstone.

tionized water powered grain milling with a line of highly efficient water-powered turbines. Up until this point, flour milling hadn't changed much for a number of centuries. Wheat was hauled to the mill in cloth sacks and ground slowly between two horizontal stones. The few historic gristmills that operate today work on the exact same principles as their forbears did two hundred years ago. A common mill would have two round horizontal grinding stones that ranged in diameter from three to five feet. The bottom bed stone sat stationary on the floor of the mill. A steel shaft came up through the bed stone to support the upper (runner) stone, which revolved on top of its mate below. Both bed and runner stones were chiseled or "dressed" with special grooves called lands and furrows—which is basically the same terminology used in describing the layout of a field for moldboard plowing.

Most millstones were covered on top and surrounded by a wooden skirt of boards to contain flour and dust during the milling process. A wooden hopper was positioned on top of the mill to contain the wheat that was destined to be ground below. Wheat kernels dropped through a regulated gate in the hopper down through the eye of the revolving top stone. The grinding process began once the wheat fell into the ingeniously designed pattern of grooves in the two stones.

It is interesting to note that the chiseled grooves radiate in a tangential pattern from the center of the millstones in order that they can readily move wheat kernels from the eye in the center to the outside circumference. These grinding channels are cut much deeper into the stone at its center point to accommodate whole kernels of wheat that are falling through the eye of the runner stone above. This is where the craft of the millstone dresser reigns supreme. The grinding channels get narrower and shallower as they radiate outward. These milling grooves are specially configured with a right-angled leading edge and a tapered trailing edge. The runner and bed stones are dressed identically with the exact same pattern of lands and furrows, which is very apparent if you lift the runner stone off the bed stone and lay it face-up next to its mate: The two stones look like identical twins. However, once the runner stone is lifted and repositioned on top of the bed stone, the patterns of the grinding channels will be diametrically opposed. This configuration of grinding channels scraping across each other is what grinds the wheat into flour.

The kernels of wheat begin their grinding journey in the deeper center grooves, where they are chopped into smaller pieces. A well-dressed sharp stone will actually scrape the exterior bran off the wheat kernel in large flakes. The remaining endosperm continues traveling outward and is pulverized by shallower cross channels (lands) into smaller and smaller pieces until it becomes fine flour at the very outer edge of the millstones. The flour is propelled outward by centrifugal force and built-up air pressure that enters the grinding stones from the eye and exhausts itself at the outer circumference.

Stone Flour Mills

Horizontal stone flour milling is a bit of a lost art these days for a variety of reasons. This is "slow food" at its best, and the finest-quality flours come from this type of milling process. These millstones rotate at a rather low 120 to 125 revolutions per minute. Flour will never heat when ground on large-diameter slowly revolving millstones. Heat is the enemy of good flour because high grinding temperatures will cause the oil in the wheat germ to go rancid. However, total output of flour in this system is rather low; a set of four-foot-diameter millstones will not grind more than three hundred pounds of flour per hour. Everything about modern industrial food production runs counter to the artisan miller of yesteryear. A modern twenty-four-inch-diameter higher-speed vertical stone mill can crank out more than double the flour of its ancestor. It may be a fact that modern flour grinders half the size of the old antique models can double the output of flour, but this shouldn't be a reason to totally ignore this time-tested technology. We have a lot to learn from yesteryear, because the principles of good milling haven't changed. The speed of feed intake is critical in any sort of flour milling, old or new. You simply must not crowd a grinder with too much wheat, because the result will be overheated, poorly ground rancid flour. Stones must be sharp and properly dressed, because dull stones will heat up the wheat and smash it instead of peeling off the bran and cutting the kernels. Another rule of thumb that holds true for both old and new flour milling is that the stones should never touch each other. Millstones need to be dressed in a slightly dish-shaped manner (deeper in the middle and level at the outer edges) in order that they may be run perfectly parallel with only a thousandth of an inch between them. Stones with little high spots that tick and rub will burn the flour and give it an off taste. Last but not least, trap iron in the form of metal bolts, screws, and nails should be removed with some sort of a magnet as the grain enters the mill for grinding. A stray piece of metal can do quite a job on any sort of millstone, old or new. Keep these old stone mills and their grinding principles in mind as we explore the myriad approaches we can take to grind our wheat into flour.

Hand Mills: Grinding Flour on a Home Scale

The simplest and cheapest type of flour grinder for home use is a hand-cranked model that can be attached to a kitchen counter or workbench. Hand mills became quite popular in the late 1960s and early '70s as the back-to-the-land homesteading movement began to take shape. *The Whole Earth Catalog*, which was the sourcebook for all things earthbound at the time, offered a whole collection of countertop-mounted flour mills. All of these hand-operated grain mills were pretty similar in construction and appearance. At first glance, the hand-cranked flour grinder resembles a meat grinder because it has a small hopper on top to contain the wheat. The main frames of these machines stand about fourteen inches tall and are usually fabricated from cast iron or steel. Just about every model of hand flour grinder is plated with tin alloy or enamel to protect the food quality of the wheat and flour. There is a cylindrical cavity that runs horizontally through the midsection of the grinder just below the hopper and a small spiral-shaped auger fits right into this opening, which is the heart of the machine. A long hand crank attaches to one end of this auger and a steel grinding wheel to the other. Hand grinders actually have two metal buhr stones that are vertically situated next to each other on the opposite end of the feed auger from the handle. The buhr stone that is closest to the inside of the hand mill works in much the same fashion as the bed stone of the antique gristmill—it remains stationary. The outer metal grinding plate is held in position by a special threaded metal rod attached to a large adjustment screw. This is the runner stone of the hand mill because it can be moved closer to or farther from the bed stone for coarser or finer flour. The grinding process is actually quite simple and straightforward: Wheat falls from the hopper into the horizontal chamber below, where it is propelled sideways to the grinding buhrs; the same auger that is carrying the wheat sideways also turns one of the grinding buhrs; wheat passes between the steel plates and is ground into flour during the process; and the flour falls straight downward into a bowl or flat pan on the countertop below.

These little grinders sound almost too good to be true. However, they do have their limitations and drawbacks. The first thing you need to know is that the crank is harder than hell to turn by hand. We bought one of these grinders from our local food co-op in 1975. It was tin-plated and came from Czechoslovakia, and the price was right at a little less than $20. We bolted it to our kitchen counter and prepared to turn wheat berries into fresh flour for baking. Well, weren't we in for a surprise. It took two hands and arms plus all the upper-body strength we could muster to crank the thing around and around. As a matter of fact, we were using so much physical force in the process that we had to be careful not to lift up the entire countertop while turning the handle. We didn't set any speed records, either. We took turns struggling with the handle, getting red in the face and sweaty; it took a good half hour to grind the three or four pounds of flour that we would need to make a small batch of bread. Much to my chagrin, the final grind of flour couldn't compare to the commercial stuff that we had been buying. It was coarse and gritty and not really well suited for making a light loaf of bread. We tried running the wheat through twice, but this seemed to improve things only slightly. The handle was much easier to turn on the second pass and the reground flour was less gritty, but coarsely ground flour became the norm in our household as we set out on our homesteading journey in the mid-1970s. Our hand-powered grain mill actually did a much better job at cracking toasted wheat for cereal and grinding softer grains like corn. It didn't matter a whole lot to us at the time—we kept using the thing because we were diehards and we were going to grind our own flour no matter what.

Steel buhr (also called burr) grinding plates needed contact with each other to perform the task of grinding wheat into flour. It's also this friction between the two plates that makes a hand mill so difficult to crank. This principle of milling is totally contrary to stone flour grinding, where the two grindstones are never supposed to touch each other. However, steel buhr stones are cheap and replaceable, which makes this not-so-perfect technology affordable and accessible to just about everyone wanting to grind their own flour. Hand-powered flour mills are still a bargain these days. The twenty-dollar hand grinder that we purchased back in the 1970s now sells for around seventy dollars. Of course, this is the bottom of the line, but one of these little units will still do the job for you. Quaker City and Corona are both readily available models. The Corona Mill is manufactured in Colombia and can be used to grind wet corn into masa as well as wheat into flour. There are other, more expensive models like the Country Living Grain Mill and the Family Grain Mill (from Germany); these will do a little better job for a whole lot more money. The top-of-the-line hand-powered grinder is the Diamant, which sells for close to a thousand dollars and comes with a large feed hopper and sizable cast-iron flywheel for easier grinding. Many of these hand-powered flour grinders can be found in the Lehman's Catalog as well as through Internet searches. There is a whole world out there of people grinding their own whole grains. One thing is certain in all of this—when you grind wheat berries through steel plates, things will wear out and get dull after a while. Stones can be dressed with chisels and other specialized tools, but steel buhrs need to be replaced. If you plan on acquiring a flour grinder with steel grinding plates, you might want to stock up on extra steel buhrs when you make your initial purchase.

Buhr Mills

My own personal adventure with farm-scale flour grinding continued with steel buhrs for quite some time. Once we had several tons of our own wheat stored in burlap bags in 1977, we knew we had to scale up to something a little bigger and faster. Luck was with us. One of my grain mentors, Milton Hammond from nearby Newport Center, had a Kelly Duplex electric buhr mill complete with sifting screens, which he used for buckwheat flour. The machine was powered by a five-horsepower motor and came equipped with seven-inch grinding plates. As a matter of fact, there were two sets of steel buhrs: one for cracking grain and one for fine grinding. We ground our wheat with the fine plates and were quite pleased. The Kelly Duplex mill worked on the same principle as every other vertical axis grinder: One plate remained stationary while the other was adjustable. I can still remember the distinctive screech of the two plates touching each other as we prepared the mill to make flour. Mind you, this still wasn't stone-ground flour. The consistency was gritty, and the final product came out of the mill much warmer in temperature than we liked. Nevertheless, we were happy because we could grind almost one hundred pounds of wheat per hour with this antique piece of equipment. Milton is now eighty-seven years old and stopped raising buckwheat several years ago. I have inherited this old electric plate mill and sifter and added it to my collection of old grain equipment.

After a while, I felt like I couldn't impose myself on Milton's generous hospitality any longer. An old flat belt-driven buhr mill came up for sale at a used-equipment junkyard across the river in Colebrook, New Hampshire. This was a considerably bigger plate grinder than Milton's; the grinding plates were ten inches across, and the hopper was big enough to hold several hundred pounds of grain. The unit was cast iron and quite heavy, and we couldn't find a make or a model number anywhere on the machine. The only markings evident were some Canadian patent numbers cast right into its main frame. We bolted the thing down just inside our big barn doors and proceeded to connect it to our forty-five-horsepower Farmall Super M tractor by a six-inch rubberized canvas flat belt. Now we could grind some grain! We used the old buhr mill for both livestock feed grinding and flour production, and even though we were using a fifty-year-old machine, we felt like we had stepped into a more modern age by increasing our grinding capacity four- or fivefold. The other thing that increased was dust, as the old grinder didn't really have an outlet spout for ground grain and flour. I devised something from sheet metal, but the dust still flew every time we operated the thing. Of course, I didn't know how valuable my lungs were back in those years of invincibility, and we never used dust masks; only occasionally would we wrap an old diaper around our mouths and noses. The best advice I can give thirty years later is take care of yourself around grain dust. It caught up with me after a while in the form of asthma and breathing issues.

The most amazing thing about these old grinders is that no one really wants them anymore. They are heavy and obsolete by modern flour and feed milling standards. The general consensus is that old buhr mills are probably best used as boat anchors. If you travel around to abandoned feed mills in our region, you will often find these old grinders still hooked to a flat belt and line shaft. These larger buhr mill grinders were even more efficient because they operated with two steel plates turning in opposite directions. But truly, there are lots of easier and better ways to grind flour and livestock grain these days, and I wouldn't go the steel buhr mill route unless I was so poor that I couldn't afford anything else. I was so fascinated with the technology of the past that I had to grind grain with this antique machinery to prove to myself that I could do it just like the old-timers. If you're as crazy and fanatical as I am about the history of agriculture and farm technology, give this method a try. Otherwise, read on to find out some simpler and more practical ways to turn wheat into flour.

Electric Mills

Sometime during the 1980s, I began to see small electric flour mills in health food stores. (On-the-spot peanut butter machines and self-serve flour grinding were the rage at the time.) There was usually a bulk container of Whitmer Montana wheat berries right next to the grinder. You turned the thing on, dumped a couple of scoops of wheat berries in the hopper, and extracted your flour from a little drawer underneath the machine. I brought some of my wheat into our local health food store once or twice and managed to plug the self-serve flour mill; the store owner told me that my wheat berries were too moist. (He didn't believe in Vermont wheat. For that matter, no one else did, either.)

I didn't think much about electric grinding until a few more years went by. In the summer of 1990, Anne and I were touring around Midcoast Maine when we happened to stop in at Morgan's Mills in the little town of South Union. We wanted to see a modern gristmill that was producing wheat flour and cornmeal from flint corn. Richard Morgan graciously showed us around and explained the process of vertical stone milling. The flour was ground in the same basic manner as our old buhr mill; the only difference was that the big round cast-iron unit he was using contained two vertical granite grinding stones instead of steel buhrs. One stone remained stationary while the other was adjustable to gauge the fineness or coarseness of the flour. We had been finally introduced to a Meadows Mill. These units varied in size from the homeowner's mini model with eight-inch grinding stones all the way up to huge commercial units with thirty-inch-diameter stones. Richard had one of the larger Meadows Mills to grind flour and cornmeal for his cornbread and pancake mixes. We did notice that he also had a smaller model over in one corner of his shop, and we ended up buying that little eight-inch stone flour grinder for two hundred dollars. A new era in homegrown flour grinding began for us. We set the thing on an old table out in our garage and put it to work. All we had to do was plug in the half-horsepower electric motor, pour some wheat in the hopper, and adjust the stones; in five minutes the hopper would be empty. Then we simply cut the power and pulled out the bottom drawer, which was full of enough flour to make a three-loaf batch of bread. This seemed almost too easy. Why hadn't we found one of these electric stone grinders ten years earlier? I guess we had to do it the hard way first to appreciate just how liberating a Meadows Mill could be.

The Meadows Mill

The company had its beginnings in 1900 when the Reverend William Calloway Meadows of Pores

Nob, North Carolina, outfitted a standard steel buhr mill with vertical millstones. The first units were manufactured and sold in 1902, and patents were issued in 1907. In 1908, a group of investors from nearby North Wilkesboro bought the business and merged it with two other local foundries to form the Meadows Mill Company. World War I created worldwide wheat shortages, which increased the demand for locally produced wheat and wheat flour all over North America. Farm-scale wheat milling had become so popular at this time that the International Harvester Company added the Meadows Mill to its farm equipment lineup in 1915. Ten years later, the postwar wheat boom was over and IH dropped the mill from its offerings, but these older stone mills are quite popular with antique farm equipment collectors and their steam and gas engine clubs. The Meadows Mills Company has continued to manufacture stone mills to this day. It's still possible to buy a brand-new unit from eight to thirty inches in diameter and put it to work grinding whole wheat flour.

Once I became an owner and a user of a small Meadows Mill, I began to notice them all over the place. Chuck and Carla Conway of the O Bread Bakery in Shelburne, Vermont, had one to grind all the whole wheat flour they used in their breads, and Fiddler's Green Farm in Belfast, Maine, used a large Meadows to grind flour for organic baking mixes. In 1995 I bought a sixteen-inch Meadows Mill that had been used at the Pine Island Bakery in Thetford, Vermont, for many years until the owners had decided to close the business. The price was one thousand dollars. This was the beginning of a new era in my wheat-farming career: I now had the wherewithal to grind larger amounts of wheat and other grains in a timely manner. It took several years to get the mill set up and working. I had to install an electrical service for the seven-and-a-half-horsepower electric motor as well as building a stand and support system for this heavy piece of hardware. Norm Collette, an all-around handyman and fix-it guy from Corinth, Vermont, journeyed up to the farm several times to dress the stones and make some of the finer adjustments to the workings of the machine. I can say with confidence that in the late 1990s, I "cut my teeth" on this old Meadows Mill. I had a few hints and suggestions from more experienced millers, but I must say that the only way to learn is by doing. I plugged the thing many times with moist wheat and by feeding it too fast. I burned the stones by running them too close together. I made coarse flour by running the stones too far apart. I put nails through it by accident. These are all ordinary mistakes that I hope you will be able to avoid by using common sense and by following the advice of an experienced operator.

Yes, I flew by the seat of my pants in the milling department for many years. I was grinding two to three hundred pounds of flour a week without too many problems. Eventually, things began to go wrong. The mill would hardly make flour anymore because the stones seemed to glaze over with something that looked like a wheat paste version of papier-mâché. I called Chuck and Carla at Obread for advice, and they recommended Hank Duncan, an itinerant millstone dresser and expert on Meadows Mills. Hank travels the whole country from coast to coast repairing and maintaining stone mills. He comes from the Piedmont region of North Carolina and had worked at the Meadows factory in North Wilkesboro for many years before he went into business for himself. I had to wait several months for him to make his rounds before he got back to Vermont. Hank is a short man with a contagious chuckle, a twinkle in his eye, and an accent reminiscent of Sheriff Andy Griffith from the 1960s TV show. We gave him supper and a warm bed and became instant friends.

Hank knows his stuff. He dismantled our Meadows Mill and went to work on the stones with his air chisel. He told us that our mill was basically a mess. The grinding surfaces of the millstones weren't even close to being parallel or true. It took him three times longer than normal to dress our stones because they were so out of kilter. It was almost as if he had to start from scratch and carve entirely new grinding surfaces. Once things got close, Hank painted the stones with food-grade red paint and reassembled the mill. Then he ran the machine, bringing the stones gently together. High spots were indicated wherever the paint wore off. He then removed the grindstones and went back to work. He did this until the mill was perfect. When Hank completed the job, he also dressed the stones on our original eight-inch mill. What a difference this made. Our Meadows Mill performed like never before after Hank worked his magic. If you end up with a Meadows Mill of any size, I would strongly recommend enlisting the services of a professional like Hank.

Maintaining a Meadows Mill

The biggest lesson that I have learned from Hank is that not all Meadows Mills are created equal. The original mills were equipped with grindstones made from a domestic flint pebble stone called North Carolina Balfour pink granite. The little chips of quartz contained within this granite are harder than diamonds, which makes for an incredibly durable and long-lasting grinding surface. According to Hank, the local granite quarry in his region began to run out of this excellent-quality rock about twenty years ago. Grindstone quality is a hit-or-miss situation in Meadows Mills that have been manufactured over the last few decades. This was certainly the case with the sixteen-inch mill that I purchased fifteen years ago: The stones are second-rate and wear down much more quickly than their pink granite counterparts. A yearly stone dressing is necessary to achieve top milling performance. Really, it's difficult to buy a twenty- or thirty-year-old Meadows Mill and be confident that the grindstones are of first-rate quality. Hank knows which years and models are the best, but the rest of us have to guess. I wanted to be sure I was getting an excellent mill, so I bought a used and reconditioned model from Hank, in the fall of 2010, as I was ready to step up to a larger grinder because of increased flour demand. Hank had a nice twenty-inch machine that he had taken in trade. I waited six months for him to recondition it and deliver it up to my farm. It cost considerably more than the sixteen-inch model I had purchased many years earlier, but the wait and the expense were well worth it because I can now grind better-quality whole wheat flour more quickly and reliably than ever before. I have kept the sixteen-inch mill to grind cornmeal. This means that I can grind gluten-free products in a dedicated mill, thereby ensuring that there are no traces of wheat in my corn products. If you plan on milling and selling flour, a mill maintenance program is essential. These machines are wearing every second they are in operation. Proper lubrication and stone dressing go hand in hand with high-quality local wheat to round out a smooth-running operation. There are many of us who mill wheat who don't have time to wait for Hank's yearly visit. These people—like Robert Beauchemin of Milanaise in Québec and Matt Williams of Aurora Mills in Lineus, Maine—dress their own millstones whenever it is necessary. One thing is certain: We need more Hanks in this world. There is a golden opportunity awaiting the individual who decides to become a stone mill specialist here in the Northeast.

Newer Isn't Always Better

Recently manufactured Meadows stone grinders are not the same machines as their twenty- and thirty-year-old predecessors. Stone grinders have gone the way of just about every other appliance out there. A new Maytag washer is not built to the same rugged specifications as the washer that we purchased (and are still using) brand new in 1976, and you could say the same thing about Meadows flour grinders. The main frame and millstone case of yesteryear's machines were constructed of durable cast iron. Contrast that with today's models, where the cast iron has been replaced by stamped and forged steel. These newer, lighter-weight grinders aren't built to the same tolerances as they were even ten years ago. A new eight-inch Meadows Mill sells for between thirteen and fifteen hundred dollars. Two of my friends have purchased these small units in the past year. In both cases, the brand-new mill needed to have its grindstones balanced, dressed, and repositioned by Hank. These machines came from the factory with defects, and the difference in milling performance was very noticeable once a number of these small imperfections had been corrected. I am so glad to have gotten to know Hank Duncan. He has taught me how to mill flour with a finely tuned Meadows Mill, and he is working on his own self-manufactured prototype stone mill right now, so stay tuned. If you can't find a good old Meadows Mill, you might be able to buy a new stone grinder from Hank.

East Tyrolean Grain Mills

The Meadows Mill isn't the only game in town when it comes to grinding locally grown wheat into flour. Some purists consider its five to six hundred revolutions per minute grinding speed to be way too fast for proper flour grinding. The thought is that high-speed flour grinding generates an overabundance of heat, which is injurious to the flour. Meadows has tried to solve this problem by equipping their larger mills with a suction fan and a cyclone. Grinding temperatures are kept cool because the flour is literally vacuumed away from the spinning stones and delivered to a cyclone to minimize dust. All of this technology is lost upon folks who want their flour slowly ground at one hundred RPM with old-fashioned horizontal millstones. You might think that this would only be possible at some old water-powered gristmill museum, but this is not

The millstone case of an East Tyrolian Austrian grinder. PHOTO COURTESY OF HEATHER DARBY

the case. There are several European small-scale electric-powered horizontal grain mills available. In recent years, many people have been buying and importing beautiful and finely crafted East Tyrolean grain mills from Austria. As seen in the accompanying photograph, these handsome wooden machines are equipped with horizontal stones and a built-in flour sifter.

Grindstone sizes range from eighteen inches in diameter for a home-scale unit to five feet in diameter for an industrial high-output model. Contrary to what you might think, these grindstones are not cut from native rock, but are made of a composite volcanic material from Greece called Naxos. Once a grindstone is cast of this extremely tough material, it is surrounded by a steel band to give it strength and durability. Best of all, these stones are self-sharpening and do not need to be dressed. John Melquist, the owner of Trukenbrod Bakery in Thetford, Vermont, has ground over seventy tons of wheat in his mill over a four-year period and has never had to dress his stones. These East Tyrolean grain mills are true works of art. They cost at least three times as much as a Meadows, but they are worth it. This is craftsmanship at its best. If you plan on buying one of these machines, start the importation process early because it can take up to six months to receive one from Austria.

Other Horizontal Flour Mills

There are several other European manufacturers of horizontal flour mills. On an organic grain growers' tour of Denmark in November 2010, I saw a number of Skiold flat stone mills on value-added grain farms and in bakeries. Located in Saeby in northern Denmark, Skiold is a major manufacturer of milling machinery for both the human consumption and the livestock feed market. The Skiold flour grinders that we saw consisted of two thirty-inch composite stones encased in a circular metal housing that was bolted to the floor. These units are heavy-duty and very well built, and as with their German counterparts the grindstones are made of a composite material. Skiold stones measure thirty inches across and are made from a very hard flint rock imported from Africa. These mills are unique in the world of horizontal flour milling: The top stone remains stationary as the bed stone acts as the runner stone, moving up or down to adjust the fineness of the flour. These mills rotate at a rather slow two hundred RPM, which allows for the production of very cool flour. An aspirated blower system takes flour away from the grinder through a series of small pipes. At full capacity, a Skiold model number 600 can produce eight hundred pounds of flour per hour. In the fall of 2010, one of these flour grinders sold for fourteen thousand dollars at the factory in Saeby, Denmark. Import fees and shipping would be added to this price, so clearly this is a serious machine intended for someone who has a lot of wheat to grind. I mention the Skiold mill to make it known that there are alternatives to the all-pervasive Meadows Mill out there. There are a number of these grinders in use here in the United States. Bob's Red Mill in Oregon has a whole battery of these machines in their production facility. If you are setting up a high-output grain-processing facility and want to mill high-quality flour with precision European machinery, the Skiold line might be something to consider.

Roller Mills and the Rise of Industrialized Flour

What about white flour? Is it possible to mill it in a low-key nonindustrial manner? White flour as we know it today was pretty much nonexistent until roller milling was invented in the 1880s. The August Wolf Company of Chambersburg, Pennsylvania, and J. M. Case of Columbus, Ohio, were two of the original manufacturers of this game-changing piece of equipment, which revolutionized the flour-milling industry at the time. I happen to own a Wolf Roller mill that measures about three feet square and stands about four feet tall, and weighs close to a ton because its frame and base are cast iron.

Belts and gears on the side of the machine drive two sets of double rolls on the top. These feed rolls are machined from high-quality steel, and measure sixteen inches long by eight inches in diameter. Roller-mill rolls are "fluted," which means they have longitudinal grooves cut into them in much the same manner as a Greek architectural column. The basic principle behind roller milling is that the wheat to be milled travels through pairs of rolls that are rotating toward each other at slightly different speeds. This action of one roll turning faster than its mate actually tears the bran off the outside of the wheat kernel and begins the milling process. Roller milling is a many-step process consisting of a series of "breaks" or passes through the rollers. The bran is removed from the endosperm on the first break. The next step is to a sifter, where the bran is removed from the remaining endosperm. Four or five more breaks and siftings will produce pure white flour. Middlings (also known as midds or shorts) are sifted out as the remaining endosperm is milled to a finer and finer consistency on each subsequent pass through the rollers.

The advent of rolling milling and the mass production of white flour in the late nineteenth century created a major rift between the artisan miller of days gone by and the factory model that it encouraged. The pace of flour milling increased manyfold as the quality declined because of the amount of heat created in the process. This was an entirely different process than stone milling—the wheat kernel was basically being peeled apart by sharp steel rollers instead of being slowly broken up and pulverized by the rock surface of a millstone. In its day, roller milling was a very capital-intensive affair. A whole battery of these monstrous machines was necessary to make white flour because the process required numerous passes through the equipment. In industrial flour-milling operations during the 1890s, roller mills were positioned side by side in a long line down the length of the mill building. The wheat to be ground began its journey in an overhead bin higher in the building. Ground wheat from the roller mill fell by gravity to a sifter on the floor below. Sifted material was then elevated to a bin on the floor above to once again fall by gravity to the second mill in line for the next break. This process of grinding, sifting, and elevating was repeated several more times until white flour was created. There is no doubt that roller milling and the industrialization of the flour grinding went hand in hand.

The days of multilevel flour mills with rows of roller mills, overhead bins, subterranean sifters, and elevators are over. Modern roller mills are self-contained units that contain multiple grinding rolls and sifters within their confines. Flour is conveyed pneumatically through internal pipes from one break and sifting to the next. I've seen several of these modern roller mills at facilities in Québec and North Dakota. These incredible pieces of equipment are made by the Italian firm Agrex. Robert Beauchemin, owner of the Milanaise Flour Mill, has the smallest Agrex roller mill made in his establishment in Milan, Québec. Even this so-called little machine is still a monster. It measures eight feet wide by forty feet long and cost three quarters of a million dollars. The Agrex mill that I saw when I visited the Dakota Pride Mill in Harvey, North Dakota, was three times bigger and cost close to two million dollars. There is good reason why artisan producers of roller-milled white flour don't exist in our part of the country. Modern small-scale wheat and flour producers face the same capitalization challenges that our stone gristmill forbears faced in the late nineteenth century.

Flour Sifters

The closest we can come to producing white flour with a stone mill is to sift what we produce. Flour sifters have been around for centuries. Some ancient European models are basically a rotating cylindrical screen. Ground flour enters at one end while coarse material exits at the other; sifted flour falls through the rotating screen. It's important to remember that sifted stone-ground whole wheat flour is not the same product as that which is roller-milled. Bran is totally removed from the white endosperm when wheat passes through a roller mill, but when wheat is ground with millstones, the bran is pulverized into particles that are almost as small as the ground endosperm. Some of the coarser particles of bran are removed by sifting, but many remain in the whole wheat flour. Monroe Stutzman, an enterprising young Amish miller from Holmes County, Ohio, regrinds bran through a hammermill and then blends it back into his whole wheat flour.

Sifting can be done on the very smallest scale in the most elementary manner with a handheld kitchen flour sifter, the same little unit that your grandmother had you squeeze for her when she was making pies. If you need to scale up to something bigger and better, there are numerous commercial sifters on the market that are fairly reasonably priced. Pictured below is a Meadows flour sifter.

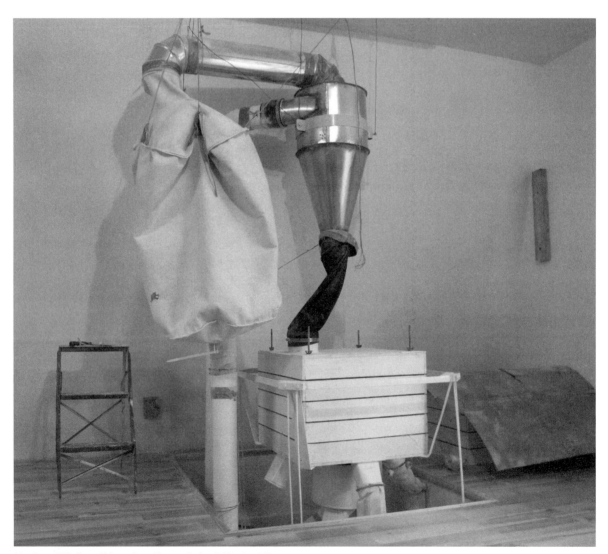

Meadows Mills flour-sifting unit at Gleason Grains, Bridgeport, VT. PHOTO COURTESY OF HEATHER DARBY

This is probably the simplest and cheapest sifter you can buy. It is basically an eccentric driven rotating box suspended from four flexible steel rods. Meadows sifters can accommodate up to nine individual sifting screens within, but four identical screens stacked one on top of the other seems to work just fine. Flour enters the top of the reciprocating box and then begins to work its way down through the stack of screens, propelled by both gravity and vibration. Thirty- to forty-mesh screens will do a nice job whitening up whole wheat flour. Coarser flour particles are separated out on each subsequent screen. Screen mesh size is measured in the number of squares to the inch—a square inch of forty-mesh screen has forty wires running in two different directions. A new Meadows sifter can be purchased for somewhere between fifteen hundred and two thousand dollars. Some farmer flour producers will sift all of their flour through larger-mesh screens (16- or 17-mesh) to remove impurities like chaff and foreign matter that escapes the preliminary cleaning process, which is like having an insurance policy that will guarantee you the best-quality clean flour. Sifters are one piece of milling equipment that can be easily home-constructed. You can buy the screening material directly from Meadows and build almost everything else out of wood. Flour sifting is truly a straightforward mechanical process.

Volumes more could be written about the process of turning wheat into flour. Whether you intend to crank a hand grinder attached to the kitchen counter or operate a large high-output mill, you will have to gain experience through practice to become a good miller. Grind the wheat, feel the flour between your fingers, and then bake with it. It's important to take part in the whole process. Even if your bread disappoints you at first, you've still accomplished something because the job was done with your own hands. Grinding and using your own wheat is the agricultural process at its best.

Wheat for Livestock Feed and Other Uses

There are other ways to consume wheat besides turning it into flour and baking with it. A number of people out there have given up eating bread and other flour products for a variety of health reasons, and refined white flour certainly gets a lower grade than its whole wheat counterpart. For those who are trying to avoid carbohydrates in their diet, processed wheat flour is off limits. However, wheat kernels can be consumed whole and cooked just like rice. Small amounts of parched and boiled wheat make a great accompaniment for a salad. In the Turkish and Mediterranean tradition, soft wheat that has been parboiled, dried, and uniformly cracked is called bulgur. With its distinct nutty flavor, bulgur is a primary ingredient in Middle Eastern dishes like tabbouleh.

We are particularly partial to sprouted wheat in our household. Good seed-quality wheat can be laid out in trays and wetted down for sprouting. We like to grind wet wheat sprouts in a hand mill for sprouted wheat bread. The ground material is very similar to bread dough. We mix it with a little sourdough for starter and then bake it several hours at a low heat. The result is heavy-textured dense bread with the sweet qualities of wheat malt. Sprouting wheat wakes up the vitamins and enzymes in the seed while it reduces the carbohydrates. Wheat can also be grown out in trays of soil to produce tender young green shoots that are harvested with scissors and squeezed through a special press into wheat grass juice. This is a healthful tonic used for various health issues. Wheat is a versatile food with myriad uses. There is life after wheat flour.

Another important end use for wheat is to feed livestock. As a cereal grain, wheat has an extremely high level of protein (11 to 15 percent) and good amino acid distribution. Many farmers consider it to be a good substitute for corn in a ration. Corn may have slightly higher energy levels, but with 14 compared with 8.5 percent protein, wheat beats corn hands-down. When the percentage of wheat in a ration is increased, protein supplements like soybean meal can be reduced. By-products of wheat like bran and middlings also make good additions to a livestock ration. Middlings are considered filler by some experts, but bran is quite useful and helpful because of its fiber and protein content. Cows, horses, and lactating sows all seem to thrive on bran.

Of all the farm animals out there, poultry probably do the best on wheat. If wheat is included in a chicken mash, it should be ground somewhat coarsely. Chickens have stones and rocks in their crops to finish the grinding process, so a mixture of cracked wheat and corn makes excellent scratch feed for them. Care must be taken when feeding wheat to dairy cows as part of their grain ration, however. Cows are ruminants with four stomachs. The rumen of a cow is basically a series of giant fermentation vats. Finely ground wheat can turn to paste in a cow's rumen, blocking everything up and halting the digestive process. This condition is known as acidosis, and it severely challenges the digestion as well as the health and well-being of a dairy cow. Because of its lower protein and higher starch level, soft winter wheat is much better for cattle than hard red spring. The key to feeding wheat to livestock is to limit the amount to no more than 25 percent of the total ration and to not grind it too fine. Pigs and chickens seem to do better with wheat supplementation than cows. When it comes to feeding wheat to our livestock we should practice moderation much as we would when feeding any other grain.

Agricultural diversity is a hidden blessing in any farming system. When we have both human consumption grains and livestock in our farming mix, we have the option of feeding a challenged wheat harvest to animals on our own farm. We keep returning to the theme of sustainability. Cereal crops, especially wheat, make sense on a farm in the northeastern United States for many reasons. We can build up the land while we are producing a crop that will sustain us. It is possible to plant next year's hay crop at the same time as this year's wheat crop. The straw by-product of our crop can remain on the farm to bed our livestock; it will in time return to the earth in the form of compost. And last but not least, if we cannot sell or consume the wheat because of weather-related quality issues, we can use it as animal feed. When it comes to growing wheat in our region, there are certainly perils and pitfalls lurking around every corner. But even if our end is result is less than perfect, a wheat crop is still well worth growing. And when we do harvest that perfect crop, we can take delight in a job well done thanks to a lot of hard work and nature's blessing.

CHAPTER ELEVEN
Barley

Barley might be described as wheat's little brother, but it is nevertheless a major cereal crop with many distinct advantages. The rebirth of wheat growing in the Northeast has received so much attention from the press and the local food movement that barley has gone relatively unnoticed, yet it has nearly the same protein levels as wheat (13 to 14 percent) and can produce phenomenal yields of two or more tons per acre. It should be considered well worth growing either as a principal crop or as part of an organic rotation. I grew my first crop of barley in 1977 right next to my first wheat crop. I remember the variety was called Herta, and I was impressed by how it thrived here on our farm in those early days. We reaped and stooked it the old-fashioned way with the grain binder and threshing machine, and our neighbor Sam Pion brought over his old John Deere 60 to provide pulley power for the flat belt that drove the old thresher. As we tossed the bundles of reaped barley into the thresher and filled burlap bags with the golden grain, Sam kept telling us that forty years earlier, everyone dreaded "thrashing" barley because it made you itch all over. The reason for this predicament was the beards or "awns" that are attached to the upper tip of each barley kernel. Indeed, barley is distinctive and easy to identify in the field because of these wispy little hairs, which protrude from the top of each stalk of grain. Barley awns grow with tiny little barbs, and it is these little hooks that irritate your skin when you get intimate with the crop during the harvest season, although itchy barley syndrome is much less a problem these days in the air-conditioned cab of a modern combine. Barley beards measure three or four inches in length and give a field the appearance of a collection of small paintbrushes. Personally, I think barley has the most beautiful growth habit of all the cereal grains.

Barley: A Brief History

Barley originated at the same time and in the same area of the Near East as wheat's ancestors, emmer and einkorn; domesticated barley descended from the wild Western Asian grass *Hordeum spontaneum*. In a similar manner to the development of wheat, genetic mutations in wild strains of barley allowed for better seed retention and less shattering. Once plants whose seeds remained well attached were selected and replanted, the domestication of barley began to progress. Evidence of domesticated barley dating back to 8500 BC has been unearthed in archaeological digs near the Sea of Galilee; barley has also been found at Neolithic sites all over the Fertile Crescent region such as Abu Hurevra in Syria and at Ali Kosh, Iran and Jarmo, Iraq, both located in the Zagros Mountains. This amazing plant quickly spread all over the Mediterranean basin and eastward into Asia. It became the staple cereal of ancient Egypt, where it was used for bread as well as beer. The Greeks also used barley to make alcoholic beverages. As a matter of fact, barley was the principal cereal grain used for human consumption in bread making in Southern Europe until the late fifteenth century. To be sure, the widespread adoption of barley for food and drink accompanied the advancement and development of important civilizations. Around the same time this bearded cereal grain was beginning to be cultivated in North Africa, it was also making its way eastward across the Asian continent to Pakistan, Tibet, and Korea. There is no other grain grown across a wider variety and range of environments than barley. In Europe, it thrives in the hot dry climate of the Mediterranean as well as in the north of Lapland at a latitude of seventy degrees. In North America, barley grows from California to Alaska in the West and from the Carolinas to Labrador in the East. It is the only grain grown in the extra-high-elevation valleys of the Himalayas. All of this versatility leads us to the conclusion that barley is a cereal crop with huge potential here in the Northeast.

Barley for Human Consumption

Barley (*Hordeum vulgare*) has a growth habit very similar to that of wheat save for a few differences. It has single-flowered spikelets that are arranged in triplets along the rachis or upper stem of the plant. The barley kernel that develops from each flower is unique because it remains covered by a tightly fitting outer hull or husk after threshing. Attached to the outer glume is the whisker-like awn that extends up in the air above the top of the plant. These beards with their barbs are what caused so much consternation and itching for my neighbor all those years ago, but these scratchy wisps also keep wildlife like deer from entering your field and chewing on your barley crop. The fact that the outer hull of the barley kernel remains so firmly adhered to the bran and endosperm underneath presents a problem to those of us who want to raise barley and put it in our soup.

Barley for human consumption must undergo a very specialized dehulling process. We will discuss the specifics of hulling later in this narrative, but suffice it to say, it's not a simple task. If you looked at a barley kernel under a magnifying glass, you might notice that the surface is mottled and ridged, which makes dehulling even more arduous. Passing barley kernels between two slightly separated stones will roll the grain around and strip off its outer husk; once the hull has been removed, the inner berry looks very much like a kernel of wheat with a deep furrow on the opposite side from the germ. Professional barley processors will polish the grain and pearl it. Pearled barley has a very white appearance because it has been steamed to remove the outer layer of bran from the kernel. Barley is probably one of the healthiest grains we can consume because it has an extremely low glycemic index. Once we eat this amazing grain, it can take as long as ten hours for our bodies to slowly break down the complex carbohydrates contained within. Little if no strain is put on our digestive and endocrine systems when we cook and eat whole barley, but curiously, even though it is healthy and delicious, whole grain barley for human consumption represents a very minuscule percentage of the ultimate end use for this crop.

Barley for Livestock Rations

As you might imagine, barley is used primarily as a feed source for livestock, and in areas where corn cannot be raised, it makes a fine substitute in animal rations. Protein levels in barley usually average between 12 and 13 percent and can sometimes reach 14 percent under ideal growing conditions. Effective fiber levels are quite high thanks to the outer husk that comes along with every kernel. Barley has as much energy as corn, but the carbohydrate in its endosperm is a bit more soluble and rapidly digested. It would seem as if the converse was true—you would think that since the starch in a kernel of barley is so much harder in texture than that of corn, it would be much more slowly digested in the rumen of a dairy cow. However, it is just the opposite. Corn is easily milled into an animal ration because of its soft texture, and the carbohydrate in ground corn has "bypass" energy that's slowly digested as it travels through the four stomachs of a dairy cow. Barley, on the other hand, is difficult to grind because it is so hard, so when a cow ingests barley meal, the carbohydrate and starch in the grain is immediately dissolved in the first stomach. For this reason, a mixture of corn and barley is recommended to round out the energy fraction in a standard dairy ration: Barley will provide the immediate energy and corn will take care of the sustained energy.

Barley is great feed for livestock other than dairy cows. Before the era of tractors in agriculture, barley was used as a grain for feeding draft horses in Europe, although, for some reason, oats were favored here in North America. Hogs also thrive on barley. Pork from the Canadian prairies has the reputation of being extremely fine-flavored and extra lean because the animals are fed barley. Barley grows much better than corn in Alberta and Saskatchewan. I remember hearing about this phenomenon when we used to cross the border to buy grain at Moulin Viens et Frères in Ayers Cliff, Québec, thirty years ago. We could purchase ground barley meal (*orge moulée*) for about two-thirds the price of ground corn, a rather finely milled product very similar in texture to coarse wheat flour. Canadian barley meal dissolved quite nicely into our skim-milk-based hog slop, and with its higher protein level, ground barley was the sole grain ingredient in our liquid pig feed. At the time, we also had the opportunity to buy rolled barley at the mill. This was purported to be a deluxe ration for dairy cows, as finely ground barley was simply too powdery and could paste up a cow's rumen and plug things up. I was so enamored by the idea of feeding rolled barley to our small herd of six cows that I loaded thirty bags of our own farm-grown barley onto our pickup truck and took them to Ayers Cliff to be rolled. This service certainly wasn't available on our side of the border, and I really felt like a big shot "going to the mill" in a different country. I'm sad to report that the whole experience was a bit of a letdown. We proudly fed our own rolled barley to our little herd of cows only to see all the kernels of barley pass through the animals and out into the manure. The experts said, "Oh, the goodness has been extracted from the grain. Check the kernels and see if there is any meat left in them." Unfortunately, our rolled barley wasn't even half digested. We tried it a few more times, and asked the mill to roll it even flatter and more aggressively the next time. The results were only slightly better. Eventually, I went back to coarsely grinding my cow barley in my old buhr mill; the chopped barley seemed to digest more completely, and we were happy once again. We also fed lots of barley to our chickens during the same era, although in retrospect I think it would have been better to have had a wider diversity of grains to formulate a more complete poultry ration. Barley is a great feed grain when combined with other sources of energy like corn.

Types of Barley

There are three subspecies of barley—two row, six row, and naked or hulless.

Six-row barley is raised primarily for animal feeding. It has six rows of spikelets, grouped as two sets of triplets attached to the rachis or upper stem of the plant. It is recognizable by the plump, somewhat rounded appearance of the grain heads. Until relatively recently, this was the only type of barley that I grew. Two-row barley has only two rows of kernels positioned longitudinally on the head of the plant. Its kernels are larger and lower in protein than those of its six-rowed counterpart. This type of barley is recognizable in the field by the somewhat flattened appearance of the grain heads. Two-row barley is used primarily for malting, as its

Two-row barley.

larger, plumper kernels contain more endosperm, which provides more fermentable starch for the malting process. The low protein of this type of barley provides a distinct advantage in the malting process because it produces a clearer, less cloudy beer. (Malting is basically a sprouting process in which the starches in the barley endosperm are changed to sugars. The grain must first be steeped or soaked in water. The feathery extension of the barley hull that attaches to the beard actually serves as a wick to help hydrate the kernel by sucking up water. The exterior hull of the barley kernel serves two purposes in the malting process: It protects the embryo of the seed, which is so important to proper sprouting; and it serves as a filter when the malt is extracted. Malting barley is big business and a science unto itself. We will discuss some of the finer points of malting and producing malting barley later in this chapter.) Last, but not least, there is naked or hulless barley. Hulless barley has been around for thousands of years as well, but since it has very little commercial value, it has been the least important of the three subspecies. Most of the breeding advances for hulless barley have been made by seed savers from indigenous cultures all over the world. These folks have been growing naked varieties of barley for their own food. Veteran seedsman and plant curator Will Bonsall of the Scatterseed Project has been cultivating and maintaining numerous varieties of hulless barley on his farm in Industry, Maine. Will even has purple varieties of hulless barley from faraway places like Nepal and the Peruvian Andes.

Hulless Barley for Home Production

Hulless barley sheds its outer husk during the threshing process, which makes it an ideal grain for the home gardener who is looking for some barley to eat directly. It is also a wonderful poultry feed: It's high in protein and much lower in fiber, because it lacks the husk covering. Since they have not been commercialized, most of these heirloom varieties of hulless barley are typically low yielding. This really isn't a problem. Who cares how much hulless grain a plot in your garden will produce? Of course more would be better, but the fact that we can grow our own soup barley on a small scale without a pile of expensive hulling equipment is reward enough. There has been a limited amount of breeding improvement work done on hulless barley in Canada, but a commercial variety named Hawkeye was released by Agriculture Canada in the 1990s. It has performed quite well in eastern Canada and offers us something we can grow on a larger scale for the human consumption market. Like all hulless grains, naked barley is not entirely without hulls when it is threshed. My experience growing AC Hawkeye has been that there is always a very small percentage of grains that slip through with hulls. Nevertheless, hulless barley is well worth growing to stock your larder with a nutritious grain that is so healthy to eat.

My Experience with Two- and Six-Row Barley

My own adventure growing barley over the past three decades has been quite interesting and telling. As I stated earlier, I began by growing six-row feed barley. My yields were quite good for many years, even with some mustard pressure in my crops. Eighty bushels or two tons to the acre seems to be the target number for a respectable crop of barley, and I've been able to achieve this a few times, especially in the drier seasons when summer growing conditions were tending toward a drought. Barley seems to thrive in these parched situations, which explains why it is such a principal cereal crop on the prairies and around the Mediterranean basin. My average barley yields have been closer to a ton to a ton and a half to the acre. I'm totally satisfied with a moderate barley harvest, as I still consider this to be a sustainable level of production for my style of mixed grain and dairy farming.

One very interesting trend has emerged in the world of barley during the past five years or so. My Québec seed dealer began offering two-row malting barley varieties in addition to the six-row feed varieties in his lineup. These barley varieties, like AC Newport, were billed as dual-purpose and could go for malting or livestock feed. I'm not sure if this was part of the larger trend toward more cash crops in Québec's Eastern Townships, but it certainly gave farmers the option to make a few more dollars from their barley crops if the grain made the grade for malting. At the same time, Canada Malt in Montreal was looking for more grain from within the province, and they also began buying large amounts of two-row barley from potato farmers in Aroostook County, Maine. I went along with the program and began planting two-row barley about five years ago. I have to say that my luck with malting barley has not been the best. I've had a few good crops, but for the most part my results have been less than perfect. My friend Blaine out in North Dakota tells me that two-row barley definitely prefers a more arid climate, and my experience has been that if my crop of malting barley gets exposed to any period of prolonged moisture, the heads and kernels simply shrivel up and don't really amount to much of anything at harvesttime. I never had this problem with the six-row types. I also have to question the dual-purpose nature of malting barley. The fact that this type of barley has protein levels that are 1 or 2 percentage points lower doesn't benefit me as a livestock producer feeding dairy cows, as I want the extra protein in my homegrown rations. This situation is particularly frustrating because local grain malting has just begun to come to the forefront here in our region, and the demand for two-row barley has jumped exponentially with the rise of local maltsters and craft breweries in the Northeast. In my own particular case, I have decided to return to growing six-row barley to use as an ingredient in my home-produced dairy feeds. I know what I can do well and what my grain needs are. I'm not a cash cropper, I'm a dairy farmer who raises grain as part of his rotation. This is not to say that raising malting barley is out of the question here in the Northeast. There are many strategies that can be employed to ensure the production of high-quality malting barley right here where we live. Again, this subject will be pursued in much greater depth in the malting section later in this chapter.

Agronomic Considerations for Barley

Whereas wheat requires extremely high levels of fertility and oats can get along with much less, the fertility needs of barley lie somewhere in the middle. Barley is the most shallow-rooted of all the cereal crops grown in our region, and for this reason it is necessary to have adequate levels of plant nutrients in the upper horizons of the soil. A moderate coat of good compost spread in the spring before planting will feed both the soil and the crop. If you use manure, it's best to apply it the previous fall at the time of plowing, but excess manure can foster the growth of annual weeds and cause a barley crop to lodge. Barley will germinate under the same cool spring temperatures as wheat. However, it is more sensitive to early-spring frosts than the wheat plant, so it is probably best to wait until the wheat is all sown before seeding barley. If you're able to get the wheat in the ground by the tenth of April, you might want to wait another week before heading out to sow your barley. Use your senses and figure out what kind of spring weather you're having. If everything is coming on gangbusters and the earth is warming up quickly, plant that barley right away, but if the April weather remains cool with lots of lingering snow flurries, you might want to hold off. A well-worked, friable, and uniform seedbed is absolutely essential for good germination and early growth. If there is any question at all as to the quality of a final seedbed for barley, fit it one more time with the field cultivator and spike drag. Those weak and wimpy barley roots need all the help they can get. Fall plowing seems to work best in our northern latitudes, and sandy or well-drained loam soils grow the best barley crops. Barley has no tolerance for wet feet, so excess moisture at any point in the growing season will stunt the plant—I've seen this happen so many times. Also, low spots in a field will not support a barley crop, and clay soils that tend toward prolonged dampness should be avoided. Lots of thought and consideration should take place when figuring out where to plant a crop of barley. This crop has pretty specific requirements—any old field just won't do.

Considerations for Winter Barley

There is also winter barley to consider. Whereas most barley grown in our more northerly regions is spring-planted, there are varieties that are planted in September. Like winter wheat and rye, winter barley germinates in the fall, establishes itself, and then goes into dormancy as winter approaches. If you can get winter barley to survive in your location, it will give you the jump on next spring's weeds, along with higher yields and an earlier summer harvest. Winter barley is quite common in the mid-Atlantic states from Pennsylvania south into Virginia. Southern and western New York State represent the most northerly reaches of its potential habitat, but this hasn't prevented farmers in other parts of the Northeast from trying this crop. Barley has the least winter hardiness of any of the cereal grains and is easily killed off by chilling winter winds and subzero temperatures. If you live in the warmer coastal areas of Connecticut or Massachusetts, you might just be successful with winter barley. If you're trying the crop at higher inland elevations in southern New England, be aware that you are gambling. The same holds true for northern New England. If the ground freezes and the snow arrives early and stays put, winter barley will most likely survive. Several of my grain grower associates have successfully harvested winter barley crops here in Vermont. The most popular varietal choice for New England has been Macgregor, which was bred in the milder regions of western Ontario near the Michigan border. I planted a small trial plot of Macgregor once with only moderate success. About a third of the stand winterkilled, leaving me with slightly less barley than I would have harvested from a spring planting.

Quality Seed and Seeding Rates

Whether you're planting barley in a home garden plot or in a large field, it is imperative to start out with high-quality seed. You want seed with at least 90 percent germ and excellent vigor. Barley is difficult enough to establish, especially if a prolonged rainy spell arrives right after planting. Planting seed that has been tested for germination and early-season vigor will give you one more leg up in establishing a first-class stand of barley. Seeding rates can vary from 100 to 140 pounds to the acre, but if the barley is to be used as a nurse crop for a new forage seeding, it's best to cut the rate back to the 100-pound level. This will give the new hay crop plenty of room to grow beneath the developing stand of barley. As a matter of fact, there is no better choice of cereal grain for seeding down, because barley has very few leaves; this lets the sunlight reach the timothy, clover, and alfalfa below. If barley is to be planted straight with no other companion crop, a higher seeding rate of 125 to 140 pounds to the acre is customary. As with all the other cereal grains, the recommended seeding rates for barley have been creeping upward over the past thirty years. One hundred pounds was standard in the 1970s; now it's 140 pounds. Somehow I think the only beneficiaries in this process of seeding rate inflation are the seed companies and dealers. If you're in doubt about how much seed to plant, consult your neighbors and university extension specialists who may have performed trials in your state. One hundred twenty-five pounds to the acre is probably a good average seeding rate that you can trust to be neither too much nor too little.

Selecting Your Barley Variety

Selecting a variety of barley was pretty simple in my earlier days of grain growing. Maybe ignorance was bliss thirty years ago, but I took what the seed dealer had to offer and that was it. The Herta barley that I planted from Stanford Seeds in Buffalo, New York, was the only thing I could get in Vermont in

1977, but while driving around southern Québec in June and July of that year, I began to see a few variety placards prominently displayed in front of some really nice-looking barley fields. One thing was certain back then: There were only two or three varieties available for planting. Each of the major Québec seed companies had their flagship variety. The one that really stood out back in the late 1970s and early '80s was called Léger. It was quite a bit taller than all of the other barley out there, with extra-large, plump heads. Developed at Laval University in Québec City, Léger became my barley of choice for almost an entire decade until it was discontinued in 1987. Around 1985, I began to notice the appearance of a much shorter variety named Chapais, which looked even thicker and higher yielding. It had been bred with extra-short straw to protect it from lodging when high fertilizer inputs were applied. Léger with its long straw did have the tendency to go flat before the combine got to the field. There were other varieties, like Bruce, that came from the Bruce Peninsula near Georgian Bay in Ontario. Varieties have come and gone over the years, but we have a lot more to choose from these days. AC Nadia, AC Newport, and AC Newdale have dominated the Québec market for the last several years, but these varieties will disappear, too, and something that is supposed to be newer and better will come along to replace them. For the past couple of years, I've been planting a Minnesota variety called Lacey that I have been purchasing from my friend Blaine Schmaltz in Rugby, North Dakota. I have to say that in my eyes, today's varieties aren't really any better than what we were planting back in the 1980s. We don't have as much choice as we think. So if you are trying to figure out what barley to plant, see what is readily available. We don't live in a part of the country where barley is a major crop, so if you find several different varieties to choose from, take some time to study their agronomic advantages and disadvantages. You'll definitely want to select a variety that yields well, but make sure it also has strong straw. Consider the end use of your barley crop. If livestock feed is what you desire, plant the variety that produces the highest levels of protein. If malting is in your game plan, a lower-protein barley with high starch levels is what you want. And last but not least, there is disease resistance to consider. Of all the supposed improvements bred into barley in the last thirty years, the ability to resist common maladies like stem rust and spot blotch have been the most important advances made by plant breeders. Factors like good weather, soil fertility, and excellent drainage play a much larger role in achieving an outstanding barley crop than variety selection.

Planting Barley Seed

I would recommend mechanical-seeding your barley with a grain drill or push-type garden seeder over hand-broadcasting. The archetypal image of the ancient farmer walking through his field, cloth sack under his arm, scattering seed by hand looks good as a block print in an old text, but it's not the best way to establish a high-yielding stand of barley. Remember that Jethro Tull invented the drill seeder in the eighteenth century; it has been the standard seeding tool for grain since the Civil War. Achieving an even stand by hand-broadcasting (even with a hand-cranked Cyclone seeder) is a lost art today. Scattering the seed is only half the battle—uniform germination won't happen unless the seed is covered with an even layer of soil. This task of burying broadcasted seed is usually accomplished by pulling a light implement like a spike drag or a tree branch over the recently broadcasted seed. On the other hand, a hand-pushed garden seeder or grain drill (both are described in detail in the planting chapter) will make short and accurate work of barley planting. Cereal seeds simply tumble down the seed tubes and are "drilled" into the awaiting seedbed. Uniformity is what we're looking for in a planting situation, and with this method, a specifically regulated amount of seed is evenly buried at the same depth across a large area. This is not to say that hand-broadcasting won't work—by all means, if you don't have access to an Earthway or Planet Jr. garden seeder, scatter it by hand and find a way to cover the seed. You will still get a barley crop. It might be thick in some areas and thin in others, but at least it's planted. Grain drills are definitely a luxury; most of these machines are equipped with a separate seed box for the grass seed, which allows both the barley and the accompanying hay crop to be planted at the same time. In the end, it doesn't matter whether you have a fancy new metal grain drill with press wheels and other modern features or an antique model with wooden wagon wheels. The seed will still find its way into the earth.

Interplanting a Hay Crop, and Use with Other Grains

Barley is an extremely versatile cereal crop that can offer us all sorts of possibilities. It is best for seeding down a companion hay crop because it doesn't get too thick or rank in its growth habit. Barley has the shortest straw as well the shortest growing season of any of the major cereals, and it usually takes only ninety days to go from seeding to ripe grain. This allows an April-planted crop to be combined in late July or the very beginning of August. By this time an underseeded hay crop will be starving for light and nutrients. The mid-summer barley harvest gives the companion hay crop plenty of time to establish itself while growing conditions are still ideal. The alfalfa, clover, and timothy will have all of August and September to gain the strength and stature needed to take them through the following winter; the remaining barley stubble will also act as a "snow catcher" to further protect the new hay seeding from being blown bare by fierce winter winds.

Barley can also be planted in a mixed-grain situation. Years ago, Stanford Seeds sold something they called the Vermont Grain Mix, which was a mixture of wheat, barley, and oats intended to be combined at the same time. Since it is at least two weeks earlier than wheat or oats, this sort of mixture never really did justice to any of the three different grains. By the time the wheat and oats were ripe, the barley was overripe and had begun to shatter its seeds and go by. Canadian field peas have the same ultra-short growing season as barley, which makes them an ideal mate for this important feed grain. Peas are usually mixed with barley seed at a one-to-three ratio. A combination of thirty pounds of peas and ninety pounds of barley is just about the right amount of mixed grain to plant an acre; mixing can take place right in the seed box of the grain drill. Dump three five-gallon plastic pails of barley across the length of the seeder followed by one bucket of peas evenly sprinkled on top of the barley. As an added bonus, peas are a leguminous crop and are able to fix their own nitrogen from the atmosphere. It is very important to inoculate them with the proper rhizobium bacteria to ensure maximum and efficient nitrogen fixation. When purchasing inoculant for peas, make sure that it is *Rhizobium leguminosarum* bv. *viciae*. Most inoculants come in the form of a peat-based black powder. To inoculate, dampen the pea seed ever so slightly, then mix in the inoculant with some sort of stirring device. Once the peas have been uniformly blackened by the inoculant, they are ready to be stirred into the barley in the drill seed box. Indeed, peas will thrive when planted with barley early in the season; the pea vines will twine around the barley straw for support. A field of headed-out barley and white flowered peas is a wonderful sight on a beautiful late-June day. Field peas usually test about 25 percent in crude protein. Mixed with barley this raises the final protein of the crop to the 18 to 20 percent level, which is more than enough protein for just about any livestock ration. As described earlier, barley is one of the best sources of sustainable cereals for animal feed. When mixed with field peas, it can go the extra mile to be the very best grain to feed a wide diversity of stock.

Some Trade-Offs with Barley

Despite all of its positive attributes, barley does present us with a number of challenges and disadvantages. For one thing, its shallow and limited root system makes it quite drought-prone when planted on a sandy soil with low organic matter levels, which is all the more reason to grow barley on high-fertility medium-loam soils. Another drawback is that barley will do poorly when grown under excessively wet conditions; an overabundance of rainfall or prolonged damp weather at any point in the growing season will stunt the crop. In these sorts of situations, grain heads won't fill normally and what started out as a pretty good-looking crop will wither up and fade away. Also, barley has the weakest straw of all the commonly cultivated cereals, so lodging can be a problem, especially after a violent mid-July thunderstorm when stalks heavily laden with grain are particularly susceptible to the damaging forces of wind-driven heavy rain. One minute you may be looking at approaching storm clouds and the most beautiful field of barley you've ever raised; twenty minutes later, disaster has struck and your barley is flat on the ground. This is part of the big gamble we call farming. Beauty can disappear in an instant.

There are, however, a few strategies that can be employed to hedge your bets and possibly avoid this sort of tragedy. Don't overfertilize your crop with any form of soluble nitrogen. Avoid chicken manure and fertilizers like Chilean nitrate; the same element that will make your barley field green, lush, and high yielding will also produce stalks that are much weaker and more prone to the vagaries of the weather. Choose a barley variety with good lodging resistance. Most seed companies provide this sort of information in their catalogs and sales literature, and usually there is a chart that lists variety offerings and their respective scores on a one-to-five or one-to-ten ranking. Consider the whole agronomic package like yield and feed value, but first and foremost, select a barley variety with excellent standability. These numbers and charts aren't totally foolproof, however, and they're not altogether accurate. I remember choosing a Canadian variety of barley called Maskot about fifteen years ago. This variety supposedly had straw that was strong enough to stand up to just about anything. The lodging score may have been correct in this case, but it didn't reveal everything about the agronomics of this particular variety. We had a series of thunderstorms just before harvest that year, and as I looked at the field from afar, it seemed that it had done pretty well. The straw was still standing after the heavy rain. But when it came time to combine the Maskot barley, there was very little grain coming into the grain tank on the machine. I got out of the cab and inspected the field more carefully. The stems of the barley were still all erect, but the grain heads had broken off at the top of the stalk and were all lying on the ground below where they would never be harvested by the combine. So much for the high lodging-resistance score. This was just another one of the many hard lessons that I have learned over the years while trying to grow grain.

Try as we might, barley straw will never be as strong or tall as wheat straw. It is so much less fibrous and lower in lignin that it will begin to break down and rot right in the field after harvest if it is exposed to repeated rains. This tendency to quickly decompose, however, makes barley straw an ideal bedding material for livestock housing situations. It is also one of the reasons why everybody and their brother constantly asks me if they can buy a tiny amount of barley straw to put in their pond to keep the algae from growing. Barley straw is a very effective algicide in an aquatic situation because it breaks down so rapidly that it increases the carbon-to-nitrogen ratio in pond water, which in turn prevents the growth of algae. Not everyone can make a business out of selling barley straw to pond owners, but if you're in the right place at the right time, you may just receive the ten dollars per bale that people are willing to pay to keep their ponds clean and beautiful. In some very rare instances, barley's weaknesses can actually be to your advantage.

Weeding Your Barley Crop

Weed control in a recently germinated barley stand is not as straightforward or easy as it would be in a similar field of wheat or oats. Tine weeding is not recommended in a young barley crop at the usual five- or six-leaf stage because the root systems of the plants won't take it. Barley's relatively weak roots set it apart from the other cereals; the plants aren't as well anchored into the soil, and even a light tine weeding will cause excessive damage to a crop. Numerous studies have been done in Denmark to determine the ultimate effectiveness of post-germination tine weeding in cereals. There are always trade-offs to consider: Will the setback to the grain crop offset the benefits of weed removal from the young stand? In most cases, the Danes have found that tine weeding is worth the effort. The field may appear bedraggled and disheveled after weeding, but it eventually recovers and outgrows the competing weeds. This simply is not the case with barley, however. Too many plants are pulled up by the roots and left to die along with the weeds.

Weed control in barley has to happen between seeding and the first signs of emerging plants, and several passes with the weeder might be needed to perform the job properly. As discussed earlier in this narrative, tine weeding is all about timing and the right weather, so hope for some sunny dry conditions during that little window of time between seeding and emergence, and dig around in your seedbed to monitor the germination progress of the barley. Ideally, you want to be out there with the weeder just before the barley shoots pop through the ground. If you're seeding forages into your barley crop, this is also the time to be laying down the seed. Broadcast your future hay crop seed and work it lightly into the soil with the tine weeder. Don't forget to balance and fine-tune the minerals in the soil of your barley field, as weeds will not thrive when calcium, magnesium, and phosphorous are balanced and in adequate supply, and micronutrient fertilization with copper, zinc, boron, and manganese will create a soil environment in which weeds fail to thrive. Options for mechanical weed control in barley are few and limited, so get out there with the tine weeder early when the sun is shining and do the best job possible before the barley emerges from the ground.

Diseases Common to Barley

Barley is the most disease-prone of all the cereals that we can plant here in the Northeast. I'm not sure why, but it seems to attract every stray virus and fungal spore out there floating around in the environment. In the previous chapter, we talked about the various rusts that can potentially plague a wheat plant. Well, barley is susceptible to every one of these rust infections and more. Scab, also known as fusarium head blight, can be a major problem in a year with prolonged wet weather during pollination, and barley kernels will end up with the same telltale pink hue that appears at the tip of the wheat kernel because of fusarium infection. The same protocol of mycotoxin testing for DON (deoxynivalenol) should be strictly followed if the least hint of pink is observed in a sample of harvested barley. Barley with DON levels over one ppm should not be used for malting. Ten ppm of DON is the cutoff point over which barley cannot be used for livestock feed.

Leaf diseases known as blotches seem to run rampant in barley. Net blotch (*Pyrenophora teres*) starts out as a small round spot on a leaf and will then turn into a larger net pattern of brown lines with yellow borders on the plant tissue. This particular affliction is most common when cool damp weather prevails for long periods of time. Spot blotch (*Bipolaris sorokiniana*) also starts out as a small dark brown spot on the leaf. As the fungal infection develops, the little dark spots turn into elongated patches surrounded by yellow. Speckled leaf blotch first manifests itself as light brown elongated spots surrounded by yellow tissue; eventually, little black raised dots will form in these spots. Stem rust (*Puccinia graminis*) also affects barley by forming dark red spore masses on its stems and leaf sheaths. Scald (*Rhynchosporium secalis*) is another common malady in barley that occurs when water droplets remain on leaves during cold weather. Little oval spots develop that are bleached or straw-colored in appearance. Powdery mildew (*Blumeria graminis* f. sp. *Hordei*) is caused by a completely different type of spore infection in which white to gray masses cover the leaf blades of the barley plant. This disease is more prevalent in cool, humid, overcast weather, and plant leaves will turn brown and then yellow. Total plant mortality is the end result of a powdery mildew infection.

These diseases and a whole host of other fungal infections can attack barley crops even in a good growing year. All it takes is a prolonged period of wet and cool weather for fungal spores to settle on a leaf and develop into something that will either stunt or kill the plant by curtailing photosynthesis. Conventional farmers with an arsenal of chemical sprays at their disposal can and will use fungicides to battle leaf diseases, but since we don't have these options in an organic system and we really don't want to use toxic rescue chemistry, we have to concentrate on strategies that are more preventive in nature. The first thing we can do is to make sure that we are starting out with disease-free, clean seed. Some of these infections, like net blotch, are seedborne. It's also important that we consider our barley crop in the larger framework of a three- or four-year period of successive crops in a rotation. Barley does much better after sod or a corn crop and should never follow another cereal, as the straw and stubble of wheat or oats can harbor all sorts of spores that have the potential to afflict a subsequent crop of barley. Another proven strategy to ward off diseases in barley is to make sure that the crop's fertility needs have been completely met. For example, low levels of phosphorous are a sure ticket to leaf yellowing. Follow the recommendations of your soil test. If you have any doubts about what minerals a crop is lacking, plant tissue testing will provide you with the answers. A foliar feeding spray will be the quickest, easiest, and best method of spoon-feeding an ailing crop. Depending on how far out on a limb you want to venture, you might consider adding extra goodies like molasses, sea minerals, and biodynamic silica (finely ground quartz crystals) to your foliar feed spray. Barley is a cereal crop with tremendous potential to produce, but also to fail. When we stop to think that it is native to the desert Near East and the arid Mediterranean basin, it is amazing that we even try to grow it here in our humid, somewhat wet climate. I would recommend *Barley Disease Handbook* published by North Dakota State University for more information about this subject. This helpful guide contains excellent illustrations and is available online.

Avoiding Lodging

Once barley has made it through the vegetative stage and headed out, it's time to think about the imminent harvest. Hopefully, your crop will progress far through the season without incident and look pretty good. By mid- to late June, hopefully, the weather will settle into a warm and dry pattern for the next several weeks, which will allow for grain fill and eventual dry-down. July thunderstorms will most likely materialize, and you can only hope that the grain will stay erect long enough to harvest it without lodging. If you've avoided overfertilizing the crop with too much nitrogen, the supporting barley straw should be strong enough to support a heavy crop against buffeting wind and rain.

Good mature compost with a low salt index is probably the best choice for giving a barley crop a little boost without comprising its straw and its standability. The relationship between crop yields and nitrogen is interesting and fraught with the same mystique as going to a gambling casino. The more nitrogen barley has at its disposal, the higher its yield potential. When a cereal crop grows fast, rank, and thick because it has been excessively fertilized, its straw, which is its underlying support system, is often compromised. All I can think of is the line from the nursery story, "I'll huff and I'll puff and blow your house down"—once a crop that is half or three-quarters ripe is lodged and lying on the ground, all of that incredible production potential is gone. There are varying degrees of lodging to consider, however. If you're lucky, you'll find your post-thunderstorm field of barley only partially lodged. It will look as if Mother Nature has paid a visit with a giant egg beater. Some sections of the field will still be standing while others have been swirled about by driving wind and rain, leaving the barley leaning at a precarious angle to the ground. But there's no need to despair in this situation. If you had the luxury of choosing your disaster, this type of barley lodging is what you might want to choose. Grain that is leaning precariously close to the ground but not actually flat can still turn into a bumper crop. You just have to hope and pray that another weather incident doesn't come along and finish the job by driving the yet-to-be-harvested barley totally flat.

Dealing with a heavy crop of barley that is nearing the end of its ripening process in the month of July will give new meaning to feelings of faith. The potential for disappointment is there arm in arm with the possibilities of an exceptional harvest. I've met some farmers who actually want to see their barley lodge a bit. Their logic is that the only way to realize the full potential of the crop is to push it with fertility inputs right to the edge of the grain beginning to fall over. These individuals are the high-stakes gamblers: If the whole field goes down, they're in trouble. Harvesting a badly lodged barley crop becomes a nightmare and a salvage operation. In this sort of scenario, the barley will all be lying in one direction with its heads driven down by the prevailing winds. The cutter bar and the head of the combine will be able to pick up most of the crop, if you pilot the machine in the direction opposite that of the lodged grain.

The direction of travel ensures that the combine's cutter bar is able to get underneath the mat of heads that are pointing toward it. Unfortunately, harvest length is more than doubled because this can only be performed in one direction. When you reach the end of each harvest pass, you must drive the combine back to the other end of the field to continue. Lodged grain is pretty well a fact of life if you're growing barley; you simply have to make the best of it and carry on. If there is only a minimal lodging, harvest is relatively straightforward, but in more severe situations when entire fields are flattened, you can only do your best and slowly limp through the process. In the end, we are still thankful for what we have coaxed from the earth.

Lodged grain comes with a host of other implications. If you have to choose a time for this unfortunate occurrence, the later it happens during the ripening process, the better the outcome will be. If a barley crop is flattened early when its kernels are only in the milk stage of development, chances are that the final harvest will be compromised. Once the stems have been bent and broken, the translocation of nutrients throughout the plant is impacted, and lighter grain with lower test weight could be the result. If lodging occurs when the barley has reached the soft dough stage, the ripening process is already well under way and chances are the quality of the harvested grain will be just fine. The efficiency of plant photosynthesis is reduced in a lodged crop of barley because plant leaves cannot get exposure to full sunlight. When those amber waves of grain are no longer upright, the entire ripening process slows down. Heads of grain that are sitting on or near the ground take much longer to dry down to the proper moisture level for harvest. The last thing you want with this set of circumstances is a prolonged period of rainy weather, as a rainy spell has the potential to promote premature sprouting and mold infection in a crop of lodged barley. Last but not least, is the possibility of damage to an underseeded crop of clover, timothy, and alfalfa. If an impenetrable mat of lodged barley sits on top of an underseeded hay crop for any length of time, smothering and mortality is usually the result. Having a seeded-down hay crop killed in the initial seeding year by a lodged companion crop of barley is a most unfortunate occurrence. So, all things considered, make every effort to encourage the growth of a strong vibrant crop of barley that has some immunity to falling over when it begins to ripen. With this in mind, you can enhance the standability of your barley crop by keeping your seeding rates moderate and your fertility inputs balanced and reasonable.

The Importance of Harvesting When Just Ripe

The ripening process for barley lasts a few weeks. Straw color gradually changes from green to faded green to golden-yellow as plants progress from milk stage to soft dough, hard dough, and eventually physiological maturity. An early barley harvest carried out in a timely manner is essential for a number of reasons. Once the grain moisture has dropped to an acceptable level, you want to combine the crop as quickly as possible to avoid exposure to rain and preserve its quality. The same system of falling number analysis used on wheat to measure pre-harvest sprouting of the grain is equally important with barley, and low enzymatic activity in the seeds of a harvested crop of barley is especially important if the grain is destined for malting. Since malting is a sprouting process dependent on high-quality endosperm, you want a full complement of starch in the kernel after it is harvested. If a ripe crop of barley has any prolonged exposure to wetness in the field, seeds will "wake up" internally and begin transforming endosperm starch to sugar in the very earliest phases of the sprouting process. So your crop might look beautiful, but the kernel moisture might still be 20 percent—too high for storage without aeration. A harvest moisture level of 13 or 15 percent would be a whole lot more sustainable because of the high energy requirements for bin drying, but since grain quality is paramount, you would have to opt for combining at 20 percent moisture. The wet grain goes in the bin, and you turn on the large drying fan. If you had the capacity to add some low-temperature heat to the air being pushed up through the barley in the bin, your job would be even easier.

A second and equally important reason for getting the barley out of the field on time is the fact that physiologically mature barley has a short shelf life. As the ripe crop dries down and its moisture levels drop into the teens, it gets increasingly fragile. The barley heads begin to bend over in a semicircular fashion, and the awns start to point downward. A ripe stalk of wheat or oats will hold on to its kernels for weeks out in the field, but unfortunately barley doesn't afford this advantage. Once the plant becomes overripe, it will begin to drop its kernels. Some varieties are better than others in this department, but when the plant gets to this gone-by stage, it seems to revert to its wild grass west Asian ancestry and lets it fruit scatter in the wind. Ripe barley will also shatter easily during harvest with the combine. In the interest of not pre-threshing the grain before it even enters the combine, adjust the speed of the reel on the cutting platform to run as slowly as possible. Just a little whack from a reel bat turning too quickly can knock ripe kernels off the plant. Barley simply won't wait—when it is ripe, it needs to be harvested. I have talked to several farmers over the years who have either stopped growing barley or have gone out and bought their own combine when they have had to wait too long for the custom harvester to show up and have lost part of a barley crop as a result.

Harvesting Barley

Barley is a nice crop to harvest because it threshes so easily. If you are growing barley on a garden scale with hand tools, it should be reaped when the plants are going from soft to hard dough. The grain is still quite erect at this time, and the kernels are well secured to the heads. If you are hand-threshing barley with a flail or by walking over it and jumping on it, you will find the process simple and relatively painless because the grain falls off the stalk so easily. A number of precautions need to be taken when combining barley, however. Barley kernels are easily damaged by overthreshing. To avoid this, drop the concave and reduce the cylinder speed of the combine; the machine should be only as aggressive as it needs to be to get the threshing done. If you find unthreshed heads coming out the back of the combine, you will have to increase the cylinder speed and raise the concave. It is very important to pay attention to these little details, because if the barley kernel is cracked or nicked during harvest, quality and germ will be lost in the process. A kernel of barley with a damaged outer husk covering will not germinate well and will fail the grade for malting. Even if barley is only going to be used for livestock feed, you still want perfect kernels to ensure that the grain will keep well in long-term storage; cracked kernels of barley will oxidize more readily and absorb atmospheric moisture in the bin. Care and quality preservation should be your number one priorities when combining a crop of barley. You should also be prepared to drop the cutter bar down pretty low out there in the field. Given barley's weak straw, it is not uncommon to cut the crop quite low in order to pick up broken, bent, and lodged stalks. Once you are filling wagons with ripe barley on a beautiful late-July or early-August day, you can breathe a sigh of relief. Barley may have the shortest growing season of all the major cereals, but it sure isn't a guaranteed crop that is easy to grow.

Post-Harvest Considerations

Barley requires more conditioning after harvest than other grains, as many of its kernels hold onto their awns after threshing. In fact, it's not unusual to find more than half the grains in a wagonload of barley still attached to four-inch-long whiskers. This condition varies from variety to variety—even the Léger barley that I grew back in the 1980s came through the combine with lots of beards on the kernels. This is more of an inconvenience than it is a particularly serious problem, as these kernels with their hair-like appendages will still flow out of wagons and travel up grain augers, albeit grain flow is not as fast. It must be that these pesky little kernels of bearded barley bothered those who came before us, too, because someone out there in the world of grain invented the debearder many years ago. Some antique seed cleaners come with their very own debearding attachment. These machines are simple and inexpensive. Grain is passed through a horizontal round wire-mesh screen by a series of steel paddles, and barley beards are literally rubbed off by the abrasive action of being knocked against the outer screen by the paddles that are moving it through the passage. When positioned on the receiving end of a seed cleaner, the debearder can also double as a precleaner because fine weed seeds like those of mustard will also fall through the screen. Stand-alone debearders are most commonly used by processors who are transforming barley into malt or food for human consumption. For those of us who are growing barley to feed to our cows or pigs, the beards really won't matter much. Once the grain has passed through an auger or two, a lot of the beards disappear, and when the barley makes its way through the feed grinder, the beards become part of the ration and are out of sight and out of mind.

Cleaning and Drying Harvested Barley

As with any grain you harvest, you must take the utmost in care with your barley once it is in a wagon or bin. Check the moisture immediately upon harvest. Thirteen percent is the magic number below which you can safely store the crop. Barley can sit at 15 percent moisture for no longer than a week before spoilage will begin to set in. There is little more disappointing than returning to a wagon load of seemingly dry barley a week and a half after harvest to find that it now has a musty odor. If you have the wherewithal, clean your barley right after it is harvested, which will remove weed seeds that can impart flavors and add moisture to your crop. If you're strapped for time during a busy harvest season, the harvested barley can be quickly conditioned by running it through a rotary screen cleaner for a quick cleaning. An 8 × 8 (eight squares to the inch) exterior screen is the recommended size for cleaning barley in this manner. A fanning mill air screen type of grain cleaning device will do a much more thorough job of conditioning barley that is fresh from the field. In my simple two-screen cleaner, I use a number 14 (fourteen sixty-fourths) screen on top and a number 7 or 8 on the bottom. If there is excess moisture to remove from your harvested barley, it is best to give careful consideration to the end use of the crop before putting the heat right to it. If the barley is destined for malting, no heat at all should be used in the drying process. Hopefully, you have a bin with an aeration floor and a big fan that will move lots of air up through the barley, and really, everything possible must be done to preserve the germ of barley that is to be malted. Feed-grade barley can be put into a grain dryer with heat; low-temperature drying is the best insurance you have to achieve the quality feed barley you will need for long-term storage. Grain temperature should not exceed 110 degrees during the drying process. To achieve lower plenum temperatures in a grain dryer, turn down the gas pressure and install a nozzle with a smaller orifice in the propane burner. All the attention to detail and extra care taken with a recently harvested crop of barley will pay you back many times over when you are ready to take the next step toward using it for feed or human consumption.

Processing Barley for Animal Feed

You needn't be fussy with cleaning your barley if it is destined to feed your cows, pigs, and chickens. The presence of a few weed seeds, barley beards, or unthreshed heads in the grain sample won't make much of a difference once it becomes part of a ration. However, a really trashy sample loaded with a high proportion of foreign material will be unpleasant to livestock and difficult to handle. Therefore, a quick pass of the newly harvested barley crop through a rotary screen cleaner is highly recommended on the way to the storage bin or silo, and a proper long-term storage moisture of 13 percent or less goes without saying at this point. We have briefly discussed the advantages and disadvantages of barley compared with corn earlier in this chapter, but a further explanation of how the starch in these two grains is utilized in the digestive tract of ruminants will help clarify the choices you have in preparing barley for animal feed.

First, the rumen, which is the first of a cow's four stomachs, acts like a large fermentation vat to process the hay and grain consumed by the animal into the energy and protein needed for bodily functions. The starch in barley is a glucan—a polymer of glucose—and barley is 64.6 percent starch, comprising amylose and amylopectin. The rumen is full of little bugs called microbes that digest ingested starch by releasing amylase. The molecular size of the starch particles is reduced by the amylase by-product of microbial fermentation in the rumen. Glucose is released in the process and used as an energy source by these same microbial bacteria. Barley starch is unique in cattle digestion in that 94 percent of it is utilized right then and there in the rumen. As a result, there isn't much left over to be metabolized farther down the digestive tract in the other three stomachs and the small intestine.

Corn starch, on the other hand, is only 74 percent degraded by microbes in the rumen, which leaves more energy available for the small intestine. Barley may be higher in protein, but it definitely has some shortcomings in the energy department. Therefore the strategy in feeding barley to cattle centers on developing an approach to feed milling that minimizes its tendency to degrade rapidly in the rumen. This is why the "experts" tried to convince me all those years ago that rolling barley into a flat flake was much better for my cows than pulverizing it into many small pieces with a hammermill or buhr mill. The logic was (and still is) that a kernel of barley processed into one flake will have much less surface area exposed to rumen microbes. As a result, its starch will be much more slowly metabolized than if it were ground into a meal, which has a much larger surface area and is very rapidly degraded in the rumen. With this little bit of animal science in mind, let's further explore the various options for milling feed barley into well-balanced wholesome rations for dairy cows and other farm livestock.

Roller-Mill Method

Let's reconsider the roller-mill method of preparing barley that I shunned so many years ago. It's pretty apparent that I didn't have an understanding of ruminant digestion at the time because I wanted my barley ground into as many small pieces as possible; when my rolled barley came back from the Québec feed mill with lots of fines in it, I considered this to be a better grind. Thirty years have passed and I've learned a few things. For one thing, the ideal flake of rolled barley is very thin and totally flattened, which allows it to travel farther along in the cow's digestive tract before all of its starch is utilized by the rumen microbes. The secret in this process is to do a good job rolling the barley.

Roller milling offers many advantages to those of us with smaller operations. These machines are compact, readily available, and relatively inexpensive to purchase. Used roller mills often appear in farm publication classified ads for as little as two or three hundred dollars, and brand-new small scale Apollo or Buhler roller mills (both manufactured in western Canada) sell for less than two thousand. These machines are elegant in their simplicity because they have a minimum of moving parts and are very easy to operate. Barley falls by gravity from an overhead hopper down between two spinning horizontal steel rolls. The action of the rolls rotating toward each other draws the barley between them and crushes it into a flake. The amount of space between the rolls determines the aggressiveness of the milling process. A small farm-scale roller mill is usually equipped with four- to six-inch-diameter rolls that measure eight to twelve inches in length. These machines are so incredibly practical because they only require four to six square feet of floor space and are powered with a very small one-and-a-half-horsepower electric motor. Standard practice on many small and medium-sized Québec dairy farms is to install one of these roller mills in a feed room between the grain storage silo and the cow stable. At the flip of a switch, an auger delivers barley from the silo to the roller mill. The grain-feeding cart is positioned underneath the roller mill. The beauty of this system is that in five to ten minutes, freshly milled barley can be fed to the milking herd at every chore.

Before you run right out and buy a used roller mill to process your barley into animal feed, there are a few things to consider. For one thing, rolls will wear out after a while, and the flutes (steel V grooves) that are cut into their steel get dull and rounded after many years of use. There's a reason why someone's used roller mill might only cost two hundred dollars. Take a look inside the machine before you seal the deal and load the old thing in your truck. It may still be worth the two hundred dollars, but be aware that you might have to remove the rollers and send them to a machine shop for re-outfitting. Depending on the diameter of the rolls and how worn down they are, this work could cost as much as a thousand dollars. Some very old roller mills have poured "babbitt" bearings to carry the shafts on the end of the rollers. Babbitt metal was the standard throughout the nineteenth century before the advent of the modern roller bearing. A mixture of molten copper, tin, and lead was poured around the shafting to form a bearing surface. This very soft metal allowed for some wear and lubrication in these old machines.

However, the lead in babbitt metal is a red flag for some people these days because of its proximity to the feedstuffs that are passing through the roller mill. I have an old Sprout Waldron roller mill that I gave away to some friends who were rolling sprouted oats for granola. They eventually returned it to me because the machine had babbitt bearings. I really don't see this as much of a problem when rolling barley for a herd of dairy cows, as the barley doesn't touch the bearings. Nevertheless, it is a good idea to know about the potential for lead contamination when using older machinery. The good news is that any good machinist or metal fabricator can replace old babbitt bearings with modern pillow block roller bearings. If you are buying an older used roller mill, be aware that you might not be able to use it immediately when you bring it home. Then again, luck may be with you, so don't be scared of an old antique. These older machines were built to last with heavy cast-iron frames and high-quality steel rollers. However you slice it, finding, buying and reconditioning a "new" old machine for your operation is an adventure in and of itself.

Tempering

Proper roller-mill operation is the key to preparing highly nutritious digestible barley grain for your cows, and good sharp rolls that run just close enough together to allow the barley to pass through are essential. Larger-diameter rollers do the best job because the barley being milled is exposed to more surface area for a longer period of time. A sixteen-inch roll is ideal, but it will take more power than a four-inch one. There are, however, still those of us out there who have apprehensions about rolled barley as dairy feed. Personally, I don't like seeing flakes of barley in the manure behind the cows. Well, leave it to the ingenuity of feed specialists from science and industry to come up with a method for processing barley—called tempering—that is superior to what I have described so far.

Tempering has become the preferred method for getting the most feed value from barley. Water is added to barley as it is augered from the bin, and it is steeped for twenty-four hours. By this time, barley that started out at 13 percent moisture is 18 to 20 percent. After the water has been drained, the barley is rolled; the finished product is softer, flatter, and without any of the fines usually found when barley is rolled in a dry state. Tempered barley is pretty close to being totally digestible cattle feed. This practice requires more preparation and the addition of a steeping tank, but it is ultimately practical in a hands-on smaller operation where time and attention to detail are the norm. Larger farms with oxygen-limiting grain storage silos can harvest their barley as high-moisture grain at this same level of humidity. The barley ferments in these special silos and can be stored safely at higher moisture levels; steeping and soaking become unnecessary in this system. The Robillard family in Irasburg, Vermont, was a pioneer in feeding high-moisture barley during the 1980s. Rolling barley at an elevated moisture level seems to be the answer for improving the digestibility of barley grain for cattle.

Sprouting and Hand-Grinding for Homestead Use

What options exist for the mini producer who harvests hundreds of pounds of barley instead of tons and who might have one cow instead of fifty? Fear not; there are ways to utilize barley as livestock feed in the absence of infrastructure like automated roller mills. The first and best option that comes to mind is to feed cattle sprouted barley. No grinding or rolling of any sort is necessary in this system. To sprout, whole barley is soaked for twenty-four hours in five-gallon pails. It is then dumped out into large plastic trays similar to those used to bus tables in restaurants, and holes are drilled in the bottom of the plastic sprout trays to permit the flow-through of water. We spent a winter sprouting grain for our herd of twenty-five cows in 1988. We had twenty-eight trays (two for each milking over a week's time) that were placed in a rack, watered, and rotated. Over the course of a week, we obtained an amazing amount of growth on the barley. Rootlets grew two inches while the barley shoots themselves got as long as three to four inches and turned green. It was like feeding a tiny amount of brand-new early-spring pasture to the cows twice a day all through the winter. The sprouting process served double duty by softening barley kernels so they could be consumed and digested without having to be rolled or ground as well as by enhancing the nutritional content of the grain. The germination process had the synergistic effect of enhancing and increasing the vitamins, minerals, and enzymes found in the barley. If your cows need a boost of tonic for health during the long winter months, try the sprouting regimen. Sprouted barley is a particularly nice tonic for a wintertime laying flock.

There are also a number of hand-operated roller mills and buhr grinders out there on the market for the very purpose of processing small amounts of feed grains for homestead livestock. The Lehman's Catalog from Kidron, Ohio, offers a number of these for sale in its pages. The brand that comes to mind first is Diamant. These buhr mills are stand-alone units made from cast iron and heavy steel. A rugged hand crank is attached to one side of the main drive shaft and a big heavy flywheel to the other for momentum during operation. A Diamant feed grinder sells for over twelve hundred dollars, but it is a once-in-a-lifetime investment. If you prefer to roll your barley instead of grinding it with steel buhrs, there are a number of hand-cranked roller mills on the market. Most of these machines are made for home brewers to grind malt for beer making, but they will double as small-scale roller mills for hand-powered low-tech feed grinding. Once you begin to explore the options for processing feed barley without electricity or diesel power, you will find that there is a whole world of inventive people out there doing creative things like hooking up bicycles to grain mills for pedal-powered operation. Homesteading with livestock is easier than ever before because of the numerous pieces of appropriately scaled equipment available to process feed grains simply and easily.

Malting Barley

Malting is the controlled germination of cereals with the intention of changing the structure and biochemistry of complex starches in the endosperm. A rather large portion of the barley grown in the world finds its way into malt production and eventually to beer making. Malting is relatively simple; the only requirements are barley, water, and heat. (Temperatures between fifty-five and sixty degrees must be maintained throughout the entire malting process.) The process begins with steeping the barley in water for two days to wash it and raise its moisture from 12 to 45 percent. This stimulates the embryo in the kernel to begin growing. By the end of the steeping process, rootlets have begun to emerge. Next comes the germination phase. Over a period of five days, the seed begins to sprout and a series of enzyme degradations take place within the kernel. The common perception is that starches are changed to sugars during the malting process, but this is an oversimplification and not quite true. Starch and protein are tightly bound together in the barley endosperm. The alpha and beta amylase enzymes that are produced during sprouting break down cell walls within the endosperm, which in turn liberates the starch fraction by solubilizing the proteins. Picture the starch in barley endosperm as a number of small and large circular bodies that are intertwined with protein. Once these molecular bonds have been broken, the barley starch can be fermented and turned into sugars by the yeasts in the beer-making process. While this enzymatic orgy is taking place, a little green shoot begins to grow inside the kernel. Called the acrospires, this is actually the future barley plant emerging from the embryo. When the acrospire contained within is somewhere close to the length of the kernel, the right amount of physical and biochemical change has taken place in the barley kernel.

At this point, the sprouting process is curtailed by kilning. Heat is applied to dry the malted barley from 45 percent moisture down to the very low level of 3 percent. The day-and-a-half-long kilning process begins with the application of 140-degree air and finishes with temperatures as high as 210 degrees. The heating stops the embryo from growing any further, and also helps the finished malt develop flavor, color, and stability. Controlling and varying the kilning temperature is a technique used by malt houses to develop their own particular trademarked characteristics. Once the malted barley has dried, the rootlets (which are called culms in the trade) are removed by passing the grain over a sieve. Culms are very high in protein and are used as a livestock feed supplement. Dried and finished malt can either be kept whole in kernel form or be ground for immediate use in breweries.

Warminster Malting

My first exposure to malting was in April 1999 on a trip to England, when Anne and I visited Warminster Malting in Wiltshire, about an hour's train ride due west of London and quite close to Stonehenge. This part of the country produces some of the finest malting barley in the world on its chalk soils. Warminster was (and still is) one of only a few old-fashioned floor maltings remaining in the UK; it's where the world-famous Maris Otter malt is produced for the brewing of cask ales. Warminster Malting was a great place to visit because we got a crash course in the basics of barley malting with quite a lot of history thrown in. Head maltster Chris Garrat picked us up at the train station and immediately brought us to the local pub for a hearty lunch and pints of ale. The tour began in the spacious cellar of the two-story stone building where the barley was steeped in large stone tanks located along the back wall. Once properly hydrated, the wet barley was transferred to the malting floor for germination. For the next five days or so, a crew of artisan maltsters kept the six-inch layer of sprouting barley properly moistened and stirred with a variety of hand and power tools. When this part of the process was complete, the wet malted barley was elevated to a series of ovens (kilns) on the top floor of the building. Chris explained to us that many years ago, wet malt was dried on grates over fires and hot coals, which is what imparted a smoky essence to many of the older heirloom beers. Once it was sufficiently dry, the malt was packaged on the top floor as well. When we were finished with our excursion through this ancient center of craft malting, Chris most kindly gave us a ride back to the station where we caught a train back to London.

This tour of a working historical malt-producing establishment in the heart of the English countryside had a profound influence on the two of us. We ended up bringing three bags of a Scottish malting barley variety called Chariot home with us to try growing on this side of the Atlantic. I know of a few other Americans who have visited Warminster Malting more recently, and the report back from these folks is that the facility has been improved and modernized, but the people at Warminster are as friendly and helpful as ever. If you ever get to Great Britain and have a genuine interest in malt, a trip to Warminster is well worth it.

Industrial Malting

Malting here in North America is an industry like any other in the field of large-scale grain processing. The business is controlled by mammoth corporations and takes place in gigantic factories that process barley and deliver malt by the train carload. Canada Malt in Montreal is probably the closest example of this sort of operation in our region. This is a medium-sized older-style operation with a material flow pattern somewhat similar to that of Warminster Malting. Barley steeping takes place in large tanks on the lowest level. The wet grain is then elevated by conveyor to a germination deck. When the malting process is complete, the material is once again conveyed upward to the top floor for kilning. Larger, more modern malting facilities have a product flow system that runs in the opposite direction: Barley begins its journey by being steeped on the uppermost level of the plant. From this point onward, all material movement is propelled downward by gravity; germination takes place at mid-level, and kilning at the very bottom of the facility. The size and scale of the industrial malting process may be impressive, but is really no different from what happens at Warminster or in your kitchen. The principles of malting are the same for everyone.

The Microbrew Beer Industry

The local food movement has been around for a decade or so, and it becomes increasingly important in our society every day. True to human nature, locally produced alcoholic beverages achieved prominence long before local flour and beans. Microbrewed beers began to come to the forefront in the 1980s when companies like Long Trail, Otter Creek, and Red Hook were established. Twenty-five years have passed since these early days of artisan beer making, and the landscape has changed over time. There are many more microbrews on the market—most of them are doing quite well—and brewpubs abound. The larger national brands of beer have lost some market share as these smaller craft outfits have gained more prominence. My neighbor Jean Couture from Troy, Vermont, can provide an even more interesting perspective on this trend in beer making and consumption as it relates to the production and use of barley malt. Jean has been a trucker for the last forty years. In 1992, he began transporting malt in two special pneumatic trailers from Canada Malt to the Red Hook Brewery in Portsmouth, New Hampshire, and Long Trail Brewing in Bridgewater, Vermont. He thought he was doing quite well moving a hundred thousand pounds of malt a day at this point, but since then, the market has exploded according to Jean. He has had to keep adding trucks and drivers to meet the demand. He now operates twelve of these specially equipped tractor trailers, delivering six hundred thousand pounds of malt per day to microbreweries from the Carolinas to Pennsylvania, Maine, and all points in between. Jean has constructed a depot in Lyndonville, Vermont, where he unloads railcars of malt from all over the United States and transships it to his ever-growing list of customers and users. He has silo storage for two million pounds of malt as well as facilities to receive shipping containers of malt from Germany and Belgium. The importance of beer in our society is not to be underestimated.

Micro-Malting

All of this massive growth in craft beer production and consumption is wonderful, but where do we fit into this trend as small-scale producers of grain in the northeastern quadrant of the country? The little bit of malting barley that we can grow on our farms would still only fill a thimble next to what is being processed and used by the microbrew beer industry. We have all kinds of artisan brewers out there, but none of them are using what I would call artisan malt, and even home brewers are happy to buy their malt out of a can or a bag. I have found this trend disappointing, because I've been interested in the small-scale production of malt for a long time, which was the primary reason why we visited Warminster Malting over a decade ago. I love malt as an ingredient in baking and dairy foods; a double-malted vanilla milk shake is one of my favorite decadent extravagances. Well, I am happy to report that "micro-malting" has finally arrived on the scene. Bruno Vachon and Rémi Verschelden started making malt in Thetford Mines, Québec, several years ago and named their company Malterie Frontenac. Their early trials were quite successful—the demand was there for what they produced from two-row barley grown in the Québec region. Thetford Mines is a rather unlikely location for this sort of establishment, as it was once the heart of Québec's asbestos-mining region, which is about one hundred miles north of where I live in northern Vermont. Asbestos mining has all but disappeared in the region, and Bruno and Rémi have retrofitted an old mine building for their malting operation.

These two very friendly individuals were gracious hosts who were both very happy to explain and demonstrate how they steep, sprout, and kiln barley on such a small scale. Malterie Frontenac uses a one-tank system to turn barley into finished malt. Each stainless-steel receptacle measures six feet wide by twelve feet long by four feet high and can hold about two tons of barley. The steeping process begins by flooding the barley with water that has been heated to the proper temperature. When the 45 percent moisture level has been attained, the steeping water is drained and the germination phase begins right in the same vessel. The sprouting grain is aerated to prevent it from souring; it's also hand-turned with a special fork halfway through the process. Once the barley is malted, kilning is accomplished by blowing high volumes of heated air through the mass of sprouted barley. This type of artisan production is beautiful in its simplicity and especially practical for this type of medium scale because the barley never moves throughout the whole process. Bruno learned his craft while doing an apprenticeship with a master maltster in Germany, and has added several very specialized and expensive analyzers to monitor the quality of the malt he produces. These two individuals are to be commended for their willingness to enter into a business dominated by giants—and for their ability to make a go of it on a scale that those in the industry have determined will not work. Bruno and Rémi also embody the inclusive and generous spirit so prevalent in the alternative and organic farming communities.

Artisan malting has taken root on our side of the border as well. Christian and Andrea Stanley of Hadley, Massachusetts, started experimenting with malting barley in their kitchen in 2009. Serious home brewers, the young couple decided to try making their own malt. Andrea did lots of research while Christian put his mechanical and technical skills to work building a prototype malting unit from stainless-steel cookware and electric heating elements. The mini malter held about a gallon of steeped barley and employed the same single-vessel process as Bruno and Rémi. Andrea and Christian have gone on to build a sustainable and well-thought-out operation that they have named Valley Malt. Their little countertop malting vessel has morphed into a large round eight-hundred-gallon stainless-steel tank, which enables them to malt about a ton of grain per batch, and Andrea has continued to educate herself by attending special courses in malting at North Dakota State University. Their artisan malt has become quite popular with home brewers and artisan breweries alike, but the most impressive thing about these folks is their generosity and willingness to share what they have learned with others.

Barley for Human Consumption

In the larger realm of barley culture and its end uses, barley grown for direct human consumption represents only a minuscule percentage of the total crop grown here in North America. Nevertheless, barley is one of the most healthful cereal grains we can eat. As mentioned earlier, the beta glucan type of fiber found in the barley kernel delivers a host of positive benefits to the human body. Medical researchers from all over the world have determined that a diet rich in barley greatly reduces plasma cholesterol and other blood lipids, and because of the unique qualities of its starch, barley consumption also helps individuals maintain lowered blood sugar levels. Although these might be good reasons to include barley as a regular part of your diet, the best reason is its great taste and texture. Barley will impart extra flavor and body to any soup whether it is vegetable- or meat-based. The longer this grain is cooked, the better the final dish will be. Whole-grain barley can be pressure-cooked or simply boiled and consumed in the same manner as rice. Slow-cooking is recommended to offset some of its chewy texture. Barley can also be flaked and used as a substitute for rolled oats or as an ingredient in granola. And last but not least, barley becomes an amazingly versatile ingredient when it is ground into flour. Flat breads made from barley flour are a standard part of the diet of people in the eastern Mediterranean from North Africa to Syria. Here in North America, barley flour is used as an ingredient in commercial baby cereals and foods. I love barley and try to include it in my diet whenever I can.

Abrasion Husking

Barley poses one big problem for those of us who want to eat it. Its outer coating or husk must be removed before we can put it in our soup or cereal bowl. The hulless varieties discussed earlier in the chapter are a possibility, but even this subspecies of barley will thresh out with a few hulls remaining on the kernels. Barley hulling is a specialized affair that requires very specific types of machinery and technology. The more common hulled grains like oats and spelt are hulled by the impact method, as the hulls of these grains fit loosely around the inner kernel. With oats and spelt, there is really no positive attachment between the exterior husk and the interior grain; such hulls are removed mechanically by actually hurling the grain against a hard surface. (Impact hulling will be described at length in a subsequent chapter.) Barley, on the other hand, has a very tightly fitting outer husk that's more like another layer of papery skin on the interior kernel. Impact hulling won't do anything for barley because it is such a hard-textured grain, with its outer skin virtually glued to the kernel. Barley needs to be hulled by abrasion. In the days of yesteryear, millers used the traditional horizontal stone mill to hull barley. The top runner stone was cranked up above the bed stone just enough to permit barley kernels to roll around in between the two abrasive surfaces. By the time barley kernels traveled from the center to the outer circumference of the slowly rotating mill, most of the outer husks would be rubbed off in the process. Further sifting and cleaning of the finished grain sample was necessary at the end of the process to remove bits of hull and cracked barley.

I have experimented with hulling my own barley for quite some time with very mixed results; most of my trials have been less than satisfactory. I've tried running barley through my Meadows vertical stone mill with no success—the millstones were separated enough to let the barley pass between them without being ground. The hulling was accomplished, but the grain got pretty smashed and chopped up because the millstones were turning too quickly. We also tried passing our barley through the Forsbergs huller/seed scarifier. Grain is forced to pass between a tapered inner cone and an outer case. The inner cone is constructed of special neoprene while the outer shell is lined with a hard rubber layer. The distance between these two surfaces can be adjusted with the turn of a hand crank. The Forsbergs huller did a very good job on spelt, although the machine's output was extremely slow—fifty pounds per hour was about as fast as it would go. Then we tried putting barley through the machine. To my surprise, the Forsbergs huller actually removed the outer husk from the barley kernels. However, the sailing wasn't exactly smooth, as the machine heated up and kept tripping electrical breakers. Barley is much harder than spelt, and it was more than this old antique could handle. This didn't stop me from continuing to hull the barley because I liked having it in my soup. You would think that after thirty-five years of breaking equipment by overtaxing things, I would have learned to stop pushing a machine when it was giving me all sorts of hints that it wasn't happy. Well, I'm sad to report that I ended up ruining my Forsbergs huller/scarifier by using it for barley. All of a sudden, little pieces of the inner rubber lining started coming out with the hulled barley. The grain was so hard that it tore up the insides of the huller. Take a lesson from me. If a machine is straining and overheating, quit while you're ahead before you damage something.

Barley-Hulling Machinery

Barley hulling and husking is a very difficult process, especially for those of us who are producing the grain on a small to medium scale. But there are a number of very specialized machines that have been designed to dehusk and pearl barley. Just about all of this equipment is industrial in scale, however, and comes from parts of the world where barley makes up a larger portion of the human diet than here in North America. There are manufacturers of barley-hulling machinery from countries like Japan, Switzerland, and Germany with names like Satake, Buhler, and Schulte. Unfortunately, the smallest machine available from any of these companies is designed to process between two and five metric tons of barley per hour. This processing capacity is more than any of us need at this point. (Perhaps one barley huller operating in a central location could service all the barley grown for our human consumption in our entire region.) The other drawback with this sophisticated hulling and husking technology is that each particular machine is designed for one specific task, and there is very little processing crossover between the different types of grains. Barley hullers are only designed to do barley and not another grain like buckwheat.

A quick Internet search for barley-processing machinery will turn up myriad inexpensive machines from China. Some of these Chinese models are designed for the smaller processing capacities that we need as artisan producers of barley. However, it has been my experience that you get what you pay for when you purchase a piece of processing equipment from China. Design tolerances for bearing, shafts, and other moving parts aren't quite what they need to be to ensure the smooth continuous performance that you will get from a more expensive German or Japanese machine. This is not to say that Chinese equipment won't do the job—if you are mechanically clever and technically savvy, one of these very inexpensive Chinese hullers will work just fine. You might have to tear the machine apart after a very few hours of operation and remachine some of its parts, but this might be a viable option when you consider that the Chinese machine cost only a few thousand dollars compared with the tens of thousands you might pay for a more sophisticated model from Europe or Japan. Another thing to consider is that the more established Western companies offer complete technical support and readily available replacement parts for their barley-hulling machinery. I'm sure that someone out there in the local food movement with a mechanical engineering background and access to a machine shop will eventually build a small-scale barley huller based on the design and operation of the larger-scale equipment that is already on the market.

How a Barley Huller Works

As I have eaten barley in my soup over the years, I have constantly wondered how a real barley huller works. A little research pointed me in the direction of the Satake Company from Japan, a worldwide leader in hulling and polishing machines for rice and barley with manufacturing facilities on every continent. I contacted their Houston, Texas, office and spoke with Thomas Kock, who kindly provided with a very detailed explanation of mechanized barley hulling.

On first glance, this machine has the same shape as an old Round Oak parlor stove—round and heavy on the bottom with a slight taper toward the top. This machine actually removes the outer husk from a kernel of barley. The VTA-5 also has the capability of removing the bran and germ from the barley endosperm, which is essentially the pearling process. Bear in mind that this is not a stand-alone piece of equipment, and it is not designed for someone to pour a bucket of whole barley in the top and catch the finished product in another bucket at the bottom. Any professional piece of hulling equipment must be situated in a mill environment where gravity-fed barley enters the top through an inlet pipe equipped with a modulating valve, which regulates the input flow of material. To accomplish this, whole barley is elevated in advance to some sort of an overhead holding bin. The machine must also be outfitted with an exhaust system powered by an aspiration fan and ducted to a cyclone dust-collection system. The exhaust fan creates the positive pressure within the huller that keeps the barley moving through the process.

The internal workings of the VTA-5 consist of a vertical central shaft that directly drives seven horizontal plates. These plates are actually stone discs fabricated from a composite rock-like material. The discs are surrounded by a perforated steel screen that encompasses the entire outer circumference of the machine's midsection. Steel resistance bars are positioned at the outer edge of each plate to provide abrasion for the barley kernels as they pass by the outer edges of the individual plates. The tiny space between the outer circumference of the plates and the resistance bars is carefully adjusted to allow barley to drop from one milling chamber to the next as the grain makes its way from the top of the machine to the bottom. As hull and bran are abraded from the barley, this waste material is pulled through the outer screen by the suction created by the aspiration fan and cyclone. Thankfully, the barley is hard-textured enough that it remains intact as it drops through the inner workings of this very aggressive hulling and pearling machine. An outlet door at the bottom of the seven-disc hulling chamber controls the throughput of barley. Internal pressure is also created during the process. A series of adjustments among the regulator valve at the top, tolerances between outer screen and hulling disc, and the outlet door at the bottom will determine the aggressiveness of the hulling process. The VTA-5 has the ability to hull or pearl barley. Depending on how the machine is set up, it can remove either just the outer hull or the bran and germ as well. I'm sure that this specific machine-intensive type of high-speed barley hulling will never be a common practice here in our region, but it sure is interesting to know how the professionals do it. A little exposure to this sort of knowledge will hopefully help some clever individual out there find a less intensive creative way to hull our regional supply of barley with the resources we have at hand here in the Northeast.

Barley is one of the most difficult but rewarding grains to grow here in our part of the country. Farmers in North Dakota, Saskatchewan, and the eastern Mediterranean don't even think twice before sowing a crop of barley. We've got a whole lot more to think about here in the Northeast. Wet weather is probably our most important concern. Unfortunately, there isn't much we can do about a prolonged spell of overly moist weather. Remember that rainfall has its advantages, too. It sure is nice that when we plant a seed, it generally germinates pretty readily. We don't have to irrigate or summer-fallow our grain fields. We also do have control over many other aspects of managing a crop of barley. We have the choices of many different varieties, where to plant, how to fertilize, and how to harvest. Nothing is guaranteed when we put a crop in the ground. We just need to visualize that beautiful sunny day in late July or early August when golden-colored barley is streaming out of the combine's unloading auger and mounding up in the wagon. It's moments like these that keep me coming back to planting barley year after year even though things don't always turn out exactly as I would like.

CHAPTER TWELVE
Oats

If there is a cereal grain that is naturally suited to the climate and growing conditions of the Northeast, it is the oat. The old-timers who were still growing grain on their farms in the mid-1970s when I first started farming were all growing oats. A decade earlier, in the '60s, grain crops of oats proliferated everywhere. When you find an old John Deere 12A or International Model 64 pull-type combine tucked away in an old shed, it was most likely used to harvest oats almost half a century ago. There is still an abundance of oat lore around from these golden days of agriculture, and stories abound about lodged grain, combine fires, and damp oats heating in a makeshift wooden bin upstairs in the barn. Wheat and barley weren't part of the farm crop repertoire in these bygone days; if you planted and harvested any grain at all, it was oats. There was (and still is) good reason for this. Oats are the most grass-like of any of the major cereals. And grasses like cool wet weather, which we have plenty of here in our region. Grasses also compete well against weeds. As a matter of fact, grasses can be weeds. If you're brand new to this grain-growing business, and you're looking to grow a first-time grain crop that is almost foolproof, try planting oats. Despite the fact that the culture of this humble cereal has been surpassed by corn, soybeans, and wheat, oats are still well worth growing, for a number of reasons that we'll discuss in this chapter.

Oats: A Brief History

Like wheat and barley, oats originated in the Fertile Crescent area of Western Asia. Some ethnobotanists theorize that oats were actually domesticated in Tartary, which was the nineteenth-century name for the Great Steppe of Central Asia that stretches from the Caspian Sea and Ural Mountains eastward to the Pacific Ocean. We do know that the modern-day oat (*Avena sativa*) is very likely descended from the wild oat plant (*A. sterilis*), which is alive and well today and has become a major weed problem in cereals on the Canadian prairies. Hardly any references to oats exist in the Bible, as they do for wheat and barley. With this fact and other evidence in mind, it is safe to assume that the domestication of oats came later.

The wild oat began as a weed in the emmer, einkorn, wheat, and barley crops of Mesopotamia. As wheat moved northward into the cooler and wetter regions of Central and Northern Europe and Asia, the wild oat, which came along as a weed, began to thrive because these climates better suited its growth habits. What we know today as the common oat was very likely selected from its ancestral wild parent in Russia and Germany where the growing conditions were much more favorable for its domestication. Since the oat plant had lower summer heat requirements and was very tolerant of excessive moisture, it soon spread even farther northward and westward into Norway, Sweden, and Scotland. Northern Europe had become the center of oat production by the Middle Ages, and oats did particularly well in Scotland with its very cool and misty climate. Scottish settlers brought oats with them to North America in the eighteenth century. Once established on this continent, oats continued to thrive and moved westward as the interior of this new land was opened up and settled. By the late nineteenth century, the center of oat production had moved to the Upper Midwest and northern plains states. Russia and Germany were still the world leaders in oat production at the time. Iowa, Wisconsin, Minnesota, and the Dakotas helped make the United States a top producer of oats as well. But although oats still grow incredibly well in the north-central US, they have become a relatively minor crop with the advent of cash crop agriculture and the disappearance of livestock-based mixed farming. Oats are easy on the land and easy to grow. This underestimated and often unrecognized cereal grain offers us lots of opportunity as we develop sustainable agricultural systems here in the northeastern United States.

The Basic Biology of Oats

I liken oats to a breath of fresh air in the world of grain growing here in our part of the country. This humble cereal plant is from an entirely different tribe than wheat or barley. Oats are from the genus *Avena*, while barley, wheat, and rye are members of the Hordae family. The growth habit and flowering patterns of oats are unique and completely different from these other grains. The inflorescence of oats is characterized by its pyramid-shaped panicle of individual flower clusters at the top of the stem. With its whorls of branches clustered around the rachis, the oat plant looks more like a little tree or brush. Whereas all of the other principal cereals form a single spike of flowers, the *Avena* genus is characterized by its very distinctive groups of spikelets that hang from the top of the plant. Oats usually have three to five whorls of spikelets that branch out in a symmetrical fashion. There is also a subspecies called side oats that has branches of flowers emerging from only one side of the rachis. Each spikelet hangs from pedicels (mini branches) and contains two or more flowers. Usually no more than two of these flowers mature into grain. The flower and future oat seed is surrounded by a papery layer called the glume. Hanging glumes shimmer in the sun and rustle in the breeze, giving oats a most magical presence. The distinctive textured appearance of a field of ripening oats brings to mind the work of a French impressionist painter like Monet. There is no other grain out there that can compare in beauty.

Growth Habits and Uses

Oats are definitely the most grass-like of all the cereals. They germinate rapidly in cool moist soil and quickly develop a very vigorous root system. Some experts claim the roots of oat plants exude an allopathic substance that suppresses surrounding weeds. During its early growth, the oat plant produces lots of leaves on a rather shortened stem. The plant will tiller profusely if not planted too thick. Because the leaves of young oat plants are characteristically broad in width and rather numerous, oats are good competitors against early-season weeds, as this crop early on develops a canopy that shades the ground and prevents weed growth. Oats grow tall. Some varieties can easily get four or more feet in height. This fact seems hard to believe when you take a look at its growth habit early in the season—it seems like oat plants spend their first four weeks growing outward instead of upward. Early production of root and leaf biomass ensures success later in the season because plants have the opportunity to gain strength and establish themselves. For this reason, many farmers choose oats to plant as a green manure and cover crop. All of this profuse early leaf growth also makes oats a great choice for forage as well as grain. When oats for grain fell from favor in the 1960s and '70s, this crop retained its place on farms for hay production. Oats make great hay when cut in the milk stage. Some seed companies specialize in wide-leafed forage oats that are bred specifically for hay production. (If you plan to grow oats for grain, avoid these leafy forage varieties—they are very low yielding.) Oats have everything going for them here in our part of the country. They grow quickly on wetter, less fertile soils than wheat or barley. Oats have their place on a livestock farm because they produce an inordinate amount of very high-quality straw for bedding. The other important cereals like barley and wheat require perfect growing conditions and lots of luck, but oats are in a class all by themselves. This crop really wants to grow in the Northeast, and it will thrive in less-than-perfect conditions. If you are a beginning grain farmer you might want to find a reason to plant oats because success is much easier to achieve with this grain than it is with anything else out there.

The Oat Grain: A Deeper Look

The oat kernel is unique in every way among the cereals. None of the other major grains bears fruit as individual florets in a large panicle like that of the oat plant. A field of oats in full splendor has the appearance of a small forest of little "grain trees." Each set of oat flowers is encased in a loosely fitting outer papery membrane called the glume. As the plant ripens, these outer glumes begin to relax their grip on the inner kernels and turn a magnificent color somewhere between beige and gold, and it is this combination of color and texture that gives an oat field that look of a French impressionist painting. The oat kernels that develop from the inner flowers of the plant remain encased within another set of membranes called the palea—which we know better as the hull. Like barley and spelt, the oat hull remains attached to grain after it is threshed. Harvested oats have a very characteristic elongated shape.

Each individual kernel is pointed at either end and has a bit of a slippery texture because of the outer hull. Floating around inside this outer casing is the inner oat kernel, commonly known as the groat. The average composition of an individual oat grain is 70 percent kernel and 30 percent hull by weight.

Oats have an entirely different outer casing from barley. Barley's husk is thin and tight fitting, whereas oat hulls are bulky, thick, and loosely fitting, which makes them much easier to remove than barley husks. For an experiment, place an oat between your thumb and forefinger and give it a little squeeze. The hull will loosen and begin to separate

Oat plant.

from the groat. Insert your thumbnail along the side of the kernel, and the oat groat will drop right into the palm of your hand. You have now gotten to the heart of the matter. Oat groats look a little like wheat kernels in size and shape. A bushel of whole oat groats weighs sixty pounds, which is exactly the same weight as a bushel of wheat. The ultimate quality of an oat crop is determined by grain color and the proportion of groat to hull. If the groats within the hulls are plump and well developed, test weight per bushel will be high and grain quality will be considered excellent. Oats with a nice white appearance and a weight per bushel of forty-two pounds or better are considered US No. 1 grade. Bushel weights for oats can vary from as little as twenty-five pounds to as high as fifty. A thirty-two-pound bushel is considered acceptable and fine for livestock feed. I'm happy if I can produce a bushel of oats that weighs thirty-eight pounds. If you are in the business of raising oats to be hulled for human consumption, a heavy, fully developed kernel is necessary to produce groats or rolled oats of any consequence. About 95 percent of the oat crop grown in North America finds its way into livestock feed. I'm interested in feeding oats to my stock, but the human consumption outlet is much more interesting and involved. We will explore both of these end uses in this chapter, but will devote much more attention to oats as people food because of the tremendous potential of this very well-adapted crop as a food source for our region.

The panicle growth pattern of individual florets hanging like fruit from the top of the oat plant provides this grain with additional advantages. Oats kernels dry more quickly during ripening because each individual grain has more exposure to sun and wind. Barley and wheat heads, conversely, which are more tightly bound to the rachis of the plant, lose moisture more slowly during their final weeks in the field. The extra hull that comes attached to the inner groat also gives oats a little bit of a leg up in the world of cereals. Oats are much less susceptible to fusarium infection because the inner grain is protected from airborne mold spores by the hull. Last but not least, oat hulls can serve as a moisture-absorbing buffer. You've got a little more leeway and a slight hedge against spoilage if you combine your oats at a moisture level a few points too high for safe storage. Once again, the hulls come to the rescue and suck up some of the excess grain moisture. This little tidbit is not a recommendation to harvest your oats at a higher moisture level and forget about them permanently, but it does mean that you might be able to condition them in a much less elaborate manner than might be otherwise used on a damp harvest of barley or wheat. Sometimes the only action necessary to take care of some damp oats is to shovel them around on a floor or throw some dry wooden fence posts into the storage bin to absorb the excess moisture. These little features are part of what makes oats such an incredible low-input crop. Whether oats are being planted by us today or by the farmers fifty or sixty years ago, this unpretentious cereal grain seems to be the natural choice here in our region.

Nutritional Benefits of Oats for Human Consumption

Crude protein of whole oats still in the hull averages around 11 percent, and the oat groat all by itself will usually measure between 12 and 14 percent. Oat protein is exceptional in quality because it is much more complete and balanced than the protein found in the other major cereals. Oats have an excellent balance of amino acids and are higher in lysine than any other grain including rice. A special legume type of protein called avenalin is found only in oats. Oats by themselves may not have a 100 percent complete amino acid profile, but they come closer to perfection than any other grain you might consume. This is why a hearty breakfast of rolled oats is so satisfying and will usually take care of your appetite until noontime or beyond.

The quality and type of fiber contained in the oat groat and its outside bran coat elevates this minor cereal to true hero status in the world of human nutrition. Oats for human consumption began their rise in ascendancy about twenty years ago with the oat bran craze in the late 1980s and early 1990s. Tons of health claims were made by all sorts of experts, but the long and the short of it was that consumption of oat bran increased metabolism, lowered blood pressure, reduced bad cholesterol, and ultimately fostered weight loss. With a lineup of potential benefits like these, consumption of oats and oat by-products skyrocketed. The craze was on. Oatmeal found its way back into the breakfast routine along with a number of other cereals that touted the health benefits of oat bran and fiber. Farmers planted more oats and the revolution was under way. After the dust had settled and the heat of the moment had passed, the truth about oat bran manifested itself—oats contain more soluble fiber than any other grain. This is the same beta glucan found in barley, but in much greater quantities. Soluble fiber promotes slower and healthier digestion because it is flexible in the digestive tract. Have you ever noticed how slow-cooked oatmeal forms an almost gelatinous paste in the cooking utensil? This same dissolved fiber forms a protective coating in the intestine, bestowing numerous benefits on human health. And while all this nutritional superiority is wonderful, we needn't forget that oats are simply great tasting. They lead the pack of cereal grains in fat content with levels in excess of 12 percent, so it's no wonder they taste so good. Here is a grain with all of the extras—high protein, healthy fiber, plenty of fat, and a hull that needs to be removed. Let's get them planted and harvested next; we'll figure out how to hull and process oats for human consumption at the end of the chapter.

Hulless Oats

There are several subspecies of oats that contribute even more to their versatility here in our region. Hulless oats (*Avena nuda*) grow and look exactly like regular oats out in the field, but are distinctive because they thresh out as whole oat groats at harvesttime. The palea or inner hull of the hulless oat is very loose fitting and is readily removed by the combine or threshing machine. I first heard of this wonderful alternative to the common oat about twenty years ago, and I bought a few bags of seed and proceeded to plant a few acres for trial. These oats seemed too good to be true. Here was the opportunity to harvest oats that could go directly to my breakfast table without all the hulling and processing necessary to transform standard oats into a product fit for human consumption. I couldn't understand why everyone wasn't growing hulless oats. I soon found out. The crop grew quite well and looked respectable out there in the field. Once the harvest was complete, however, I found that this so-called miracle oat was rather low yielding. Another major disappointment was the fact that these oats did not thresh out completely clean. At least 10 percent of the hulless oats retained their hulls. This made for some extra-chewy breakfast cereal, and spitting out the hulls got a little tiresome after a while. Another drawback to this crop was its smaller kernel size. Hulless oats could not compare in size to plump whole oat groats produced from traditional oats.

My friend Michel Gaudreau in Compton, Québec, had had the same experience. In the early 1990s, Michel built a processing facility to produce rolled oats. He began with hulless oats, thinking they would be a simple way to make oat flakes. After several years of trial and error, Michel invested in hulling equipment and returned to growing regular oats, as larger oat kernels make better oat flakes. My adventure with hulless oat production offered me one final lesson that contributed to my final decision to stop growing them. I didn't realize that tiny, almost microscopic hairs remain attached to hulless oat kernels after threshing. I found out just how irritating these hairs can be to your skin and respiratory tract when we ran our crop of hulless oats through the seed cleaner that first year we grew them. The old-timers cursed barley for making you itchy, but these hulless oats were even worse. Long-sleeved shirts and lots of body covering offered a bit of relief and protection during seed cleaning, but, all in all, this was not a pleasant task. My days as a producer of hulless oats for human consumption ended almost as quickly as they began.

I don't want to totally condemn hulless oats as a viable crop for the Northeast, though, as they have their place and can be successfully raised and processed by just about anyone with a minimum of investment. Eastern Ontario farmer George Wright has built a very successful farm enterprise growing hulless oats and selling rolled oats at the Ottawa farmer's market. George claims that the secret is in variety selection, and he grows a relatively new cultivar from western Canada named Gehl that lacks those little hairs that make you so itchy. George also saves and selects seed only from plants that are totally hulless. So please don't let my rocky and inauspicious start with hulless oats deter you from trying to grow them, as varieties have improved over the years. I should also mention that hulless oats make great animal feed for chickens and baby pigs. The absence of the extra fiber from the outside hull contributes to an easily digested feed grain with high concentrations of both protein and energy. Hulless oats have also been used as a component along with wheat and peas in mixed cereal grains grown for on-farm dairy rations. You can't beat the extra energy and high fat levels of hulless oats for livestock feed. Seed selection and new variety development continue in earnest in Canada, so keep your eyes peeled for ever-improving varieties of hulless oats that might work well on our farms here in the Northeast.

Winter-Grown Oats

Last but not least, there are winter-grown varieties of oats. Winter-seeded oats will not survive in the Northeast; they are generally planted in warmer areas like Virginia and points south. These oats are seeded in late September and October; they go dormant in December and begin growing again in late February or early March, depending on latitude and climate. The well-known Buck oats that people seed for luring deer are winter oats. There is absolutely no reason to even try winter oats in the Northeast. If you are planting oats in the fall for a cover crop, you will be amazed at how long into the winter they will survive before they are killed dead by prolonged freezing temperatures. Even where I live in one of the most frigid localities in northern Vermont, lush green fall-planted oats will persist until late December. When temperatures finally drop into the teens and single digits in late December or early January, cover crop oats will finally turn brown. Many vegetable and grain growers who are exploring no-till options of farming will plant right into this thick mat of dead oats in early spring: Weed competition is virtually eliminated because the ground is covered and protected by the oat mulch. Oats are the grain of choice for many reasons and offer a long list of options to almost everyone out there who is farming in our part of the country.

Finding and Choosing Seed

The first order of business when growing oats is to find the seed and variety that will best suit your needs. Before you begin, ask yourself a few questions to determine the parameters involved. *What will be the end use of my crop? Are these oats for human consumption or livestock feed? How much straw do I hope to harvest with my oats? What kind of soil fertility levels am I working with? Are there any soil drainage issues? Do I have any yield goals?* Once you've considered your goals and needs, you can proceed to source seed with a clearer sense of purpose. Now it's time to peruse every cereal seed catalog and variety trial report that you can lay your hands on. Within these publications, you will find tables that list varieties and their traits. If you are studying variety trial information, make sure that multiple years of trials are included from replicated test plots. When doing this sort of pre-planting research, more is better when it comes to information. If you are growing oats for animal feed, yield, test weight, and straw quantities will be the most important traits to look for, but if you are growing oats to process into groats and flakes, test weight and grain plumpness are much more important than yield. Further inspection of all this data will reveal that you also have choices for grain color and days to maturity. For human consumption, you want the whitest oats you can get. For livestock feeding, a yellow or red oat will be just fine. If the oats you plan on growing are to be used as a nurse crop for an undersown hay crop of timothy, clover, or alfalfa, an oat variety with fewer days to maturity will better serve you. The sooner the oats are harvested and the straw is baled, the quicker the hay crop underneath can obtain some sunlight and really begin to grow. Oats are so adaptable that any old variety will probably grow for you. However, if you want perfection and the best chance at success, look for the variety that best suits your needs.

Sometimes the perfect variety will elude you. Suppose none of the local dealers stock what you are looking for. What now? You have a choice—either accept what is available and live with it or bring something in from far away. When you buy seed from another region of the country, however, be prepared to pay quite a bit extra for shipping. This is especially true if you only need a bag or two for a garden plot. Carriers like UPS or FedEx charge dearly for transporting bags of seed from faraway places like Albert Lea Seedhouse in Minnesota. More often than not, shipping charges will far exceed the catalog value of the actual seed. Pallet shipments are much more cost-effective. You might pay two hundred dollars or more per pallet for common carrier delivery of seed from Minnesota, but when you consider that you are receiving forty bags, at fifty pounds each, the actual freight cost is only five dollars per bag. Many seed distributors have sweetheart deals with trucking companies and can offer special pallet rates if you buy a large order of seed. Whatever you do, don't wait until the last minute to buy your seed—start looking in December or January for the seed you will be planting in April. The best varieties sell out early and you might be stuck planting something you don't really like. Winter is a great time to prepare for spring planting. Once you have procured your oat seed, you will be able to plant when the ground is ready in April.

Often, the oat seed you want will be right there under your nose. Check the classifieds in your local farm paper for seed. You might just have a neighbor who has some cleaned oats for sale. This was certainly the case for me in the late 1970s. Several people directed me to Clarence Huff, a seventy-five-year-old seedsman from Compton, Québec. Clarence grew about forty acres of Gary oats a year and processed them all into seed, which he sold very reasonably. Gary oats were the favorite old Eastern Townships variety at that time, as they grew to almost chest height and yielded quite well. Gary was my oat of choice for more than ten years until it faded from the scene in the early 1990s. I bought seed from Clarence until he sold me his combine in 1985. He became one of my early mentors who taught me a lot about growing and cleaning cereal grain, and now I sell oats to my neighbors and other farmers just like Clarence did. As grain growing becomes increasingly popular here in our region, there will be more and more farmers who will be able to raise, clean, and sell seed. I encourage you to buy seed locally if you can. You might have to drive fifty miles or more, but you will still be purchasing locally grown and adapted seed from someone who loves the land as much as you do. I find this sort of act of local commerce far superior to buying from a dealer. Humankind has been growing and saving seed for millennia. Let's carry on the tradition.

Choosing Your Variety

What about specific varieties of oats? you might ask. A quick Internet search will turn up hundreds of choices from places as far away as North Dakota and Saskatchewan. To avoid confusion and frustration, I usually don't look any farther than Québec and eastern Canada for oats to plant. I absolutely loved the Gary oats of yesteryear because they were such a great straw crop. They also tended to lodge quite readily, however, and for this reason breeders began releasing shorter varieties with stiffer straw. The last time that I ever saw Gary oats offered was in 1995 from Maine Potato Growers Association in Presque Isle, Maine. Other older American varieties like Robust and Rodney have also disappeared from the scene, but I have had very good luck growing several older Canadian oat varieties that are still available. AC Aylmer and AC Rigodon are both heavy test weight, white-colored varieties that work well for milling and flaking, while Robust and Ogle are American feed varieties that have stood the test of time and are still available for planting. Bia, a newer variety from Co-op Féderée in Québec, has done very well in Heather Darby's University of Vermont grain trials. It really doesn't make much sense to bombard you with a list of the one hundred or more oat varieties that are available currently. Varieties of oats will come and go, and it is my hope that the relevance of this book will outlive them all. Rest assured that good cereal plant breeding will proceed in the future and that excellent varieties of oats will continue to be available to us as growers.

The Culture of Oats

Oats love cool, moist weather, which makes them an ideal crop for very early-spring planting. Early planting also provides a recently sown oat crop with the extra water it needs over and above its sister crops of wheat and barley. It's not unusual to see farmers seeding oats in late March or very early April when there are still winter snowdrifts remaining up against the edges of fields—cold temperatures and spring snowstorms seem to give oats even more reasons to flourish. With this early-spring planting requirement in mind, pre-plant fall tillage is almost a must for oats. As described numerous times already in this narrative, moldboard plowing in October or November of the previous season is the way to go if you are planting on sod. If you are following a corn or soybean crop with oats, spring harrowing with a heavy disc or other medium-weight secondary tillage tool will suffice. Oats don't require the same finely worked seedbed that is necessary for wheat or barley, but if you are following grain corn with oats, some sort of after-harvest stalk chopping and fall tillage to incorporate the corn stover will put you way ahead in the spring. Sometimes this isn't possible in the late fall because prolonged rain and snow may have made soil conditions too wet for tillage, but if you can incorporate the crop residue into the earth, soil biology will have a head start breaking everything down.

When oats follow soybeans, we have an entirely different situation, as soybeans leave very little crop residue behind. This lack of biomass going into winter makes harvested fields of soybeans very prone to erosion. So if there isn't enough growing season remaining to establish some sort of soil-holding green cover crop after soybeans, stay out of the field with the plow and harrow in the fall. Spring tillage will be just fine for oats following soybeans. The thoroughness of seedbed preparation for a crop of oats will also depend on whether you are seeding straight oats or oats with an underseeded hay crop. If oats are to be planted all by themselves, a couple of passes with the disc and the field cultivator will be enough to establish an adequate seedbed. However, if the field is destined to be a hay field after the oats are harvested, careful preparation and smoothing will be necessary to provide good seed-to-soil contact for the little clover, alfalfa, and timothy seeds that will be growing along with the oats. You will also want a relatively smooth surface for the hay field that will follow the next year. There is little worse than harvesting hay on a rough field that is full of bumps and ruts—both tractor operator and haying equipment will be needlessly rattled if you seeded down your oats on poorly prepared land. Like anything else in farming, it is important to consider future growing seasons and crops while you go about the business of planting this year's crop. Sometimes, shortcuts work just fine. Precise seedbed preparation isn't necessary if oats are being seeded all by themselves because the field will eventually be tilled again. More care needs to be taken, however, if a forage crop accompanies the oats.

Oats and Fertility

Oats have the lowest fertility demands of all the cereals, aside from buckwheat. Supplemental nitrogen and potassium are almost never required—oat plants will scavenge the leftover fertility from plowed-down sod or the corn crop that preceded them. A minimal level of phosphorous is needed for adequate root development and plant metabolic function. Too much nutrient input is more damaging to an oat crop than not having enough inherent fertility, actually, because excessive levels of elements like nitrogen will foster rank vegetative growth, which leads to plants with weak straw and a high probability of lodging. There is nothing more disheartening than watching a field of extra-thick dark green oats being flattened by a July thunderstorm. If you plan on spreading some fresh stable manure on your future oat field, do it the season previous to planting. Whether you spread on sod and plow it down or spread on top of plowed ground, an August or September application of raw manure will give the soluble salts contained in the material time to mellow and digest a little before next spring's oats begin to grow. The timing of a spring application of compost on oats is not as critical, especially if the product is well made and mature. A light topdressing worked into the soil just prior to seeding with a field cultivator or spike drag will work perfectly.

Ideally, the highest-yielding and best-quality oat crops are produced on soils with high levels of inherent fertility. High-organic-matter soils with plenty of minerals and diverse biology are the ideal environment for just about any crop including oats, and plants will grow tall and green with strong straw and high grain yield. Quality fields like these are usually reserved for heavier-feeding crops like wheat and corn. The humble little oat usually gets second or third best when it comes to choosing a field location for planting. But the beauty of oats as a crop choice is that they will still yield moderately well in these less-than-perfect locations with little or no supplemental fertilization.

Adding nutrients to the soil in advance of planting an oat crop is a bit of a double-edged sword. Extra potassium and nitrogen supplied to a growing crop of oats increases plant mass and yield potential as well as the chances for lodging. The secret is to apply just enough inputs to give the crop a bit of a boost without having it fall over. You can tell quite a bit by the color of growing plants. A dark green color is indicative of adequate nitrogen levels, while pale green and yellow plants are exhibiting chlorosis—which means nitrogen is lacking. Most of us don't have the luxury of a surplus of manure to spread on the land prior to planting oats, and we just have to be content with plowing down some sod the fall before. This should be all that is necessary to do reasonably well with oats—the grain that doesn't ask for much.

Checking Your Seed for Quality

The importance of planting high-quality seed goes without saying when planting any grain crop, and it's no different for oats. Upon receiving your seed, take a good look at the analysis tag sewn to the top of the bag. It may be a blue certified seed tag or a simple white piece of paper. Here you will find the results of several official tests that have been performed on the seed—germination should be 90 percent or better, and noxious weed seed content below one-tenth of 1 percent. Open the bag and have a good look at what you're buying. The oats should glisten with brightness and be white or yellow in color. Stained, dark-colored oats are an indication of prolonged exposure to wet weather. Scoop some seed in your hands and hold it up to your nose. If you sense a musty or moldy smell, you won't want to plant these oats. Last but not least, check the test weight. The oats you intend to plant should weigh a minimum of thirty-two pounds to the bushel; a thirty-six- or thirty-eight-pound bushel is even better. Denser seed provides more food to a young oat seedling. You only get one chance in the spring to plant your oats, so make sure well in advance that you are starting out with high-quality seed. Weather and a fickle climate will offer enough perils and pitfalls in the approaching season; planting the very best seed you can find will give you one less thing to worry about at the beginning of the growing season.

Planting Your Oats

Oats can either be broadcasted and incorporated with a light harrow or seeded with a traditional grain drill. Hand-scattering seed will most likely be the method of choice for smaller garden plots of oats. Broadcasting is a lost art and far more difficult than it looks. As with other grains, the skill is in getting an even distribution and covering of the oat seed. Some people are really good at this task. Before you begin the planting process, weigh out the right amount of seed for the number of square feet that you intend to cover. The archetypal farmer depicted in old woodcuts scattering seed from his cloth sack has given way to the Cyclone hand-cranked seeder and the tractor-mounted spin-type fertilizer and seed broadcaster, and you can make pretty short work of planting a quarter acre of oats with a Cyclone seeder strapped over your shoulder. The Herd Seeder from Kasco Manufacturing in Shelbyville, Indiana, is the tractor-powered version of the hand-operated spinner/spreader. These wonderful little machines are versatile and inexpensive, and can be bolted to the front or back of any small tractor or all-terrain vehicle. The seed spinner is driven by either a battery-powered electric motor or a small PTO shaft. Whatever method you use to broadcast your oats, make sure that you get the distance between each seeding pass correct. When seed is spun onto a field, it usually covers the ground more heavily directly under the spinner and more lightly at the outer extremity of the spread pattern. This makes it more important than ever to overlap each spreading pass. One secret that I have learned from operating the Herd Seed Broadcaster is to go back over the field a second time with a route of travel that puts the tractor and seeder halfway between each of the first passes over the field. A thicker, more even stand of oats will result because more seed will be applied to the areas that received less seed on the first pass.

Once the seed has been applied, it's time to incorporate it into the soil. You really don't want to bury it more than an inch deep. If you're working in a garden or a small field plot, a hand rake or an old bed spring pulled by a rope will do the job simply and easily. For larger acreages, a spike drag or disc harrow set in the shallow position will put just enough soil over the oats. This isn't precision farming, so don't be too worried if you see a few remaining oat seeds on the soil surface. The next thing to figure out is whether or not to roll the ground to firm up the seedbed. Packing, as it is called, has its advantages and disadvantages. Germination might improve as a result of better soil-to-seed contact, but early seed weed growth might be encouraged as well. It has been my experience that oats planted in the early spring don't need to be packed in with a roller; there is usually plenty of moisture around at this time of year to get oats to sprout. If you are planting grass seed along with the oats, a final rolling of the field will definitely promote germination of the little clover, timothy, and alfalfa seeds that were sown on top of the final seedbed. A little common sense and attention to detail will make all the difference if you decide to plant your oats using the broadcast method.

Oats

Oats in early flower. PHOTO COURTESY OF HEATHER DARBY

Seeding Oats with a Grain Drill

The easiest and most straightforward way to plant oats is with a grain drill. We have already discussed the finer points of grain drill operation, but let's review a few things that pertain to the peculiarities of seeding a crop of oats. Oats are lighter and bulkier than all the other cereals except spelt. With this fact in mind, don't be surprised if you end up using what appears to be a lot more seed per acre. The seed volume adjustment lever will be a lot farther open for oats than it will be for wheat and barley. When I first began growing oats in the 1970s, the standard seeding rate was three bushels (about one hundred pounds) per acre. Recommended seeding rates have experienced inflation over the past three decades like almost everything else. I have seen seed catalog recommendations for oats as high as 140 pounds per acre, but I've found that 125 pounds works pretty well for me. If you're seeding grass along with the oats, drop the seeding rate back to one hundred pounds to the acre. This will help the future hay crop compete with the oats. When planting oats, be prepared to refill your grain drill twice as often as you would if you were planting wheat. This becomes perfectly obvious when you consider that a bushel of oats weighs only half as much as the same volume of wheat. If your grain drill is equipped with adjustable seed cups above each delivery tube, you will want to enlarge the seed outlet opening to allow the much larger, bulkier kernels of oats to pass through unobstructed. I can't speak for every grain drill out there, but this means putting the seed cup adjustment lever in the second notch on my International grain drill. If you're planting field peas along with the oats, now is the time to add them to the mix at a rate of three parts oats to one part peas.

The only other caveat that I can offer about planting oats is to make sure that your seed is well cleaned and totally free of extraneous chaff, as chaffy oats may not flow evenly through the seed box and down into the seed delivery tubes. Dirty grain will hang up and bridge, causing uneven and diminished seed coverage. This situation can be particularly troublesome because you won't even know that you are having a problem before it is too late. Half a field may be planted before you realize that the seed hasn't been flowing evenly. This sort of situation is difficult to remedy by going back over what you just barely planted because you won't really know when the problem began and how much area to replant.

The last thing to consider is whether or not to roll the field after you are finished. A newer grain drill with press wheels behind each set of seed discs offers a few more options than the older models. The beauty of press wheels is that they only pack a very narrow band of soil directly over the planted seed. Since press wheels are adjustable for down pressure, you also have the option to match the right amount of packing force to the soil conditions at planting time. This will give the sprouting oats a bit of an early growth advantage over the weeds. Planting oats with a seed drill is paradise in motion for me on a warm spring day in early April.

Robust Early Growth

Once planted, oats will continue to demonstrate their prowess and home-team advantage. No other cereal, including wheat, will germinate more quickly under some of the worst early-spring conditions imaginable. Oats like cold weather, and they need more moisture in the beginning of the season than the other cereals. Six inches of wet spring snow on top of recently germinated oats in mid-April seems to only make them turn a darker shade of green once it has melted. There is some truth in the old adage of snow being "poor man's fertilizer," especially when it comes to oats in April. Oats are extremely robust in their early stages of growth, which makes this crop easy to differentiate from the other cereals. The leaves are broader in width and almost seem to reach out horizontally across the soil. The characteristic grain drill pattern of young oat seedlings in six- or seven-inch rows disappears very quickly as the plants tiller outward and form a ground-covering canopy. Annual weeds are smothered quite quickly in a rapidly growing field of oats. There is also the possibility that I mentioned earlier of unique weed-suppressing root exudates that are emitted from young oat plants. This phenomenon remains a theory in need of clinical trials, but there is no doubt that early-growing oats will beat the weeds better than any other grain crop except maybe buckwheat. The beauty of a newly seeded field of oats is apparent in how little you have to do once planting is completed. Stand back and watch this amazing cereal grain grow!

Early-Season Weed Control

Because of their strong roots and robust foliage, oats respond very favorably to early-season weed control with a tine weeder. Cereal grain tine weeding requires strategy and good observational skills to be effective and successful. If you seeded down your oats by putting grass seed down with a grain drill at the time of planting, you will probably want to stay out of your oat field with the tine weeder, because by the time the oats are at the five- or six-leaf stage and ready for the weeder, the little seedlings of timothy, clover, and alfalfa are just popping up, and weeder tines tearing through a field of young oats at this stage of growth don't differentiate between weeds and young forage seedlings. A different strategy is required if you want to plant forages and still use the tine weeder. Delay planting the grass seed until just before you are ready to head to the oat field with your Einbock or Kovar. By adopting this tactic, you will be doing double duty with your tine weeder. Young annual weeds like mustard and lamb's-quarters will be uprooted at the same time the freshly scattered grass seed is being mulched with a light layer of soil. In my personal experience, oats seem to suffer less weeder injury and setback than any of the other cereals. Several years ago I had a small field of oats that developed a rather thick carpet of lamb's-quarters underneath the crop. My Kovar weeder was occupied in some distant cornfield and unavailable to use in the oats. I called my vegetable farmer neighbor, Gerard, and asked him if he would be willing to attack the weeds in the bottom of my eight-inch-tall oat crop with his brand-new Einbock weeder. He obliged and came right up to do the deed. I was a little worried because the Einbock is probably the most aggressive tine weeder on the market. After he finished the job, the understory of weeds was severely set back and the oats were quite bedraggled and a bit flattened. A week later, though, the oats looked like nothing had ever disturbed them and the weeds were gone. We harvested a very nice clean crop of oats from the field later in the season. The motto of the story is not to worry about damaging a crop of oats with early-season tine weeding. This amazing plant is so strong and so vigorous that it takes care of its own needs very adequately.

The Growth Habits of Oats

Oats have an interesting growth habit as they progress through the vegetative stage into flower and finally into seed. Although this is one of the tallest of the common grain crops, oats remain short and bushy for what seems an extra-long time. The plants keep putting on more and more leaves as they grow outward as well as upward. This extra leaf surface contributes to the plant's ability to gather energy from the sun's rays, which eventually turns into high levels of fat in the oat kernel. The abundance of vegetative growth also makes oats a great choice for hay and forage crops. Indeed, oat hay and silage are popular choices with dairy farmers and beef producers who are looking for heavy yields of annual high-quality forage. Until recently, the standard procedure for making oat hay has been to cut the crop in the milk stage just after heading out. Contemporary research on the relationship between harvest dates and energy levels in oat hay has found that cutting the crop a little later in the soft dough stage of maturity produces the best-quality forage. Oats spend a good long time putting on lots of growth in the vegetative stage. Most varieties will grow as tall as three or four feet before there is any sign of flowering. By the time the plant is loaded with panicles and in full flower, an additional foot of height is possible. The tall stature of the oat plant is a blessing and a curse simultaneously. High yields of hay and straw are possible, but so is the chance for lodging.

Oats have about the same number of days to maturity as wheat, and both these grains take about two weeks longer than barley to mature. It takes a little more than a month for an oat plant to progress from flower to physiological maturity. After heading, the flowering panicles will first fill up with a white liquid form of starch called the milk. As with every other ripening grain, this internal milky starch will eventually begin to thicken and transform itself into soft dough. Kernels in the soft dough stage will continue to lose moisture as they become ripe grain in the hard dough stage. Although the oats are considered to be physiologically mature at this point, grain moisture can still be as high as 20 percent. Now it's time to patiently wait for final dry-down and harvest. The oat plant undergoes a very distinct and beautiful series of color changes during this reproductive stage of development. The seed head panicles that first appear at the tops of the plants are whitish green in hue in relation to the remainder of the oat plant, which is dark green. From this point onward, the ripening oat plant will begin to fade and change color. Deep dark green morphs into light green when the plant is in the milk stage. Once milk turns to soft dough, any remaining light green pigment disappears into a rather drab yellowish brown shade. Oats finally develop the bright golden-yellow color for which they are so well known during the final weeks of the ripening process. Careful inspection of a field of oats in the very last days before final harvest will reveal that the seed heads and upper stem portions of the plants will be golden-colored while the straw at the base of the stem might still be somewhat green in appearance. This green straw will slowly turn golden as the oat grain at the top of the plant drops to a final moisture level below 13 percent. Patience and vigilance are necessary at this point. You've been watching these oats ripen for the last month and you want to go out and combine them right now, but it's best to wait until the entire oat stem has changed color from green to gold before beginning the harvest. This is especially true if the weather is expected to stay sunny and dry. The oats will be better off still attached to the plants in the field as opposed to sitting in a wagon in a high-moisture state.

Harvesting Oats

Oats are an absolute joy to harvest if they remain standing until they are dead ripe. This is my favorite crop to watch being gobbled up by the combine. All those branching panicles interlock with one another to form a tightly knit mat of grain that holds together very well as it passes over the cutting platform and up the feeder house of the combine. This ease-of-harvesting feature serves us equally well when oats are cut by hand sickle or grain binder. Once the straw has been severed, this crop likes to stick to itself. I found this to be particularly advantageous when we used to reap oats with the grain binder thirty years ago. The reaped oats traveled gracefully across the platform canvas and up between the elevator canvases with nary a hitch, and bundles of grain were packed and tied very nicely before being ejected by the machine. When walking through the field after it was reaped and stooked, there was no evidence of any loose or escaped oat plants lying on the ground. This certainly wasn't the case with reaped wheat or barley. After harvest, I would always find plenty of stray wheat and barley stems that had somehow slipped out of the grain binder. But oats had the advantages of extra height and branching foliage, which make them an ideal crop for old-fashioned harvesting. There are certainly many reasons why oats were so popular here in the Northeast in the past.

There are, however, a few perils and pitfalls to consider when harvesting a crop of oats. Ripe oats stain and discolor very readily when exposed to rainy and wet weather conditions. Bright white or yellow oats will darken to a black color if they are wet down by a good rain. Once again, we return to the man-against-nature scenario at harvesttime. If rain is imminent and your oats are still a little green on the bottom with a 20 percent moisture level, it is best to harvest them before they get soaked. Drying will have to be completed with forced air in a bin or added heat in a grain dryer. If you are feeding oats to your own stock, however, all of this extra effort might not be worth it. On the other hand, if your oat crop is destined for milling or even for high-end horse feed, white oats will command a much higher price.

Lodged oats are the other major challenge that can undo an otherwise prolific crop. When a field of oats gets flattened by a mid- to late-season thunderstorm, you can only hope that the crop was far enough along in its development that grain fill is not affected. If the oats have already made it to dough stage when they go down, chances are that they will ripen just fine. Downed oats that have just barely headed out won't fare as well because they will be less likely to ripen well and will also have to spend a longer time down on the ground. There are varying degrees of lodging to consider as well. A lodged stand of oats that is still up off the ground at a ten- or fifteen-degree angle will continue to ripen because the plants will still be able to translocate nutrients through straw stems that are not totally crimped and broken. Partially lodged oats will still be receiving sunshine for photosynthesis as well. The biggest setback handed to us by Mother Nature in this type of situation is that lodged oats will not dry out as readily as a standing crop. The rate of ripening will slow down considerably, and grain discoloration is pretty well a foregone conclusion. The reality of a downed oat crop will certainly bring out the philosopher in anyone of us who has had this experience. You'll have plenty of time to live and relive as you slowly combine your oats from only one direction. This sort of harvesting process takes two to three times longer because the combine isn't threshing as you drive it back to the other end of the field to take the next pass. Your only consolation is that at least you are picking up the oats by driving the machine in the direction opposite to that of the lodged grain and running the header as low as you dare to gather the flattened crop. All you can do is keep reminding yourself that at least you're getting something from the harvest. The oats may be dark and unfit for milling, but there are oats coming out of the field. Farming is full of disappointment and heartbreak, and you have to be ready for it when it comes. Hopefully, 20/20 hindsight will help you learn from your mistakes and do some things differently next season. Maybe you'll apply less nitrogenous fertility or choose an oat variety with stronger straw. Whatever the outcome, we simply have to take the good with the bad and move on to harvest the next field.

Once your oats are combined and out of the field, you follow the same set of practices you would for harvesting any other grain. Moisture levels should be monitored immediately. If the oats are over 15 percent, you will need to dry them as quickly and efficiently as possible. Since oats are so much lighter and fluffier than all of the other cereals, drying by bin aeration works amazingly well. Once again, oats are at the top of the list for ease and simplicity when it comes to post-harvest conditioning. A quick cleaning with the rotary screen is highly recommended before you put oats away in any sort of a storage bin. Removal of excess chaff, weed seeds, and other foreign material will allow for better aeration and speedier drying. Because of their larger size and bulky nature, oats will require slightly larger inner and outer cleaning screen sizes than barley or wheat. Whereas an 8×8 outer screen and a 4×4 inner cone will suffice for these two other grains, oats are best cleaned with a 6×6 outer screen and a 4×4 inner cone. It usually takes two to three hours to change a set of screens on a rotary cleaner. Keep this in mind if you are growing barley and wheat in addition to oats. You might want to harvest and process these crops first before moving onto your oats. Once your oats are dry, stable, and stored safely in a bin, it's time to take a deep breath, relax, and go bale the straw.

Processing Oats for Animal Consumption

Many livestock producers like to add a small percentage of oats to a ration to bulk things up and provide some extra energy and fiber. The process of preparing oats for animal feeding is relatively simple and straightforward. Oats require only very minimal processing—they can be fed straight out of the bin to ruminants with upper and lower teeth like sheep and goats, as these animals have the ability to chew and more completely digest the oats. Chickens and other poultry can also process whole oats when fed as scratch feed, because all those little stones carried in the chicken's crop will help grind up the whole oats. Monogastric farm animals like pigs perform much better on ground oats, but can survive on whole oats if they are softened by pre-soaking them in warm water or skim milk. But oats have always been associated with the horse, and most of the oats grown on farms prior to the widespread adoption of tractor power in the 1930s were harvested for horse feed. Most horsemen like to crimp oats with a simple roller mill before feeding them to horses. The rolling action breaks the hull and flattens the inner endosperm, which contributes to improved digestive efficiency for horses. Horses can eat whole oats, but will definitely do better when the grain is rolled. The same holds true for cattle; many low-input beef operations feed whole oats to their cows. This is not a bad practice, as the extra fiber contained in the oat hulls slows down passage through the rumen and helps to produce essential body heat during the digestive process. Dairy cows, however, are a little fussier than their beef brethren and require oats that have been either rolled or ground in a hammermill. The consensus out there these days, though, is that oats aren't really a great dairy cow feed, and most traditional dairymen will tell you straight up, "There isn't a whole lot of milk in oats." This may be true if high production is your only goal. More milk will most certainly be produced if cows are pushed on large amounts of corn and soybeans. However, if grass-based sustainable dairying is your cup of tea, small amounts of oats in a dairy ration are just the ticket because the fiber in the oats will balance nicely with a high-forage diet of hay and pasture. Oats definitely have a place in livestock operations.

Implements for Processing Oats

Most oats fed on farms these days are processed with the same roller mill that was described in the previous chapter on barley. Whole oats pass between two tightly fitting steel rollers and emerge from the mill crushed and flattened. Some fines will be produced during the rolling process, and these can be fed right along with everything else. It seems like oats and roller mills were made for each other, as the soft-textured, bulky-hulled grain is easy to roll and requires very little input power for the process. A one-horsepower electric motor may be all you need to power a roller mill that will take care of a small herd of cows. This sure beats starting up a sixty- or seventy-horsepower tractor to power other types of feed grinders. As discussed in the barley chapter, fresh oats can be rolled directly out of storage and immediately fed to a herd of cows at each chore. *Fresh* and *simple* are the two key words to remember when choosing the small on-farm roller-mill option for feeding oats to livestock in a small farm operation.

Hammermills and plate mills can also be used to grind oats for farm livestock feed, which will be discussed in more detail later. This type of feed-grinding technology requires more power than the roller mill and produces a whole lot more dust and fine feed. The most commonly utilized farm implement for feed grinding is the grinder-mixer—basically a hammermill attached to a mixing tank. Oats, or any other grain for that matter, are metered into the top of the hammermill and beaten through a perforated concave screen by a series of revolving thin pieces of metal called hammers; the ground oats are carried by a horizontal auger into a vertical mixing tank. Hammermills can be outfitted with a number of different screens with varying-sized holes. The smaller the screen, the finer the oats will be ground and the more power it will take. Screens with tiny holes will produce very finely ground oats that are best fed to baby pigs. Medium-ground oats work much better for poultry feed. If you plan on adding oats to a dairy ration, choose a hammermill screen with larger holes (half an inch or better) to ensure that the oats are chopped and coarsely ground. You really don't want to feed oat flour to dairy cows and paste up their rumens. The beauty of processing oats in a grinder-mixer is that other grains and minerals can be added and mixed in during the process. This allows for the creation of a balanced ration put together from a number of different grain sources.

Plate Mills

Plate or buhr grinders are the oldest, most time-tested method for grinding oats or any other grain into grist for animal feed. These antique farm machines have been described several times already, but a simple review will explain why they can work so well to process oats. Two metal plates with rough surfaces are positioned side by side on a vertical axis. One disc remains stationary while the other turns freely on a horizontal shaft. Grain passes between the two plates and is ground either finely or coarsely depending on the space between the two surfaces. Plate mills are the ideal choice for the low-tech homesteader who is feeding a small number of farm animals. There are several brand-new models on the market that are hand-operated with a large turn crank and heavy flywheel to provide momentum during the grinding process. These machines are readily available in the Lehman's Catalog from Kidron, Ohio. If you happen to procure an older buhr mill, you will find that most of these machines were equipped with a large pulley and powered by a flat belt. (Most older tractors from the 1950s and '60s had special pulleys to run flat canvas belts.) If this is not an option, an electric motor or stand-alone gasoline engine can be modified to power an old buhr grinder. If this project is beyond your mechanical capabilities and comfort zone, I would recommend hooking up with an antique steam and gas engine enthusiast. These individuals are everywhere in our region; they have shows and expositions all the time. You might just find someone grinding grain with one of these old units at a steam show.

Grinding oats with a plate grinder isn't very complicated. There are only a couple of things you will have to know to operate and adjust the machine. First and foremost, the grinder has to turn at a speed that is fast enough to carry the grain through. Feed throughput depends on plate revolution and the amount of horsepower supplied to the grinder. If you are running at low RPM and low power, you will only be able to trickle oats through the machine. Fineness of the grist is determined by the amount of space between the buhrs. If you are grinding oats for dairy cows, back off the plates and let the machine just chop the oats. A finer, more flour-like grind can be achieved by positioning the grinding plates closer together. Plate adjustment is regulated by a large cast-iron hand wheel on one end of the grinder's main shaft. There isn't much of an antique market for these old machines. If this sort of process thrills you, it is something you can do for very little money.

Just remember to not be the same fool I was all those years ago when I ground my grains in this manner. Wear ear and eye protection along with a dust mask to protect yourself from the noise, dust, and flying particles produced by these beloved antiques.

Processing Oats for Human Consumption

Oats have everything going for them. However, I will have to stop singing the praises of oats when it comes to processing them for human consumption. You need to remove the outer hull from the inner groat to make oatmeal or oat flour, and dehulling is a very machine-intensive process. The sad reality of this situation is that you leave the realm of the simple and straightforward and enter the dimension of the complex and complicated when you go from growing and harvesting oats to hulling and flaking them. There are very few if any options for removing hulls from whole oats in a low-tech fashion. Some very clever low-key technologist I came across has put free and downloadable plans on the Internet for a hand-crank-powered oat huller—basically a reconfigured version of the countertop-mounted Corona flour mill described in detail in the chapter on wheat. The stationary steel buhr inside the grinder is relined with a layer of hard rubber to provide abrasion instead of grinding action when oats are passed between it and the adjustable outer steel buhr. The hand mill is not cranked down as tightly as it would be to grind flour. Instead a small space is left between the outer buhr and the inner rubber surface; the oats roll around as they pass through the mill, and most of the hulls are removed. The end product of this elementary process is a collection of loose hulls, broken kernels, oat groats, and some whole oats that escaped hulling. Here is the heart of the matter when it comes to hulling oats. It doesn't matter whether you are hulling on your countertop with a hand crank or in a big Quaker Oats mill out in Iowa, the edible oat groats need to be separated from all of the other material in the hulling process. The other problem you'll encounter when hulling oats is grain breakage. Oats break easily because they are soft and somewhat fragile. There is a fine line and balance that must be achieved between aggressive hulling and grain breakage. We can certainly live with a few stray hulls and broken kernels if we are processing oats for our own breakfast cereal. Unfortunately, if we want to provide oat groats or flakes for the growing market out there for locally produced grain products, we need to seek perfection in the process.

Hulling and Flaking Oats

Setting up a hulling and flaking line is costly and very machine-intensive, and I can speak from personal experience on this subject. I have grown high-quality oats for years and have always wanted to manufacture my own signature brand of rolled oats. In 2007, I transported five tons of my own oats across the border to Michel Gaudreau's oat mill in Compton, Québec, for hulling and flaking. Michel had begun building the infrastructure about five years earlier and had finally perfected a system in which he could reliably produce food-grade oat groats and flakes from field-run oats. But importing and re-exporting my own oats to and from Québec was a real logistical and bureaucratic nightmare. I had to procure a Canadian tax ID number and pay the 12 percent Canadian government GST (Goods and Services Tax) on my own oats, which were only coming into the country for processing. Then I had to file an FDA Prior Notice with US Customs in order to bring my rolled oats back home. Add to all of this a special permit to drive my farm truck on Québec highways and it becomes obvious very quickly why my little oat import/export adventure only happened twice.

One additional lesson that I learned after going through this whole process is the concept of "shrink." My five tons of whole oats very quickly became two tons of rolled oats when they left the mill. Three tons of small oats and hulls remained behind as part of my payment to Michel for the processing. Despite all of these drawbacks, my effort to produce a Vermont-grown oatmeal was rewarded with huge success in every manner except in the financial column. Localvores loved the rolled oats and were willing to pay higher-than-market price for them. I may have gained lots of glory by pulling this project off, but it certainly didn't pencil out as a sustainable business model when the final income was balanced against the costs involved. However, I am the perpetual optimist and dreamer. I vowed right then and there to set up my own oat-processing line here on my farm in Westfield, Vermont. I set out to learn as much as I could about oat processing and began to acquire the knowledge and machinery necessary to make this dream come true. Four years have passed and a lot of money has been spent, and at the time of this writing I am almost ready to begin producing my own rolled oats. I will share my knowledge and experiences with you so that you can understand the specialized equipment that is required as well as the basic principles behind hulling and flaking oats.

The first person I called to find out more about small-scale oat processing was Andy Leinoff from Cabot, Vermont, who founded Eric and Andy's Homemade Oatmeal in 1991. The little company manufactured and distributed Vermont-grown oat flakes until 2000. A rather detailed account of Eric and Andy's decade-long experience with oat processing appears in chapter one. After nine years of struggling with supply and production issues, Andy shut the business down in 2000. A lot of the specialized machinery that he used in the hulling and flaking process ended up being sold to Michel Gaudreau, who was just beginning to build his oat business at the time. The antique Wolf roller mill used for oat flaking stayed behind and eventually made its way to my granary. Andy told me in no uncertain terms: "If you want to turn whole oats into oatmeal, get ready to spend a lot of money."

He also referred me to his former consultant and supplier of equipment, Rick Gilles from Codema LLC of Maple Grove, Minnesota. Rick is a kindly midwesterner who has spent most of his life in and around grain-milling facilities. He just happened to be going on a business trip to Québec at the time and offered to stop by my farm to discuss the details and specifics of installing a potential oat-processing line. Codema makes their own hullers in addition to representing grain-processing machinery manufacturers from all over the world. Right about this same time, I also spent part of a day with Michel Gaudreau. He gave me a very detailed and extended tour of his oat-processing facility in which he had invested close to a million dollars. Among Andy, Rick, and Michel, I learned that this was very serious business indeed. You needed a whole lot more infrastructure than a huller and a roller mill to make oat groats and flakes. Most sane people would have gathered all of this information and abandoned the project right then and there, but I was not to be deterred. I decided to continue onward despite the fact that I would need to acquire a fairly extensive collection of very specialized and expensive machinery. Each of these pieces of equipment could be purchased used at a cost of between ten and twenty thousand dollars, and brand-new machines were priced in the forty-thousand-dollar range. What follows is a step-by-step description of the actual hulling and flaking process and the technology necessary to do the job.

Mulling and Flaking, Step by Step

Oats are best hulled by the impact method, while abrasion hulling works well for barley because it is such a hard-textured grain with a very tightly fitting outer husk. Oats are much softer and have a very distinct hull that is chaff-like and loose fitting. In the process of impact hulling, oats and other grains are thrown by centrifugal force from a flat spinning disc against a surrounding band of steel. Grain from an overhead hopper is slowly metered onto this very specialized veined impeller, which is very similar to the revolving spinner located at the bottom of a cone-type fertilizer spreader. Forsbergs machines are still manufactured in Thief River Falls, Minnesota, and there are quite a few out there that can be purchased for between five and ten thousand dollars. The impeller on a Forsbergs huller rotates between 1,250 and 1,700 revolutions per minute, and the steel veins on the impeller disc and the outer band against which the oats are thrown will wear and will need to be replaced occasionally. The impeller should be operated at the lowest RPM possible to get the job done without causing too much oat breakage. The Codema huller, however, looks very similar to the older Forsbergs model, but it is equipped with a few more bells and whistles and does a much more thorough and complete job. Codema hullers sell new for around twenty thousand dollars and are quite popular in large commercial oat-processing mills. The Forsbergs 15-D huller is more like a piece of farm machinery; its operating capability is a bit more limited and less exacting because it was designed to dehull grain for livestock consumption. This was always a popular machine with midwestern hog farmers who needed to produce and grind oat groats for baby pig rations.

Aspiration

The initial removal of the outer oat hull is only the first in a series of steps that make up the complete hulling process. Once the whole oats have been flung by the impeller against the outer edge of the hulling chamber, a mixture of oat groats, unhulled oats, and loose hulls is vacuumed away by suction into a cyclone. Aspiration is the next step in this process. This chaffy mixture of oat groats and hulls is subjected to a regulated air blast, which blows the lighter loose hulls away from the heavier material as it passes through the aspiration chamber. Aspirators must be adjusted carefully—if the air blast is too strong, good usable grain will be blown out with the hulls. Commercial industrial-grade aspirators look like big metal boxes with fan units attached to the exterior. Rick Gilles from Codema provided me with some fancy glossy literature depicting a whole line of aspiration equipment including ductwork to carry the airborne hulls away. I asked him how much one of these units cost, and he calmly told me just over twenty grand. On top of this fact, there was still one more piece of equipment to procure before the hulling line was complete. My head was beginning to spin at this point. How would I ever be able to hull my own oats on my farm?

The Roskamp Huller

At this point, Rick asked me if I had ever heard of a Roskamp Champion Oat Huller. Apparently, this older piece of feed-milling machinery had once been the standard for the industry. In 1938, John Roskamp of Waterloo, Iowa, built one of the first commercial-scale oat hullers available at the time. The Roskamp Huller Company was founded and soon began providing oat hullers as well as roller mills and other equipment to the feed-milling industry. This machine is a rather large hulk of very heavy cast iron. It is a hybrid piece of equipment, incorporating features from both impact and abrasion-hulling technology. Oats are fed into a hopper midway up the side of the machine, whereupon they are carried upward to the hulling chamber by a vertical screw auger. Oat hulls are removed as they pass between two very specialized discs rotating on a horizontal shaft. One disc is an abrasive backing plate while the other is outfitted with eight protruding "buttons" that actually rub the hulls from the groats. The distance between these two plates is controlled by a large lever located on top of the hulling chamber. If the oats are not hulling well, the gap between the hulling disc and the backing plate is closed down. If the oats are being smashed up and broken, on the other hand, the space is made larger.

Once the oats have been hulled, the mixture of groats and loose hulls is augered horizontally to the other end of the machine, where it is subjected to an air blast from a built-in aspiration fan. Herein lies the beauty and ultimate advantage of the Roskamp Oat Huller. This is a two-in-one machine because it not only hulls the oats, but aspirates them as well. In fact, it's actually a three-in-one piece of equipment because the oats that fall through the aspirator drop into a Carter disc cleaner at the very bottom of the unit. The aspirated oats reverse direction as they are carried back across a pan at the very bottom by another horizontal auger. The Carter indent discs revolve vertically through this stream of moving oats, and the little indent pockets in the revolving discs actually pick out and separate the oat groats from the unhulled oats that are being augered along underneath them. Once the groats are trapped in the indent pockets, they are lifted upward and tossed sideways into another small horizontal auger. From here the groats travel to an outlet at the bottom of the machine located underneath and alongside the aspiration fan. The remaining oats that aren't grabbed by the Carter indent discs travel back to the hopper end of the machine, where they're picked up by the original vertical infeed auger to begin the hulling journey all over again. Oats that pass through the Roskamp unhulled are recycled around and around in a circular pattern until they are finally hulled and separated out by the built-in Carter disc cleaner.

Despite its amazing ability to do a multitude of hulling tasks, the Roskamp Oat Huller is a crude machine by modern oat-milling standards. It is designed more for the livestock feed industry than the human consumption market. Nevertheless, Rick Gilles from Codema had a line on one of these machines that he was about to take in trade from a pet food and birdseed company in Ontario. He agreed to rebuild and recondition the machine and sell it to me for eighteen thousand dollars. Much to the chagrin of the rest of my family, I said yes to his offer. Once again, I was sailing into uncharted territory and had no inkling of the steep learning curve that was ahead. My repainted and reconditioned Roskamp Oat Huller arrived by truck about six months later, and I was amazed at how heavy it was when I went to unload it from the freight truck with my tractor. We then set about elevating the machine sixteen feet up through a trapdoor to the second floor of my granary. We accomplished this feat of levitation by using a neighbor's forklift and a couple of heavy-duty chainfalls. The fun was just beginning. It was time to spend even more money on this project. Like all large industrial machinery, the oat huller was powered by a three-phase electric motor. I had to take another step backward and install a special "rotophase" converter to turn my single-phase electricity into three-phase power that would be compatible with my new cast-iron hulk. This project—complete with wiring, motor starters, and switches—set me back around five thousand dollars. The next task was to provide an escape route for the oat hulls that were to be blown away by the aspiration fan, which involved having special ductwork custom-made at a local sheet metal shop. Add on another three thousand dollars once this was installed. Last but not least, a large cyclone dust collector to receive the aspirated oat hulls had to be hung from the side of the building twenty feet up in the air. By the time a special steel hanging frame was fabricated and the cyclone was hung with a crane, another five grand had slipped through my fingers. At a total cost of twenty-nine thousand dollars, my rebuilt Roskamp Oat Huller was very rapidly putting me in the poorhouse before I had even hulled a single oat. My "twenty/twenty" hindsight advice about oat hulling or any other type of machine-intensive grain processing is to enter this sort of realm of the unknown with your eyes wide open. Be aware of the extra and incidental costs involved in setting up a complete processing line. After all, Michel Gaudreau had spent close to a million dollars of borrowed money on his oat-processing installation. What was I thinking? Stay tuned to my story, because the plot thickens.

About six months after the Roskamp was delivered, I had finished the installation and was ready to start the thing up for a trial run. Our cash reserves were pretty well shot by this time, and I was anxious to see the thing perform. We turned on the three-phase converter and then tripped the switch for the twenty-five-horsepower motor that powered the huller. The whole building started to vibrate and rumble as the machine sprang to life, and the wail of the aspirator fan made it sound like an airplane was taking off right inside my granary. We opened up a small sliding gate in a pipe and let the whole oats drop from an overhead bin directly into the feed hopper of the Roskamp. Sure enough, the machine did work. We adjusted the damper of the aspirator fan so that only hulls were being blown out into the outside cyclone and down into a wagon. The Carter disc separated the groats from the unhulled oats as promised. We quickly filled a fifty-five-gallon drum on the ground level below through a delivery pipe with a steady stream of mostly oat groats. Everything went pretty much as planned, but we certainly didn't achieve perfection. There were quite a few whole oats mixed in with the groats. The operator's manual claims that the best the machine can do is to hull about 95 percent of the oats, and I could live with oats in my groats, but I really wasn't pleased by the amount of breakage I saw in the finished product. We tinkered a little more with the hulling chamber adjustment lever and were able to lessen the oat breakage somewhat, but we still had plenty of broken-up oat groats at the end of the process. I have to remember that this is a feed mill machine. Perhaps I am demanding too much from such an antiquated and outdated piece of equipment.

The Paddy Table Separator

I have learned several important lessons from operating my Roskamp Oat Huller and I will share them with you. First and foremost, you need to start the hulling process with very heavy and plump oats. Don't bother with anything less than thirty-eight-pound-test-weight oats. Clean your oats as aggressively as you dare to produce the heaviest grain possible. Using bottom screens with larger slots and turning the wind up on your cleaner will allow you to successfully accomplish this task. Be prepared for lots of lighter, number two grain and pin oats to divert into livestock feed. Larger-sized heavy oats are hulled and aspirated much better than lighter fare. The other important consideration is that even if you are using a new and improved modern machine like a Codema, unhulled oats are always going to slip through the hulling process and end up in the groats. Have no fear; there is an easy and expensive solution to the problem. You need one more essential piece of equipment—a paddy table separator—to complete your full line of hulling machinery. The paddy table is quite commonly used in the rice industry to separate unhulled or "paddy" rice from hulled product. This machine works equally well on oats and performs the same function of separating unhulled oats from groats. Paddy tables come in a variety of sizes and shapes, from small Japanese models that measure four feet square to larger European units that are fifteen feet long and six feet wide. These machines have as many as seven stacked decks. Each contains a number of cells or small compartments. Paddy tables must be securely fastened to the floor because they shake back and forth fifty or more times a minute with as much as an eighteen-inch stroke. Oats and other hulled grains travel down through the series of decks and inner cells, and separation of hulled from unhulled grain is accomplished by friction. No air is used in this process as it would be in a fanning mill or gravity table. Groats are slipperier than whole oats and will travel to one side of the vibrating deck while the oats travel to the other. Final quality control in oat hulling happens on the paddy table separator.

After Rick Gilles explained the principles of paddy table separation to me, I knew that this tool would have to be the final jewel in my oat crown. I meekly asked him how much one of these wonderful units would cost. He pulled out some more glossy pamphlets to show me pictures of an "economical" lower-priced machine made in Colombia of all places. It was a high-quality copy of the German and Swiss machines made by Schulte and Buhler. The price tag was forty-two thousand dollars; European models sold for over eighty. Needless to say, there was to be no paddy table in my granary at that point. I asked about used machines and was informed that once in a great while some large oat processor in the Midwest or western Canada traded in a paddy table for a brand-new machine. I was going to have to wait and experiment with other methods of oat separation. So we experimented with two different approaches to solving this problem. We tried running our Roskamp-produced mixture of groats and oats over the gravity table and through the air screen cleaner with a variety of screens and wind settings. I didn't have much luck. Then I contacted Matt Williams in Aroostook County, Maine, who was also using a Roskamp Oat Huller. Matt informed me that he passed his whole-oat/groat mixture over his gravity table multiple times to achieve good separation. Of course, he had a much newer and more modern multiple-air-blast Oliver gravity table. My ancient 1946 Forsbergs gravity table wasn't going to cut the mustard.

When you are pioneering in the field of grain processing in this part of the country, you will eventually reach a point of no return where you have to stop and take stock of your situation. You've spent a ton of time and money and still have a long way to go to reach your goal of producing food-grade grain products for human consumption. What do you do now? Do you quit or plod blindly onward? Lack of infrastructure is perceived to be the principal impediment standing in the way of New England–produced local grains, but I would go farther and say that the largest problem is the absence of resource people who can help us learn how to use the specialized processing machinery that is so essential to making the dream of local oatmeal come true. I had finally arrived at this point of no return. I wasn't totally discouraged, but I did realize that I wasn't going to be selling my own brand of Butterworks Farm oatmeal anytime too soon. I decided that I would save up some money and wait for a used paddy table to come along. Rick Gilles had several prospects for trades in the coming year. For the time being, I knew that I could hull oats with my Roskamp and produce groats that were good enough for us to eat at home, though not for sale to the public. We didn't mind spitting out a few hulls from our morning cereal. Patience was more important than anything else at this point.

Thankfully, help came along just when it was needed most. In the fall of 2009, the Vermont Housing and Conservation Board announced an infrastructure grants program designed to build in state capacity for grain processing and animal slaughtering. This grant announcement was part of the VHCB's Farm Viability Program, which helps existing farms remain vibrant and profitable by helping them design and carry out realistic business plans for future growth. I contacted program manager Ela Chapin and enlisted my oat project in the grant competition. This opportunity could not have come along at a better moment. The board actually provided trained business consultants to assist all of the grant applicants in the writing of professional business plans for their proposed enterprises, and a very knowledgeable consultant from Ontario who was familiar with grain processing was hired by the project to provide very specific information to myself and several other applicants who were working on various grain projects. Writing the business plan really helped me narrow down my goals to the point that my hulling and flaking project could materialize and proceed into the future. I decided to concentrate on hulling oats before proceeding with the flaking.

In April 2010, I found out that I had been awarded a grant and now had the funds to buy a used paddy table; I just had to find one. The prospects of buying one from Codema were quite distant. Rick Gilles informed that some trade-ins were in the pipeline, but still a way off. I found some small Italian models that sold brand new for around fifteen thousand dollars, but these were rice machines and there were no models working in North America to inspect. I then found a used Schulte paddy table in Manitoba for the same price of fifteen thousand; I almost bought it, but decided against it when I took a closer look at the photos that were sent to me by email. The deck portion of the machine was very rusty. The owner assured me that running oats over the table for a day or two would remove the rust by friction. But somehow, the idea of processing a food-grade product in a rusty environment didn't thrill me. At this point, I called back Rick Gilles at Codema and told him to put me on the waiting list. It took another year, but finally my "new" used paddy table arrived on a dreary cold day in late October 2011. The machine had been traded in the previous July by a large oat processor in Manitoba, and it was now all mine for a price of eight thousand dollars, which included shipping. My patience was finally rewarded, and I have the generosity and vision of the Vermont Housing and Conservation Board to thank for this wonderful gift.

This brings my personal oat-hulling and -flaking adventure just about up to the time of this writing. The paddy table has been installed; it is huge and heavy and looks rather daunting. The three-phase power is hooked up, and the machine is bolted to the ground floor of my granary—I can't believe how much floor space this unit takes up. We've fired it up and watched it run, but have not put any hulled oats through it yet, though this will happen soon. I have high hopes for producing high-quality oat groats with my new acquisition. From this point onward, my narrative about oat hulling and flaking will be more based on research than actual hands-on experience.

The Process of Turning Oats to Oatmeal
Once you've successfully produced oat groats, you are still only halfway to having oatmeal, but given my own personal taste in breakfast cereal I'd just as soon stop the whole process right here. I prefer a chewy breakfast porridge made from slow-cooked oat groats over the standard rendition of pasty oatmeal. The best oat cereal begins by soaking groats or rolled oats in water and whey the evening before. If you've done a less-than-perfect job of removing the hulls from your groats, you get one more chance to make things right at this point. When you pour water over your groats, the loose hulls and unhulled grain will float to the top, and you can skim off this less desirable grain and feed it to your chickens. Several rinses and soakings might be necessary. By morning, the oats are soft, lower in phytates, and ready to cook. I prefer my oats cooked in milk, probably because I'm a dairy farmer, but water works just as well, and I throw in some raisins or other dried fruit for variety. Maple syrup and yogurt also both go quite nicely with a bowl of oat groats or oatmeal. But we all take oat-based breakfast cereals for granted; rolled oats are an inexpensive and readily available staple in the American diet. Who would have thought that the process of turning field-run oats into breakfast food would be so complex and involved? When you buy oat groats or rolled oats out of the bulk bin at your local store, you hardly ever find a stray unhulled oat, and this is testament to what a good job the oat-milling industry does here in North America. Let's hope that we as farmers can learn some lessons from the big guys and produce a homemade, homegrown product that tastes even better because we did the work from field to cereal bowl ourselves.

Toasting Oats
The primary limiting factor in oat processing is the high fat content of the crop. Oil levels in oats range between 5 and 8 percent. Once the outer hull of the oat has been removed, the fat contained in the groat will begin to oxidize and eventually become rancid. The oat industry sees this as a shelf-life problem and has developed procedures like toasting and steaming to stabilize their products. Toasting oat

groats is actually an ancient tradition that originated centuries ago in Scotland; the Scots brought this tradition with them when they immigrated to Nova Scotia in the early nineteenth century. In fact, one of my favorite historical grain-processing museums to visit is Balmoral Mills in Nova Scotia. Located on the north coast near Tatamagouche, this water-powered gristmill museum processes local oats just as the original Scottish settlers did when the mill was constructed in 1874. After the oats are hulled between two large millstones, they are toasted over a slow-burning fire in a specially constructed kiln adjacent to the mill, and then thinly spread out on a perforated cast-iron floor above a hardwood fire in the basement of the stone building. Scottish oats are famous for their smoky flavor, and these oats are as exceptional as anything you would bring home from Scotland. All of which is to say that long before food scientists even existed, our forefathers knew they needed to stabilize oats for keepability, and they succeeded in producing excellent-tasting oats.

Oat Planking

Industrial-scale oat processing and flaking is complicated and exacting. People go to school and get degrees in food engineering to become experts in this process. I envision myself as more of a "farmer flaker" and not as a highly trained oat-processing technologist. For this reason, I will keep my explanation of the oat-flaking process simple, understandable, and hopefully possible with our limited nonindustrial resources.

To begin with, oat groats generally need to be tempered with some sort of live steam heat to soften them before rolling. But steaming serves a multitude of other essential functions as well. For example, the activity of the lipase and peroxydase enzymes that will eventually promote oxidation and rancidity is virtually eliminated by subjecting oat groats to direct steam heating for periods of one hour or more. Some of the proteins contained in the oat groat are denatured at the same time, however. The simplest way to carry out this tempering process is to hold the oat groats in a large stainless-steel box and bombard them with piped-in high-pressure steam that is produced by a boiler. High-pressure steam is a very concentrated form of energy. It is quite a bit hotter than boiling water. To put this in perspective, think about how badly you can burn your hand when it passes too close to the spout of a boiling teakettle. At sixty pounds of pressure to the square inch, steam temperature is close to three hundred degrees Fahrenheit. Since it is under pressure, steam will propel itself through pipes right into the specially constructed oat box, where it is dispersed into the recently hulled groats. This process can take up to an hour because the whole mass of oats in the holding bin needs to be warmed to over 150 degrees. Once the oats are softened and their moisture level has risen to 16 or 17 percent, they are ready to become flakes as they pass through the roller mill.

I could explain roller milling in terms of physics and altered oat groat cell structure, but I think it's easier to simply state that a pliable soft-textured oat groat is flattened into a flake as it passes between two revolving steel or cast-iron rolls. The food science engineers will use terms like *shear* to describe the disruptive process in which the cell walls, starch granules, and protein bodies of the oat are transformed when it is subjected to the pressure of the roller mill. To put it in layman's terms, we are trying to squash the oat groat as flat as possible and still have it hold together in one piece when it becomes a flake. This is basically the same process that would take place in a feed mill or with a small roller mill positioned outside a grain silo full of barley; the only difference is that we don't worry about the extra "fines" that are produced when milling oats or barley for animal feed with a roller mill, and the fine grain by-product just gets mixed in with the ration. But when oat groats are rolled for human consumption, any fines that are produced represent less overall yield in terms of flaked oats. Flaked oats (oatmeal) are not usually sold with a pile of extra oat flour in the bottom of the bin or bag. An oat groat sufficiently softened by steaming will generally hold its shape during the flaking process.

Flaking with a Roller Mill

Almost any old roller mill found on the farm or in a feed mill can be used to process oat groats into flakes. Machines with larger-diameter rolls will do a much better job because they inherently have more milling surface area to perform the actual flaking process. Larger rollers have an increased circumference, which will in turn keep the flake under pressure for a longer period of time than a set of smaller rollers would. A well-maintained roller mill with good sharp rolls that are precisely adjusted is essential if you want to turn out a good-looking oat flake. There are plenty of used roller mills out there that can be purchased quite reasonably—but before you do so, remove the top of the machine and carefully inspect the rolls for nicks, wear, and other imperfections. After you finish with a visual inspection, run your fingers over the grooved surface of the rollers to determine how sharp everything is. The longitudinal grooves and valleys should still have sharp pointed edges. If everything is rounded, the rolls are worn. This isn't the end of the world; the rolls can be sent out to machine shops that specialize in remanufacturing roller-mill rolls. The process is quite expensive, however, and it's not uncommon to spend a thousand dollars or more for this type of remachining. If the remainder of the machine is in perfect order with good bearings, shafting, and tin work, roller reconditioning just might be worth it. Sometimes, luck will prevail and you might pick up a machine that is in excellent condition for an excellent price. Buying a new roller mill isn't totally out of the question, either. For three to five thousand dollars, you can set yourself up to roll oats and other grains. There are plenty of new imported machines from China that sell for half the price of models manufactured in the United States, but my advice in this department is to buy a US product. In general, you get what you pay for.

The Mechanics of Flaking

Now that we've discussed the required machinery and the theory of flaking, let's get right to work and try flaking some oats. I plan to set up the antique Wolf roller mill given to me by Andy Leinoff to do the job, and I have a ten-horsepower electric motor to power this one-ton hulk of steel and cast iron. The first order of business will be to determine the pulley sizes on the mill and the electric motor so that the mill will operate between 250 and 450 RPM. This is a simple geometric problem—most electric motors revolve at 1,750 RPM, which is somewhere between four and seven times the required shaft speed of the roller mill. A four-inch pulley on the drive motor shaft should be matched by a sixteen- to twenty-eight-inch pulley on the roller mill to achieve the proper RPM for operation. I plan to roll my oat groats in my shop, which is attached to my barn and farmstead milk plant. We have high-pressure steam piped around the building for heating and milk pasteurization, but we will need a warm room with steam access to temper and roll our oats. I bought a specially constructed stainless-steel steam box from Codema that will be positioned over the top of the roller mill. The plan is to pipe live high-pressure steam directly into this overhead vessel until the oats are warm and soft enough to be rolled properly. The oats will then fall by gravity through a chute and gate valve into the Wolf roller mill below. Moist oat flakes will be delivered from the bottom of the roller mill onto a flat belt conveyor, which will transport them to some sort of a drying unit to reduce their moisture from the 17 percent level back to a more stable 13 percent.

As far as I can tell, the post-milling treatment of the wet oat flakes will be the most complicated and difficult part of the entire process. These flakes are quite fragile when they are in this state and must be handled with the utmost in care to prevent breakage and shattering. This is why they will fall onto a belt conveyor instead of being moved away by a normal grain auger. Professional oat millers (aka the big guys) use a very expensive specialized piece of equipment called a fluidized bed dryer to gently move oats away from the roller mill and dry them at the same time. This is a very energy-intensive process in which the wet oat flakes are suspended in the air and moved along by a powerful hot-air blast that blows upward from a trough underneath. By the time the oat flakes reach the end of the trough, they are dry and ready for packaging. I have a vibratory conveyor unit that I use for dry beans that I hope to refurbish for oat flake drying. Any fines that break off the flakes in this final process are usually sifted out and used for oat flour.

Flaking Oats on a Home Scale

It's time to get realistic, folks. Who in their right mind except a grain-processing nut like myself would ever want to jump through this many hoops and spend this much money to make oatmeal? Is there any way to make oat flakes that is a little less complicated and lower-key than all the antics I just described? Thankfully, the answer is yes. If you want your own oats for home consumption, the hulless varieties are a whole lot easier to deal with. You might end up with a small percentage of oats that don't thresh clean, but you will still have fewer hulls than if you were working with regular oats. Once you've gotten to this point, making oat flakes is straightforward and easy. Grab a copy of the Lehman's Catalog and buy one of their little hand-cranked table model rollers, which cost less than a hundred dollars and will roll just about any type of grain. I know several people who leave their hand hullers clamped to their kitchen counters all winter long. When they want to make oatmeal, they roll their groats the evening before prior to the soaking process. There's no need to steam or soften the groats, because the cereal is being eaten fresh; broken groats and fine particles go right into the same pot and are cooked along with everything else. If for some reason, you are a fanatic like me and want to pre-steam your groats before they are hand-rolled, there are simple ways to make this happen. Spread the groats out on a fine stainless-steel screen placed on top of a large diameter pot of boiling water. Once your steamed groats have been hand-rolled, they won't need to be redried because they will be going directly into the cooking pot. It's good to know that small-scale oatmeal production is indeed possible for those of us who insist on doing the whole process ourselves.

If you just must do this process in a fashion that is a little more scaled up, there are two individuals who have pulled it off and who are actually selling homegrown oat flakes to the public at large. Matt Williams of Linneus, Maine, in southern Aroostook County is producing very nice rolled oats without going through the tempering and steaming process. He has found that his customers actually like the taste of raw oat flakes. In fact, Matt's oats are in such high demand and sell so quickly that rancidity and shelf life have really not been problems for him, so he saves a whole lot of aggravation and money by skipping the steaming process. In addition, Michel Gaudreau of Golden Grains in Compton, Québec, has developed a hybrid tempering process that serves him quite well in his hulling and flaking operation. Michel tempers his oats by passing them through a salvaged propane-fired coffee roaster. The oats are toasted as they tumble through a revolving four-foot-diameter drum that works in much the same manner as a rotary screen seed cleaner. The roaster is set on a slight incline to keep the oats moving through, and heat is provided by a large propane flame. The oats are then steamed as they move slowly through a stainless-steel auger equipped with several ports for steam injection. After exiting this special auger, the oats drop right into the roller mill for flaking. Michel has built his own homemade version of a fluidized bed dryer to lower the moisture of his oat flakes once they exit the roller mill. He uses a series of infrared heat lamps suspended over a twenty-foot-long vibratory conveyor to radiate heat onto the oat flakes below. I am always amazed by his use of electric heat in this part of the process—we would never be able to afford it on our side of the border, where our electricity costs in excess of fourteen cents a kilowatt hour. Electricity is four cents a kilowatt hour in Québec thanks to the James Bay Hydro Project. Michel has designed his specialized dryer-conveyor to sift his flakes as they move from one end to the other. All of the fine material drops through perforations and ends up as oat flour. The finished flakes are delivered off the far end of the conveyor directly into bags, which are sewn shut and palletized for shipment.

I would imagine that by this time you really don't want to hear much more about oat hulling and flaking, and we've pretty well exhausted the subject. If you never intend to try any part of this process, at least you have more of an idea of what is involved, but if you're like me and bound and determined to go from the field to the cereal bowl, I hope that you will at least have enough basic information to proceed. Rest assured that there are plenty of very generous people out there to provide you with advice and help if you decide to hull and flake your own oats. It's a small club, and there is plenty of room for more members. The local foods market is wide open for homegrown oat groats and flakes, too. The majority of localvore consumers won't mind paying you a fair price for the extra work these products require. I'm also sure that these folks would be quite forgiving and would not be bothered by an occasional stray hull in their oatmeal.

Advances in Oat Breeding

Oats have been on the decline in the United States for the past fifteen years. In 1998, oat acres in the United States totaled about fourteen million; ten years later, in 2008, only two million acres were being planted. Acreage planted to other cereals has also declined, but not as precipitously as oats. Indeed, America's love affair with corn and soybean production has been heating up as cereal culture has fallen by the wayside. The introduction of shorter-season varieties of corn and soybeans with improved cold tolerance has allowed these crops to move into the northern plains states like North Dakota, and grain corn production has even become popular farther to the north in Manitoba. When every acre is seen as a profit center, low-income crops like oats are replaced with today's new moneymakers—corn and soybeans. This is unfortunate. The diversity and flexibility that come with having cereal grains in a rotation is lost, and long-term soil health suffers as well. Fewer and fewer animals are being kept on the land as cash crop agriculture becomes the norm, so if there are no beef cows that need bedding, why produce a straw crop like oats? As the tradition of growing oats disappears on the plains and in the Upper Midwest, this wonderful undemanding crop becomes even more important here in the Northeast. Despite all of the newer super-duper varieties of corn and soybeans, oats still remain the most well-suited and sustainable crop that we can grow here in our region.

The general decline in oat culture has not gone unnoticed by the milling industry and plant-breeding community. A group of individuals from Agriculture Canada, the USDA ARS (Agricultural Research Service), North American Millers Association, and Prairie Oat Growers has formed a collaborative to advance the lowly oat and save it from extinction as an important cereal crop. The Collaborative Oat Research Enterprise (CORE) is a global research partnership dedicated to the "increased production and consumption of higher quality oats with increased nutritive value." The organization consists of twenty-eight institutional members as well as many other individuals who are plant breeders and industry people. The few oat breeders who do come from the United States hail from the Dakotas, Idaho, and Montana. Thankfully, Canada still has a number of active oat breeders, many of whom are supported by both Agriculture Canada and oat-milling trade associations. The two Canadian oat research centers closest to us are located at Laval University in Québec City and at the Agriculture Canada Research Station in Ottawa, and oat varieties developed at these two locations will generally do quite well in the Northeast.

I've had the pleasure of having several conversations in person and on the phone about oat improvements with Jennifer Mitchell-Fetch, who works for Agriculture Canada in Winnipeg, Manitoba. Jennifer and her contemporaries have been concentrating their efforts on developing improved oat varieties with better nutrition, standability, and test weight. Apparently, oat varieties from Scandinavia are lower yielding, but are known for extremely durable straw that helps resist lodging. One variety in particular, Swedish Crown, has been crossed onto high-yielding North American varieties to improve standability, and Jennifer claims that she can breed oats with straw so tough that the modern combine will have difficulty cutting it at harvesttime. "Percent plumps" is another important factor in the world of oat breeding. The oat-milling industry wants a greater percentage of larger-sized kernels coming off the field to improve hulling and flaking efficiency. The amount of fat in oats has also come to the attention of this collaborative of oat breeders, and heart-health-conscious nutritionists have requested that all new oat varieties have a fat content of less than 7 percent. Many popular varieties have fat levels near 8 percent. More progress has been made in breeding for disease resistance in oats than any other trait; the incidence of barley yellow dwarf and crown and stem rust in oats has really declined in the past few years.

The final and most amazing advancement that has occurred in oat- as well as other plant-breeding efforts in the last few years has been the introduction of genetic marker technology. Molecular biologists are able to mark the genome of specific traits in an oat plant and watch their transfer under the microscope. This is not the same genetic manipulation that takes place in biotechnology—plant traits are still passed down through traditional breeding techniques. Genetic marking speeds up varietal improvement considerably because plant breeders can now verify heredity in the next generation by looking for specific marked genes under the microscope. Before this technique was developed, plant breeders had to spend another year growing out their crosses to see if the traits they wanted had been transferred to the offspring.

Oats definitely do not get the attention that wheat, corn and soybeans receive, but it is good to know that there are people out there who are working hard to make improvements to the crop. This works to our advantage in this part of the country. When we plant a new oat variety, we can give thanks to the tireless efforts of those plant breeders who are working behind the scenes. If you can think of any excuse at all to work oats into your crop rotation, by all means adopt this crop. Nothing will grow any better or more naturally for you than oats.

CHAPTER THIRTEEN
The Winter Cereals: Rye, Spelt, and Triticale

Fall-seeded winter grains have a very special place on an organic farm. These cereal crops have many advantages not found in their spring-planted brethren. Along with winter wheat, rye, triticale, and spelt are all planted in mid- to late September when the land is drier and time isn't quite so short. There is a bit of a lull in grain activities right about this time. Hopefully, all of the spring cereals are all harvested and in the bin. Soybean harvest is a still a few weeks away in early October, and corn usually isn't ready for picking until late October or early November. The pace of September work is a little less frenetic because you've got a little more time to prepare fields for planting than you do in April when it seems like everything needs to go into the ground at the same time. Soil conditions are also more favorable for tillage and seeding at this time of the year—those persistent wet spots that don't want to dry up in mid-April are all dry enough to seed at this time of the year. If everything is going perfectly, you've had time to fall-apply lime, compost, and other fertility materials earlier in the month of September. The season is beginning to wind down and the earth is getting ready to draw its forces inward again in anticipation of the coming winter. Now is the time to add the organic amendments that need to be assimilated by the soil life for next year's growing season. The green leaves on the trees are fading and turning to yellow and gold. One of my favorite things to do at this beautiful time of year is to prepare fields and plant them to winter grains on a warm sunny autumn day as I look out at the blaze of color that is lighting up the landscape.

Winter grains also have the ability to beat the annual weeds that compete with spring-planted crops. Wild mustard, lamb's-quarters, and other bothersome weeds will germinate along with a fall-planted cereal in late September, but will not survive through the upcoming winter. When temperatures tumble into the teens and single digits in late November or early December, the fall-planted winter grain will go into dormancy while the weeds freeze to death. It's almost like the arrival of winter acts like a natural herbicide, sparing the grain crop while it eliminates the annual weeds. If the winter season is kind and the dormant grain makes it to spring unscathed, you will have a green grain crop that is already at least four inches tall when the snows finally recede in late March or early April. These well-developed six- or seven-inch rows of cereal plants have a three-week to one-month jump-start on any cereal grain that is seeded in the very early spring. This same spring advantage holds true for early-spring weeds like mustard. Winter grain grows rapidly the following April, quickly establishing a canopy over the bare ground between the rows of young plants, and spring weeds won't wake up and grow because spring tillage is not necessary in an already established field of grain. This jump-start translates into better plant stands and much higher yields. Winter grains are harvested in the dry weather of mid-July, at least three weeks before spring grain is combined, and it is not uncommon for winter grain yields to be double those of spring-planted cereals. I have sung the praises of winter wheat in an earlier chapter, but rye, spelt, and triticale are also minor winter grain crops that can have a very important presence as part of a diverse crop rotation on any organic farm in the Northeast.

Rye

Rye (*Secale cereale*) is a member of the wheat tribe (Triticeae). Although this fall-seeded cereal is closely related to wheat and barley, it has more in common with less demanding crops like oats. Rye has often been labeled the "poverty grain" because it will produce moderately well on thinner, low-fertility soils that will not support a crop of wheat or barley. It is one of the most popular and versatile fall-planted cover crops on organic and conventional farms alike. Germination is rapid even at lower soil and air temperatures well into the month of October. Once rye is up and growing, it produces lots of leafy biomass before fall dormancy sets in later in November. Well over half the rye planted in the United States is plowed down for green manure the following May.

If you had to choose one word to differentiate rye plants from their wheat cousins, it would be *slender*. It's true that rye straw is wispier and much taller than wheat. A field of rye can grow to a height of nearly seven feet when planted on rich soil, which makes the crop very prone to lodging at harvesttime. Rye is somewhat comparable to wheat in its manner of growth. It has two to three flowered spikelets; its flowering glumes are long-awned. However, its flowering head is much longer and considerably thinner than that of wheat. These compressed grain heads have glumes and appendages that are so firmly attached, very little chaff is produced in the threshing process. Wheat, on the other hand, produces lots of chaff when its grains are removed during harvest. Rye also has a unique characteristic in which the individual grains in the ripening head are partly exposed instead of being entirely enclosed in the glumes like wheat. Rye kernels are much longer and more slender than wheat's. The berries of this grain are pointed on the embryo end where they attach to the spike. The longitudinal crease that is so common in a kernel of wheat is barely noticeable in rye grains. Threshed kernels of rye are easily discernible by their long, narrow shape and their grayish green color. Rye averages about 10.6 percent protein and 1.7 percent fat. It is very easy to grow and can provide us with an alternative grain for flour milling and bread baking.

The Origins of Rye

Although ethnobotanists are not entirely sure about the origins of rye, they do know that its domestication happened much later than wheat's. There are several theories about the origin of the wild rye that still grows in central and western Turkey. Two different varieties have proliferated in modern times. *Secale montanum* is found in Southern Europe and nearby Asia. *S. antolicum* grows farther east in Syria, Armenia, and Turkestan. Either one or both of these plants first appeared ages ago as weeds in the wheat and barley fields in Southern Asia. In much the same manner as oats, the wild rye moved northward and began to thrive in the cooler and wetter environments of Eastern Europe and Western Asia. The rye plant as we know it today most likely was domesticated from its wild ancestor in Eastern Europe somewhere between the Austrian Alps and the Caspian Sea. There is no mention of rye in any literature of Europe until the Bronze Age, which dated from 1800 to 1500 BC. The Greeks were certainly not acquainted with it, but there are references to rye in Roman literature, and Pliny described rye as one of the major staples for the barbarian tribes the Romans had conquered to the north near the Rhine River.

Rye was much better suited to the damper and wintrier climate of what was to become Germany, Poland, and Russia, and has always been known as "the bread of the northern peoples"; it's still a predominant bread grain in this part of the world. (Hearty European dark breads like pumpernickel are rye-based.) Rye eventually found its way to Britain and traveled to North America with the first colonists who settled in New England. It turned out that rye was much better suited to the climate of the New World than wheat. After several years of losing wheat crops to the Hessian fly, settlers in the Massachusetts Bay Colony switched over to rye, which they found to be much more reliable. Rye and "Injun" bread became a common staple made from a mixture of flint corn and rye. But despite the fact that rye is easy to grow and thrives with minimal management on low-fertility soils, it has never really become a dominant grain here in North America. There was certainly an increased demand for rye flour in the late nineteenth century with the arrival of waves of German and other Eastern European immigrants, but by 1905 Thomas F. Hunt in his *Cereals of America* was describing rye as a cereal in decline. Fortunately, rye is still very much in demand, especially with the advent of artisan baking in the last decade. There is definitely a place for rye in the burgeoning local grains movement here in the northeastern United States.

Choosing Rye Varieties

What varieties of rye should we plant? Spring varieties of rye do exist, but they are quite rare. I saw trial plots of organic Gazelle spring rye on a tour of the Dickinson, North Dakota Agricultural Experiment Station several years ago, but the crop was short in stature and rather uncompetitive with summer weed growth; seed costs were also exorbitant. At a dollar or more per pound, Gazelle spring rye wouldn't fit the budget of most farmers. I came away from my visit pretty well convinced that regular old winter rye was the best agronomic and economic choice for grain growers here in the Northeast. The time to buy rye seed is in mid-August, after the harvest is in and the seed has been cleaned. Most farm fertilizer and seed suppliers stock rye seed for the planting of fall cover crops. The seed is relatively inexpensive because it is usually of local origin. Many potato farmers grow rye as a soil-building part of their crop rotation. In many cases, a potato farmer will use his harvest of rye seed to offset his fertilizer bill from the previous spring, so if you happen to live near a potato farm that grows rye, you can save yourself some money by buying directly from your neighbor.

Variety choices for winter rye are just about nonexistent in this part of the country because most farmers keep planting their own seed year after year, but chances are pretty good that the variety you will be planting will be Aroostook. This cultivar was selected from the old variety Balbo in 1981 by the USDA Soil Conservation Service in New York. Aroostook is a very low-yielding but extremely winter-hardy variety that has worked quite well in the harsh climate of northern Maine's potato country. There is also a Canadian variety called Musketeer that is supposed to have both good yield potential and excellent winter hardiness, but I have never seen it offered in this part of the country. Unlike the other common cereals, which are self-pollinators, rye reproduces by cross-pollination. With this fact in mind, it is understandable why it would be so difficult to maintain varietal purity in crops of rye. Most of what we know as common rye is likely a hodgepodge of several different varieties. Most of the rye out there is diploid, which means it has fourteen chromosomes. There are a few tetraploid varieties, which all have twenty-eight chromosomes. Many years ago I decided to try planting a tetraploid cultivar called Tetra Pektus. Its kernels were said to be 60 percent larger than normal rye. Tetra Pektus was supposedly the miller's choice because it made better flour, so I drove way up to northern Maine and paid dearly for enough of this "wonder seed" to plant a five-acre field. I am sorry to report that the results were disappointing. Hopefully, you can learn from my experience and not make the same mistake. The seed was several years old and did not germinate very well; I found out years later that rye doesn't hold its germ over multiple years as well as other grains. I also had no idea that Tetra Pektus was a tetraploid rye and that it was different from the more common types. Tetraploids and diploids are genetically incompatible and cannot cross-pollinate. Given that rye needs to cross-pollinate to bear fruit, a mixture of these two genetically different types of rye in the same field will yield dismally because flowers that are not pollinated will never make grain. My best advice when it comes to obtaining rye seed to plant is to buy good clean seed with excellent germ from a reputable source. Inspect the seed before you seal the deal—poke your nose right inside the bag and give the grain a good sniff. It should smell fresh and clean without any hint of mustiness. Give it a good visual inspection for weed seeds, excess chaff, and little ergot fungal bodies. (These look like large jet-black kernels of rye. Ergot is a real problem in rye and will be discussed later in this chapter.) I have found that rye seed produced in drier regions of the country like Minnesota and North Dakota has a considerably lower content of ergot contamination. When it comes to rye seed, choose quality before variety.

Preparing Your Soil for Planting Rye

Soil preparation for rye is simple and straightforward. You needn't overwork the land, but some secondary tillage will be necessary to achieve a weed-free seedbed. Of course, if you're plowing down sod, you'll need to take some extra care and time to cut up the furrows with the disc and fit the soil with a cultivator or spike drag. A four-inch layer of good friable earth is all you need for good seed-to-soil contact. Sometimes, rye will follow another annual crop like a cereal or beans, and if this is the case only minimal tillage is required—one pass with the disc or cultivator to incorporate the residue of the last crop will suit. Rye really isn't fussy. If the leftover crop residue isn't too thick, rye can be seeded no-till right into the remaining stubble. This has become a common practice after corn silage has been chopped in late September. A heavy no-till grain drill equipped with special seed discs is pulled through the corn stubble without any advanced tillage or soil preparation of any kind. The rye will germinate just fine, and the savings in time, tractor fuel, and soil health are monumental.

Rye likes a dry soil much the same as any other cereal grain, so if your new seeding is prone to prolonged periods of standing-water inundation, you might want to think about finding a more suitable location. Lighter loams work best. Top-notch fertility isn't a requirement, either. Rye will produce moderately well on acid soil with low mineral and organic matter content. However, if the ground is up to snuff, you can expect a bumper crop. A light coating of aged manure or compost coupled with some mineral supplementation will make the difference between an average and a great crop. Rye is a great crop for a beginning grain farmer because it so flexible and forgiving.

Planting and Early Growth

Rye is definitely the least demanding and most forgiving of all the fall-planted cereals, and can be planted anytime between the last week of August and the middle of October. This isn't the case with the other winter cereals, especially wheat, which has a much narrower window of opportunity when it comes to seeding dates. Late-August-planted winter wheat can get too tall in the field and actually be smothered by winter snows during dormancy, but rye can stand more abuse than wheat. Some farmers with cattle will even carefully graze a well-established field of rye just before dormancy sets in. The trick is to not overgraze the rye and to keep animals off the field if conditions are wet and muddy. Ideally, the top couple of inches of soil should be frozen to support hoof traffic, but lush green rye will provide highly productive pasture for dairy or beef cows when everything else has long since been grazed.

Despite its abilities to withstand early or late planting, you will get the best results when rye is seeded during the third week of September. You can use a grain drill or broadcast seed by hand or machine. One hundred and fifty pounds of seed to the acre planted with a drill will guarantee you a nice thick stand of rye, but seeding rates can be increased by twenty-five to fifty pounds to the acre for broadcasted rye because not every seed will get covered with soil and germinate. If you are planting rye as a cover crop to be plowed down the following spring, a two-hundred-pound-plus seeding rate will ensure a plentiful supply of green biomass for incorporation in April or May of the next season.

Rye is easily distinguishable from the other fall cereals because its recently emerged seedlings are reddish in hue for up to a week before they turn green. When viewed from a distance, sometimes it's hard to tell if a field of rye has even come up yet, because the purplish red rows of young plants blend right in with the soil. When rye germinates, it throws out a whorl of four temporary roots instead of the three found in the other cereals. This phenomenon gives the crop quite a bit of seedling advantage at a time of the year when the days are getting shorter and the amount of sunlight is diminishing. Rye is well known for its ability to hug the ground and spread laterally, and it will continue this growth for several weeks and begin the tillering process before it goes dormant at the time of final freeze-up. Rye is probably the toughest of all the fall-planted cereal grains—a well-established thick stand that reaches three to five inches in height is almost certainly guaranteed winter survival. I have seen rye planted at the very end of October or beginning of November just barely sprout before the arrival of winter, then go on to be a bumper crop the next season. We should never underestimate the ability of this extra-tough crop to defy all of the odds and thrive despite deteriorating late-autumn weather.

The Dormant Period

Rye enters a state of suspended animation for the winter. This dormancy is quite miraculous and unique to winter cereals. Once the ambient temperature drops below freezing for an extended period of time, rye plants simply stop growing and go into hibernation. Leaves and stems remain green throughout this period. Plant color might fade to a lighter shade of green, but the rye is still very much alive in its winter sleep mode. During this time, rye and other winter cereals undergo a process called vernalization, which can be best described as the chilling effect that turns on the genetics responsible for next summer's flowering and eventual seed production. Without vernalization, a rye plant will not make fruit. This explains why rye seed planted in the spring will not flower and make seed. When I visited Washington State University's wheat-breeding department, I saw firsthand how winter cereal seeds were placed in special time- and temperature-sensitive freezers to be artificially vernalized so they could be planted in greenhouse studies.

Once rye has entered dormancy, you want the ground to freeze a bit followed by a nice blanket of snow. That little bit of frost before permanent snow cover will protect dormant rye plants from being attacked by snow mold, which is a foliage-consuming fungus that thrives in unfrozen soil at temperatures between freezing and forty degrees. The last thing you want is a heavy snowfall on top of ground that has not yet frozen. If this does happen, the snow can act like a blanket insulating the dormant rye, and the end result could be an onslaught of snow mold, which will attack the green rye leaves and leave them white and near death. Let's hope this doesn't happen to any of us. If you do notice some bleaching in your rye when the snow recedes in the spring, dig around the base of the plant. If the main stem is still green, the afflicted rye will send up new leaves. Your crop might be stunted and slowed down a bit by this, but at least you will still have some rye to harvest in a few months.

Vegetative Growth and Cultivation

Rye is the first of the winter grains to break dormancy in the spring. There's something rather heartwarming and encouraging about the sight of bright green rows of healthy rye plants poking up through the receding snow of early spring. Once the sun comes out and the temperatures climb into the forties, rye will begin to grow again. While you're racing against time to plant spring wheat, oats, and barley, the rye is off and running, and it is not uncommon to see it grow to a foot tall by the first of May. If an underseeded hay crop is in your plans, you will want to frost-seed into the wintered-over rye as early as you can get out there with some kind of broadcaster. It's best to scatter seed when the ground is still frozen and at least partially covered with snow. To accomplish this task, you can either walk through the field with a hand-operated Cyclone seeder strapped over your shoulder or drive around with a power seeder attached to a small tractor or ATV.

If your field of rye is intended for a cover crop plow-down, some observation and decision making will be required. By the middle of May, rye will be transitioning from the vegetative leafy stage of growth to its reproductive phase. When this happens, a more pronounced central stem will appear and plant leaves will become less lush. Make sure you incorporate your rye crop before the plants begin to show signs of stem formation, as plant cell structure changes to a less nitrogenous, woodier type of cellulose during this process. When rye with stems is plowed down, it will contribute much less nitrogen to a spring soil. As a matter of fact, if you wait too long to incorporate a rye crop, the plow-down can actually *remove* soil nitrogen to assist in breaking down the tough plant residues. The final thing to consider when using rye as a spring-incorporated cover crop is the type of tillage used in the process. Rye is tenacious stuff and will do everything in its power to reappear and survive. This is especially true if you use a heavy disc or field cultivator for the job. If you don't want volunteer rye in your next crop, choose a moldboard plow for this job and bury the stuff under six inches of soil. Rye makes a much better plow-down for a row crop like corn or soybeans than for a spring cereal. Since row crops are usually cultivated, wild rye is pretty easily removed. I have to admit that I've learned about this the hard way by finding lots of volunteer rye in a crop of wheat or barley. Rye kernels are almost impossible to remove from other cereals during the cleaning process. Cover crop rye needs careful management and consideration in late April and early May.

The Reproductive Phase

Heading and anthesis—the scientific term for pollen shed—is the next stage of development in the life of a rye plant. As the month of May progresses, the vegetative stage of rye growth gives way to the reproductive phase. That amazing dark green color that you have been observing since the snow's departure begins to fade. The first major hint that the rye crop is moving onward is the appearance of a lone leaf stretching upward above the rest of the plant. This is the flag leaf, and it signifies that the plant has reached the boot stage. If you peel the flag leaf down the stem of the plant, you will find the makings of the future seed head down in the boot. Several days later these heads emerge with awns attached on top. What once looked like a field of tall grass now looks like a field of grain with beards waving in the breeze. By this time, the rye has taken on a very distinctive gray-green color—ripening rye has a hue like no other cereal. It is also the most precocious of all the winter grains, heading out weeks before winter wheat. In the warmer southern areas of the Northeast, the first appearance of rye heads and beards can come as early as the tenth of May, but for the rest of us who farm farther north, this usually happens between the fifteenth and the twenty-fifth of the month. A week or so after the rye heads have emerged from the boot, the pollen will begin to fly. Rye produces profuse amounts of pollen, which is quite noticeable during the anthesis period. Lots of pollen is necessary because rye flowers are fertilized by cross-pollination instead of the usual internal self-pollination that takes place in other cereals. All of this extra pollen floating above a field of rye is an insurance policy of sorts. It is imperative that a flowering rye spikelet receive its pollen from another plant.

Once a field of rye has reached anthesis, there is no turning back. Grazing and cover cropping are no longer options because the basic cellular structure of the plant has transformed from leafy and tender to tough and woody. Once heads of rye appear, the crop can be safely harvested for hay or straw without regrowth of any kind. If rye (or any other grain for that matter) is mowed down during the vegetative stage, it will grow back and try to make a seed head. Some fruit and vegetable growers will grow rye and harvest it early as "green straw," because once a recently headed rye plant is mowed and dried, the green-colored straw is much more biodegradable when it is used for mulching vegetable and berry crops. Farmers who employ this practice of early rye harvesting are also pleased by the absence of viable weed seeds in their future mulch.

The Rodale Institute of Kutztown, Pennsylvania, has been experimenting with rolling down young headed-out rye plants for the no-till planting of row crops like corn and soybeans. A machine called a roller-crimper flattens the rye and creases the stem every eight inches. Corn and soybeans seeded directly into this mulch of rolled-down rye are able to germinate and make good growth in a weed-free environment. Mechanical cultivation is totally unnecessary in this system. The only requirement for success is a good thick crop of rolled-down rye to totally shade the ground and prevent weed growth. A heavy corn planter equipped with special wavy coulters will cut right through the rye mulch to plant seeds in the protected soil underneath. This innovative practice is testament to the versatility of rye as more than just a cover or grain crop. However, no-till planting into roller-crimped rye does have some limitations, especially in more northerly locations where rye anthesis comes later. It is possible to miss the ideal window for corn planting when you have to wait until very late May or early June for rye anthesis.

Ripening and Harvest

Once a rye crop has undergone pollination, it will spend the next six to eight weeks making grain. During this time plants will grow to a height of five feet or more. The process is exactly the same as any other cereal. Fertilized flowers will first produce a very liquid milky starch that eventually thickens into soft dough, but as ripening continues plant stems and leaves lose their vitality much earlier in the process than other cereals. A field of rye may look golden and ripe from a distance, but a careful inspection of seed heads will indicate that maturity has progressed only to the soft dough stage. Patience is essential if you want to harvest high-quality rye. You would think that because rye heads out two weeks earlier than winter wheat, it would be harvested two weeks earlier, but that isn't the case. The normal early dieback of leaves and stems delays the onset of hard dough development and final maturity for at least a couple of weeks.

Rye is generally ready for combining about the same time as winter wheat. I've learned this lesson about rye maturity the hard way over the years. In the early 1980s, my harvested rye became musty smelling after a month of storage. It sure seemed ripe at harvesttime, but nevertheless the grain started to spoil. I consulted with veteran potato farmer and rye grower Burt Peaslee, who told me that rye was never ready to combine until after spring barley harvest. I took that lesson to heart and have followed his advice ever since. Don't rush your rye harvest unless you have grain-drying facilities. Another reason to harvest wheat and barley crops first is to prevent cross-contamination of these harvested grains with kernels of rye. Combines are very difficult to clean between crops, so harvest your rye last. You can chew kernels or hold them in your hands to determine how wet they are, but only an actual moisture test will tell you if your crop is really below 13 percent. Once you've waited this long to take the combine to the rye field, the crop is generally tangled and a bit lodged by virtue of its height and slender wispy straw, but fortunately rye seems to tangle and intertwine instead of going totally flat like a crop of oats. Your field might look like an eggbeater has been through it, but everything will be all right if you lower the combine's cutter bar enough to pick up all the grain.

Ergot Infection

Rye is pretty well unaffected by most of the common plant diseases that plague the other cereals. Black stem rust and orange leaf rust can be a problem, although I have never seen any evidence of rust in my rye during the past thirty years of cultivating the crop. Rust spores usually blow in during the month of July, long after rye has headed out and begun to make grain. Rye does have one major disease problem, however. It is extremely susceptible to fungal infection from ergot (*Claviceps purpurea*). Although ergot can invade the seed heads of wheat and some of the common forage grasses, it is particularly problematic in rye. As with every other fungal invasion in a cereal crop, airborne ergot spores enter the floret of flowering rye at the time of anthesis. The spores mimic pollen grains while they colonize the stigma or male part of the flower. A fungal mycelium then develops that destroys the ovary of the flower and eventually connects with the vascular bundle of the plant that is responsible for plant nutrition. Once this ergot growth has access to the rye plant's circulatory system, it grows right along with the developing kernels of rye. The final result is a large black ergot body (also known as a sclerotium) that can be seen protruding from the infected rye head. As I stated earlier, rye is one of the few cereals whose kernel is not entirely covered by the husk. This particular type of semi-open kernel physiology lends a helping hand to the ergot body as it parasitizes the rye plant. The result is very noticeable: The ergot sclerotium can grow quite large as it protrudes upward and outward from the rye head at a forty-five-degree angle.

The bad news is that these large black growths are toxic and poisonous to humans and farm animals. Ergot bodies harden up and thresh out just like ripe grain. They are a bit larger than a rye kernel and are not easily removed by a fanning mill grain cleaner.

It seems like small amounts of ergot infection are almost a given in our humid climate. If you only find a little ergot in your clean rye sample, you're probably all right—you can remove it by hand as you pour the rye into your flour grinder. Larger infections of ergot pose more of a dilemma, as the ergot fungus produces alkaloids that can pose challenges to the circulatory systems of mammals. Several years ago we had a very serious late-summer bout of ergot in the mature seed heads of pasture grasses here in Vermont. The end result was an outbreak of gangrene and bloody hooves in some cattle that had been grazing these grasses. Fortunately, the affected animals all recovered once the problem was diagnosed and the ergot-infected pastures were removed from their diets. The best advice that I can give about ergot infection in a rye crop is to be vigilant. Walk your fields in advance of harvest to determine how much ergot is actually growing out of the rye heads. If you see an inordinate number of little black growths waving around in the breeze with the rye heads, it might be time to make some post-harvest plans for your rye crop. A gravity table will remove a high proportion of the ergot. First and foremost, try to plant clean seed on ground that is relatively free of ergot contamination.

Uses for Rye

Once your crop of rye has been harvested, cleaned, and safely stored, you have a few options for its use. I like to grow rye as a source of cover crop seed. A homegrown source of rye seed for fall planting of cover crops can save you a lot of money. Since so much of the seed available from farm suppliers is loaded with ergot, I figured that I could do just as well raising it myself for my own use. Sometime back in the early 1980s I let one of my cover crop rye fields grow up to grain. I found the experience to be quite rewarding. I had my own seed to use later in the summer and a bit of a surplus to sell to other farmers and gardeners. I usually manage to plant between ten and fifteen acres of rye each season and sell my surplus, as newfound interest in soil conservation and cover cropping on farms has created a very good demand for rye seed in the past few years. If you are harvesting weed-free rye with a properly adjusted combine, you might not even need to clean the grain before replanting it. A small amount of ergot in the seed won't be a major problem, especially if the rye is being sown as a cover crop. My advice, however, is to be totally up front about what you are selling. Let the customer know if the seed contains any ergot. If you've got the wherewithal to clean your rye seed, I would recommend it. Ergot spores are everywhere and can infect any crop of rye even if it was planted with clean seed. Whatever the situation, at least you'll know that you started with the best seed possible.

Rye really isn't much of a feed grain; its bitter taste and heavy texture render it almost unpalatable for most classes of livestock. Pigs, chickens, and cows would rather consume barley and corn if given the choice. If you must feed some rye because your grain supplies are limited, it is recommended that rye should make up 25 percent or less of the total

An ergot body attached to a head of rye.
PHOTO COURTESY OF SID BOSWORTH

ration. If an animal doesn't have anything else to eat, rye will simply have to do. Rye is much more edible if it is sprouted first, but the true value of rye for livestock use lies in the quality of its straw. Rye straw makes fine bedding because it is long and soft. It fluffs up readily and breaks down quickly in a compost pile. Rye straw was an important export from Vermont in the nineteenth century. After the rye was being gently threshed with a special threshing machine that left the straw long and unbroken, bundles of straw were shipped to New York City for horse stall bedding. Francis Angier has told me stories of his grandfather packing rye straw onto a special lake steamer for the long journey down Lake Champlain and the Champlain Barge Canal. So if all else fails and you are looking for some income from your field of rye, there is always the straw to sell. Rye may yield a bit on the low side with harvests of half a ton to a ton to the acre of grain, but more straw will be produced to the acre than with any of the other cereals.

Rye is most profitable as a crop for human consumption. If making rye flour is your intention, do everything in your power to produce the very best-quality grain. A timely early harvest will protect rye from in-field sprouting and ensure the high falling number necessary for good baking characteristics. (For a detailed explanation of falling numbers, see the wheat chapter.) You will also want to be certain that your rye is entirely free of ergot. A double cleaning with a fanning mill and a gravity table is a good place to start. Any grain that doesn't make the grade for milling can be used as seed. Rye acts completely different from wheat when it travels through the millstones; it seems to pulverize into a fine powder without much resistance. The germ and bran of a rye berry are very tightly bound to the endosperm, so as a result, the flour doesn't separate into its individual components as easily as wheat would do. Rye flour is naturally high in fiber with levels of over 8 percent. The fiber source in rye consists of noncellulose polysaccharides, which are known for their high water-binding capacity. The high levels of this unique fiber give rye health-promoting properties not found in any of the other cereals. When whole-grain rye bread is consumed, the human body has a much lower insulin response than it would for other grains like wheat. Rye is also a wonderful food for diabetics, as the larger starch particles found in the rye bread actually slow the rate of starch digestion into sugar. In addition, rye lacks the type of gluten present in wheat. It is, however, high in another type of gluten called gliadin that allows leavening agents to do their work properly. When sourdough culture is mixed with rye flour, it goes to work digesting fiber and the bitter taste present in the grain. We have found that sourdough works much better than yeast to make a really good loaf of rye bread. Rye is probably the most sustainable bread grain we can grow here in the Northeast. Its fertility requirements are low and it is a natural here in our climate. I love the chewiness and the robust slightly bitter flavor of this minor grain.

Spelt

Spelt (*Triticum spelta*) is a hexaploid species of wheat that has the same winter growth habit as rye. It is a very ancient grain whose origin dates back to between 4000 and 5000 BC, just like emmer and einkorn. Ripening spelt has a very distinctive appearance out in the field: It is much taller than wheat and a little shorter than rye in stature. Spelt heads have multiple florets like wheat, but are much thinner and a bit longer. Each floret contains two kernels that are larger than a wheat berry and pointed at one end. As seen on page 27 of the color insert, spelt looks completely different from any of the other cereals. Its dark brown color and individualized husk-covered florets give this grain the look of antiquity. The genetic makeup of spelt is virtually the same as it was three thousand years ago. For this reason, it has become quite popular with nutrition-conscious individuals who want to avoid modern wheat and the allergies that accompany its advanced genetics. Otherwise, spelt has gone relatively unnoticed for the last several thousand years—when you bake with spelt flour, you are using the exact same grain that was used to make matzo in the Old Testament. The other unique (and slightly troubling) feature about spelt is that the grain remains in the hull after it is threshed, and harvested spelt is about 20 percent hull by weight with a fluffy, loose texture. Once it is in the bag or the bin, this grain continues to look like it is from an earlier time in the history of civilization and agriculture. Spelt is prized for its superior nutritional qualities—it is higher than wheat in protein, vitamins, and minerals. It has a nutty mild flavor and contains a type of gluten that is totally different from wheat's. The fact that spelt continues to grow in popularity each year makes it a good choice for a winter cereal with income potential here in our region.

Origins of Spelt

The jury is still out on the exact origins of spelt, although some archaeologists claim that it originated in what is now Iran around 5000 BC. Recent DNA evidence points to two separate origins for spelt, one in Asia and one in Europe. It is most likely a hybrid of regular bread wheat and emmer, and could also be a cross between wild goat grass and emmer. The grain seems to have made its way to Southern Europe during the Neolithic era, which lasted from 2500 to 1700 BC. By the Iron Age (450 to 15 BC), spelt had become the main wheat species in southern Germany and Switzerland, and this area of Central Europe eventually became the breadbasket of spelt production. Spelt arrived in North America with Swiss and German immigrants around 1890. Most of these folks settled in Ohio, which became the center of American spelt production for a few short decades. The crop was known as spelz in the Census of Agriculture and farming publications at the time. As these Swiss and German farmers became more assimilated into American culture, spelt production and consumption died out pretty quickly and was replaced by wheat. By 1920, spelt was almost extinct as a North American cereal. It wasn't until the 1960s that it was resurrected as a grain crop by W. B. French, a seedsman from the town of Wakeman in Huron County, Ohio. After noticing some spelt growing in a neighbor's field, French procured a small amount of seed and began to propagate it. By the 1970s, he teamed up with Ohio State University agronomist Hal Lafever to improve common spelt for modern-day production. A breeding program released several new varieties that were quite well suited to the Great Lakes region of the Midwest. Since this time, Ohio has become the center of North American spelt production; somewhere between one hundred and two hundred thousand acres of spelt are grown in the state each year. Spelt is an integral part of the crop rotation practiced by the Amish in Ohio. Instead of planting soybeans after corn, Amish farmers seed spelt in the fall after their corn has been harvested. Grass seed (alfalfa, timothy, and clover) is frost-seeded into the spelt the following spring.

The organic farming community discovered spelt in the late 1980s, and demand for it began to grow as health-conscious consumers began to look for alternatives to wheat. The crop also proved to be quite adaptable across many different climate zones: Spelt culture spread into Michigan, Ontario, New York State, and Pennsylvania. I first began to see it in Québec around 2000, and I bought some seed from the Quirion brothers in Barnston, Québec, and planted some on my farm in northern Vermont. The crop thrived here, too, and several other farmers in northern New England have also started growing spelt with great success—proving that this ancient grain has its place here in our region as well.

Fertility Considerations

Spelt is much less exacting in the fertility department than winter wheat. A good harvest can be had with 25 to 50 percent less nitrogen input than for a similar crop of wheat. Soil preparation is about the same as it would be for any other winter cereal, and a well-prepared seedbed that is smooth and well packed works best. Spelt can be seeded anytime between the middle of September and the first part of October. The third week of September seems to be ideal for those of us in the more northerly areas of the region. If you are growing spelt in southern New England or New York, an October planting date will work just fine. For some unknown reason, spelt seems to thrive after a crop of corn silage. Minimal soil preparation is necessary because very little crop residue remains after corn has been chopped for silage. One pass with a disc is all that is necessary before spelt can be seeded. The Amish out in Ohio really know what they are doing following corn with spelt. Soil drainage is not as critical for spelt as it is for some of the other grains. You certainly won't want to plant spelt in a field that is constantly soggy, yet it will stand a lot more soil moisture than almost any other fall-planted cereal. There are definitely a number of advantages to planting a crop with five-thousand-year-old genetics—Swiss and German farmers weren't adding any fertilizer to their spelt crops in Europe three hundred years ago.

Considerations for Planting

Spelt seed is not as readily available as winter rye or winter wheat, and you won't find it as part of the regular inventory at most seed dealers and farm supply establishments in the region. If you do want to plant some spelt in September, start looking for seed in early August when it is being harvested. Most dealers with any connections should be able to order it in for you. Lakeview Organic Grain in Penn Yan, New York, always carries spelt seed. Your best bet will be to buy seed from another farmer, and the price will certainly be much lower if you can buy seed from another individual as opposed to a commercial establishment. The most common variety of spelt available to us is called Oberkulmer. Some varieties of spelt contain traces of wheat genetics, but Oberkulmer is true common spelt. It seems to thrive across a number of different climate zones from Pennsylvania all the way northward to Québec. I have been growing Oberkulmer spelt for close to ten years and have always been happy with its productivity and winter hardiness. Because it remains in the hull, spelt seed is bulky and very light in weight. A bushel of spelt still in the hull only weighs twenty-eight pounds, and seeding rates can approach two hundred pounds to the acre. So if you are planning on planting any amount of acreage, be prepared to handle a lot of seed, and don't wait until the last minute to procure it.

The most difficult aspect of growing spelt is seeding the crop. Standard practice is to use a grain drill just as you would for any other cereal. The problem is how to get two hundred pounds of seed evenly distributed on an acre of land. It's hard to get enough seed into the ground, and you won't find calibration numbers for spelt on the seed rate charts of most grain drills. If you can't find another grower to consult with, you'll have to calibrate your own grain drill. I have certainly learned how to plant spelt the hard way by loading up the grain drill with a set amount of seed and going out there to try it, but when I run the drill down on seed, I look at the acreage meter and find that I have only planted one hundred pounds to the acre instead of two hundred. Hopefully, you'll be able to learn from my experience and not have to go through all of this extra work when it comes time to plant your spelt. If you find that you have only seeded at half the necessary rate, you will simply have to leave the grain drill at its present setting and drive back over what you just planted a second time. I find that with my International 5100 drill, I need to adjust the seed-control lever to its wide-open position. I then reach underneath the seed box and open up the seed exit gates at the top of each seed tube to the widest position. This allows the rather large spelt seeds plenty of free travel through the seed-metering gears and down the seed-delivery tubes. There is no other seed out there as bulky as spelt, and it doesn't flow readily through most seeding equipment. Thus seed flow needs to be totally unimpeded to ensure that this very unusual-shaped grain can make its way through the grain drill and into the earth below.

Sometimes you'll get lucky and end up with spelt seed that has been partially dehulled for easier planting. If this is the case, your drill will have to be closed back up to accommodate seed that is much closer to wheat in its shape and density. A bushel of hulled spelt weighs sixty pounds, which is exactly the same as a bushel of wheat. Spelt must be hulled carefully and gently if it is to be used for seed. If the hulling is too aggressive and the spelt seed gets smashed and chipped in the process, it will not germinate. Plant hulled spelt at a rate of 120 pounds to the acre.

The Growth Process

Once your spelt is planted, there isn't much left to do except wait for it to germinate. My experience with this grain has been quite positive: It germinates quickly and grows with fervor in the early fall. Make sure to treat your spelt crop exactly the same way you would a new seeding of rye or winter wheat. If you've got extra top growth and ground conditions are dry enough, consider giving the spelt a light grazing in late October. Spelt enters winter dormancy in exactly the same manner as any other winter cereal. Vernalization by cold winter temperatures is required for the crop to produce seed the following summer. Once winter has passed, you carefully inspect your crop with high hopes that you have been spared by the vagaries of winterkill. It has been my experience that spelt doesn't come through the winter quite as green as rye would. Nevertheless, it seems to turn green with the arrival of warmer temperatures and some spring rains during the month of April. Spelt produces plenty of vigorously growing, wide green leaves as it moves into the month of May, and is a fairly long-season crop. It heads out and enters anthesis at least three and sometimes four weeks after rye. It's always such a pleasure when the spelt heads finally emerge from the flag leaves. By this time, the grain has gotten quite tall as it approaches four feet in height. Once grain fill begins, spelt straw begins to turn a very special color that can only be described as a mixture of purple and brown. Spelt keeps reminding us that it is a very ancient grain every time we gaze out at a ripening field. There is nothing else out there that is anything like it in color, growth habit, or grain quality.

Harvesting Spelt

Spelt takes longer to ripen than any of the other winter grains. It will need to remain in the field a week or two longer than rye or winter wheat; in most years, spelt seems ready for harvest about the same time as spring wheat. Part of the beauty of this ancient grain is that it is easy to harvest. Although its straw is pithy and quite hollow, spelt usually remains standing in the field with only minimal lodging. Spelt is a covered grain, which means that its kernels retain their husk covering through the threshing process, and because it is such an ancient plant, the ripe husk-covered grains of spelt are not very tightly bound to the rachis of the upper stem. Therefore threshing is quite easy—very little power and abrasion are required to remove the grain from the remainder of the straw. This makes spelt an ideal grain to grow in your garden because hand-reaping and -threshing will work quite well at harvesttime. You could say that this ancient grain that has changed very little over millennia is ideal for the sickle and flail. If you lay some spelt out on a tarp on a threshing floor, you will be surprised at how easily the kernels fall off when walked upon or banged with a wooden flail. The rest of us who will be harvesting spelt mechanically with a combine or a threshing machine will need to keep the fragility of this ancient crop in mind when we head to the field with our equipment, as it is very easy to overthresh spelt by running the combine cylinder too fast or bringing the concave up too close to the rasp bars.

Spelt needs to come out of the field still in the hull. You won't find combine adjustments for spelt in the back of your operator's manual because this is not a mainstream crop in this country. (This would be an entirely different story in Germany or Switzerland.) Start out by setting your cylinder and concave for oats. The chaffer and sieve in the back of the machine might have to be opened a little wider because spelt in the hull is much larger than oats. Once you have begun combining and clean grain is flowing into the tank, stop and inspect your sample for signs of overthreshing. If you see lots of hulled spelt kernels mixed in with your sample, back off the cylinder speed, lower the concave, and give it another try until you get better results. Spelt kernels are easy to see: They look almost exactly like wheat berries, but they are lighter in color and slightly longer. It is very important to keep the grain in the husk because this is what protects it from future damage as it travels through augers and the post-harvest cleaning process. Spelt is much softer than wheat and is very easily damaged when it is moved around. Remember, there will always be a few stray hulled kernels of spelt in the mix, as perfect threshing is damn near impossible. All we can ask of ourselves as growers of spelt is to do the best harvesting job we can with the machinery we have. Attention to detail and precise combine adjustment will achieve the best results no matter what crop you are harvesting.

How do you know if your spelt is dry enough for harvest and keeping? Once again we find ourselves in uncharted territory. You won't find spelt on the calibration chart of your moisture tester, so I usually use the settings for oats when I test a sample of spelt for moisture. Deciding when to combine your crop of spelt is a pretty subjective matter, as there are lots of factors to consider. Has the straw totally turned brown in color? How moist do the spelt berries feel in your mouth when you chew them? What are the weather conditions like at the moment and what is the forecast for the next couple of days?

If rain is imminent and you have a dry day or two, you might just want to go out there and harvest the crop before it gets really soaked. Spelt's biggest advantage is its hull, which not only protects the inner kernel but can also serve as a moisture buffer. It is very easy to blow air through this very light and fluffy crop. Several years ago, I had a combine break down right in the middle of spelt harvest. It took all day to repair the machine; we were ready to get back to work about five o'clock in the afternoon. Rain was in the forecast for the next day, so we went to the field and gave it everything we had. Normally, we cease combining when the sun goes down and the dew begins to fall, but the dew seemed to be holding off because rain was on its way. There was a little dampness in the air, however. We kept the cutter bar quite high to minimize the amount of straw that had to travel through the machine. We combined until eleven PM when we finished. I was a bit worried that maybe the crop might be too high in moisture, as we had a rather large truckload of spelt to deal with. I inserted three screw-in aerators into the twenty-two-foot-long truck body full of spelt. We backed the truck into a shed and plugged in the aerators. In the end, the spelt never heated and the quality was perfect. We pulled the truck out two weeks later and were able to put excellent-quality bright-colored spelt through our cleaner. If we had waited until after the rain to harvest our crop, the quality would have been reduced I am sure. All of this is to say that deciding when to harvest any grain crop is always a hard call to make. Usually, there isn't much time to fret, especially if weather is on its way. Under these sorts of circumstances, weigh your options and consider your infrastructure, then you make your move. This is definitely part of the excitement that makes farming a bit nerve racking—and so much fun at the same time.

The Value of Spelt for Livestock Feed

Up until this point everything about growing spelt has been relatively easy, but things start to get a little more complicated after our spelt is harvested, cleaned, and in the bin. Spelt makes great livestock feed and if this is what you plan to do with it, the easy path will continue. According to animal nutrition value tables, spelt has the same relative feed value as oats. Protein averages between 12 and 15 percent; there is plenty of fiber as well. Most farmers who feed spelt will tell you they like it a whole lot more than oats because the hulls are almost as valuable as the inner grain. All that extra fiber from the hulls acts as a buffer to "cushion" and balance the protein and energy found in the grain. With its low bushel weight and bulky nature, spelt nourishes livestock in a much gentler manner than high-potency grains like corn and soybeans. When you consider the unique nutritional qualities of spelt alongside its ancient genetics, it is the ideal grain to feed to cattle that are on low-grain, high-forage diets. Spelt is equally beneficial in the human diet. The only problem that we encounter in the human consumption department is that we can't eat and digest the hulls like a cow can. If we want to make bread or pasta out of spelt, we must find a way to remove the hulls first. Dehulling spelt is not a simple process. The slight inconvenience of spelt hulls surrounding the kernels of such an amazing and nutritious whole-grain food stops many of us "wannabe" spelt growers and farmer processors right in our tracks.

Dehulling Spelt

The ideal solution for those of us who want to dehull our spelt would be to have a local processor who could custom-hull our crops. If you live near the Finger Lakes region of New York State, Klaas Martens will do the job for you at his Lakeview Organic Grain facility in Penn Yan; Klaas has invested well over a hundred thousand dollars in specialized German spelt-hulling and -cleaning equipment, and for a fee he will process your spelt. The machinery performs best when the crop is low in moisture and has been precleaned. Higher-moisture spelt doesn't hull as well and travels through the system rather slowly. Several years ago I took a couple of tons of spelt in large tote bags up to Michel Gaudreau in Compton, Québec, for processing. The price for custom-hulling was close to forty cents a pound, and Michel kept the hulls and spelt cleanings. After paying this high price, losing part of the crop, and putting up with the hassle of crossing an international boundary twice, I decided pretty quickly that I wouldn't try this route a second time. However, if you can find someone to hull your spelt for a reasonable price and who is close at hand, by all means take advantage of the opportunity. If you're not fortunate enough to have access to custom spelt hulling and you're still bound and determined to feed yourself and others with your homegrown spelt, you will have to acquire some hulling machinery and learn how to use it. There are numerous machines that can do the job. Some are hundred-year-old antiques, while others are newer, very expensive pieces of equipment that come from Northern European countries like Germany and Denmark.

I will give you a brief overview of spelt hulling in general and the different methods for actually removing the outer hull from the inner kernel. We have discussed both abrasion and impact hulling in the two previous chapters on barley and oats, and interestingly, abrasion hulling seems to work a little bit better on spelt. The actual spelt kernel is quite soft and easily damaged during the hulling process, so it's important to be gentle when operating any sort of huller. Light abrasion seems to work better than the impact method because fewer kernels are broken and damaged.

If you are dealing with very small amounts of spelt for home consumption, it is possible to build your own hand-cranked dehuller by modifying an existing Corona or other brand of counter-mounted steel buhr type of grain mill. The actual modification was designed and patented for free use in the public domain in 1989 by Allen Dong and Roger J. Edberg.

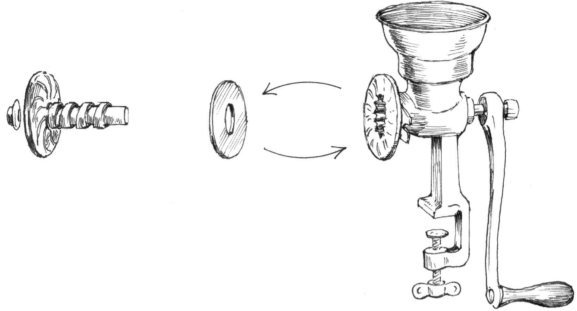

Hand-cranked flour grinder converted to a hulling unit.

The plans are on the Internet for public use and can be found at www.savingourseeds.org/pdf/grain_dehuller.pdf. An updated and more modern version of the homemade huller with step-by-step instructions is also available at www.bilagaana.com/dehuller/Sunflower%20Dehuller.html. To build a homemade spelt dehuller, you simply remove the stationary steel buhr from the main body of a Corona or any other hand-cranked flour mill and replace it with a large round metal washer and rubber disc. Reinsert the handle, feed auger, and runner buhr and loosely adjust the mill to crack grain as opposed to grinding flour. The abrasion that is created between the runner buhr and the backing plate will remove most of the hulls from your spelt. This is crude technology at best, but it's simple and cheap. The rubber disc will wear out frequently and have to be replaced and reglued to the backing washer once in a while. The mixture of hulled spelt berries and loose hulls will also need some kind of simple aspiration, which can be accomplished by standing in front of a box fan or in a breezy doorway and pouring the mixture from one five-gallon pail to another. This certainly isn't precision technology, but it does present a doable possibility for someone with a small amount of spelt to hull. A basic illustration of this homemade grain huller appears above.

Monroe's Mill

Some of the earliest mechanical spelt hullers worked on exactly the same principle as the hand-powered model I just described. I took a trip in July 2011 to Holmes County, Ohio, for the express purpose of studying some of the low-tech spelt-hulling operations run by the Amish community. On this trip, I had the pleasure of spending the better part of an afternoon with Monroe Stutzman, whose spelt huller is pictured in the insert. The machine was basically a large vertically positioned stone grinder encased in a rugged cast-iron housing. The interior stones measured about two feet in diameter and were positioned just far enough away from each other to rub the spelt hulls off the inner kernels, but not grind them. This machine was manufactured by the Engelberg Huller Company of Syracuse, New York, and had received patents in 1906 and 1907.

Monroe operated his huller at about eight hundred RPM, and the mixture of hulled spelt and loose hulls was then run through a two-screen fanning mill for preliminary cleaning. A twenty-eight or thirty sixty-fourths screen was used on top and a ¾ × 8 slotted screen on the bottom of the fanning mill. After this, the cleaned grain was put over a gravity table for density separation. The final step was a pass through a series of five quarter-inch mesh hardware cloth screens, each positioned at a forty-five-degree angle and stacked one over another. This very clever little fabrication was Monroe's simple and efficient method of paddy table separation. As the spelt fell down through the stack of tilted screens, unhulled grains would tumble to the bottom of each deck while the hulled spelt made its way through the quarter-inch openings. Five screens gave the tumbling spelt five different opportunities to separate during the process. Monroe's small mill embodied human ingenuity at its very best. For a very minimal amount of money, he was able to construct a smooth-running efficient spelt-hulling operation from very inexpensive and discarded equipment. The Amish of Ohio are masters at growing and processing spelt, and are very friendly folks who are more than willing to share their knowledge and expertise with the rest of us who are newer to this business. If you are really serious about hulling spelt on any sort of scale, I would recommend a trip to Holmes County, Ohio, and a visit to Monroe's mill.

Using Antique Hulling Equipment

In my travels around the Midwest, I've had the good fortune to visit more than one spelt-hulling operation. Ken Zimmerman of Craigville, Indiana, owns Fruited Plain Seeds and specializes in processing field-run spelt into milling-quality grain on his farm in the northeastern part of the state south of Fort Wayne. Like Monroe Stutzman, Ken uses older equipment to produce a very high-quality product in a simple and straightforward manner. Ken has gained the reputation of being able to extract more usable milling spelt from a ton of raw grain than anyone else in the industry. He does it all with ancient outdated equipment in an old horse barn on a back road out in the middle of nowhere. After precleaning the spelt with a screen cleaner and a Carter disc, Ken runs it through a Roskamp Champion Oat Huller like the machine I described in the oat chapter. I've tried hulling spelt several times with my Roskamp only to produce a smashed-up and partially ground product that is totally unsuitable for human consumption. Ken's big secret is that he has found a way to selectively slow down the RPMs of the hulling disc on his Roskamp machine without compromising the speed of the aspiration fan or the Carter disc in the bottom. Roskamp hullers normally rotate at eighteen hundred RPM for processing oats, but Ken has been able to preserve the integrity of the spelt he hulls by slowing the machine down to between thirteen hundred and a thousand RPM. He also showed me how to improve the separating function of the Carter disc at the bottom of the huller by adjusting the opening in the inspection door. For some unknown reason, when this little steel flap is propped partially open instead of remaining closed, more hulled spelt is propelled into the clean grain auger by the rotating discs. Once his spelt has been hulled, Ken's process is pretty much the same as Monroe's. The seed first travels by elevator to an air screen cleaner and then to a gravity table. Ken can get his spelt so well hulled that paddy table separation is not necessary at the end of the process. Ken emphasized the importance of placing a strong magnet on the inlet end of the Roskamp Champion. All it takes is one little nail, screw, or some other piece of trap iron to enter the metal workings of the huller to cause thousands of dollars' worth of damage to the machine. My take-away lesson from visiting Fruited Plain Seeds was the value of variable-speed hulling. When I purchased my Roskamp huller, I didn't have a clue about this feature. I was aware that the Roskamp Company actually did make variable-speed machines, but I didn't really know why or how much difference this feature could make in the hulling process. It is possible to retrofit a Roskamp huller, but it would be very expensive and would require lots of machining and fine-tuning. So once again, please learn from my mistakes. If you plan on acquiring one of these old machines, buy one that has the variable-speed feature.

European Hulling Equipment

Some farmer processors of spelt have decided to forgo retrofitting older US-made machines in favor of specialized European machinery. Most of this equipment comes from Germany and Denmark, where spelt—or dinkel, as it is called in Germany—is a major commodity crop. I have only seen one of these machines in action, when on a recent summer trip to Ohio I journeyed to Dean McIlvain's Twin Peaks Organic Farm in West Salem to see his spelt-hulling operation. Dean grows several hundred acres of spelt and processes it for the human consumption market. We were fortunate that his mill was working full-tilt on the afternoon that we arrived at his farm; it was mid-July and he was emptying last year's crop out of a large grain silo in preparation for the coming harvest. The spelt began its journey by traveling up a very tall elevator leg whereupon it was delivered by gravity to his processing building. First stop was at a large air screen cleaner. Once the field-run grain was cleaned, it was then delivered by another, shorter internal bucket elevator to a Franz Horn Dinkelschäler located high above our heads on an overhead steel deck. We climbed a ladder to watch the machine in action. The Dinkelschäler DVC-2 consists of two separate components: a hulling chamber and a unit to separate hulled spelt from unhulled grains. This machine is designed specifically for spelt, and it is basically a small rotary hammermill with T-shaped revolving hammers and a specially designed circular wire-mesh screen. The spelt is delivered to the intake spout of the huller by a variable-speed auger. Once inside the screen, it is dehulled as the revolving hammers push it through the wire mesh. As the product leaves the huller, loose hulls are pulled off by vacuum through a pipe. The hulled spelt then falls onto a specially designed vibrating circular cone-shaped screen that separates finished product from grains that escaped hulling on their first pass through the system. I had never seen anything like this unit in any of my visits to other grain-processing facilities. Dean told us that it was actually patterned after separation technology used in the coffee industry. When I studied the machine, I could see that it measured about five feet in diameter and was slightly dished toward the center point. Hulled spelt was delivered by a pipe from above, and the entire screen reciprocated back and forth quite vigorously. Hulled spelt made its way to the outer circumference of the cone and exited via the spout visible. Unhulled spelt ended up at the center of the cone, whereupon it was delivered back to the hulling unit for a second pass through the system. The final step in the process was a pass through an Oliver gravity table. This sort of technology is for the serious high-volume producer of spelt only—Dean's machine was capable of processing close to a ton and a half of spelt per hour. He paid twenty thousand dollars for it fifteen years earlier, and the same Dinkelschäler DVC-2 costs well over fifty thousand today. Service and parts can be a problem because Franz Horn is located in Sagau, Germany. Here in our neck of the woods an installation like the one I just described would be a wonderful asset if it could be operated as a custom service on a regional basis to serve all of us smaller spelt producers. It would be worth driving several hundred miles for this kind of service.

Processing Spelt for Human Consumption

After you've made all this extra effort to remove the hulls from your harvested spelt crop, it's time to reap the ultimate reward and actually eat the stuff. Spelt is much closer to soft wheat in hardness and texture, but it will bake into a fine loaf of bread regardless. The softness of the inner grain contributes to the difficulty of hulling because the kernels are very easily damaged in the process. Chipped and broken kernels of spelt are acceptable if the sample is destined to be ground into flour immediately. However, if you are processing spelt to sell as whole berries, you will definitely want to do a first-class hulling job: Whole grains definitely keep a lot longer than broken ones. By virtue of its extra-soft texture, spelt is much easier to grind than hard wheat, and it takes a whole lot less power to grind spelt into fine silky whole-grain flour. True spelt will lend a yellow crumb color to freshly baked bread because the grain is exceptionally high in carotenoids, which are a source of B vitamins. The nutritional advantages of spelt don't end there, and there is a growing segment of natural foods consumers who want to use spelt as their sole source of carbohydrates. These folks will tell you that spelt is superior to wheat in every way, because it is nutritionally balanced and contains higher levels of vitamins, minerals, and fats. For example, spelt averages 2.8 percent in fat compared to 1.8 percent for wheat. It is more easily digested by the human body because of the structure of its starch; spelt fiber also has excellent water solubility, which allows for much better absorption in the intestinal tract. Protein levels in spelt, with its ancient genetics, are comparable to those found in wheat, but its amino acid levels are much higher and more diverse. Baking good bread with spelt requires a bit of practice that can be gained with experience. Spelt flour is more difficult to mix than wheat because the surface of the dough remains moister in the process. Most artisan bakers use sourdough and a sponge system to skillfully craft excellent loaves of spelt bread. While I'm singing the praises of spelt as a superior food, I cannot forget to mention its most important attribute—taste. Spelt has a nutty flavor and extra-light texture that is pure pleasure for the taste buds. Slather some butter on a slice of freshly baked whole-grain spelt bread and it's easy to understand why this relatively unknown ancient grain has made such a comeback in the last twenty years.

Triticale

Triticale is the only cereal grain that we can't trace back to pre-Christian times in the Fertile Crescent of Western Asia. It is a modern stabilized hybrid that has resulted from crossing rye and durum wheat. The first deliberate attempts to cross wheat and rye were made in Scotland by A. S. Wilson in 1875, and since these two grains were genetically different, the resulting crosses were self-sterile and not of much value to anyone at the time. In 1918 at the Saratov Experiment Station in Russia, a number of natural hybrids between wheat and rye inadvertently crossed in the field and were able to produce viable seed. These crosses were selected and seed was saved. Work on this project also continued in Sweden and at the University of Manitoba in earnest in the 1930s. The new grain was named triticale, which was a combination of the two Latin names for wheat and rye—*triticum* and *secale*. These first crosses were tall, highly sterile, late maturing, and produced only small amounts of shriveled grain. There was certainly a lot for these plant breeders to do, and much of this work was performed at the International Maize and Wheat Improvement Center (CIMMYT) in Mexico. It wasn't until 1969 that the first commercial triticale variety was released to the public. At the time, triticale was dubbed a miracle crop and greeted with great fanfare by the agricultural press, as this new man-made grain was seen to have great potential as a human food crop. All of this original hype and excitement might have been more of a setback than a benefit to triticale, because bakeries tried it with less-than-perfect results. I remember that the Fassetts Company here in Vermont sold a triticale loaf for a very short time. Triticale lacked the gluten found in wheat, and very soon after its initial splash it was relegated to the lowly status of a minor feed grain.

Triticale as Part of a Livestock Ration

The initial breeding of triticale was accomplished by doubling the chromosomes of the sterile wheat/rye cross. When there are more than two sets of chromosomes, it's known in genetics as polyploidy. This breeding work combined four sets of seven chromosomes from wheat with two sets of chromosomes from rye for a total of forty-two (6 × 7) chromosomes in triticale. The genius behind this cross comes from the strengths that both of these cereals have to offer: Wheat brings grain quality, yield, and disease resistance to the mix, while rye supplies the vigor and hardiness. Triticale self-pollinates like wheat instead of cross-pollinating like rye; its seeds are longer than wheat's and plumper than rye's. Grain color ranges from the light tan of wheat to the gray-brown hue of rye. Spring-planted varieties do exist, but winter cultivars are much more common. Protein levels in triticale can range as high as 20 percent, with the average being closer to 16 or 17. Lysine, which is an important amino acid in livestock nutrition, is significantly higher in triticale as well, all of which adds up to an excellent feed grain for livestock. Palatability is reasonable also. Stock, especially swine, don't relish this grain as much as they would corn, but they will eat it, especially when it is mixed with other grains in a ration. The same holds true for cattle. Poultry probably do the best of any farm animal on a high-percentage triticale diet. Whatever the case may be, the inclusion of triticale in a grain mix with its high lysine and 17 percent protein will allow you to cut back on expensive protein inputs like soybean meal. Plant breeders have continued to work on this grain since its introduction in the 1970s, and yields and grain quality have improved to the point that triticale is now a viable feed grain and well worth growing on an organic farm in our part of the country.

Use as a Forage

Triticale has also become an important forage crop for beef and dairy producers. It is fairly well adapted to a number of different climates over a wide geographic area. Triticale thrives equally well on the Great Plains from Oklahoma to North Dakota as it does in the Northeast. It can be a good source of both late-fall and early-spring pasture, and I have found spring triticale to be an excellent "seed down" crop for new plantings of hay. The procedure is simple: Mix spring triticale seed fifty–fifty with inoculated Canadian field peas and load them into a grain drill. Fill the grass seed box on the grain drill with your chosen hay mixture. (On lighter, well-drained soils, I like to plant a mixture of timothy and alfalfa with a sprinkling of meadow brome. On heavier soils, I plant medium red clover and timothy.) Proceed to the field as early as possible and plant just as you would for a crop of wheat or barley. For some reason, the triticale–pea combination actually complements the growth of the young grass and legume seedlings. The triticale and peas grow quickly into a very tall, thick mass, which is mowed for hay or silage in late June or early July. The best-quality forage is achieved by mowing at the preheading boot stage of development when the triticale is still in the boot: Forage protein levels can reach 17 percent, and hay yields are tremendous. It's not uncommon for this Peakal crop to grow to four feet in height, and it doesn't take long to pile up a lot of round bales or chopped silage with this kind of volume of feed harvested from a field. The best news of all is that the newly seeded hay crop down there on the bottom is usually six inches to a foot tall by the time the nurse crop of triticale and peas is mowed. New hay seedings have a major advantage in this system because they still have two or three months of growing season remaining to thicken. The competition from the nurse crop is eliminated early enough in the summer to allow the newly seeded hay stand plenty of time to establish itself, and another cutting of hay can usually be made in late August or early September. Truly, triticale is a tremendous forage crop that can be used in a multitude of different management systems.

Use as a Fall-Planted Feed Grain

This man-made cross of wheat and rye has equally excellent potential as a fall-planted feed grain crop. Its grain protein levels are higher than wheat or barley, and yields will often exceed a ton to the acre. Triticale has a fifty-pound-per-bushel test weight, which means that it has a good density and concentration of nutrients. Cultural methods for triticale are exactly the same as they would be for any other fall-seeded grain crop like winter wheat, rye, or spelt. Plant 120 to 150 pounds of seed to the acre in late September and hope for good winter survival. Make sure you procure seed that is well adapted to our climate. There are numerous substrains and regionally adapted varieties of triticale, and I have found that a lot of the forage varieties that are available for planting in the Northeast aren't really suitable for grain production. A lot of this seed comes from much warmer places like Kansas and Oklahoma. These varieties will germinate quite well in the fall and look fantastic going into winter, but more often than not winterkill will finish off these beautifully thick and green stands of forage triticale. There is little more depressing than going out there in March and finding that your new triticale seeding that had so much potential has died and turned brown. My advice is to find triticale seed from North Dakota or the Canadian prairies. I happened to be in North Dakota at my friend Blaine Schmaltz's in mid-April when the snow was melting back from the edge of his triticale field. The crop hadn't made a whole lot of growth the previous fall. It was tiny and had been pressed into the dirt by a heavy snowpack. The little triticale plants had the purplish color of young rye seedlings, and Blaine wondered if the crop was going to make it. He told me later in the summer that this crop of triticale had exceeded all his expectations. It rallied and grew like blazes to give him a fifty-five-bushel-to-the-acre harvest in early August. Blaine loves the stuff, and finds it makes great livestock feed when substituted for corn in a ration. It also grows in climates and locations where corn won't grow. Triticale has potential indeed.

I have tried planting triticale three times in the past twenty years. I'm sorry to report that I have never harvested a suitable crop. Winterkill has claimed it every time. I'm sure that buying seed from warmer places like Pennsylvania and Kansas didn't increase the chances for crop survival on my cold and windy hilltop in northern Vermont. In typical farmer fashion, I'm not ready to quit trying yet. I'm going to give it another go with seed from Blaine. It sure would be nice to mix some triticale into some of my homemade grain rations. I have grown winter wheat, rye, and spelt with great success. These crops will continue to add variety and diversity to my mix of planting choices. I am bound and determined to do well with triticale. It grows well in western New York and southern New England. Just because I have had difficulty growing triticale here on "the rooftop of northern New England," don't be afraid to try it where you live.

CHAPTER FOURTEEN
Soybeans

The soybean (*Glycine max*) stands alone in its importance as a food crop for humans and livestock. No other grain plant produces as much protein per acre. Soybeans average close to 40 percent in protein and 20 percent in oil content. Protein quality is exceptional as well, and soybeans contain a fairly balanced profile of all the necessary amino acids that are required to support life. To supply compete protein, however, they must be cooked or heated to destroy a trypsin inhibitor contained in the plant's chemical composition. This is why soybeans are roasted for livestock rations and boiled before we can eat them. One of the reasons why soybeans produce flatulence and are so difficult for us to digest is because their carbohydrate portion is composed of nondigestible sugars called oligosaccharides, which can cause abdominal discomfort in monogastric animals. If soybeans are fermented into products like tempeh and miso, these soluble carbohydrates are broken down and rendered more digestible. The supreme advantage of this wonder crop is that it is a legume; it can extract its own nitrogen from the atmosphere just like alfalfa and clover. Soybean roots produce little round nodules with the help of the bacteria *Rhizobium japonicum*, a biological mechanism that allows the plant to "fix" its own nitrogen from the air. Soybeans are in a totally different class than the cereals in the grass (Gramineae) family. Broad-leafed flowering plants known as angiosperms, they grow more like little trees or stalks of green. The good news about the soybean is that it will grow and produce large amounts of edible protein in all parts of our region. Soybeans have always been acclimated to the warmer southern locations, but now thanks to modern breeding and selection, there are numerous soybean varieties that will perform well in the colder areas of northern New England and New York.

Soybean Origins

The soybean was first domesticated in northeastern China around 1100 BC. Its ancestor was a wild vine (*Glycine soja*) that produced tiny hard seeds, which required lots of preparation to be used as human food. In fact, this plant stills grows wild throughout Korea, China, and other parts of Eastern Asia. Once this wild vine had been transformed into an upright annual leguminous plant whose fruits were larger and easier to harvest, the soybean spread like wildfire through all of Asia and eventually to Japan and other islands in the Pacific Ocean. A whole variety of high-protein foods from basic bean sprouts to tofu and tempeh were developed from the soybean as its culture spread throughout the Far East. By the early eighteenth century, soybeans had made their way to Europe on sailing ships returning from the Orient, and French scientists took a fancy to the little bean when they discovered that it contained very little, if any, starch. Suddenly there was a high-protein, low-carbohydrate food that could be consumed by individuals with diabetes. The very

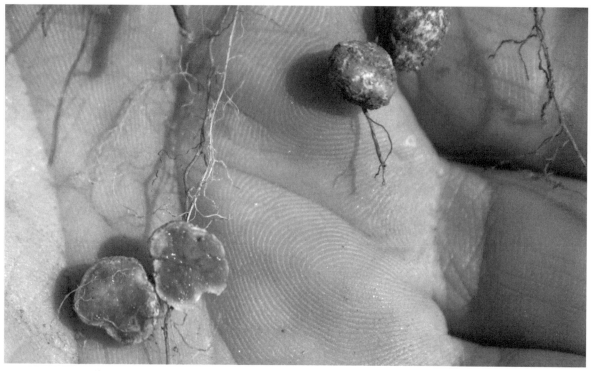
Soybean nodule cut in half to illustrate a healthy pink interior. PHOTO COURTESY OF SID BOSWORTH

Nitrogen-fixing nodules affixed to the root system of a soybean plant. PHOTO COURTESY OF SID BOSWORTH

first record of the crop in North America comes from a sailor named Samuel Bowen who returned to Savannah, Georgia, with soybeans he had procured in China. Around 1765, Bowen began cultivating small amounts of soybeans and exporting soy sauce to Europe. Benjamin Franklin also took a fancy to soybeans and was instrumental in introducing them from France to Philadelphia in 1804. Soybeans remained a curiosity through the remainder of the nineteenth century and into the first few decades of the twentieth; the land grant colleges and agricultural experts of the time promoted them as a forage crop to be mowed for hay or plowed down for green manure. The experts knew that this crop was full of nitrogen, but few, if any, efforts were made to harvest its fruit. Henry Ford was a staunch vegetarian and America's first champion for the widespread industrial usage of soybeans. Between 1932 and 1933, he spent over a million dollars on soybean research. His efforts were mostly concentrated on finding uses for soybean oil in the manufacture of automobiles. By 1935, the Ford Motor Company was using soybean oil in paints and for fluid in its shock absorbers. Ford was also a pioneer in developing the first soy-based plastics.

The Rise of Soybean Production Post–World War II

Soybeans were definitely a neglected crop until World War II. Up until the early 1940s, the United States imported 40 percent of its cooking oil from abroad, but the disruption of trade routes that resulted because of the war encouraged the widespread adoption of soybean culture for the domestic production of vegetable oil. Henry Ford's research from ten years earlier paid great dividends, and a domestic soybean oil industry was born. American agriculture in general changed drastically after World War II. As the standard of living increased, people began eating more meat, and the soybean oil meal by-product from the oil extraction process quite handily supplied the newfound protein requirements of the burgeoning beef, poultry, and pork farming sectors. The true potential of this mysterious bean plant from Asia was finally realized. A sixty-pound bushel of soybeans yielded eleven pounds of high-quality oil and forty-eight pounds of 42 percent protein meal. No further processing was required to improve the digestibility of the soybean oil meal because the heat generated during oil extraction was sufficient to transform it into a complete protein source. This parallel development of a vegetable oil industry coupled with increased meat production was a match made in heaven. Soybeans really began to come into their own as an important crop in the 1950s. Up until this time, most of the soybeans grown in the country were concentrated in the southeastern states of Georgia, Alabama, Mississippi, North Carolina, and Kentucky. As the demand for oil and meat increased after the war, production began to move into the Corn Belt of the Midwest—soybeans just happened to be an ideal crop rotation mate to corn. The archetypal corn–soybean rotation that dominates American agriculture today was born during the 1950s.

Soybeans have come a long way since the early days of commercial production. Once output began to climb, the genius of American agricultural ingenuity went to work on the crop during the late 1950s and early '60s. Row spacing got narrower and yields increased with the introduction of new varieties. Combine grain platforms, which up until this time were used exclusively on cereals like wheat and barley, were modified to work down closer to the ground to "shave" the low-hanging pods of soybeans. By 1965, soybeans had become America's number three crop, after corn and wheat. At first soybeans were used exclusively in the manufacture of cooking oils, margarine, shortening, and salad dressings, but as the crop increased in economic importance, university researchers found that soybeans also worked well for industrial applications. Soybean oil eventually found its way into paint, inks, varnishes, caulking compounds, and numerous other products. Henry Ford would be proud, as the United States was the world leader in soybean production through the 1970s, dominating the market with a 75 percent share until the latter part of the decade, when worldwide shortages of protein encouraged new production in Argentina and Brazil. South America is now a major player in the international soybean commodity business; sadly, thousands of square miles of the Amazon rain forest have been cleared for soybean farming. No other agricultural commodity has undergone as much change in as short a time as the soybean. In just sixty years, this little bean plant from Asia has transformed from total obscurity to major international importance. However, over 90 percent of the soybeans grown in North America are now genetically engineered to withstand the killing effect of Monsanto's Roundup herbicide, and increased organic and non-GMO soybean production is needed now more than ever as we find ourselves in an environment that has become totally dominated by multinational chemical and seed corporations.

My Search for a Short-Season Variety

My own experience as a grower of soybeans was late blooming. When I first began growing wheat and barley in 1977, soybeans weren't on my radar screen. My impression was that this was a Corn Belt crop that thrived from Ohio westward to Illinois and Iowa. A totally homegrown livestock ration was out of the question at that time, because we had to purchase additional concentrates to increase the protein level of our ground animal feeds. At first we bought bags of soybean oil meal from our local grain dealer to mix with our farm-produced barley and oats, but by the early 1980s we found that we could buy individual bags of specific protein premixes at the Viens et Frères feed mill in Ayers Cliff, Québec. Once a week we would journey northward to buy forty-kilogram (eighty-eight pounds) bags of high-protein chicken and hog concentrate. We could only guess what these supplements contained for ingredients. There was certainly a variety of breed-specific minerals and vitamins in each bag. The extra protein came from sources like fish meal and tankage—which is a nice word for dried slaughterhouse waste. We really weren't worried about what was in stuff because these "Shurgain" feeds allowed our animals to produce more meat and eggs than ever before. We supplemented our dairy cow rations with corn gluten and linseed meal. (Corn gluten was a by-product of the corn starch industry; linseed meal was what was left over after flaxseeds were pressed into oil.) None of these commercial products was very organic, but there really wasn't anything else available at the time. Organic certification was still five or six years away, and we were very happy that we had found a way to enhance the value of our farm-produced cereals. We considered growing other high-protein crops like field peas, lupines, and flax, but didn't have the machinery or know-how to make it happen. Throughout this whole early era of my farming and homesteading career, I had the desire and dream to produce my own protein. If there were only an ultra-short-season soybean available, I wouldn't have to buy all these extra products from the commercial marketplace.

It was right about this time that Johnny's Selected Seeds from Albion, Maine, began offering Fiskeby V soybeans in their catalog. This was the latest and most modern of a series of ultra-short-season Fiskeby cultivars that had been bred and developed by Swedish plant breeder Dr. Sven A. Holmberg. Beginning in the 1930s, Dr. Holmberg made it his life's work to bring the soybean to Sweden, and between 1939 and 1940 he went on an expedition to Japan's northernmost island of Hokkaido in search of soybean germplasm that might be adapted to southern Sweden. He returned to the village of Fiskeby in his native Sweden and began the process of trialing and hybridizing the hundreds of northern Japanese soybean varieties for which he had obtained seed. In 1949, Holmberg released his first commercially viable soybean variety named Fiskeby III, which yielded between twenty-three and thirty-five bushels to the acre. He remained dedicated to northern soybean production for the rest of his life, working in Northern Europe, the United States, and Canada. In 1954, he made a second visit to East Asia to collect more breeding material; his final trip to Japan and Siberia followed in 1970. It was right about this time that he released the crowning achievement of his life's work in breeding—the Fiskeby V soybean. Sven Holmberg deserves much of the credit for bringing soybean culture to southern Canada and the northern latitudes of the United States. He passed away in 1981 at the ripe old age of eighty-seven.

I began to notice more and more evidence of northern soybean production during the mid-1980s. In 1983, we took a trip to Ottawa in eastern Ontario to visit friends, and on the return trip, we stopped at a farm near Cornwall and purchased a thousand pounds of raw unprocessed soybeans from a farmer. The variety, called Maple Glen, had been bred and developed at the Agriculture Canada Research Station near Ottawa. The beans were big, plump, and beautiful. We had to bring them to a local bakery and roast them in a large convection oven in order to make them palatable for our cows. It was right about this time that I fell in love with the idea of growing soybeans for myself. If it could be done in Canada to our north, we should be able to do it here on our side of the border. I began to do some research into the subject and found that the Maple series of soybeans had many varieties that differed in days to maturity. Maple Arrow took 132 days; Maple Glen, 130 days; Maple Isle, 121 days; and Maple Presto, 107 days. The provinces of both Québec and Ontario published variety and maturity guides that indicated how these various cultivars performed in their particular climatic subregions. It took me a few more years before I mustered up the courage to actually try growing soybeans on my farm. In 1989, I planted half an acre of KG 20 to see if they would actually mature on my hilltop. Interestingly enough, KG 20 was descended from a cross between McCall from the University of Minnesota, Hardone and our old friend, Fiskeby III. Much to my delight, the experiment was successful. The soybeans ripened nicely, and I made plans for a larger acreage the following year.

Soybean Maturity Zones

Soybeans differ from most of the other common grains in that they are a photoperiod-sensitive plant. Cereals like wheat, barley, and oats have an internal clock that tells them to flower when a certain number of days have passed since germination. Corn goes from vegetative to reproductive when a set number of heat units have accumulated, but soybeans are totally different. They are daylight- and photoperiod-sensitive. Plants begin to produce flowers when the hours of daylight start to wane after the summer solstice on the twenty-first of June. Each soybean variety is adapted to transition from vegetative to reproductive at a predetermined number of hours of sunlight. Since the hours of daylight increase as you move northward, northern varieties will usually flower earlier at greater day lengths than soybeans that are acclimated to shorter days farther south. Soybean maturity is more a matter of latitude than any other factor. If a northern variety is moved south, it will get the photoperiod message to flower much earlier than it would in its normal environment because the days are shorter. Contrarily, if a southern variety is moved northward, flowering will be delayed because of the longer days. If flowering is held back too long, the soybean plant won't have enough growing season to finish ripening before killing frost finishes the process. With this phenomenon in mind, agronomists have developed a North American soybean maturity zone map and a variety classification system based on latitude.

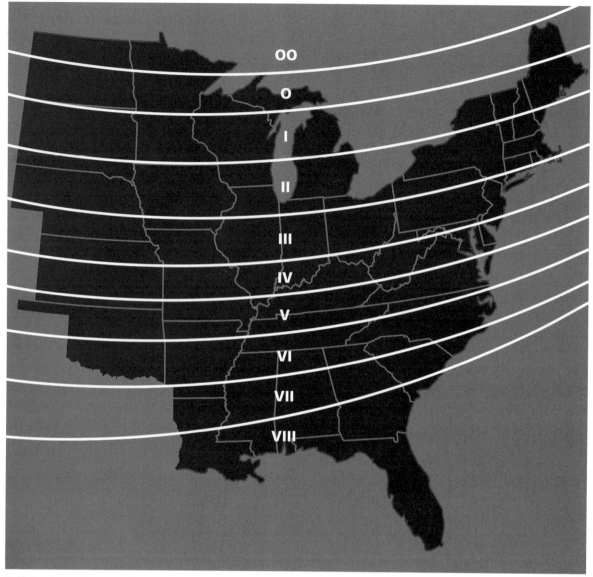

North American soybean maturity zones. PHOTO COURTESY OF HOW TO GROW GREAT SOYBEANS, ACRES USA

There are eleven maturity ranges that stretch from Group 000 in the far north to Group 8 in the southernmost part of the United States. There are actually two more soybean maturity groups that stretch across Mexico and the Caribbean. Group 10 grows in the tropics. The lines on this map are arbitrary and not exact; variations do exist because of cooler and warmer local microclimates, which might result from a mountain range or proximity to a large body of water. Nevertheless, each soybean maturity zone is approximately 150 miles from north to south. This is about how many miles it would take to notice a slight difference in day length from one location to the next. A quick glance at the map puts the northeastern United States in groups 0, 1, and 2. My experience tells me that these determinations are a bit too generous. In far northern New England, we tend to have better luck with Group 00 varieties, but if you are nervous about a short growing season and an early frost, plant Group 000 soybeans for peace of mind. The good news these days is that there are hundreds of varieties to choose from in these maturity ranges. It was a different story back in the 1980s when there were only a few ultra-short-season Canadian options available.

Things to Consider Before Getting Started

Soybeans are probably the one grain crop that will tolerate late planting without severe yield loss. Because flowering is triggered by photoperiod instead of the number of days since germination, a mid-June-planted soybean will still flower and make fruit despite its late start. Yields will be a bit lower because the vegetative period will be shorter and the resulting plant smaller. A mid- to late-May-planted soybean has the potential to produce more pods because the plant will have had two to four more weeks to gain in size and stature before flowering is triggered by day length. For every three days of delayed planting, flowering will be pushed back by one day. Canadian agronomists don't use maturity zone designations to measure soybean maturity. Since most of the soybeans grown in Canada are Group 0 and lower, corn heat units are used to determine the suitability of a cultivar to a specific region. So if you are trying to grow soybeans in the northern United States, you will have to consider your latitude as well as the number of growing-degree units in your area. The best news of all is that soybeans are flexible and forgiving. Late planting will still produce a crop, and while it might not be bin busting, at least your field will produce something. Heat and sufficient moisture are also essential for a good harvest. If you are brand new to growing soybeans, it will be best to err on the conservative side when choosing a variety. Plant something with a little shorter season to ensure there will be ripe beans to harvest in early October.

Choosing the Right Variety

Soybeans exhibit all sorts of characteristics, which make them one of the most diverse agricultural crops we can grow. There are short varieties and tall varieties. Some plants are quite bushy while others are more erect with fewer branches. Seed size and color can vary greatly as well. All of this diversity is to your advantage because it gives you lots of choices. The first order of business is finding a variety that will flower early enough to ripen in your location. Consider your circumstances and microclimate. Do you have a long growing season or do you live in a frost pocket? Will you be planting your soybeans on a cool high-elevation hilltop or on a warmer river bottom flood-plain? Once you have determined your soybean maturity zone, you can refine your variety selection by planting an early or later cultivar from within that group. For instance, if you are planting a Group 0 variety, you have the choice of an 0–1 soybean—which flowers and matures earlier—all the way up to an 0–9 variety, which could flower a week to ten days later. If you have any doubts about getting your crop to fully ripen, it's best to err on the side of caution and choose a soybean with a lower maturity group rating. If you're farming up on the windy hilltop, you might choose a Group 00–1 soybean. If your field is located down in the valley and protected from autumn frost by river fog, you might choose a Group 00–9 or even a Group 0–1 variety. I have very successfully grown some Group 000 varieties in my very cool northern Vermont location. These plants are short in stature and produce ripe beans by the third week of September. Although yields are higher in the later-flowering varieties, respectable harvests are possible from the Group 00 and Group 000 varieties. To achieve a good yield, plant these cultivars at high populations in narrow rows with a grain drill. If you are planting Canadian seed, you will want to familiarize yourself with their system of measuring corn heat units. The very shortest season Canadian soybeans require 2,250 to 2,300 CHUs (Canadian Heat Units). This is about what would be required to ripen a seventy-five- to seventy-seven-day corn. A midseason Group 0–9 soybean might be designated a twenty-six-hundred-CHU bean, which is the same as an eighty-five-day corn. Once we get up above three thousand CHUs, we are working with ninety-day-plus corn and Group 1 and 2 soybeans. Be conservative. It's better to harvest a few less beans from a shorter-season variety than to have no beans at all because the frost came earlier than you had planned.

Choosing for Foliage and Plant Structure

Once you've determined what maturity soybean to plant, it's time to choose what sort of foliage and plant structure you'd like to see in your field. When you peruse the variety description tables in most farm seed catalogs, you will notice that soybean varieties are described as either narrow-leafed or bushy. Traditionally, soybeans have always been quite broad-leafed. This type of stature has worked quite well for those of us who plant our beans in rows that are thirty inches or wider. These types of soybeans grow outward as they grow upward, and by the time a true bush-style soybean plant reaches two feet in height, its leaves are usually stretching halfway across a thirty-inch row and nestling against those of the next row over. Once the leaves close in the row, an impenetrable canopy is formed that won't allow sunlight to reach weeds growing below. There is no better strategy for weed control in soybeans than a completely closed-in row. Since the advent of herbicides, soybean breeders have been developing plants with narrower leaf expression. These plants are tall and skinny and do quite well when planted in narrower seven- and fifteen-inch rows. The logic behind planting soybeans in these narrower rows is to cram more plants into an acre and increase yield. Narrow-row soybeans are generally seeded with a special grain drill or "fold down" corn planter, but this situation presents the organic soybean grower with a bit of a dilemma. For us, thirty-inch rows generally work best because they can be cultivated two or three times until the plant leaves reach all the way across the rows. Unfortunately, there are fewer and fewer wide-canopy varieties being bred and released into today's marketplace. It is especially difficult to find bushy Group 00 and Group 0 soybeans; the best you might find is a semi-bushy variety that might grow partway out across the row. Group 1 and 2 varieties are much more likely to fit the bill, but are much too late maturing for many of us to grow. I have tried growing soybeans in both seven- and fifteen-inch rows with only limited success. Weed control is difficult. I finally solved this problem by switching to thirty-inch rows and a late Group 0 variety that forms a reasonably good canopy. It is possible to grow narrow-row soybeans organically. If you can keep the weeds at bay for the first two to three weeks, these types of soybeans will form a canopy much earlier in the growing season, and less mechanical cultivation will be required—saving time and energy. (Weed control and cultivation will be discussed at length later in this chapter.)

Considering Fruit Characteristics

The actual bean or fruit of the soybean is as variable in size and appearance as the plant is in its growth habit. As a grower of soybeans, you have many choices when considering what you would like the fruit to look like. Before you select a specific variety, determine the end use of your potential harvested crop. If you are growing soybeans for livestock feed, the decision-making process will be relatively easy, as just about any soybean variety will work. If you can find an extra-high-protein variety, that will be even better. Soybean protein levels usually run in the 42 percent range. Some seed companies like Prograin in St. Cesaire, Québec, offer specially bred feed varieties with protein levels of 48 percent. These cultivars are a little lower yielding, but will actually produce more pounds of protein per acre. If you're feeding these beans to your own livestock,

Early Riser seed corn ears ready for shelling.

Corn beginning to canopy the row.

Picking ear corn for seed.

Ear corn being dried for seed.

The author with Early Riser corn beginning to tassel.

An ear of corn ready to dent.

Combining hybrid field corn in late October.

Some weed-free corn in early June.

An ear of corn at milk stage.

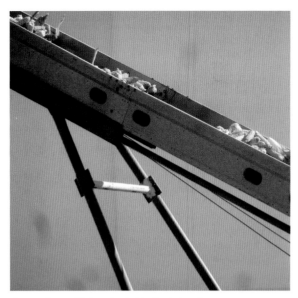
Ear corn for seed being elevated to the drying silo.

Combine grain tank filling with corn.

Picking ear corn with a New Idea Uniharvester.

First signs of silk.

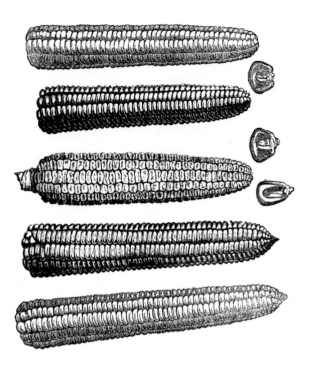
Types of corn, from top to bottom: flint (top two), flour, dent (bottom two). PHOTO COURTESY OF FEDCO SEEDS

A ripe ear of Early Riser open-pollinated corn.

Drying corn.

Combining tangled open-pollinated corn.

A full crib of ear corn.

Wheat going from milk to soft dough stage.

Red Fife ready for harvest.

Red Fife heirloom wheat at milk stage.

Red Fife getting ripe.

Round bales of winter wheat straw for winter bedding.

Harvesting a thinner stand of soft white winter wheat.

A modern variety of hard red spring wheat recently headed.

A Meadows Mill home-style flour grinder.

Bearded wheat.

Crossing two varieties of wheat in a protective sheath.

Inspecting hand-threshed wheat prior to harvest.

Bagging unit at a commercial-scale flour mill in Ontario.

The chew test. PHOTO COURTESY OF STEVE LEGGE

A horizontal stone wheat grinder at the Maud Foster windmill in Boston, Lincolnshire. PHOTO COURTESY OF ANNE LAZOR

Awned barley just after heading.

Barley field beginning to turn color and lodge.

An upland field of barley at harvesttime.

Individual heads of six-row barley.

One of my first barley crops in the early '80s.

Paddy table separator for sorting hulled and unhulled oats.

A Roskamp Champion Oat Huller.

Oat hulls dropping from an overhead cyclone.

Oats in the milk stage. PHOTO COURTESY OF BETHANY DUNBAR

Oat panicles.

Winter rye past due for harvest.

First pass through an oat field with the combine.

A field of rye in the late winter.

A nice harvest of plump white oats.

A spelt huller at an Amish facility in Holmes County, Ohio.

A well-established field of winter grains in the late fall.

Spelt ready to harvest.

A modern German-made spelt huller.

Stooks of winter rye. PHOTO COURTESY OF ANNE LAZOR

The author in waist-high soybeans. PHOTO COURTESY OF BETHANY DUNBAR

Soybean plants with pods.

Field-run beans heading to the roaster.

A portable soybean roaster.

Planting soybeans in thirty-inch rows in late May.

Bird's-eye view of the soybean-roasting process.
Right to left: Wagon to roaster to grain dryer for cooling.

Harvesting a plentiful crop of soybeans in October.
PHOTO COURTESY OF LISA ROBINSON

Hand-pulling dry beans at harvest.

Hand-feeding a bean combine.

A weed-free field of dry beans at mid-growth stage.

Black turtle beans ready for the cleaner.

Hand-sorting Marfax beans on a bean belt.

Sunflowers in full bloom.

A field of windrowed flax.

Flax beginning to form purple flowers.

Harvesting flax with a combine windrow-pickup attachment.

My best crop of sunflowers, summer of 1999. PHOTO COURTESY OF ANNE LAZOR

A pull-type grain windrower in a field of harvested buckwheat.

Buckwheat in full bloom.

Making animal feed with a grinder-mixer.

Research trial plots.

a high-protein variety is a great idea. However, if you're selling commodity organic soybeans to a marketplace that pays by the ton, a higher-yielding, lower-protein crop will earn you more money. If you plan on producing soybeans for the human consumption market, your options are a bit more limited. Most tofu and tempeh makers want large soybeans. Soybean seed can range from two to four thousand seeds per pound. Choose a variety toward the low end of the spectrum if possible, one with twenty-three or twenty-four hundred seeds per pound. These seed count numbers aren't written in stone. Environmental factors like a drought during pod fill can reduce fruit size, and we have to constantly remind ourselves that farming is a gamble and that we need good weather for perfect results.

There is also a very specialized market out there for natto soybeans. Natto is a fermented Japanese soy food made from very tiny soybeans. These soybeans usually number about four thousand seeds to the pound and are very easy to grow. But the most important requirement for human consumption soybeans is hilum color. The hilum is the little dot or speck that you will find on the side of an individual soybean. It is basically a tiny scar that represents the point of attachment between the soybean and the pod. When beans are developing inside the pod, plant nutrients travel through the hilum to the ripening fruit. Depending on the variety, a soybean hilum will vary in color from clear and yellow to gray, brown, and black. Soybean processors want beans with clear or yellow hilums because they don't want dark specks in their products. Study the variety description charts carefully before purchasing seed, as there is always a column for hilum color.

Choosing for Human Consumption

Choosing a variety to grow and sell into the human consumption market is no easy task. Many processors will specify a particular variety that meets their own special performance requirements. Before jumping off the cliff and venturing into this market, it would be a good idea to contact your potential buyers and other farmers who are already growing specialty beans for advice and direction. For years the Vinton 81 soybean was the favorite soybean for many soy milk and tofu processors. The fact that Vinton is a Group I–9 variety made it impossible for farmers in the northern part of our region to raise higher-value specialty soybeans. Thanks to the efforts of Canadian plant breeders, there are now a few more varieties in the Group 0 range that will work well for food processing, but once we drop down another maturity zone into the Group 00 range, there are fewer choices. Generally speaking, longer-season full-canopy soybeans produce larger soybeans that best meet the needs of soy food manufacturers. After years of trying numerous soybean varieties, I have finally found a dual-purpose cultivar that works well for human consumption as well as livestock feed. The variety is called Dares and is produced and sold by Co-op Fédérée in the province of Québec. Dares is a Group 0–8 soybean. I'm pushing my growing season a bit by planting it, but I have been able to successfully ripen it for the last several seasons. Choosing the right soybean for your farm and marketing needs can be fun and interesting and a bit frustrating when you try something that doesn't perform as you imagined. Once you find the right variety, you'll want to stick with it until something better comes along.

Planting Soybeans

Like all beans, soybeans like to be planted in warm, reasonably moist soil. The minimum soil temperature should be fifty degrees, and all risk of frost should be past. Standard practice for soybean seeding is to wait until after corn planting has finished. Depending on your location north to south, this could be anytime after the fifteenth of May. Soybeans are a whole lot less sensitive to cold growing conditions than dry edible beans—for example, you would never plant a kidney or yellow eye bean in the middle of May, but a soybean can take it. In my northerly location, I usually try to have my soybeans in the ground by the twenty-fifth of May. Later planting during the first half of June is acceptable because soybean flowering and fruiting is dependent on photoperiod instead of a set number of days. Despite this fact, I still find that the earlier a soybean is planted, the more time it has to put on vegetative growth before the advent of flowering. The result is a bigger, more robust plant that will produce more flowers and pods and thus yield more. Early soybean planting is a must, especially if you are growing a variety whose maturity is marginal in your area. This is certainly the case for me. Since I am planting a Group 0–8 soybean, I need to seed early in order to give this variety plenty of time to mature and dry down at the other end of the growing season in October. If I were planting Group 00 or Group 000 soybeans, I could afford to put off planting until the first or second week of June without any adverse effects on yield.

Soybeans require prolific moisture for good germination. A soybean seed must absorb 50 percent of its weight in water before it begins to sprout. The last thing you want is a dried-out seedbed right after planting. You can never tell if it's going to be dry or wet once late May and early June roll around. If things start to dry out a bit, you'll be glad you got those soybeans planted on the twentieth of May when there was still some moisture left in the soil. If you're on a dairy or beef farm, you'll be glad to have finished soybean planting by the third week of May so you can move on to some early first-cutting haying. The race is on. You've been going like blazes since April planting cereals and corn, and soybeans are no different from any of the other crops. There's no time to relax until haying is finished.

Preparing Your Soil

Soybeans should be planted in a well-worked seedbed that is firm and moist. The fact that soybeans aren't the first crop to be planted in the spring gives you plenty of time to work the soil completely and thoroughly. Most people will follow a corn crop with soybeans. Logic has it that a leguminous crop like soybeans will replace some of the soil nitrogen that has been removed by the corn. There is also a symbiotic relationship between the roots of young soybean plants and the chopped-up decomposing corn stover that has been incorporated into the soil in advance of the bean crop. No matter what crop precedes a new seeding of soybeans, it's always a good idea to work in last year's crop residue as early in the growing season as possible. An April discing and chopping of last season's cornstalks or cereal stubble will provide soil microbes with plenty of time to begin breaking things down. You also might want to think about spraying some kind of microbial inoculant on the soil to aid in the decomposition process. If you've got the time and the wherewithal to plant a cover crop of oats in early April on what will be your soybean field, by all means do it. The oats will grow for four to six weeks and provide green plant food for the soil food web just prior to soybean planting in late May. Even if you don't cover crop in advance of planting your soybeans, a bit of tillage every two weeks during April and early May will help you attain a perfect seedbed when it comes time to finally plant your soybeans later in the month. One pass with the disc and a second with the field cultivator should be just about right. The perfect seedbed should be well worked down to a depth of six inches and firmed with a cultipacker if the soil is too soft and fluffy. Excellent seed-to-soil contact is essential to attain good soybean germination; some growers will roll and firm a field of soybeans right after they have been planted. Soybeans are fussy and require the very best soil conditions to germinate completely and properly.

Choosing Your Seeding Method

Soybeans can be planted in a variety of different ways. If you're a small-scale garden grower, you can carve a furrow with a hoe, plant a soybean every inch or two, cover the seed with soil, and tamp with your hoe. A push-type garden seeder like a Planet Jr. or an Earthway will allow you to plant larger bean patches in a timely fashion. When it comes to mechanically planting larger acreages, there are two basic ways to put seed in the ground: You can plant either narrow rows of beans with a grain drill, or wider rows with a corn planter. This brings us back to the choice between bushy foliage soybeans for wide rows or thin erect plants for narrow rows. Each seeding method has its positive features and its drawbacks. Corn planters are very accurate, and seed spacing in the row is usually quite precise. Germination is almost guaranteed because these machines place soybean seeds at a uniform depth in well-firmed soil. However, corn planters are a bit more cumbersome to navigate in the field, necessitating the careful layout of rows. Some soybean growers would rather plant with a grain drill, which doesn't require as much adjustment or maintenance as a corn planter. Drilling soybeans is much less technical and exacting: Just set the seed adjustment lever, fill the hopper, and start planting. There is no need to worry about headlands or laying out a field in thirty-inch rows.

Planting soybeans in seven- and fourteen-inch rows (with a grain drill) is also known as solid seeding. This method of soybean culture is a bit more free-form than wide-row planting, plus it's quick and easy. Simply operate your grain drill just as you would if you were planting wheat or barley. Although you may cover more acres in a day, grain drills do have a few major drawbacks. Dialing up a specific plant population is much more difficult with this type of planting, as the fluted seed gear in the bottom of the grain drill hopper is nowhere near as precise in metering seed as the delivery system of a corn planter. Older planters use revolving seed plates while newer machines use finger pickups and vacuum to place an exact amount of seed in the ground. Corn planters can be adjusted to drop anywhere from two to twenty seeds per row foot. For example, if you have determined that you want a population of eight seeds per foot of row, the seed sprockets on a corn planter can be set to make this happen. A grain drill won't give you this kind of accuracy. Germination might also not be as good with a grain drill because soybean seeds won't be planted as deeply or firmly in the soil. You can fret about how to plant soybeans for a good long time. The truth of the matter is that nice crops of soybeans can be established using either seeding method.

I planted my first KG 20 soybeans with my old John Deere Van Brunt Model FB grain drill in 1990. This was a 1950s-era machine equipped with single disc openers on six-inch centers. It had done a great job planting barley, oats, and wheat for me in previous seasons, but soybeans were something new for both me and the machine. I consulted the seeding chart on the soybean seed bag to get a better idea of how to go about the process, and according to the population tables, I would need to sow between two and three seeds per foot of each six-inch row to attain the recommended 185,000 plants per acre. The next step was to try the machine on a hard surface like the road to see how many seeds it was actually laying down. I found this mock planting project to be more theoretical than actually possible. As I pulled the grain drill across the yard with the tractor, seeds came pouring down the tubes, hit the turning single discs, and then went rolling around on the ground. After several attempts at various seeding rates, it quickly became obvious that the precision planting I was hoping for was just not going to happen with my 1950s-vintage grain drill. It seemed as if the soybeans had a mind of their own in this matter. Each six-inch row contained a different number of beans, and it was going to be quite difficult dialing this old machine down low enough to drop only three seeds per foot. There was no setting for soybeans listed on the seeding chart under the drill's cover, and I also found out that single disc grain drills only do a marginal job burying round seeds like soybeans and peas. When an oblong seed like wheat or oats drops down a seed tube, the single disc opener places it in the seeding trench with little or no trouble. But soybeans act differently when they hit the revolving single disc opener of a grain drill. It's much harder to bury a soybean with the single disc because it acts like a little rolling marble once it emerges from the seed tube. The combination of the rolling action of the concave disc and the round shape of the seed permits the opener to lift the seed out of the earth instead of burying it at a uniform depth. The opposite action takes place when soybeans are drilled through double disc openers: As the seed falls between the two flat discs, the rolling action of the metal blades grabs the soybeans and pulls them down to a uniform depth. Despite all of these obstacles and setbacks, I still managed to plant my first crop of KG 20 soybeans with my ancient grain drill. The germination wasn't perfect, but I still had plenty of bean plants in the field. Maybe my germination problems were a blessing in disguise. The drill actually planted a lot more seed than 185,000 plants to the acre. The fact that the germination was a bit reduced probably saved me from planting too thick. I was certainly pretty inexperienced back in 1990 when I tried grain drilling soybeans for the first time, but I still managed to get a lot of experience and did get a twenty-bushel-to-the-acre crop in my first year. These sorts of lessons are invaluable, and we all need to go through these first attempts in each of our personal journeys.

The Evolution of Grain Drill Use for Soybeans

Fortunately, grain drills have come a long way since the 1950s. Beginning in the '60s and '70s, farm machinery manufacturers began producing heavier grain drills equipped with features that made soybean planting possible. The single disc opener gave way to the double disc, which worked a whole lot better for burying soybeans at a uniform depth. Newer machines were constructed of heavier materials with down-pressure springs that provided for better soil penetration across a wide array of soil types. It was now possible to plant soybeans one and a half to two inches deep with a grain drill. The addition of adjustable seed cups below the hopper box was another feature that improved grain drill performance with soybeans. Often, the fluted seed gears on older drills would damage soybeans because there wasn't adequate space for the seed to pass underneath the gear on the way to the drop tube below; larger soybeans could be cracked or partially ground up in the process. Now with the flick of a three-position lever, this space underneath the seed gears could be made large enough to comfortably accommodate large round seeds like soybeans.

I purchased a 1970s-vintage International Model 510 grain drill in 1993 after several years of marginal soybean-planting results with the older John Deere model. At the time, it seemed like I had just emerged from the Dark Ages into the Renaissance. The International drill was twelve feet wide and had double disc openers as well as adjustable seed cups, and it did a much better job putting the seed in the ground. Seed spacing was still an issue with this drill, however. I found it difficult to close the seeding-control mechanism down to a low enough rate to follow the planting recommendations on the seed bag. After six or seven years of struggling with trying to attain the proper population of soybeans, I traded again in 1999 for an even better and more modern grain drill. The next generation of grain drill after the International 510 was the Model 5100 Soybean Special. I still own this machine and hope that it lasts me for as long as I am still farming. By the late 1980s, when this drill was made, farm equipment companies had finally added all the bells and whistles necessary to transform an ordinary seed drill into a dual-purpose cereal and soybean planter. My new Case IH 5100 was outfitted with a hard rubber press wheel that followed each double disc opener. Not only were soybeans well placed in the soil, but now they were firmed and pressed by a narrow little roller with five possible positions for adjusting down pressure.

The most revolutionary improvement developed by the agricultural engineers was the addition of a two-speed drill shaft to allow for more accurate seed spacing. The drill shaft runs underneath the entire length of the hopper, providing motive force to each seed gear. This particular shaft is chain-driven by a seven-tooth sprocket and turns at the same speed as the end wheels of the grain drill. Soybean Special grain drills have the option of powering the drill shaft with a fourteen-drive sprocket, which cuts the RPMs of the drill shaft in half and in turn sends fewer soybean seeds down the seed tube. Precision soybean seeding is much more in the realm of possibility with this sort of setup; indeed, dropping only three seeds per foot of each seven-inch row was nearly impossible before the Soybean Special came along. There are a few important details to keep in mind when you're planting with the fourteen-inch sprocket. Remember that you are only planting half the amount of seed per acre that is indicated on the calibration chart, usually located under the hopper cover. For example, if lever position number ten is supposed to plant one hundred pounds of beans to the acre, you will actually only be seeding fifty pounds with a half-speed drill shaft. You will have to find the setting for two hundred pounds to the acre to achieve a one-hundred-pound seeding rate. The other minor inconvenience about a Soybean Special grain drill is the fact that sprockets need to be changed and drive chains need to be either lengthened or shortened when you change back and forth between cereals and soybeans. This seems like a small price to pay, however, for the flexibility of being able to accurately plant both soybeans and cereal crops with the same machine.

More Considerations for Seeding Soybeans

If you decide to plant soybeans with a grain drill, choose a variety with a narrow leaf canopy that thrives when planted at higher populations. Quick emergence and good early-season vigor are also traits to look for if you decide to go this route. Most seed catalogs have variety rating charts that will provide you with this sort of information. Weed-control options are much more limited when soybeans are seeded with a grain drill. The strategy behind this methodology is to plant your soybeans thick and hope that they grow quickly after germination, beating out the weeds by forming a leaf canopy as early in the season as possible. Mechanical weed control in drilled soybeans is possible on a very limited basis and will be discussed a little later in this chapter. Soybean plants have the ability to "flex" and compensate for yield when planted at lower populations. When fewer plants are present, the existing plants will form additional branches and yield more beans. Soybeans can be planted at rates as low as 150,000 plants to the acre when seeded in thirty-inch rows and up to 225,000 when planted in six- or seven-inch rows with a grain drill.

Another important fact to consider when seeding soybeans is that beans tend to vary in size. A pound of large-sized soybeans will contain somewhere around twenty-three hundred seeds, while average-sized soybeans will have a twenty-eight-hundred-per-pound count. Keep seed size in mind when you are trying to calibrate your grain drill and figure out a seeding rate. Although you may be advised to plant a certain number of pounds of seed per acre, you'll do a much better planting job if you know your relative seed size and determine how many seeds to plant per foot of row. Most seed companies will provide seed count information on the bag or variety tag. My Soybean Special drill has a calibration chart under the hopper cover that tells me how many pounds of seed are required per acre for soybeans, ranging from twenty-two hundred all the way up to three thousand seeds to the pound. There is also information on the number of beans needed per foot of row to attain various end populations, and you have the option to plug every other or every third seed run on a grain drill to plant soybeans in fourteen- or twenty-one-inch rows. Grain drill calibration becomes a mathematical exercise. Consult the seeding chart on the drill to make an educated guess at where you might set the seed flow adjustment lever to come up with your desired population. Now go try the drill out on a driveway or some other hard surface to get an idea of how many seeds are actually being planted. Don't forget to increase the number of seeds per row foot by two or three if you plan to plant in fourteen- or twenty-one-inch rows. You may have to try several times before you get the seed setting right. Don't be alarmed if you err a little bit higher or lower than the recommended population. It's probably better to seed too thin than too thick. Too many bean plants to the acre can self-compete for water and nutrients and lodge later in the season. Once your grain drill is properly adjusted, soybean seeding will be a breeze. Load the drill with seed and calibrate it properly, and planting soybeans is no different than seeding wheat, barley, or oats.

My Switch to the Corn Planter

After fifteen years of planting soybeans with a grain drill, I switched to a corn planter about five years ago. I figured that since I had been developing my skills planting corn over the years, I could certainly master the art of seeding soybeans in thirty-inch rows, but weed control had become more and more difficult for me in my grain-drilled beans. Planting beans in rows would give me a lot more opportunity to cultivate the crop for a longer period of time. I was still using a six-row International plate planter at the time, and I bought some plastic soybean seed plates, studied the owner's manual, and gave it a try. The first thing I learned was that planting soybeans with a corn planter was not the precision process I thought it would be. Soybean seed plates have larger cells that are meant to drop two or three beans each time they revolve over the seed outlet hole. The first time I planted beans with the old six-row International planter, I didn't put enough seed in the ground. I must have been bashful about cranking up the population, and the final plant count ended up somewhere between six and seven beans per foot of row when it should have been at least twelve. If you're using a plate planter to seed soybeans, the seed plates have to spin pretty quickly in order to drop enough seed to attain twelve plants per foot of row. The gearing on the drill shaft of the planter will be turning at a rather high RPM, which can put extra stress on drive sprockets and chains. (You might compare it to riding your ten-speed bicycle in high gear all the time.) Normal preseason corn planter maintenance that includes replacing worn sprockets and chains will keep your machine running flawlessly so that once you start planting soybeans you won't have to stop and fix something when you should be seeding.

Once I switched to a John Deere 7000 finger-pickup planter for corn seeding, I had to learn a whole new planting process for putting my soybean seed in the ground. It's not as easy as just changing plates and planting rate sprockets. The little corn seed clip belt planting mechanism will not pick up and deliver a round seed like a soybean. Instead, you must install special soybean cups down in the bottom of the planting unit. I had heard about these cups for years, but really didn't know anything about them. Thankfully, I was able to procure some used ones from another farmer. There really wasn't much to these little devices. The John Deere soybean metering cup looks like a small inverted plastic bowl that spins on a short stub axle. They are actually rather crude in design. Seed enters from the top and is delivered out the side of the spinning cup to the seed tube below. If you are looking to plant twelve to fourteen plants per foot of row, it's very easily accomplished with these units.

The key to success in planting soybeans with a corn planter is to choose a bush-type cultivar and make sure that you plant enough seed. Once I began using a corn planter to sow my soybeans, I found that it was necessary to plant a Group 0 instead of the Group 00 variety. I worried a bit about pushing the limits of my growing season, but everything turned out just fine. My yields increased because I was able to do a better job with weed control in the thirty-inch rows, and past experience taught me to not be shy about planting enough seed. Theoretically, nine beans per row foot should be the right number to achieve a plant population of 150,000 soybeans per acre. This might be all right for the warmer fuller-season areas in the Corn Belt where thinly seeded soybean plants will develop more fruiting branches to compensate for lower population levels, but here in our part of the country, we need all the plants we can get. Make sure to plant at least one seed for every inch of row—having a few extra plants is actually a really good idea, because some will be lost in the cultivation process. I was very pleasantly surprised with the results of my transition to thirty-inch-row soybean culture. Germination was fantastic, because the seeds that I planted were placed firmly in the ground at a very uniform depth. The plants grew quickly and canopied the row by early July; in the end, my yields were improved as well. This was the first time I had ever consistently harvested a ton or more of soybeans per acre across a number of different field environments. After ten or more years of hesitation and fear of the unknown, I am convinced that the corn planter is a better choice than the grain drill for seeding a successful soybean crop in our region.

The Importance of Seed Inoculation

The very last and most important detail to consider when planting soybeans is seed inoculation. One of the reasons that soybeans are such a wonder plant—as I said before—is that they fix their own supply of nitrogen from the surrounding atmosphere. With the help of the very specific bacteria *Rhizobium japonicum*, soybeans develop nodules on their roots that make the process of nitrogen fixation possible, but this bacterium is nonexistent in soils where soybeans have never been planted. It's always a good idea to inoculate soybean seeds before planting, even if they have been grown in the field in previous years. Inoculation is cheap insurance, and inoculant is available in both dry peat-based powder form and also as a liquid. If you are certified organic, check with your supplier to find out if the particular brand they sell is legal under the National Organic Program. Some inoculants are fabricated from genetically modified bacteria, but most manufacturers will gladly provide you with a non-GMO affidavit.

I like to double inoculate my soybean seed, which gives me the peace of mind that the proper rhizobia will be there in the soil to do their work. Soybeans need to be dampened ever so slightly to allow peat-based inoculants to stick to the seed. I start the process by pouring half a bag (twenty-five pounds) of seed into a large widemouthed tub. I then sprinkle a tiny amount of a liquid inoculant called Cell-Tech on the seed and stir it up. Careful attention to detail is very important in this part of the process. Start out with just a few drops, because if the seed gets too wet, it can become too soft and be damaged by the inner workings of the planter. It's always easier to add a few more drops of Cell-Tech than it is to take it away. If for some reason the seed gets too wet, add a few more dry soybeans to the mix. Once the seed is sufficiently dampened, pour on some dry peat inoculant and stir it all together. You'll know when everything is adequately mixed, because the soybeans will appear uniformly blackened. When you're satisfied with your work, gently pour the mix into your grain drill or corn planter. Once the machine is loaded with seed, give the soybeans in the hopper one more stir to evenly distribute the inoculant. Now you're ready to plant.

Soybeans are very delicate little fruits that are easily damaged by rough handling. The actual germ of the soybeans is very near the surface right behind the hilum, and if beans are banged around or mishandled they can lose some of their germ as a result of rough treatment. Don't drop or throw bags of soybeans around like bales of hay. One Canadian seed dealer told me that every time a bag of soybeans was dropped or tossed, germination declined. So remember to be gentle with your soybean seed when you are moving it around before planting. Once soybean seed has been wetted down and inoculated, it should be planted right away. I've made the mistake of leaving inoculated soybeans in both a grain drill and a corn planter overnight, and upon resuming planting the next day, I discovered that the seed didn't want to come out of the planter. It had swelled overnight and bridged inside the planter box because it was all stuck together. The only remedy to this situation was to empty the seed back out and remix it with some drier stuff. I also made the mistake of wetting down my soybeans with a little water and maple syrup once. The mix got so sticky and hung up in the grain drill that I had to empty everything out and start all over again. The seeds that we are planting are precious. They are full of potential life and promise and must be treated with care and respect. I've learned these lessons the hard way.

Early Growth and Cultivation of Soybeans

Soybeans planted in warm moist soil will germinate quickly. A nice warm gentle rain after planting is ideal. If you've got lots of small rocks in your soybean field, rolling with a cultipacker will improve the final harvest because it will push small stones down into the soil out of the way of the combine cutter bar. A timely rolling will also serve to promote good soil-to-seed contact, especially if the beans have been planted with a grain drill. Pre-emergence weed control with a rotary hoe or tine weeder will eliminate many of the early-season weed seedlings that would be able to get a foothold between planting and emergence a week later, and a pass over a newly-planted soybean field with one of these weed-control machines will expose the white-haired roots of infantile weeds to the drying rays of the sun. Of course, timing is everything. All you can do is hope for a sunny warm day four or five days after the field has been planted.

It is also very important to monitor the progress of germination during this time period. Go out in the field and dig around every day to see how far the young soybean shoots are from popping up through the soil's surface. Soybeans are dicotyledons, which mean they are a two-sided seed that has two embryonic leaves or cotyledons at the time of germination. When a soybean—or any other bean, for that matter—pokes its growing point up through the surface of the soil, its two symmetrical seed halves are attached to either side of the emerging stem. Very soon after emergence, the plant's first two unifoliate leaves will unfold below the two cotyledons. This is probably one of the most critical periods in the life of a soybean plant. If one or both of the cotyledons is knocked off the emerging stem at this stage of development, the young seedling might not survive. The very best scenario for a damaged soybean seedling would be a 10 percent yield reduction. For this reason, take the utmost care when the growing point is just below the soil surface as well as during the plant's first several days of growth, when the cotyledons are still attached to the stem. You will want to avoid mechanical cultivation with the rotary hoe or tine weeder when the growing point is just below the ground and also during the first couple of days of growth. Soybeans are particularly susceptible to damage from light tillage and cultivation at this point in their development, so make sure to stay out of the field with the iron until the two cotyledons have fully expressed themselves and the first unifoliate leaves have unfolded at the first node of the plant. After this it becomes safe to go in there with the tine weeder or rotary hoe and really stir things up a little. In fact, once a soybean has made it through these first couple of critical days, it becomes a rather tough little plant that can withstand the stresses and strains of mechanical cultivation and wheel traffic, and it becomes time to eliminate those first weeds especially if you have planted your beans in narrow rows with a grain drill. These very young soybean plants don't mind being driven over by tractor wheels. At this point, the plants are also quite well anchored with the beginnings of a taproot. This actually allows for a fairly rigorous mechanical attack on the weeds without damaging any of the emerging soybean crop. Take advantage of this situation, particularly if your beans are solid-seeded. Once the soybeans grow a few more inches in height, driving over them will be more difficult and the possibility for crop damage increases.

Early-Season Weed Control

Soybeans will grow quite quickly if the weather is warm and there is plenty of moisture available. If the last week of May or the first week of June comes off cold and wet, the soybeans will stand still while the weeds take off. There really isn't much we can do about weather, but weather patterns are predicted weeks ahead of time these days. If you put stock in this sort of thing, it might be a good thing to consider before planting soybeans on the twentieth of May. You might even want to hold off until the first week of June. Then again, if you take the chance and get soybeans going early, you will be rewarded in the end. The choice is up to you. If you do experience a slowdown in the early growth of your soybean crop, you would do well to consider what sort of weed pressure you have. Walk your fields to see what is competing with the tiny, recently emerged soybean plants. What are the predominant weeds? Are they annuals like mustards and lamb's-quarters, or are grasses the problem? Different strategies are needed for different types of weeds. For example, light cultivation with the rotary hoe or tine weeder works best on the annual weeds. Unfortunately, you need warm and dry conditions to uproot and dry out these weeds so they can be cooked by the sun. These machines won't do anything for grasses that are deeper-rooted and more firmly anchored in the soil; you'll need a row-crop cultivator to knock back such grasses before your young soybean crop gets choked out. I had this experience in 2010. My soybeans were up about two inches, the weather was cool and cloudy, and the grass was invading fast. We mounted a three-point-hitch Danish tine cultivator on the tractor and very slowly and carefully made our way through the young soybeans. As the day progressed, our confidence with the machine increased and we were able to speed up the cultivation a little bit. It was cloudy and cool, but the soil was dry, which was just what the crop needed. The grass was set back and the soybeans took off the following week when the sun returned. We actually went back in with the tine weeder the next week when the beans were four inches tall.

Because of their taproots, soybeans can stand up to tine weeding until they are six inches tall. The tines will vibrate around the stems of the beans, pulling out the weeds while leaving the bean plants. The soybeans might appear a bit bedraggled after weeding, but they will recover and be much better off for it. If you are working with solid-seeded beans planted with a grain drill, you can charge through your stand with the weeder and tractor until the beans are six inches tall; a few plants might be sacrificed in the process, but most will survive and thrive. This will be your last chance to cultivate your soybeans, so if you have lots of emerging grass in a solid-seeded bean crop, you're in trouble—the weeder isn't going to do anything for the grass, and your crop will suffer. This is the primary reason why I switched to planting soybeans in thirty-inch rows. I can keep knocking back the grass with a row-crop cultivator until the soybeans get knee-high or better. Successful early-season weed control is essential in getting your soybeans off to the very best start.

The Vegetative Growth Stages

Soybeans go through a number of vegetative growth stages on their way to bearing fruit as a mature plant. Once the cotyledons have fully expanded after germination, the first set of unifoliate leaves appears at the first node of the plant. Just as much activity is going on belowground as above. The bean plant's taproot begins to lengthen, and the formation of root nodules for nitrogen fixation begins; very soon afterward, the first actual trifoliate leaf appears at the second node of the soybean plant. The plant is officially at this V-1 stage of growth when the three leaflets in the trifoliate cluster are fully developed and no longer touching one another. The plant stem continues to lengthen, and a third node develops that supports a second set of trifoliate leaves. At this V-2 stage of growth, root nodules actively begin fixing nitrogen, and the plant's lateral roots begin growing rapidly through the top six inches of soil profile out into the plant rows. If you've got your row-crop cultivator shovels running very close to the rows for precision weeding, now is the time to adjust them back away a bit to avoid root pruning. Soybean vegetative growth continues for at least another month through the V-6 stage, where seven more nodes and six more sets of trifoliate leaves develop as the soybean gains size and stature through June and into early July. By stage V-5, the axillary buds in the top stem axils are beginning to develop into flower clusters, and the number of nodes that a plant will develop is fixed. At stage V-6, the unifoliate leaves at the very bottom of the plant stem have dropped off and left a noticeable scar. Row-crop cultivation is possible throughout most of the vegetative period; two and sometimes even three passes through the growing soybeans might be necessary to keep the weeds at bay.

The biggest limitation in plant cultivation is the bushiness factor. As the sets of trifoliate leaves grow and elongate out into the rows, there is less and less room for the tractor tires to drive through without damaging the foliage. Once the rows are shaded by the plants, the canopy has been established and weeds are no longer a threat. As the plants get bigger and bushier, cultivator shovels are pulled farther and farther away from the plants, which allows you to increase tractor speed and to perform the cultivation duties much faster. The most important rule to remember about row cultivating any sort of bean plants is to stay out of the field with machinery when the leaves are wet from dew or rain: Bean plant leaves are very susceptible to fungal diseases like anthracnose that are easily spread by machinery contact with wet plant foliage. Wait until things have dried out before entering the field with the cultivator.

Insects and Diseases in Soybeans

Disease is extremely rare here in the Northeast because soybeans are a relatively recent addition to the crop lineup in this part of the country. The Midwest and the Southeast, however, which have been major soybean production areas for more than fifty years, are plagued with all sorts of problems that we only read about in magazines and seed catalogs. Indeed, there are a number of fungal and viral infections that can wipe out a crop of soybeans in pretty short order. Among them are bacterial blight, brown spot, downy mildew, and rhizoctonia stem rot, which are all brought on by extreme weather conditions like excessive dampness, coolness, or too much heat. The inoculants for these various maladies must also be present in the environment. The two most common afflictions of the soybean in the Corn Belt are phytophthora root rot and soybean cyst nematode (SCN). Phytophthora is a soilborne fungus commonly associated with the stand reduction in alfalfa crops. It is a major problem in the Midwest, where it causes seed rot in recently planted beans as well as root die-off in more mature plants. The soybean cyst nematode is a little soil organism (*Heterodera glycines*) that attaches itself to the roots of soybean plants and forms little cysts that parasitize the plants. Plants afflicted with SCN will experience moderate to severe yellowing and chlorosis, and the crop will be stunted and fail to thrive. Nematodes persist in the soil for two years. And while phytophthora root rot and soybean cyst nematode are quite rare here in the Northeast, they do exist to our north in Québec where soybeans have been cultivated for many years. We have as yet to see any major evidence of any of these maladies in our region, but that's not to say they might not become an issue in the future.

White Mold

The one soybean disease that is quite common in these parts is white mold, also known as sclerotinia stem rot. White mold doesn't usually show up in a field until midseason or later when the plants have grown tall and bushy and have begun to flower. Tan or white lesions will first appear at stem nodes and eventually girdle and strangle the plant. A cottony fungal growth around the stems follows the first sign of lesions. You won't be able to see white mold from a truck driving by a field, though; you have to get out and actually walk up and down the rows to find it. White mold is caused by the fungus *Sclerotinia sclerotiorum*, which thrives in thick soybean foliage in moist humid conditions with little air movement. Eventually the fungus will fruit into large black structures called sclerotia, which cling to infected stems and invade the pith of the plant. White mold is a soilborne fungal organism and it certainly is no fun; even worse, once it's present in a field, it has the ability to stick around for seven years. Deep plowing of infected crop residue can help bury it and slow it down some, but the sad part is that you won't be able to grow a host crop like soybeans or sunflowers in that particular location for a full seven years. I've only seen a white mold outbreak once in my twenty-plus years of growing soybeans, and it was in solid-seeded beans planted on a river bottom field that was prone to prolonged periods of thick morning fog. Chances are that if you are new to soybean growing, white mold will not be a problem for you, but the fear of it alone is enough to convince me to keep planting soybeans in thirty-inch rows where good air movement and field ventilation can go a long way toward preventing an outbreak in the first place.

The Soybean Aphid

The only real insect threat to soybeans is the soybean aphid (*Aphis glycines* Matsumura). This exotic little bug hails from Asia and was first positively identified here in North America in 2000. Soybean aphids have been moving eastward from the Midwest for the last decade and have only very recently been identified in the Northeast. These little creatures will first invade a field of soybeans in early June. By the time the plants have gained considerable stature in late July, the aphids have been through as many as fifteen generations. A walk through the rows and an inspection of the leaf undersides will reveal masses of little yellow-brown aphids attached and sucking away at the plants' vascular system. The normal mode of action in a situation like this is to start worrying immediately that all of the life is being sucked out of your crop by these little devils, but on further inspection you might also notice an elevated population of ladybugs feasting on the aphids. Some farmers will want to spray some sort of organically approved pyrethrum insecticide to kill the aphids, but this is a costly practice in both dollars and plant condition. By this time, the soybeans have closed in the rows, and a drive through with a spraying apparatus will damage the plant foliage probably more than the aphids. My own inclination in this circumstance is to stand back and let the ladybugs eat the aphids. Pyrethrum spray may be organically approved because it comes from a plant source instead of a test tube, but it is still poison in my book. Plus, field spraying will upset the balance of nature because it will kill the ladybugs and other beneficials in addition to the aphids. Some people will still opt to spray, but you can spend at least forty dollars an acre to spray a crop of soybeans and the aphids still might return for another couple of generations. I saw my first-ever soybean aphid infestation in 2011 and chose not to spray—and I stand by my decision. The soybeans survived quite nicely despite the aphids.

Fruiting and Maturation of Soybeans

Soybeans will continue their vegetative growth after the onset of flowering and pod formation. By stage V-7, plants are fifteen to seventeen inches tall. At this point, stage R-1 (early bloom) begins with the appearance of tiny little flowers between the third and the sixth node of the plant. Flowering proceeds upward and downward from here. If the soybean plants are extra bushy and have axillary branches, flowering will begin here a few days after the main stem has begun to set blossoms. Vertical root growth also increases rapidly at this point in the soybean plant's development. Vegetative growth will continue through the V-10 or V-12 stage, which means that the height of an average soybean plant could more than double once flowering begins.

The Bloom Stage

The soybean plant is at full bloom in the R-2 stage. There will be an open flower at each of the two uppermost nodes of the main stem where a fully developed leaf appears. At this point, the soybean plant is a powerhouse of activity. The plant is seventeen to twenty-two inches tall and has attained about half of its total height. Lots of dry matter accumulation is still happening in the vegetative tissues, but the plant is beginning to shift its metabolism over to the production of pods and seeds. Nitrogen fixation is happening at a frantic pace to feed the rapidly growing plant. Roots are growing deeper; the lateral roots have pretty much made their way entirely across the thirty-inch rows. This is an amazing time in the growth and development of the soybean plant—leaves and stems are lengthening and growing wildly at the same time flowers are thinking about producing seedpods. This is indeed a miracle plant.

The Growth Stages

July and August are the big months for soybean growth and development. Those little bean plants that were so short and small back in early June will eventually grow to thigh, waist, or even chest height by the end of August. There's something magical about the whole process. Soybeans that are healthy and happy, growing on soils with adequate and balanced fertility, will have the most amazing dark green color. The richness of this green hue and the plant's incredibly rapid growth habit come compliments of those little round root nodules that remove nitrogen from the atmosphere and synthesize it for the plant. The soybean is a little nitrogen factory growing there in our fields.

Growth stages R-3 (beginning pod) and R-4 (full pod) both take place in August and early September, when the little purple flowers that first appeared in the crotch between the leaf stems and main stems begin to morph into tiny quarter-inch pods. The progress of pod development follows the sequence of flowering. The first pods you will see will be toward the bottom of the main stem where the first flowers appeared. This first evidence of future fruit in early August is a good sign, and means these rapidly growing plants still have a way to go before final maturity, but will be producing soybeans soon. Once these little pods grow to three-quarters of an inch long, the plant is considered to be in the full pod or R-4 stage. Pod fill commences right after this. Now is when you hope for the very best growing conditions, because this is the most stressful time in the life of a soybean plant. Everything is happening at once. The very top of the plant is still vegetative, producing flowers and baby pods. Lower down on the stem, pods are growing and beginning to fill. Dry matter accumulation has reached a crescendo as nutrients are being translocated from other parts of the plant to the developing fruit. This is the time when a waist-high soybean plant that is loaded with developing pods on the bottom might just grow another six inches to a foot in height and set three or four more clusters of pods. The potential for extra yield is tremendous, but a mid-August dry spell could really throw a wrench in the works. A soybean plant will start to abort pods if it runs out of water or nutrients. The best you can do is stand back and marvel at what is going on out there in your fields. These plants that were only four inches tall two months ago are now close to four feet tall. They are dark green in color and loaded with developing pods of soybeans from top to bottom. Hopefully, you have enough inherent fertility and humus in your soil to deal with some weather-related stresses. Pray for some rain, but not too much, and keep your eyes on the crop.

The Reproductive Phase

Seed formation begins in late August and continues well into September. By this time, soybean plants have attained their maximum height and are no longer putting on any vegetative growth. Agronomists classify this period in the soybean's reproductive phase as R-5 (beginning seed) and R-6 (full seed). All the plant's resources are now being directed at finishing the fruit. Nitrogen fixation begins to decline and root development ceases, and little green soybeans first begin developing inside the oldest pods at the bottom of the plant. Throughout the month of September, soybean pods will enlarge and ripen sequentially from the bottom to the top of the plant. As this process finishes up in the latter half of the month, hope and pray that a killing frost will hold off until all of the pods are filled with plump green soybeans.

Maturation

Beginning and full maturity are classified as the R-7 and R-8 stages of soybean development. A quick glance at your soybean field toward the end of September will tell you that maturity is right around the corner. First, you will notice that the soybean plants have lost their deep green color. Plant leaves will first fade into a lighter green shade as nutrients are carried away to the developing pods. All of a sudden, you will look over at your field and notice that the leaves are beginning to turn yellow. At first, the yellowing will be splotchy and limited to a few bottom leaves. You may even notice that some of these bottom trifoliates have begun to fall from the plant. Once this process begins, it will only be a matter of a few more days before the entire plant turns yellow. Things are winding down and finishing up at this point. Physiological maturity occurs when the seeds in the pods have accumulated 100 percent of their dry matter. Bean pods have now turned from green to yellow or brown, and in a matter of days all of the remaining yellow leaves on the soybean plants fall to the ground. Overnight, the soybean field will have the appearance of row upon row of stark soldier-like plant stems standing in formation loaded from top to bottom with little clusters of brown bean pods. The green beans inside these pods have turned to a beige creamy brown color, which is typical of the soybean as we know it. Once the plants have reached this stage of full maturity, the beans contain all the necessary components to reproduce as a seed, but they are still rather swollen with a moisture content of 60 percent. It will take at least another five to ten days of field-drying before these recently mature soybeans will lose enough moisture to drop below the 15 percent level for a safe harvest.

Soybean Harvest

Waiting for soybeans to dry down and ripen in late September and early October is a time when I worry a lot, as extended periods of sunny, warm, dry weather are rare at this time of the year. Here in the mountains of Vermont's Northeast Kingdom, we are apt to get an early snowfall anytime after the first of October, and while rain on drying soybean plants is acceptable, six inches of heavy, wet snow could be a disaster. The agronomy textbooks that are written out in the Midwest might tell you mature soybeans will dry down to suitable harvest moisture in five to ten days, but the situation is a bit different in this part of the country. When I watch the clouds build up on the mountains to our west in early October, I am well aware that anything more than two sunny days in a row at this time of the year is a rare gift. Autumn is fast approaching, and the beautiful dry weather that we had for harvesting cereals back in mid-August is long gone. At this point, in retrospect, it seems like growing and maturing the crop of soybeans was the easy part; it becomes imperative to get them dry and out of the field. I don't mean to paint too dismal of a picture here—soybean-harvesting weather always seems to arrive sooner or later. It has been my experience that we always seem to get that nice spell of warm, dry weather sometime between the tenth and the twentieth of October. And while I'm not recommending that you be obsessive-compulsive all the time, you might want to be when it comes to soybean harvest. Make sure that all your harvesting machinery and other infrastructure are all set and ready to go; when the weather won't let you into the field, spend your time preparing the combine, the grain dryer, and the storage bin. Those three or four dry days in a row will arrive eventually and when they do, you need to be ready and raring to go. You really want to get those beans out of the field before the weather gets any worse. Harvesting grain corn at this time of year is nowhere near as stressful as soybeans—corn carries its fruit three to four feet off the ground where the ears can continue to dry and escape the vagaries of the weather. Soybeans, on the other hand, carry their fruit only inches from the soil's surface.

Home-Scale Hand-Harvesting

The whole process of harvesting soybeans is nowhere near as worrisome if you are working with a smaller garden plot. Hand-harvesting soybean plants gives you all sorts of options not available to larger growers who need to cover acres and acres with large machinery. Also, soybeans don't have to be bone-dry to be hand-pulled. This mode of harvesting may be time consuming, but it is very thorough and complete, and if you are careful you won't be leaving any beans in the field. Most varieties of soybeans bear their bottom pods only inches from the soil surface, so combine cutter bars have difficulty cutting the plant off below the bottom pods and many times will leave this lowest-hanging fruit in the field. If you are carefully pulling soybean plants out by the roots, not only will you be able to harvest every last bean pod, but high-moisture harvest is also not a problem. Lay your harvested soybean stalks out in an airy dry place like an old barn or shed to finish drying down to the proper level for threshing. Once everything dries out, the soybeans can be removed from the pods by running the whole plants through a small portable thresher or by laying everything out between two tarps on a threshing floor. Get a group of people to dance on top of the upper tarp and the threshing will be easily accomplished; or you could hit the harvested bean plants with a flail. It's probably best to keep the beans between the two tarps so they don't ricochet all over the place from the impact of the flail. Once threshed, the mixture of soybeans, pods, and other plant material can be poured in front of a fan or in a breezy doorway to accomplish separation and final cleaning. You will be surprised how many soybeans can be harvested from a quarter acre without the benefit of large-scale expensive equipment.

Harvesting with a Combine

For those of us who will be harvesting our crop of soybeans with a combine, it's important to remember that soybean harvest is totally different from a cereal crop like wheat or barley. Older pull-type combines and some of the earlier self-propelled machines were never designed to harvest soybeans in their day because the crop was a relatively minor one in the 1950s and early '60s. In 1990, I remember struggling to harvest my first crop of KG 20 soybeans with the 1958 Oliver 25 self-propelled I owned at the time. I couldn't position the cutter bar low enough to harvest the lowest pods on the stalks, and it was frustrating to see all of the unharvested bean pods still hanging on the stubble that remained in the field. The grain head on this older machine was designed to cut cereals at a height of eight inches to a foot off the ground. When I lowered the head all the way down to ground level, the guards ended up pointing right into the earth. Let's just say it was a neat trick trying to operate this old grain head low enough to cut the soybean plants without running the mechanism right into the dirt. In 1992, I stepped up to a 1970s-vintage John Deere 3300 combine. Since it had been primarily a barley machine, my new piece of equipment was outfitted with a "Quick-tach 100" grain platform that really didn't do any better job harvesting soybeans than the old Oliver had done two years before. I finally purchased a thirteen-foot John Deere 213 "flex" head for my combine in 1994, which was a dual-purpose platform that could be used for both cereals and soybeans. The removal of five special bolts in back of the cutter bar turned the head into a flexible platform that could easily follow the contours of the field surface. The 213 head was also much longer front-to-back than my old grain head, and the increased length in this dimension allowed for fore-and-aft positioning parallel to the ground. When I set the head down on the ground, the guards remained level and no longer pointed downward into the soil. I likened the thing to a Norelco shaver with the floating head that cut your whiskers, but not your skin. The performance of the 213 flex head in my 1994 soybean crop was flawless; it did everything it was supposed to do, and I had somewhere around fifteen or twenty acres of soybeans that year. I finished the harvesting job in two short October afternoons. Once again, I had to ask myself the question, *Why did I spend all those years struggling to make older equipment work on a shoestring budget?* Flex heads are standard equipment these days. As a matter of fact, you'd be hard-pressed to find a grain-only platform. Smaller flex platforms in excellent condition like the John Deere 213 are getting more difficult to find because they were made in the 1970s and '80s. They do exist, however. You might have to look far and wide, but eventually you will find one. A good used smaller flex head for any model of combine can cost between three and four thousand dollars. Unfortunately, it's one of those scenarios where you have to pay to play. If you want to do a reasonably good job harvesting soybeans, the first thing you need is the right cutting platform.

When to Harvest?

Once you've got the right combine and flex head for the job, it's time to take a closer look at field conditions. The ideal situation would be a nice field of soybeans totally free of weeds, but this is certainly not the norm for most of us. Weeds are a fact of life if you're not using herbicides, and they can be a bit of a problem because they slow down the drying process of the soybean crop. Large annual weeds like redroot pigweed and lamb's-quarters remain green and wet well into the fall harvest season, and when these plants are threshed along with the soybeans, their juices can stain the lighter-colored soybeans and make them unsuitable for the human consumption market. If you have heavy weed pressure in your soybean crop, you might just have to wait for a few heavy frosts to kill and remove some moisture from these large juicy plants—it's amazing what a night in the low twenties can do for a large lamb's-quarters plant.

Several extra-cold evenings followed by warm sunny days is all it takes to prepare a field of weedy soybeans for harvest, but you will want to check the moisture level of your soybean crop. Try to gather a representative sample of soybeans from different locations in the field and on the plants themselves, because soybeans ripen unevenly. The pods at the bottom of the plant generally ripen first, while the last beans to develop at the very top of the plant ripen later on. Ultimately, you will have to decide when to put the combine in the field. Soybeans can be combined at 20 percent moisture or even a little higher, but this is not recommended, as damp pods and moist beans don't thresh well. Higher-moisture soybeans are much softer in texture and can be easily damaged in the threshing process. Fifteen percent seems to be the magic number for harvest moisture: Once the moisture of standing soybeans drops to 13 percent or lower, field losses from shattering are an increased possibility. Beans also tend to split and crack if they are harvested in too dry a condition. Splits will work for livestock feed, but not for seed or for the human consumption market. There is no set prescription for when or how to harvest a crop of soybeans, but once you're into the month of October, the realm of weather predictability begins to decline. The best advice I can give is to carefully monitor field moisture and weather conditions, and to harvest your soybeans as early as possible. If a ripe crop remains in the field for an extended period of time and experiences continual wetting and drying, the bean pods on the plants will be much more prone to shattering and splitting open from the impact of the combine reel.

Preparing the Combine

The combine needs to be set and operated differently for soybeans than it would for cereals like wheat, oats, and barley. When adjusting and preparing your machine for soybean harvest, start at the front and work your way to the back. The first thing that deserves attention is the cutter bar. Since conditions are damper in October than August, a sharp cutting mechanism is more important than ever to efficiently mow down soybean plants. Replace worn knives with brand-new serrated-edge ones. Newer-style section knives are much easier to change because they are bolted instead of riveted to the cutter bar. Guards should also be straightened or replaced if they are bent or damaged. To ensure good cutting action, hammer down the hold-down clips so the cutter bar makes good solid contact with guard surfaces. Check the skid plates under the flex platform for wear. This underside of the head rides right down there on the soil surface and must be in good condition to operate properly. All of the aftermarket farm equipment accessory catalogs offer polyethylene skid plate protectors that can be easily bolted to the bottom of any flex head; these poly skid plates will reduce friction and improve harvest performance in soybeans. Last but not least, check over the combine reel for fatigued metal parts and broken teeth. The reel is what guides the crop into the machine. It must be in topnotch working order to deliver soybean plants to the combine feeder house. It seems as though the reel on my combine always waits until soybean harvest to self-destruct and slow down the show. Each wooden bat is connected to a round steel rod that rides in a corresponding metal bushing, and metal on metal wears thin after a while. If you are harvesting weedy soybeans, vines and stems will wrap around these metal parts and put lots of extra stress on the components. I have an older "parts head" from which I can rob pieces to repair my machine when these breakdowns occur. Some people don't like clutter, but this is the best reason I can think of for never throwing anything away—every farm should have a "bone yard" to house its collection of old iron that might just come in handy some day. A properly working and maintained flex platform is a must if you want to do the best job possible harvesting your soybeans.

Suggestions for Combine Adjustment

A nice crop of soybeans is a joy to combine. If everything is operating properly, you will be able to hear the beans rattle as they are threshed by the cylinder right below your feet. A high-yielding harvest can be so loud that you can hardly hear yourself think as you drive your combine through the field. Consult the operator's manual for your machine as a starting point for field adjustment. The general rule for combining soybeans is to be as gentle as possible—threshing needn't be anywhere near as aggressive as it would be for a cereal crop. Whereas cylinder speeds in wheat can be as high as twelve hundred RPM, soybeans are easily removed from pods at speeds of five hundred RPM and less. Plant moisture and field conditions will have the final say in how fast to run the threshing cylinder. If for some reason you need to harvest your soybeans at moisture levels in the high teens, you might have to increase the cylinder speed. Check the grain tank; if there are a lot of unthreshed pods, you will either have to increase cylinder speed or wait a few more days for things to dry out a little more. The same holds true for the concave setting. Drop the concave down to five or six instead of the one or two setting that you would use on a crop like wheat. There should be ample room between the cylinder and concave for the crop to pass through.

Remember that soybeans are fragile and that the germ is right there behind the hilum. When harvesting this crop, the general combining rule of using no more threshing force than is absolutely necessary applies more than ever. A few soybeans are bound to crack and split during combining, but careful attention to detail will help you harvest the crop in the best possible condition. This in turn will allow you to command a premium price in the marketplace, as processors of soy foods will reject beans that have been mechanically damaged during harvest. The shoe of the combine works pretty much the same for soybeans as it does for any other grain. The chaffer and sieve screens might be opened up just a little more than they would for wheat or barley, and the cleaning fan is usually cranked up to its highest speed because soybeans are quite heavy and need all the air they can get to arrive at the grain tank in clean condition.

These suggestions for combine adjustment are only meant to be general guidelines. A good combine operator is constantly tweaking the machine as conditions change throughout the day. Soybeans will be easiest to harvest if the sun is out, a drying breeze is blowing, and the moisture content of the crop is somewhere between 13 and 15 percent. Although we would all like to order up these weather conditions for our soybean harvest, we would do well to realize that this is the ideal and not necessarily the norm. If our soybeans are at 17 percent moisture and torrential rain is imminent in a day or two, we might just have to combine the crop a bit more aggressively and get the crop out of the field before the fall monsoon season sets in.

Storage and Post-Harvest Treatment of Soybeans

Soybeans are generally combined into trucks or gravity grain wagons just like any other grain crop. However, beans have a bit of an advantage over the cereals when it comes to post-harvest stability. A wagon- or truckload of wheat or barley between 15 and 20 percent moisture will start to heat and spoil within hours, whereas harvested soybeans with elevated moisture levels can sit there for days before heating begins. Depending on how moist your soybeans are, you've got at least a couple of days before you have to take some sort of drying action. We do have the weather on our side at this time of the year in the Northeast, as cool temperatures will help hold off spoilage. I've kept high-moisture soybeans stored in a gravity box for up to a week with no ill effects—but don't push your luck, because I've also had soybeans in the same situation begin to turn fuzzy with mold. Be proactive and get some air onto your crop. Install a couple of screw-in aerators in each wagon, which will give you weeks of slack. Better yet, put your soybeans in a bin with an aeration floor and turn the fan on. Whatever the case, careful monitoring of crop moisture is still essential right after harvest. If you don't have access to a moisture tester, the simple chew test will suffice. If your soybeans are rubbery and chewy, you'd best be making plans to dry them. If the soybeans crack between your teeth, there isn't much to worry about in the short term. It's always a good idea to get an official moisture test so you know where you stand with your harvested crop. After investing so much time, energy, and hard work, it would be a shame to see your soybeans go moldy sitting in a truck or a wagon.

Roasting Your Soybeans

Soybeans that will be used as livestock feed have some special considerations. At 38 to 42 percent protein, they are perfect ration complement to lower-testing grains like corn and barley. The only problem is that beans by themselves are bitter and rather indigestible, and feed specialists recommend feeding no more than two pounds of ground raw soybeans per day per head. There is also the trypsin inhibitor factor to worry about. If soybeans are heated, the enzyme trypsin is liberated, which makes all of the soybeans' amino acids available. Complete protein that can then be utilized by livestock is the result.

The simple solution to improving the bitter flavor and indigestibility of raw soybeans is to roast them. This process is usually accomplished with a propane-fired revolving drum roaster, which can be seen on page 29 in the color insert. Fortunately, there are now a number of itinerant roasters who travel from farm to farm in our region processing people's soybeans. Denis Boucher from Swanton, Vermont, for example, comes to our farm every October to roast our beans for us. He sets his machine up in our farmyard and taps into our big grain dryer propane tanks for his gas, although some of the other roasters out there bring their own propane and electrical generators to power the process. We simply bring the soybeans wagon by wagon to the roaster. The beans travel up an auger and drop into a slightly inclined rotating cylinder that has a huge gas flame shooting down its center. Careful adjustment of the drum angle and the flow of beans through the drum is essential in order not to burn and blacken the soybeans. The ideal final roasting temperature for beans exiting the far end of the drum is somewhere between 290 and 310 degrees. Depending on the amount of soybeans you have to process, the roasted end product can be augered into either a stationary grain dryer or an aeration bin for steeping and cooling. Common practice is to let the soybeans remain hot for an hour before turning on the air to cool them. This so-called steeping process determines the amount of bypass protein in the roasted soybeans. If the beans are cooled too quickly there will be more readily soluble protein—which is less desirable in livestock feed rations—but if the beans remain hot too long, bound protein can result.

Soybean-roasting day at our farm is a big deal, and the whole process can be a bit nerve racking until all parts of the system are working smoothly in unison. There are augers, wagons, trucks, tractors, and grain dryers everywhere that all must be coordinated. The burner in the roaster has the whooshing sound of a small rocket ship. Once the beans are flowing through the system, you can stand back and enjoy the whole process. There is smoke and steam billowing up from the roaster, and the smell of toasting beans in the air is indescribably delicious. You keep reaching down for handfuls of roasted beans to snack on as they exit the roaster. The hot freshly roasted soybeans sure are better than anything you can buy in a little bag at a health food store.

Most farm-scale portable roasters will process around ten tons of soybeans per hour. Prices for this service will vary between twenty-five and forty dollars per ton depending on who supplies the propane. It usually takes the better part of a day to set up, roast our thirty to forty tons of soybeans, and dismantle. It's also amazing how much gas can be consumed when you are firing a propane burner with a rating of several million BTUs. And while on-farm soybean roasting might seem like a lot of work, it is well worth it. The soybeans that you have produced for feeding your livestock aren't really worth anything until they have been roasted, as the bitter flavor of raw soybeans is transformed into a nutty taste that is relished by all classes of livestock, and soybean protein becomes complete for a highly palatable feed additive to use on your farm. To top everything off, the soybean harvest remains home for processing because the roaster comes to the farm.

The Shrink Factor

The concept of "shrink" is something you discover when you start roasting soybeans. For the sake of an example, let's say that the soybean roaster comes to your farm and puts thirty tons of your beans through his equipment. The hot beans are augered into a drying silo and aerated with a fan for several days to cool them down to the ambient outdoor temperature. But you might be surprised several weeks later when you go to sell your crop that the roasted beans now might weigh twenty-seven or twenty-eight tons. Thanks to the heating and cooling process, your crop can lose as much as 10 percent of its weight and mass. The shrink factor is the primary reason why experienced producers of commodity roasted soybeans don't overroast or oversteep their harvests. However, there is an unseen advantage in shrinking soybeans. The primary substance that is driven off in the roasting and cooling process is water. Soybeans can lose as much as 4 or 5 points of moisture during roasting, which can work to our advantage if we've got a quantity of high-moisture beans that need drying. As a matter of fact, many growers will combine their soybeans at 17 percent moisture in anticipation of a 4-point drop in moisture once the crop has been through the roaster. This sort of situation can be a blessing or a curse depending on whether or not the soybean roaster can arrive at your farm in a timely fashion. You've got a week or a little more to let those 17 percent moisture soybeans sit there in wagons before they begin to spoil. If it is at all possible, try to coordinate the arrival of your soybean roaster before you put the combine in the field. If he is two weeks from arriving at your farm, you might want to delay your harvest or make some other arrangements for drying your crop. Personally, I like to finish drying my soybeans with the roasting process. I end up with 13 percent instead of 9 percent moisture roasted soybeans.

Considerations for Food-Grade Soybeans

A whole lot more care must be taken if your harvested soybeans are headed for the food-grade human consumption market. Hopefully, you've chosen a larger bean variety with a clear or yellow hilum, and you've waited for the weeds to get frosted and dry up before combining to prevent any sort of staining. Once you are in the field, the combine must be operated at the least aggressive setting possible to avoid damaging the beans in any way. Theoretically, you can reduce the speed of the threshing cylinder to between three and four hundred RPM. The concave can also be lowered even more to the number eight or nine setting. Check your combine's harvesting performance. You can keep backing off the settings until you begin to see unthreshed pods. Then tighten things back up just a whisker. You will also do a better job if the soybeans are closer to the 13 percent level. If you've got the time, clean and process your soybean crop before putting it into storage. You will do less damage to the soybeans at this time of the year than later on in the winter when temperatures begin to plummet, as soybeans can crack and break if they are processed under frigid conditions. A simple fanning mill or air screen cleaner should be all that you need to clean your soybeans. If you have access to a gravity table, you will be able to make your soybeans look even nicer. Small stones that are the same size as the soybeans are probably the most important type of foreign material that must be removed from this food-grade crop. These little rocks are hard on soy-processing equipment and even harder on the teeth of the people who will eventually consume the soybeans as a processed product. A gravity table will usually deposit these little rocks up against the far right-hand wall of the vibrating deck. Forsbergs Inc. makes another piece of specialized equipment called a destoner that is constructed and operates in a similar fashion to its gravity tables. Destoners can be found in parts of the country where dry beans are grown and processed.

Growing food-grade soybeans is a labor of love that isn't for everyone. Because the demand for organic grain for livestock rations has increased over the last few years, the price differential between food-grade and feed-grade soybeans has declined to less than one hundred dollars per ton in many cases. If this is the case, you have to ask yourself if all the extra care, infrastructure, and effort it takes to produce soybeans for human consumption are really worth it. The good news is that more and more small local enterprises that produce soy milk, tofu, tempeh, and other products are popping up everywhere in our region, and these outfits would rather buy soybeans from local farmers than from Canada or the Midwest. This new development in the local food movement bodes well for all of us who want to feed our neighbors with our farm-produced grains, and most of these small local soybean processors will pay more than the organic commodity marketplace for high-quality local beans. This also works extremely well for those of us who are just getting our feet wet with food-grade soybeans, because these small enterprises would rather buy soybeans by the ton or half ton instead of the tractor-trailer-load. Small soybean growers delivering limited amounts of beans directly to a microprocessor half an hour or less from their farm is a win–win situation for everyone involved. If the local food movement continues to thrive, local northeastern-grown soybeans will continue to have an ever-more-important place in the scheme of things.

Organic Soybeans in an Ever-Increasing GMO Environment

I'm sure that the advent of genetically modified crops in the agricultural sector isn't news to any of us. But let's just have a bit of a recap. Roundup Ready soybeans were the first commercially available transgenic crop to hit the marketplace in 1996, where the Monsanto Company released a number of soybean varieties with inserted genes that gave these plants resistance to the killing effect of its broad-spectrum herbicide Roundup. Formerly, Roundup had been sprayed with the intention of killing everything that grew in that particular spot. The technology seemed so simple and benign at the time, and I didn't think much about it because I would never use herbicides anyway. And while Roundup had been touted as a miracle herbicide that broke down rapidly in the environment and had no negative side effects, most of us in the organic world didn't give it much thought until several years later when we began to see transgenic crops begin to totally dominate the marketplace for agricultural seeds. In a few short years, three-quarters of the soybeans grown in the United States were Roundup Ready; by 2011, transgenic dominance of the soybean crop in our country had climbed to 93 percent. Roundup Ready corn was first introduced in 1998 and now makes up almost 90 percent of the North American corn crop. Genetically modified crops are here to stay these days—and there is lots of cause for concern in the organic farming sector.

How do we keep from getting run over by this GMO freight train that is barreling down upon us? These plant varieties that are stacked with all kinds of inserted traits are perfectly legal and all-pervasive out there in the environment. Corn is a lot more worrisome than soybeans because it is cross-pollinated by airborne pollen that might just come from a neighboring field of GM corn. Fortunately, there is little if any risk of airborne pollen contamination in soybeans, because they are self-pollinators. This is one of the few bright spots out there in this new world of transgenic farming that has taken conventional agriculture by storm. But there is still plenty of cause for concern, and we can't rest on our laurels or let our defenses down, as GMO soybeans are everywhere. They are in any product on the supermarket shelf that has soybeans in its list of ingredients as well as in all conventional livestock grain and the meat and milk from the animals that consume this grain.

As organic producers, we don't want anything to do with the transgenes of conventional agriculture, and we must be aware of the potential for contamination and be constantly on guard to prevent it. The onus is on us, not them, to develop a farming strategy that steers us clear of GMOs. We have to be continually mindful that the origin for most of the contamination that occurs in organic soybeans comes from the inadvertent mixing of conventional and organic crops, and we need to be acutely aware of every opportunity for this sort of cross-contamination. Some of it we can see right before our eyes—like when the traveling roaster brings his machine to our farm after he has just roasted conventional GMO beans the day before at another farm. It's up to us to open the cleanout trapdoor on the bottom at the delivery end of the roaster and remove all of the regular soybeans that remain. Some contamination from GMOs, however, is less obvious and hidden from our view. We need to buy clean seed from reputable sources. Most of the big conventional seed companies offer GMO soybeans in their catalogs. These larger companies generally have some excellent non-GMO variety offerings and are well worth dealing with. Make sure the seed company has an IP (identity preserved) program for its non-GMO seed. This means that they have programs in place to segregate GMO and non-GMO seeds in order to guarantee the varietal purity of what they are selling you. If you have the opportunity to purchase a good variety from an all-organic seed company, by all means do it. But the very best thing you can do to guarantee the purity of the seed you plant is to save your own seed.

The last and probably the most common source of soybean contamination happens when organic crops are transported by commercial grain haulers. If you are sending your soybeans away to any sort of buyer, ask the transport driver to show you the clean truck affidavit that is required by organic certifiers. Before you load any of your soybeans in someone else's truck, you might want to climb inside the truck body and have a look for yourself to see if everything is clean. We've come a long way learning how to grow, harvest, and clean this amazing crop that has the potential to produce so much protein per acre in a few short months, but we need to do everything in our power to keep our crops pure.

CHAPTER FIFTEEN
Dry Beans

Dry edible beans (*Phaseolus vulgaris*) originated in Central and South America. This leguminous plant worked its way northward over time and was being grown here along with corn and squash by Native Americans when the first colonists from Europe arrived. But dry beans never became a staple for the early settlers like Indian corn did. Green beans were grown as a fresh vegetable in kitchen gardens of the time; the only dry beans grown were the string beans that were left to ripen for next year's seed. In fact, there is no record of any commercial production of dry beans before the 1830s. Story has it in Liberty Hyde Bailey's 1907 *Cyclopedia of Agriculture* that Stephen Coe of Yates, New York, brought a single pint of beans from eastern New York back to his farm in the western part of the state in 1836, and planted and raised them for three successive growing seasons. His seed increases allowed his son, Tunis H. Coe, to harvest a crop of thirty-three bushels in 1839, and the crop was sold to H. V. Prentiss of Albion, New York, who was the only person around who could be persuaded to buy such a quantity. This transaction is thought to be the very humble beginning of the dry bean industry in western New York—and the whole country, for that matter.

Dry Bean Production in the Late 1800s

Crop acreage really began to climb after 1846 when the wheat weevil ravaged Genesee Valley grain crops for several years in a row, and beans proved to be a viable alternative to wheat. After this, the fledgling New York State bean industry remained stable, but did not experience any increases until 1861 when the US government began buying beans to feed the soldiers in the Union army. Dry bean production increased rapidly during the years of the Civil War. When the war was over, government demand for beans ceased, but the troops returned home with a newfound taste for beans. The late nineteenth century was the heyday for dry bean production all over the country. Acreage increased rapidly all the way westward into what was to become the Corn Belt, and Michigan soon outpaced New York as the country's leading producer of dry beans. The traditional ham and baked bean suppers that were so famous in New England originated at this time. According to the Twelfth Census of Agriculture compiled in 1899, Maine had 10,252 acres planted in dry beans that year. Vermont had 2,404 acres and New Hampshire had 2,982 acres, but New York led the eastern United States with 129,298 acres planted to dry beans.

Stories abound about the legacy of the dry bean industry in our part of the country. In a recent interview, eighty-eight-year-old Winston Way—who retired long ago as UVM's agronomist and crop specialist—described the two railroad bean depots that once existed in Grand Isle on the Champlain Islands. These were little wooden buildings with upstairs bean storage bins that were situated right next to the railroad track for shipping local beans to distant markets. Groups of local women were hired to sit in a room underneath and handpick beans for final cleaning; dry beans dropped by gravity from the overhead storage bins onto specially designed conveyors called bean belts. I've also heard stories of many farm families who raised an acre or two of dry beans to sell to their village neighbors who didn't have the wherewithal to produce food for themselves.

Maine was and still is the leading New England state in dry bean acreage. Production peaked in the 1960s when B and M Baked Beans of Portland was canning beans for national distribution and Kennebec Bean of North Vassalboro was packaging many different varieties of beans in its little one- and two-pound State of Maine Brand plastic bags. Both of these companies were institutions in their day, but are now out of business. Dry bean production in western New York, unfortunately, has given way to cash grain corn and soybean farming and is now a shadow of its former self. But despite all the doom and gloom and decline in the industry, the vestige of dry beans lives on here in our region, and there are still individuals who grow this crop all over the Northeast. Acreage is on the increase again as the local food movement has begun to demand dry beans from local sources.

The Benefits of Growing Dry Beans

Dry beans have been our most naturally adapted protein crop here in the Northeast for well over a century. The soybean may be the latest addition to our lineup of crop choices, but it is the dry edible bean that has stayed the course over time, and while soybeans may have the potential to produce tremendous amounts of protein per acre, their growing season can be rather long and fertility requirements out of reach for some of us who are trying to farm on marginal soils in cooler areas. For example, the very shortest-season Group 000 soybean cultivars take at least 105 to 110 days to ripen and mature, but a typical dry bean will ripen in as little as 80 days. When it gets to be the third week of September and you are hoping and praying for one more week of growing weather without frost to finish maturing a soybean crop, dry beans will already be ripe and ready for harvest. This is agricultural security at its best.

Most of us will be growing dry beans to feed ourselves and our neighbors. This humble crop is so much fun to grow because there is such a diversity of bean varieties, each with its own particular characteristics and unique flavor. The most common northeastern varieties are soldier beans, yellow eyes, and the Jacob's cattle bean, and have probably been the most popular varieties used for making traditional baked beans. Dark red kidneys are more common in western New York with its longer growing season, whereas light red kidneys are more suited to northern New England's shorter growing season. Soybean lovers might disagree, but I think dry beans are much more interesting and pleasing to the palate. They are certainly easier to cook and can be used in a variety of different ways from baked beans to chili, soups, salads, and Mexican dishes. Dry bean plants are much smaller and more compact than soybeans, and they lend themselves well to hand-planting, cultivation, and harvest in a garden situation. When we grow dry beans, we are feeding ourselves first and foremost—this isn't a crop that easily lends itself to large-scale commercialization, especially with organic practices in the absence of herbicides. Dry beans have been such a popular crop over the years because they have allowed all sorts of people to raise considerable amounts of homegrown high-quality, complete protein.

Determinate and Indeterminate

Dry edible beans will grow in a wide variety of soil types, but they like a well-drained loam the best. Good drainage is the secret to doing well with this crop. Prolonged moisture and saturated soils at any time during the growing season will stunt a field of dry beans. There are actually many similarities between dry beans and soybeans—they are both legumes that can fix atmospheric nitrogen by means of root nodules. Fertilization takes place within the flower itself, which makes the dry bean a self-pollinator just like its distant cousin the soybean. Dry bean flowers form at the junction of the axil and main stem, but are much larger and prettier than soybean flowers. The agronomic term used describe a dry bean flower is *papilionaceous*, which translates to "butterfly-like." (*Papillon* is the French word for "butterfly.") Bean flowers are usually purple and white in color and have two distinct rounded petals that look like little wings. The leaves of a dry bean plant are trifoliate because they are expressed in groups of three.

Soybeans and dry beans differ considerably, however, when it comes to plant stature and the number of days required to produce mature fruit. Dry beans seldom grow above knee height. They are classified as either determinate (bushy) or indeterminate (vining or trailing). A determinate bean plant grows to a certain point, after which it flowers and then sets its pods all at once. A bush bean grows outward into the row just as much as it does upward, and most of the common New England large-seeded bean varieties like soldiers and yellow eyes fit into this determinate or bush category. The line between vegetative and reproductive growth is much less distinct with indeterminate dry beans. Much like soybeans, this type of plant will keep producing new flowers for several weeks after the beginning of pod formation and will need an extra week or two to ripen. Indeterminate beans are usually much more vine-like and are quite a bit taller than their determinate brethren. Many varieties of black turtle beans fit into this indeterminate group. There is also a third group of dry beans that bridge the gap between determinate and indeterminate. Semi-indeterminate plants are somewhat bushy and short, but they will send out short vines with a few more flowers and extra pods.

A good crop of black beans will grow to thigh height and will be rather viny in nature. Much like the soybean, a black bean can have a ripening pod at the bottom of the stalk and a new flower at the top simultaneously. Dry beans are famous for their incredible diversity. There are thousands of different varieties and subvarieties that have developed over time in a wide array of geographic regions and climates, and locally adapted cultivars of a particular type of dry bean are quite common because individuals select seeds from plants they like. In a matter of a few generations, a new and slightly different dry bean can develop. The greatest thing about growing dry beans is the personal and local nature of the crop. The big corporate seed giants are not involved in breeding and producing dry bean seed because the potential for profits is just not there. We, the people, have control of this crop. It feels so good to be able to grow this much high-quality vegetable protein on a sustainable scale here in our own region.

Fertility Considerations

Hopefully, you took a soil test the previous fall and applied adequate mineral fertility to make up for any deficits. Balance the macro elements like calcium, magnesium, phosphorous, and potassium with the addition of rock powders and organic fertilizers. Substances like gypsum, limestone, rock phosphate, and sul-po-mag might be needed to correct imbalances and supply nutrients. Micronutrients are equally important. Sulfur, boron, zinc, copper, and manganese should be supplied in minute amounts, which can be accomplished by blending them with the sul-po-mag or gypsum. Good, well-made compost is the icing on the cake, as it will feed and encourage soil microbes in addition to providing a buffering action for the added mineral salts. Just because beans are a legume and can fix nitrogen from the air doesn't give us license to shortchange their fertility needs. Dry beans will certainly grow without any inputs, and you will most likely harvest some sort of crop. However, if you want to grow quality dry beans that yield well and are resistant to fungal diseases, pay attention to your soil's needs. Photosynthesis and nitrogen fixation are much more efficient in the presence of the right minerals and trace elements, and microbes and other soil life depend on these substances as well for proper function. A soil that is properly balanced with micro- and macronutrients will have an amazing digestive capacity. This translates into dry bean plants with top-notch photosynthetic function. The end result is superb plant health and production, and your final harvest of beans will be an incredibly nutrient-dense food that will nurture you and everyone else who consumes them.

Planting Dry Beans

Dry beans are a warm-season crop. Soil temperatures should be at least sixty degrees and stable before seeds are sown, which usually translates into late-May or early-June planting times. If cold and damp weather persists into the month of June, dry beans can be successfully planted until the end of the month, which is to say there is plenty of time to prepare a perfect seedbed with adequate fertility. For the most part, you will be planting your crop on a very well-drained and fertile piece of ground that could have been planted to wheat or some other cereal in early or mid-April. I have found that a six-week cover crop of oats before dry beans is a recipe for success with the crop.

Work the ground up and apply compost and other necessary amendments as early as possible in the growing season, and plant about 150 pounds of oats to the acre with a grain drill. This is a much better strategy than leaving the ground bare and watching the weeds grow through April and May. In many cases, dry beans will follow the previous year's crop of corn, and if the corn was harvested for grain, there will be a considerable amount of stover left in the field. The best-case scenario in this situation would be to have incorporated the stalks late the previous fall, and an early-April discing and soil stirring will re-add the necessary air into the soil to promote further decomposition of last year's corn stover. Early April can be an extremely busy time if you are racing around trying to plant oats, wheat, and barley in a timely manner, but whatever time investment can be spared and devoted to readying a field for dry bean planting two months later will be paid back many times over later in the season. Cover crops and preseason tillage are highly recommended.

A nice warm late-spring rain after planting followed by sunny warm weather in the seventies will promote rapid germination; during these sorts of conditions, I've seen dry beans pop through the soil in four days or less. When a crop of beans literally blows out of the ground like this, it establishes itself quickly and gets the jump on the weeds. Of course, this is the ideal situation, and things seldom work this way. I speak from experience on this subject. When I first started planting dry beans fifteen years ago, I didn't worry too much about timing. I simply planted them and they germinated without much trouble. My beginner's luck gave me the confidence to plant more and more dry beans as each year passed. In 2007, I rented a very nice flat eleven-acre field with beautiful sandy loam soil. Up until this point five acres had been the largest amount of dry beans I had planted in any one year, but bolstered by my past successes and encouraged by growing consumer demand, I took the plunge and decided to plant all eleven of these newly rented acres to dry beans. Right around the first of June, the weather came off hot and dry for a few days. We finished fitting the new ground with the field cultivator on a Saturday, and I spent all day Sunday planting six different varieties of dry beans. The weather was extremely hot and the dust was flying. In my mind, planting conditions couldn't have been any better. By evening when I had finished, the heat and humidity of the day had worked itself up into a severe thunderstorm. An inch of new rain on very warm sandy soil seemed like a guarantee for rapid germination of the just-planted beans, but by Monday, a new cold rainy weather system had arrived and settled in for the better part of a week. I started looking for sprouted beans on Wednesday and was surprised to find no germination as of yet. By Friday, there were still no beans up. Now I began to worry. The following week, seven or eight days after planting, a few plants had emerged. Something was indeed wrong here. UVM agronomist Heather Darby came to look at the field a few days later, and dug up some seeds with a shovel as part of her investigation. Many seeds had sprouted, but the young bean shoots had turned into "crooknecks"—they had not been able to break through the crust and had turned 180 degrees and grown back downward—while other seeds had rotted right in place. This unfortunate incident was a major disappointment and a valuable lesson to me at the same time. In hindsight, I should have ripped the field up with a cultivator and replanted. With germination at about 50 percent, I prepared myself for half a crop. All of this is to say that a prolonged spell of cold, wet weather is the worst thing that can happen to a field of recently seeded dry beans.

This and one other experience has taught me that planting dry beans by the calendar is foolhardy, and we must wait until the soil is above sixty degrees and the weather is stable. I've come to believe that the tenth of June might be a more reliable planting date than May twenty-fifth or June first. When you consider that dry beans can go from seed to mature fruit in less than ninety days, even a planting date of June fifteenth will work. Dry beans planted in the middle of June should still be getting close to ripe by the middle of September. Several years ago, Lynn Hazen, a bean grower from North Hero in the Champlain Islands of Vermont, planted his yellow eyes in early July, and Lynn harvested a respectable crop of beans in early October of that year. If you happen to be a hay producer as well as a bean grower, putting off your bean planting until the end of the first week of June will give you some time to put up some high-quality early-cut hay before heading to the bean field for planting. Flexibility and a "nose for the weather" paired with a little old-fashioned good luck are necessary ingredients for success when it comes to figuring out planting dates for dry beans. Take your time, consider all the variables, and make an informed decision.

Finding Good Seed Stock

Planting high-quality seed is just as important as planting at the right time. Dry bean seed is expensive and difficult to find, especially if you are looking for some of the lesser-known heirloom varieties. Once you've begun to actually produce dry beans, you can and should save your own seed from your very best plants. If you're just starting out, you'll have to bite the bullet and buy seed. Just about all of the smaller seed companies offer dry beans. These seeds must be considered specialty items because prices can run from five to eight dollars a pound. The sixty pounds of bean seed that it takes to plant one acre can turn into quite a major investment. I've been able to circumvent these high catalog prices by buying dry beans directly from several producers in Maine. Tony Neves's Freedom Bean Company of Albion and Patty Qua of Dexter have both been sources of very high-quality food-grade dry beans that have worked equally well for seed. Western New York is also a good place to source dry bean seed. Mark Callan of Caledonia, New York, is a dairy farmer who specializes in producing dark red kidneys and black turtle beans. Even though Mark grows close to five hundred acres of dry beans on excellent soils in an ideal climate, he buys all of the seed he plants from Idaho. Mark is very concerned about spreading disease by planting seeds that are carrying anthracnose and other fungus infections; apparently, dry bean seed can appear perfectly fine, but still be carrying disease. Seedborne disease quickly translates to soilborne disease. Once a field is infected with bean blight or some other fungal organism, dry beans cannot be planted there again for many years. Idaho is ideal bean seed country because of its hot, arid climate. When it's time to harvest, the irrigation water is turned off and the beans finish maturing in full sunshine uninterrupted by rainfall. The state of Idaho has a very rigorous certified seed program for dry beans. Everything is laboratory-tested for disease before it is released to the market. I have been very happily buying my black turtle bean seed from Mark Callan for the last few years. The larger-seeded New England dry bean varieties are a little more difficult to procure in Idaho. If you are saving your own seed or buying it from another bean producer, you might at least want to have it tested for disease before planting it.

Inoculating Bean Seed

Since dry beans are a legume capable of fixing nitrogen from the atmosphere, the seeds should be inoculated just prior to planting. Bean roots need to be supplied with their own specific bacterium (*Rhizobium phaseoli*) in order to form root nodules. This is the same inoculant that you buy for your garden beans at a farm and garden center; it comes packed in tiny plastic bags that might be enough for a half pound of seed. Source your inoculant well before planting season begins because it is quite difficult to find in larger packages. You can ask your seed supplier to special-order it for you, but sometimes it is unavailable here in the East. I have been special-ordering my dry bean inoculant from Agassiz Seed and Supply in West Fargo, North Dakota, for the past five years. These large packages will inoculate a thousand pounds of seed. These larger packages work well because they provide you with enough of the peat-based material to triple- or quadruple-inoculate your bean seed.

It's best to wait until you are in the field with your planter to inoculate your dry bean seed. I use an old washtub or similar plastic vessel to do the job. Dampen your seed with just enough water to allow the peat-based powder to stick to it, but be careful not to overmoisten the seed. Start with a few drops first—you don't want the seeds to swell from being too wet. Remember that beans are dicotyledons, which means the seed comprises two halves that can break apart if it gets too soft from too much water. Once a seed breaks in half, it won't germinate. Pour some inoculant out of the bag onto the dampened seed and stir it around gently with your hand. When the seed is sufficiently blackened, it's ready to pour into the planter. Once again, be gentle with the seed. If you're not satisfied with the amount of inoculant on the seed once it is in the planter box, add a little more and give it another thorough mixing. Carefully gauge the amount of seed you plan to plant. Once it has been wet down, coated with inoculant, and in the planter box, it will no longer be fit for making baked beans or anything else for human consumption. Ideally, you'd like to finish planting with only a minimum of leftover seed.

Preparing the Soil for Dry Beans

Beans need a well-worked seedbed of friable soil, so it's important to work the field a few extra times if necessary to ensure that it is smooth and that most of the sod or last year's crop residues are sufficiently buried. Large amounts of surface crop residue can interfere with tine weeding after the beans germinate—you don't want to be dragging cornstalks over tender young bean seedlings with your weeder. Standard practice is to plant beans in thirty-inch rows with an ordinary corn planter. A hand-pushed garden seeder will work well in a smaller-acreage garden situation, but grain drills aren't recommended for seeding dry beans because it is difficult to accurately establish the proper population of plants to the acre with this type of equipment. The tendency will be to plant more seeds than are necessary per foot of row; bean seed can also be damaged as it passes through the seed gears of the drill. Even if you simulate the row spacing of a corn planter by blocking the right number of seed tubes across the drill, it is highly unlikely that the seed will be planted deep enough or firmed adequately.

You can't beat a corn planter for putting beans in the ground. If you are using a plate planter, make sure to find some special bean plates. These are usually made of cast iron and have very large cells that let more than one bean drop through at a time. Cast-iron plates work much better than the newer plastic plates for planting beans. I was fortunate enough to buy a whole stack of these antique dry bean seed plates from a retired red kidney producer in New York State. Smaller beans like black turtles and great northerns will readily pass through a standard soybean plate. If you are planting with a finger-pickup planter like a John Deere 7000, you will have to install some special dry bean cups to do the job. Beans should not be planted too thick. Five to six plants per foot of row is just about right. Both the seed plate and bean cup planting systems are designed to drop more than one seed at a time, which can result in uneven seed spacing: You might see two plants close together, and then four or five inches before the next to double-dropped seeds. It took me years to figure out how to remedy this situation, but it's quite simple. Increase your ground speed. Standard planter speed is usually three to four miles per hour, but if you increase tractor speed to six miles per hour, multiple dropped seeds will spread themselves out more evenly in the row. Be sure to calibrate your planter before heading to the field by driving it on a hard-packed surface like a road or a driveway. Count the number of beans per foot of row and adjust the seeding rate sprockets upward or downward to achieve the desired population of six plants per foot. Set the planter depth to somewhere between an inch and a half and two inches deep. Adjust the packer wheels to firm the seed properly. Once you begin planting, stop and check the length of the markers to make sure that each planter pass is thirty inches away from the last pass. Now you're ready to plant away. If you don't have a monitor on your planter to let you know when a row plugs or stops planting, watch the ground behind you as you lift the planter at the end of each pass to see if there are few seeds dropping out on the surface. Good planter operation requires vigilance and constant attention to detail.

Growth and Cultivation of Dry Beans

Once your dry bean seed is in the ground, all you can do is hope for favorable weather conditions. Continued warm temperatures and some timely light rain will do a world of good. Ideally, you want to see little bean shoots poking out of the ground in four or five days. This period between planting and emergence is critical for weed control. Take a walk out to your bean field each morning with a shovel and dig around a little bit to monitor the germination process. When the little bean shoots are half to three-quarters of an inch from breaking through the soil surface, it's time to do some pre-emergence weeding, which basically amounts to a light scratching tillage of the soil surface in the newly planted bean field. You can accomplish this with a tine weeder adjusted to its most aggressive setting or a spike drag smoothing harrow. This is your last chance to knock out those weeds that can establish themselves right in the row between the bean plants. As you know, weed kill will always be much better when the sun is shining; hot, dry conditions will help "cook" those unearthed little weed rootlets in the white hair stage. I generally use my McFarlane spike-tooth harrow for this job because it is gentle and aggressive at the same time. I travel at a forty-five-degree angle to the rows to ensure the best possible disturbance of the soil's surface. When the job is done, the ground is worked up to a depth of about half an inch. The dust mulch that is created by this very shallow tillage helps to kill further weeds and conserve moisture for the sprouting bean seed below. Tine weeders and rotary hoes are much less aggressive, but also work well for this job, although you might have to double back and take a second trip over the bean field to get the best results. Rotary hoes are probably the best choice for this activity if the soil is on the wet side, but it really doesn't matter what sort of implement you use for this job of pre-emergence cultivation—the most important thing is that it gets done. If you can't find a spike drag, weeder, or rotary hoe, try an old bedspring or small tree towed behind the tractor. If you're working in a garden plot, this soil disturbance can be performed with a light raking action. You need to get the jump on the weeds before they get the jump on you.

Germination

If everything is going perfectly, you'll begin to see the first signs of germination five or six days after planting. Beans are extremely tender and vulnerable at this stage of early development and should not be disturbed by any cultivation machinery. This is also the time to find if you have a woodchuck problem, as these creatures like to come out from their underground holes next to a stone wall or field edge and snack down on the recently emerged bean shoots. Hopefully, you won't have this problem, but if you do notice that some major munching has taken place, you will have to take some sort of action. It can very quickly turn into an "us or them" situation. I don't want to sound brutal, but war will have to be declared and the woodchuck population reduced. The seed coats fall off the bean spikes in a day or two as the first set of trifoliate leaves begin to grow from the plant's first node. If you've done a good job of pre-emergence blind cultivation, you should be looking at row upon row of tiny bean plants that are totally free of weeds.

Early-Season Weed Control

Once the first set of true leaves has appeared, it's time to get back to weed control. When they have leaves, the young beans plants are much tougher and can easily stand up to the rigors of mechanical cultivation. This is the ideal time for tine weeding—the young bean plants very quickly begin to sink a taproot, which helps them withstand the pulling action of a spring steel weeder tine bumping up against their stems and leaves. We've talked a lot about what an act of faith tine weeding is in other crops, and it's no different in dry beans. Before heading into the field with your weeder, make sure it is a sunny day and that all of the dew and morning moisture has totally evaporated from the bean foliage. Stay out of the bean field when it is wet; as I've mentioned, human and mechanical contact with wet plants can easily spread an anthracnose fungal infection throughout a healthy stand of dry beans. If you manage to get out there with the tine weeder very early just after the appearance of the first true leaves, be careful and observant. Most weeders can be set to scratch heavily or lightly. Try the light settings first to make sure that you do as little damage as possible to the young plants. This early-season weed control is the key to success with a dry bean crop. The soil is still loose and friable from the earlier pre-emergence cultivation, and tine weeding will yank out the next flush of weeds and expose their rootlets to the baking action of the sun. As long as the young bean plants are not being broken off or pulled out, tine weeding is the best action to pursue. Don't be too concerned if the dry beans look a bit bedraggled—the absence of weeds is the most important thing at this point.

Bean plants grow very quickly, especially if the weather is warm and rainfall is adequate. The tine weeder can be used until the plants reach four to six inches in height. As the plants get taller and stronger, you can set the tine weeder more aggressively. This is usually done by pulling the top lever toward the rear of the weeder. This in turn pulls the tine teeth forward and positions them at a more upright angle to the ground below. Maintain that observation and vigilance. If you see leaves flying or damaged plants, you'll either have to stop or back off on the tine angles. There are no steadfast rules in this business. If you've got the ability to make two or three quick trips through the bean field with the weeder, by all means do it. Tine weeding is much faster and easier than row-crop cultivation. Every trip that you can make through the field with the weeder will lessen in-row weed pressure later in the growing season.

Tips on Cultivation

Dry beans have a vegetative growth pattern very similar to their relative the soybean. New leaves and stem nodes appear every few days in the best of growing conditions; these plants grow fast. It's not unusual for dry beans planted the first week of June to be eight to ten inches tall by the first of July. Growth is outward as well as upward as these plants bush out into the rows. Row-crop cultivation with a shovel or rolling cultivator should start when the bean plants have reached four to six inches in height. If you start too early, there's a risk of burying the young bean plants. The same rules that apply to cultivating other row crops like corn also apply to dry beans. The first cultivation can be deeper and closer to the plants in the row. Shovels and sweeps need to be pulled back away from the bean plants as they spread outward and begin to send their roots laterally. Common sense and astute observation are the keys to effective row-crop cultivation. You want to wipe out as many weeds as possible as early in their life cycle as possible. It's a whole lot easier to cut off an annual weed like lamb's-quarters when it is half an inch tall as opposed to a foot. All that early-season spike dragging and tine weeding pay big dividends later on in the summer, and the same holds true for in-row cultivation. The first pass with the cultivator is the most critical. Slow down a little and try to get as absolutely close to the row as possible without damaging the bean plants. If you don't get that weed on the first pass through the field, you certainly won't be able to eliminate it later on. Pay attention to the bean foliage and the tractor tires. If you find that you are actually running over leaves and damaging plants, it's time to regroup and develop another strategy—a cultivating tractor with narrower tires might be all that you need. Once the bean leaves have spread out a third of the way into the row space, opportunities for row cultivation rapidly decline. Some growers like to use disc hillers on their last trip through the field. This operation stunts in-row weeds because it throws quite a bit of soil up against the lower stems of the bean plants. The last bit of cultivation advice that I can provide may sound a bit redundant, but it's so important that I will repeat myself once again. Stay out of the field when the plants are wet with morning dew or a recent rain shower. Beans are extremely prone to leaf diseases that can spread like wildfire in high-moisture situations.

The line between vegetative and reproductive growth is much more distinct in dry edible beans than it is in soybeans. Dry beans will put on rapid vegetative growth for the first four to six weeks. Flowers will begin to appear at the junction of the leaf axil and the main stem sometime around the middle of July. By this time, the vegetative growth process is pretty well finished. Unlike the soybean, which keeps growing after it begins to flower, dry bean plants are about as big as they are ever going to get by the first of August. A quick peek under the leaves will reveal rather large double-lobed purple and white flowers, which soon turn into tiny pods. If you want a meal of string beans, now is the time to do the picking. Once a field of beans has begun to flower, it's time to pull any and all equipment from the field, as tractor tires and cultivator iron will injure plant leaves and tender flowers. Mid- to late-season cultivation can also prune back lateral plant roots; this can limit water and nutrient uptake during the pod-fill stage of reproductive development.

Pests and Diseases

Dry beans can be susceptible to disease and insect pressures. However, we are very fortunate to live and farm in this part of the country where beans are a relatively minor crop, because disaster is quite rare. If you practice good crop rotation and manage your soil fertility well, your problems should be few and far between. It doesn't matter if you are growing a small garden plot or a larger acreage, dry beans should not be grown in the same spot for at least four years. Anthracnose (*Colletotrichum lindemuthianum*) is the one fungal affliction you should be able to recognize and take steps to prevent in your bean crops. Prevention can be as simple as planting clean seed, as most anthracnose infections come to a field on the seed. The fungus can attack plants early or late in their development. Affected plants will have brown or black spots on their leaves or pits in the stems and the cotyledons. Bean growers have been known to walk their recently germinated fields pulling up diseased plants. It is highly recommended that these rouged plants be completely removed from the field and burned to prevent further spread of the fungus. Anthracnose can also attack bean pods later in the season. Splotches will appear on the fruit. Throughout the course of ripening, the disease will gradually work its way through the pods and onto the seeds contained within, but once a bean seed is dry, the fungus goes dormant and is unrecognizable to the naked eye. This is how seed can become infected and transmit anthracnose when planted the following year. Anthracnose also becomes a soilborne pathogen when bean straw from afflicted plants is returned to the soil through tillage.

Bacterial blight can be a problem in dry beans. Several types of blight are carried on the seed. The symptoms of common or fuscous blight are brown necrotic lesions on plant leaves surrounded by a bright yellow zone. Halo blight infection is characterized by prominent greenish yellow zones or halos that appear on bean leaves early in the season. But bean blights are totally preventable by buying excellent-quality disease-free seed from faraway places like Idaho that have rigorous seed-quality-assurance programs.

I have found insect challenges to be even less of a problem in dry beans: The only insect that has ever impacted my beans is the bean weevil. This dark-colored beetle is about a quarter inch long and drills little holes in stored beans with its short, broad snout. The larvae of the Mexican bean beetle feed on the tissues between the veins of the leaves, but thankfully I have never seen one of these little creatures in my extreme northern location. In addition, experienced bean growers don't like to plant a crop on plowed sod because root-damaging insects like white grubs and wireworms take up residence on the roots of grasses. These little creatures can decimate whole sections of a newly planted bean field by feeding on plant roots. But rest assured that if you pay close attention to balanced soil fertility and crop rotation, bugs shouldn't be much of a concern in your dry beans.

Late-Season Weed Control

Once dry beans have flowered, it's time to stand back and leave the rest to Mother Nature. By this time, plant foliage should be reaching out toward the middle of the rows. A thick canopy is the only late-season weed control we have as organic growers, and while I'd like to think that this seemingly impenetrable mass of bean foliage would be enough to guarantee us a weed-free field for the upcoming September harvest, unfortunately this is not always the case. Those same late-July and early-August rainstorms that supply moisture to bean plants in the midst of filling pods with fruit also encourage the growth of annual weeds that escaped the row-crop cultivator on its last pass through the field. By mid-August, it's not uncommon to see a whole new crop of pigweed and lamb's-quarters poking up through a field of dry beans that appeared to be totally weed-free two to three weeks earlier after the last cultivation. These unwanted plants have the potential to grow quite tall and bushy through the remainder of August and into early September. More often than not, I've had towering four- to five-foot-tall "weed trees" in my dry beans at harvesttime. These annual weeds won't severely impact the final yield of a dry bean crop, but they will make harvest extremely difficult. Most beans are either pulled up by hand or "lifted" out of the ground by special harvesting machinery. A four-foot-high lamb's-quarters plant is doubly troubling because its extensive root system will interfere with bean pulling as well as make it hard to find the bean plants underneath. An overabundance of weeds in a field of ripening beans also impedes the dry-down process.

Dry bean growing is no different from any other agricultural activity: Challenges and setbacks lurk around every corner. This reality of late-season weeds is reason enough to start out with small and manageable dry bean plantings. The best strategy for dealing with late-season annual weeds is to hand-pull them when they first appear in late July, which means walking through the field or garden either alone or (better yet) with a small army of people to yank out these unwanted intruders. An annual weed like a pigweed or lamb's-quarters plant will pull out easily when it is just beginning to stick up above the beans, because its root system is still relatively infantile and fairly minimal. If you wait too long to remove the invading weeds, however, the root-balls will be much larger and bean plants will pull out, too. If you want to avoid "harvest hell" in September, gather up your friends and localvore bean eaters and have a weed-pulling party in late July or early August.

Reproduction and Ripening

Most varieties of dry beans take about twelve weeks to go from seed to final fruit. Bean plants spend their first six weeks in a vegetative state putting on leaves and plant mass to nourish and carry the fruit during the next six weeks to final harvest. Changes in the crop are gradual and subtle. A field of dry beans appears verdant and dark green at the time of flowering. Once the flowers turn into small string beans, the reproductive process is slow and steady over the next six weeks. Bean pods lengthen and fill out during the first week or two, and soon grow beyond the crispy snap bean stage to a tougher and more leathery consistency. Little bumps along the edges of a five- or six-inch-long bean pod indicate the presence of developing seeds within. The bean plant itself has now transitioned into directing all its photosynthesis and nutrient flow to the growth and development of these five or six little seeds inside each pod. The first visible sign of change in a field of fruiting dry beans is a gradual change in color from deep dark green to lighter and paler shades.

By the first of September, the bean pods have reached maximum size and plumpness. If you like shell beans for soups and salads, now is the time to harvest some pods and shell them out for eating. The beans themselves are full-sized and rather soft, which makes them ideal for freezing or a late-summer soup. As the month of September wears on and the crop ripens, the light green leathery pods will turn creamy white in color and become much more brittle, which means physiological maturity is fast approaching. Bean leaves also begin to lose their green pigment during this time. That field of dry beans that appeared lush and dark green six weeks earlier is now more than half yellow and a little tired looking. Clusters of ripe bean pods become visible as the foliage dies back. When the beans inside the pods have dropped to 50 percent in moisture and about half of the bean leaves have turned yellow, the crop is considered to be ripe. Harvest will still have to wait for a little while until the beans themselves have dropped to 20 percent moisture. The dry-down process is totally dependent on the weather. A lot can happen in a few short days if the sun is out, the wind is blowing, and the temperatures are warm. As the beans continue with the final ripening process, the pods hanging from the dying plants will actually start to rattle as they dry out and become more papery. The beauty of this crop lies in the fact that we are able to grow all of this quality protein for human consumption in twelve short weeks. We can harvest our dry beans in September when the weather is still stable and the temperatures are warm enough to promote good in-field drying. Soybeans may have the potential for more yield, but they need to remain in the field for at least three additional weeks, which exposes them to the vagaries of early snow, incessant rain, and colder temperatures. So while dry beans aren't easy to harvest, they do offer us the security of a shorter growing season and harvest under reasonably good weather conditions.

Dry Bean Harvest

Up until harvest, dry bean growing is a fairly straightforward process. Field preparation, planting, and grow-out are subject to specific parameters and procedures that require attention to detail and hard work. But dry bean harvest is a different story altogether. Getting a large crop harvested in good condition and out of the field without major losses is a daunting task that is dependent on favorable weather conditions, specialized bean equipment, and good field conditions. Trouble-free harvesting of dry beans seems to be totally dependent on scale. If you're a garden grower of this crop and harvesting by hand, the process is simple and straightforward. All you need is some sunshine, a supple back, and a bit of time to hand-pull your dry beans. When the beans are rattling around in the pods and most of the plant leaves have turned yellow, it's time to start pulling. Bend over at the waist, grab the stem of the bean plant at ground level, and give it a little tug. When you've got a handful of bean plants in a little bunch, stack them in the field or put them on a wagon or in a barn for further drying. Traditionally, pulled dry beans were placed between two fence posts that were pounded into the ground out in the garden patch. The posts were situated about twelve inches apart and bundles of dry beans were alternately nestled roots-first on either side of the improvised bean drying rack. Single drying poles were quite popular as well. In this system dry bean plants were stacked radiating outward from the central post. However the pulled bean plants were stacked, the end product was a four- to five-foot-tall column of bean vines exposed to the drying forces of sun and wind, and the tightly packed-in bean plants in these stacks formed a thatch that repelled rainfall. Once the drying process was complete, the bean plants were hauled away to a stationary bean thresher for the final separation of the beans from the rest of the material.

Hand-harvesting of dry beans hasn't really changed much in the last hundred years. We still have to bend over at the waist to pull up the plants. Bean drying poles and racks are no longer extensively used because we now have more modern materials like plastic tarps and sheeting to protect our pulled plants from rainfall and heavy dew. Bean plants can be loosely placed in small four-foot-diameter fluffy stacks for further drying, if you cover the stacks with inexpensive small plastic tarps every night and on rainy days. Just uncover the bean piles every dry morning after the dew has evaporated. This process can be a bit of a chore, but it is well worth it. Field-drying can be rapid if the weather cooperates, and you conserve labor and time by moving the bean plants only once or twice. You can also bring your pulled bean plants under cover for further drying. An old barn with a large door at either end of a central drive will work quite well because the central passage is usually a bit of a wind tunnel. Bean plants can also be stacked on old hayracks and wagons for drying. Try to keep your wagon-stacked bean piles fairly small and fluffy to permit good aeration. Wagons can be pulled out of sheds on nice days and kept inside when it is damp and raining.

Hand-Threshing Dry Beans

Dry beans are probably the only crop that easily lends itself to hand-threshing. An old-fashioned hand flail is the tool of choice for many old-time bean growers aged eighty and on up. Bean pods will become very brittle and prone to shattering open once the entire plant has dried out to 15 percent or less in moisture. For those of us younger than eighty, a fifty-five-gallon drum with its top removed makes a fine threshing chamber. Large cardboard and plastic soap barrels can also be obtained free of charge from most dairy farms. Hand-threshing is simple and easy—just grab two or three whole bean plants by the roots, lower your little mini bundle down into the barrel or drum, and whack it against the sides. Beans will come flying out of the pods and fall to the bottom of the drum. Inspect what's left of your bundle for unthreshed pods. If you find any, whack it some more until all the pods have split open. The mixture of dry beans, chaff, and broken-up pods in the bottom of the barrel can be sifted through half-inch-mesh hardware cloth for a preliminary cleaning. The final cleaning can be accomplished by pouring the beans and light chaff from bucket to bucket in front of a fan or in a windy door opening.

Because harvesting and threshing are such simple matters for small-scale garden growers of dry beans, it is certainly possible to plant and harvest several hundred feet of dry beans totally by hand without having to invest in any specialized bean equipment. Self-sufficiency is indeed possible if you are willing to do the work.

Bean Threshers

Dry beans can also be shelled out and winnowed in a one-step process if you have access to an antique dry bean thresher. Even though they haven't been manufactured for seventy or eighty years, bean threshers can still be found around the countryside stashed away in old sheds and barns. If you find one, it can usually be purchased for a couple of hundred dollars. These old wooden relics came in large and small sizes. All of these machines were powered by a long flat belt that ran from a large steel pulley back to a power source like a tractor or a stationary engine. Bean threshers operate in much the same manner as a grain thresher except they turn more slowly and have two cylinders instead of one. The threshing cylinder on a bean machine is equipped with spikes or pins instead of the usual rasp bars found on a grain model. Spike-tooth threshing is much gentler and easier on beans than rasp bars. The primary cylinder on a bean thresher turns no faster than two hundred RPM, and the combination of slow cylinder speed and spike-tooth abrasion ensures that the dry beans won't split in half as they go through the process. The second spike-tooth cylinder on a bean thresher turns a little bit faster; if a bean pod doesn't get split open by the first cylinder, the second cylinder is there to finish the job.

The remainder of the bean-threshing process is identical to grain threshing. Winnowing is accomplished by a fan in the bottom that pushes air up through a shoe of two reciprocating screens. The only difference between the cleaning units of a bean thresher and those of a grain thresher is in the types of screens used. Bean screens have much larger round holes than grain screens, and larger bean threshers are conveyor-fed while smaller models are hand-fed. Remember—bean threshing is dusty, dirty, and dangerous. This is especially true if you are hand-feeding little bundles of beans into a whirling spike-tooth cylinder. Even though it is only turning two hundred RPM, those steel spikes have no mercy for human flesh and bone, so please, *be careful*. Paint a red mark or put an imaginary boundary on the in-feed portion of the thresher past which your hands should never venture. Be alert and wear a mask for respiratory protection. Remember that your beans need to be bone-dry if you hope to get top performance from a bean thresher. A well-maintained and properly operating bean thresher can make very short work of a lot of dry beans. The bean straw by-product makes good stock feed and bedding.

Mechanical Bean Pulling

Dry beans grown on a larger commercial scale are lifted up and out of the dirt with special bean knives and pulling machinery. The two-row horse-drawn bean puller, for example, dates back to the early twentieth century. This implement consists of two side-by-side tapered steel knives that angle obliquely inward. These bean-cutting blades travel horizontally along the ground and are attached to the rest of the framework by means of two mounting shanks. The knives are adjusted so their fronts tip slightly downward into the soil and their tail ends are slightly upward, which allows the bean puller to slide along beneath two rows of beans and lift the plants by their roots right out of the soil. The two rows of pulled beans are merged together in the process because the tail ends of the blades angle toward each other in the center. The basic bean puller of yesteryear was rapidly modified for tractor use when horses faded from the scene in the 1930s and '40s. Mid-mount cultivators for row-crop tractors were very easily outfitted with two-, four-, and six-row bean-pulling units.

Mechanical bean pulling is an early-morning job. You have to be out there in the bean field at the crack of dawn when the plants are still soaked with dew, because if dry plants are lifted and dragged out of the ground, pods will shatter and beans will spill all over the place. The best time for bean pulling is between five and ten o'clock in the morning in mid-September. This job is neater and more precise in theory than it is in actual practice. Everything is soaking wet from dew, and a good deal of dirt is pulled up and moved with the bean plants. As a result, the pulled-up plants are often caked with damp soil. The windrow of dry beans often looks like a sorry, bedraggled mess. This is where another piece of specialized bean equipment called a windrower comes to the rescue. Like pullers, windrowers also come in two-, four-, and six-row configurations. The Innes bean windrower pictured in the accompanying photograph is hitched to the rear of a tractor with a two-row mid-mount bean puller.

This is a simple little machine whose two steel cylinders are outfitted with rubber mounted side-delivery hay rake teeth. The metal drums revolve over the top of the dirty pulled beans; the front drum lifts the just-pulled bean plants out of the loose soil and tosses them backward to the next drum, which gives them one more bit of agitation. This process shakes the dirt from the beans and leaves them behind in a nice fluffy row where they can dry for the rest of the day. Standard practice is to leave the windrowed beans out for a day to dry. The next morning, the windrower is back in the field to move the beans a second time. A small revolving canvas apron-type cross-conveyor is added to the back of the windrower, and each windrow of

Innes dry bean windrower.
PHOTO COURTESY OF JACK LAZOR

Close-up of beam lifting teeth on Innes windrower.
PHOTO COURTESY OF JACK LAZOR

beans is lifted once again and thrown back onto the cross-conveyor to be moved sideways. Three sets of windrows are lifted back up and merged to form one large windrow for final harvest by a bean combine later in the day. More soil is removed from the bean plants during this secondary windrowing.

The Dry Bean Combine

In commercial production, these large windrows of dry beans are usually picked up and combined in the afternoon of the second day after pulling. If drying conditions are exceptional, the process can take place on the same day the beans are pulled. A dry bean combine is a very special piece of equipment designed to pick up windrows of dry beans and thresh them all in one pass through the field. These machines were first made in the 1930s and '40s in Saginaw, Michigan, by the Bidwell Bean Thresher Company. Charles Bidwell started making bean threshers in 1848 in Albion, New York, and later moved to Michigan. Early bean combines were powered by their own gasoline engines and drawn by small tractors or large teams of horses. The first Bidwell combines weren't much more than a glorified bean thresher on large tractor tires with an old-fashioned canvas feed apron on the front to pick up beans in the field. According to Klaas Martens of Penn Yan, New York, Bidwell fell on hard times after the Cuban trade embargo went into effect in 1960: Once Cuban demand for dry beans was eliminated, the market experienced a major contraction. The Bidwell Company shut its doors not long afterward. But in 1961, Bob Bergstrasser began retrofitting and repairing Bidwell bean combines at his service station in Stanley, New York. Bobs Equipment soon began when the little repair shop began fabricating its own machines and improving on the Bidwell design. The brand-new Bobs bean combine was born by combining the best features of the Bidwell threshing chamber with the superior design of the windrow-pickup head from another brand called Pioneer. As the dry bean industry continued to decline in central and western New York State, Bobs Equipment Company disappeared from the scene.

There are still plenty of these machines out there in the bean-growing regions of New York, however. They are advertised in regional farm papers like *Country Folks* at prices ranging from one to five thousand dollars. A Bobs pull-type bean combine is a hulk of a machine equal in size to a large cabin cruiser. There's a lot of metal and fine workmanship there for what little you will have to pay for one of these fine old machines. A Bobs Model 44 measures thirty-five feet in length from the pickup in front to the straw spreader in back and close to twelve feet to the top of the clean bean elevator. These machines are constructed of galvanized sheet metal and many wooden parts. They are so big and cumbersome that they are carried on thirty-eight-inch-diameter tractor tires. All Bobs combines were PTO-powered from the tractor that pulled them through the field.

Bean combines are pulled along a windrow of dry beans. Revolving pickup teeth near the ground grab the pulled bean plants and flip them up onto a perforated-metal inclined deck, and a series of wooden slats with special teeth that are connected to a revolving bell crank move back and forth to carry the beans upward into the throat of the machine. Bean combines have two spike-tooth cylinders just like a bean thresher. The secondary cylinder turns a bit faster than the primary one to ensure complete threshing. Threshed beans drop onto a special bean pan and then onto a set of reciprocating double cleaning screens. One of the reasons a bean combine is so long is to provide adequate space for good separation between beans and chaff. Clean beans are conveyed across the bottom of the combine to an elevator that transports them upward to the grain tank. The grain hopper that sits up on top of the combine is a tetrahedron-shaped steel box that tips sideways to empty clean beans into a wagon or truck. This particular unloading setup is unique because it avoids the auger unloading found on conventional grain combines. Since augers split and damage beans, their use is minimized on a bean combine. When you're sitting in the seat of a tractor operating a bean combine you're towing along behind, there is a lot to keep track of. You are constantly feathering the hydraulic remote level to fine-tune the height of the pickup header. The beans need to move evenly up the feeder ramp to ensure continuous nonslugging threshing. The reciprocating screens in the shoe need to be monitored constantly to make sure they are not getting plugged with the excess dirt that always comes along with mechanically pulled dry beans. Dampness and dirt are the enemies of efficient bean combining, so it's a good idea to have another person present to closely monitor the inner workings of the machine, allowing you to concentrate on moving the whole operation forward. If all systems are in order and luck is on your side, a well-tuned bean combine can harvest ten or more acres on a nice warm September afternoon.

All of this machine-intensive large-scale bean pulling and combining sounds rather glamorous. The idea of tons and tons of dry beans pouring into the grain tank of a Bobs bean combine being slowly drawn across a field of windrows is almost too good to be true. But the actual reality of bean harvesting on an organic farm differs greatly from this idyllic scene. Organic farmers usually have some weeds in dry beans. The conventional growers who use the pullers and combines that I have just described have been using herbicides in dry bean production since the 1950s, but knife-type bean pullers don't work well in weeds or grass. The knives and shanks get plugged up and cease to work when you get into weeds, and material simply won't flow through like it is supposed to. All it takes is a couple of lamb's-quarters plants to hang up on the shanks, and after another foot or two of travel a ball of weeds, dirt, and bean plants is dragging along under the tractor. You stop and back up to give it another try—or maybe you have someone walking alongside with a pitchfork trying to help by manually pulling stuff through. Frustration will begin to build: You've got acres and acres in front of you to finish, but you're stuck in one spot. The reality of being an organic farmer and a beginning bean grower hits hard. You either have to figure out a way to keep your field relatively clean of late-season weeds without herbicides or park the tractor and puller and start hand-pulling.

Growing Beans at Butterworks Farm

My own particular adventure with dry bean growing has been full of challenges and lessons. I planted my first crop in 1995. I tried light red kidneys and soldier beans. The field was relatively weed-free, but the population was a little on the low side because I didn't know how to set my planter properly. We never got that first crop harvested because I had not yet acquired any bean equipment and the task of hand-pulling all those beans was just too daunting at the time. I started to educate myself about larger-scale dry bean production and began looking for the right equipment. In January 1996, I saw an ad in the *Country Folks* farm paper for a complete line of dry bean machinery for sale right here in Vermont. To my surprise, Robert Stuart had been growing ten or more acres of dry beans on the floodplain of the Winooski River down behind the Burlington International Airport. He was leaving his position as the manager of the S. T. Griswold farm in Williston to pursue other interests, and I bought his whole dry bean outfit for five thousand dollars. The machinery lineup included a bean combine, tractor-mounted pullers, and a windrower. As usual, I was partially blinded by enthusiasm. The equipment was ancient—the combine was a Bidwell Model 38 that had been made back in the early 1950s. It had an apron-style feeder canvas and was powered by a large flat-head six-cylinder Continental gasoline power plant. I remember pulling it home with my pickup truck, causing a traffic jam at the busy five-corner intersection in Essex Junction,

Vermont. The puller was equally old and clunky. It mounted out on the very front of a tractor and was raised and lowered by its own hydraulic cylinder. The windrower was powered by a hydraulic motor instead of a PTO shaft. Both of these machines had been manufactured by the Heath Company of Fort Collins, Colorado, which has long since gone out of business.

I attempted to put all of this equipment to work in September 1996 when my three acres of dry beans were ready for harvest. This is when I began to learn about the differences between theory and reality. The puller was a four-row machine, but I had a six-row planter. I tried and tried to pull my beans—with only marginal success, because beans were getting covered with too much dirt. Pods were popping open and beans were scattered all over the ground, and we ended up hand-pulling and field-windrowing. When it came time to try the combine, it seemed like more beans were dropping out the back with the straw than were going into the tank. The machine leaked beans everywhere. We quickly learned that the best way to operate the bean combine was to park it in one place with a plastic tarp stretched out underneath to catch all of the stray beans that ricocheted out of the thing. The bean combine quickly became a stationary thresher driven by its own power plant instead of a flat belt. We limped along hand-pulling dry beans and putting them on hay wagons for transport to the now stationary bean combine. It was now very apparent to me that I was a real neophyte in the dry bean business. I might talk to as many old-timers as I could find and I read all sorts of books and manuals, but this was no substitute for experience or better equipment. Despite all these setbacks, we were able to harvest our own beans and I remained as determined as ever to make the process work better.

The next year I managed to borrow a tractor-mounted bean puller for my Farmall Super C. Tony Neves from the Freedom Bean Company in Maine finished up his harvest early and was kind enough to loan me his puller. We had a few problems negotiating some weeds in the beans, but were able to get them pulled without too much trouble. By this time, the old Heath windrower was beginning to get tired. It seemed like dry bean harvest was a constant reminder of just why there weren't very many people growing this crop. I had plenty of help and advice from bean growers in New York and Maine, but pulling, windrowing, and combining beans were difficult enough for them and they were spraying their weeds with herbicides. Still, I wasn't ready to quit. The market for dry beans was insatiable in the late 1990s and the whole local food movement hadn't even been born yet. And so I persisted with my passion of raising dry beans year after year, trying different harvest strategies each season. In 2003, I tried using my grain combine with the soybean flex head. I found out very quickly why this was not standard practice in the dry bean industry: Dry beans have longer pods than soybeans and hang down much closer to the ground. Even with the flex head operating right down at ground level, bean pods were getting cut in half and never making it into the combine for threshing. I slowed the threshing cylinder down as low as it would go and dropped the concave all the way down. Despite these adjustments, the harvested beans in the grain tank looked like they had been through a war. There was a high proportion of splits and cracked beans because of rough treatment from the cylinder rasp bars. The one exception to the rule was the black turtle bean, which performed a lot more like soybeans. The plants themselves were much taller than the larger-seeded varieties of dry beans. The smaller black bean seeds seemed to stand up much better to conventional rasp bar threshing. Once I learned this little secret, my acreage of black beans began to increase while the heirlooms declined. The black beans were quite popular and easy to grow, and by 2005, my yearly bean acreage had stabilized to three acres of Jacobs cattle, soldiers, yellow eyes, and light red kidneys and as much as seven acres of blacks.

It soon became apparent that I needed better machinery that actually performed well out in the field. I finally broke down and bought a newer Bobs Model 44 dry bean combine from Herbert "Bussy" York in Farmington, Maine. Not long after that, I picked up an Inness windrower near Skowhegan, Maine. The state of Maine still has a small and vibrant dry bean industry; there are growers and former growers all over central Maine. I found most of these people through my acquaintance with Tony Neves, who now grows dry beans near Albion. Tony was state tax commissioner at one time, but has now retired to a full-time career of dry bean farming. The dry bean community that remains in Maine and New York State is very friendly. These folks have been incredibly helpful to the newcomers like me, and are generous with their time and willing to provide lots of free advice to those of us with little or no experience. Tony and Helen Neves even made the two-hundred-mile trip over from Albion, Maine, to my place to help me learn to how to do a better job pulling beans. If you are looking for old bean equipment, a couple of calls to any of these individuals in New York or Maine will provide you with the leads to find what you need. One of the greatest things about a career in agriculture is the relationships that are forged among farmers—that genuine love of farming transcends all sorts of other barriers that divide people. It doesn't really matter if you use organic or conventional practices to grow your crops. Everyone has something to teach.

After fifteen years of trying to grow dry beans, I have developed a system that seems to work well for me. I have been able to successfully pull, windrow, and combine crops a number of times, but I am always dissatisfied with the performance of the harvest as it is taking place. There is so much waste from shattering and other field losses, and I just hate seeing all those beans scattered all over the ground. People in the business tell me that this is a fact of life that must be accepted with mechanized dry bean harvesting. We have gone back to hand-pulling our bean plants because we can do a very precise job with little or no shattering. Sometimes we make windrows right out in the field while other times we load the bean plants onto wagons. Even though I now have a twenty-five-year-old combine instead of a fifty-year-old machine, bean losses are reduced if the combine is parked in one spot and the bean plants are brought to it. I am always surprised by how many stray beans end up on the plastic tarp spread out underneath. Choosing to hand-pull my beans has also forced me to take a closer look at the scale of my dry bean enterprise. Three acres is about as much as we can handle here on our farm. As bean harvest approaches, I begin to put out calls for help to localvore consumer groups. We have been most fortunate to receive help pulling beans from the City Market/Onion River Co-op crop mob in Burlington and from the students of Sterling College in Craftsbury Common, Vermont. The old adage of many hands making light work sure rings true when these sorts of reinforcements arrive at your bean field: A group of ten or fifteen helpers can pull up two acres of beans in a short afternoon. Black turtle beans have turned into our largest dry bean crop because they can be readily harvested with a normal rasp bar grain combine and flex head. Some field losses from shattering do occur, but they are minimal and acceptable, and we are able to produce tons and tons of dry beans with only a fraction of the work that is required for the larger-seeded heirloom varieties. If you find that you have the same indomitable passion for dry bean growing that I do, just make sure to start out with a small enough acreage that you can comfortably handle. You will produce more and better-quality beans on a small intensively managed plot than you will with a larger acreage that gets ahead of you.

After the Harvest

Unlike most of the other grains, beans can be kept at 18 percent moisture for quite some time before they begin to degrade and spoil, which gives ample time to dry and process the crop. There are numerous low-tech methods to finish drying small amounts of beans harvested from a garden-sized plot. The easiest method is to build a two- to three-foot-square box out of boards. Nail a fine hardware cloth screen to the top of this receptacle, lay a small box fan on its side inside the wooden frame, and spread your damp beans out on top. The air blowing up through will dry the beans in pretty short order. If you find yourself with several hundred pounds of dry beans, store them in loose-weave burlap bags so they can breathe—storage in standard woven plastic grain bags will cause your crop to sweat and spoil rapidly. You can quickly improvise a makeshift dryer by setting out two parallel rows of beans in burlap bags about eighteen inches apart. Place a common box fan at one end between the two rows of bags, and close off the other end with another bag of beans. Then lay more bags of beans across the top to bridge the space between the two rows of bags. The idea is to form an air tunnel by tightly nestling the bags of beans into one another; the box fan will pump air down the tunnel and out through the bags. This arrangement is cheap and easy and will serve to stabilize moist beans, slowly lowering their moisture to an acceptable level for longer-term safe storage. Larger amounts of dry beans can be stored in wagons or large tote bags with screw-in aerators pulling air through to accomplish the drying.

Issues with Staining and Underdevelopment

Another major problem with dry beans is staining and underdevelopment. Not every bean that comes out of the field will be perfect. Usually the whiter the bean, the more chance it has of being off color and underweight. Darker-fleshed beans like kidneys and Jacobs cattle are much less prone to staining than their white-fleshed brethren like soldiers and yellow eyes. Staining is caused by prolonged contact between the bean pod and the wet soil. An extremely wet September is the worst enemy of a ripening field of dry beans. These plants, by their very nature, have low-hanging pods that come in contact with the soil. When a bean stains it not only loses its creamy white color, but also changes from being smooth and glossy to a texture that is rough like sandpaper. Unfortunately, passing your beans through an air screen fanning mill cleaner doesn't remove these second-rate specimens. Weed seeds and chaff can be removed with the fanning mill, however, and standard procedure in cleaning large-seeded dry beans is to use a twenty-nine or thirty sixty-fourths-inch round-hole top screen and an eighth-inch by three-quarter-inch cross-slotted bottom screen to remove the splits. Commercial establishments will then run the clean beans over a gravity table to sort them by density; the rough-textured stained beans are generally a little lighter in weight and are separated out on the gravity table. Most brands of gravity tables can be outfitted with a special wire-mesh bean deck. Often the number two and number three quality beans from gravity table separation are run back through the machine repeatedly until only the worst-looking ones are left. The culls are then roasted or cooked for animal feed. Probably the most disappointing thing about producing a crop of dry beans is the amount of yield reduction that takes place all through the harvest process. Even though your stand of beans may actually yield fifteen hundred pounds to the acre, you may end up with twelve hundred or less by the time the crop has been pulled, combined, and cleaned several times. Beans shatter and fall on the ground during harvest. Then they shrink a little if drying is needed, and after that, a good thorough post-harvest cleaning can take care of another 10 to 15 percent of the total volume. If you still have lots of passion for dry bean growing after this, you're really in it for the long haul!

The "Pick Out" Process

Once dry beans have been cleaned, they still have to go through the final "pick out" process. There will still be malformed, discolored, split, and generally poor-looking beans that escape the fanning mill and the gravity table. Large commercial bean processors accomplish this final cleaning with electric eye technology that costs hundreds of thousands of dollars, but the rest of us will have to do this job by hand with the help of our own sharp eyes. Bean belt conveyors are standard fare in the industry. These units consist of a hopper and an endless belt that carries a small stream of beans along to be hand-sorted. The simplest of these units is a treadle-powered small canvas belt that is intended for one person to use. The one-person bean belt is an antique well worth owning for the small bean grower. The beauty of this handcrafted item is that it's foot-powered and compact enough to fit in your kitchen or in front of your woodstove on a cold winter's night. The operator is in total control of its speed and capacity. If you need to stop the belt, simply stop pumping the treadle with your feet. Reject beans and small stones are easily picked out with your fingers, leaving an end product ready for the bean pot. It is powered by an electric motor and requires at least four pickers seated on benches to keep up with the flow of beans on the conveyor belt. Six people make an even better picking team. There are little receptacle boxes along each side of the belt to receive the culls. If the electric motor were changed to a variable-speed drive, this belt would work even better. This is high-capacity picking. Bean pick-out is time consuming and labor-intensive and really cuts into the final profit of this crop.

At first glance, dry beans appear to be a high-value crop with the potential for large profits. With a yield potential of a thousand to fifteen hundred pounds to the acre and an average wholesale price of two or more dollars per pound, this crop seems well worth the effort. However, a little experience out in the field hand-pulling weeds and bean plants coupled with the potential for field losses and time-consuming post-harvest processing paints a different picture. Still, I'm going to keep growing dry beans because I like how they look with their variety of colors and designs. Beans come in so many interesting sizes and shapes, and they are nutritious and taste good. There is nothing better than sitting down to a burrito, black bean soup, or a pot of baked beans that you grew yourself.

CHAPTER SIXTEEN
Oilseeds

Until relatively recently, the production of homegrown vegetable oil wasn't even a consideration here in our part of the country. Oil for cooking and salad dressing was something that came in a drum or bottle from some distant location. Thirty years ago, my favorite vegetable oil was an organic unrefined corn oil that we bought in five-gallon jugs from our local food co-op; it was golden yellow in color and had the most amazing nutty taste. This oil did not work well for frying, however, because it was thick with solids and started to smoke as soon as it was heated. It sure did enhance a salad, though. I remember dreaming about making oil from my own corn, but a little research into the subject provided me with the reality check I needed to put that dream up on the shelf. Corn is only 2.8 percent oil by weight. The oil is found in the germ of the kernel, which must be removed and pressed separately. This was certainly more than I could ever envision doing at the time. My favorite unrefined corn oil soon disappeared and we moved on to other vegetable oils like safflower, canola, and olive oil. I also tried growing flax for oil and animal feed in the mid-1980s, but that experiment was also a flop because of poor germination and weed pressure. Soybeans (with an oil content of 20 percent) were also out of the question at the time. It looked like cooking oil was one of those things we were just going to have to buy.

Lessons from Québec

It seems that most of my inspiration for innovation has come from north of the border. This was certainly the case for oilseed production. Michel Gaudreau of Compton, Québec, grew a couple of very nice sunflower crops in 1996 and 1997. I really hadn't considered growing sunflowers before this, but I was ready to now. I called Michel for some advice and proceeded to order some Pioneer short-season sunflower seed for the 1998 growing season. There was a bit of a steep learning curve, but I did manage to get the crop planted and harvested. (I will discuss the specifics later in this chapter.) I knew that the seeds would need to be pressed into oil, but beyond that, I had no clue how it was done. In the summer of 1998, I made a few phone calls in search of an oilseed press. This was long before the days of Internet searches for me. Somehow, I managed to contact a fellow in Minnesota who was a sales representative for a German company that manufactured expeller presses. I became a bit discouraged when I found out that the very smallest and cheapest model of oilseed press sold for fifteen thousand dollars. However, I did find out that the company had recently sold a machine to Loïc Dewavrin, who owned Les Fermes Longprés in the little hamlet of Les Cèdres just west of Montreal.

I called Loïc immediately and found him to be very friendly and helpful. He had been pressing lots of different oilseeds that he had been growing on his thousand-acre cash crop farm, and was also in the process of transitioning his entire corn and soybean operation to certified organic. Several years of experience with oilseed production and processing had taught Loïc and his two brothers that sunflower oil was the most practical product for them to produce in this climate. They had formed a little oil-production enterprise called Les Huiles Naturelles d'Amerique. Loïc agreed to press the three thousand pounds of sunflower seed that I harvested from my two-acre trial plot that year, and so I bagged up this crop and transported it in my pickup truck to their farm 130 miles away. The operation was immaculate and very impressive. Several weeks later, I drove back up to their farm and picked up the sunflower oil, which they had professionally packaged in special green-hued square-shaped one-liter oil bottles. I also brought back the sunflower cake to use as a protein source in animal rations. This was the beginning of a long and fruitful friendship with Loïc and his brothers Thomas and Come. I had a much larger and more abundant crop of sunflowers the next year in 1999, and brought somewhere between five and seven tons of sunflower seeds to Les Cèdres that year for pressing. The oil was very well received here in Vermont. We saved a twenty-liter bulk jug for our own use and sold the rest to a number of health food stores and food co-ops throughout the state. Loïc has continued to teach me the finer points of sunflower production over the years, and the fact that I had a good case of beginner's luck with my first two crops of sunflowers kept me in the game for many years. Unfortunately, there have been some poor crops, too. But sunflowers and oilseed production in general have now caught on all over the Northeast. A core group of individuals has begun to grow sunflowers and other crops like canola and flax. Not every year is a good one for these crops, but they will grow rather well here with a little bit of luck and cooperation from the weather.

Sunflowers

Like corn and dry beans, the sunflower (*Helianthus annuus*) is native to North America, and was first domesticated by Native Americans in Arizona and New Mexico around 3000 BC. They cultivated the sunflower for its seeds, which they pounded into meal for cakes, mush, and bread; the oil from the seeds was rubbed onto their skin and hair. Spanish explorers first encountered these strikingly beautiful plants early in the sixteenth century on their forays northward into what was to become the American Southwest. By 1550, sunflowers had been brought back to Spain and Mediterranean Europe for use as an ornamental flower, and the culture then spread eastward to Egypt, India, and Russia. In 1716, the English patented a process for squeezing oil from sunflower seeds. The Russians, however, deserve credit for turning sunflowers into a food crop.

Russian Influence

Olive oil was the natural choice for cooking in Southern Europe because the olive tree was so well suited to the warm, arid climate of the Mediterranean basin. But Russia was not blessed with the same climate, which meant that it had to import oil from the south. When Russians discovered that copious amounts of oil could be pressed from sunflower seeds, the crop took the country by storm. By the eighteenth century, sunflowers were being extensively cultivated in Russia as an oilseed crop. Peter the Great was a great champion of sunflowers, and the Russian Orthodox Church forbade the consumption of all oils except sunflower during Lent. There is historic evidence of commercial oil production in Russia as early as 1769. Russian farmers and plant breeders should be given credit for selecting and improving sunflowers and turning them into a field crop. Yields and standability increased at the same time as a sunflower oil industry developed. By the nineteenth century, Russia had become a major exporter of sunflower oil to Europe; two million acres of sunflowers were being grown there at the time. Some of the sunflower varieties that we grow in our gardens today, like Mammoth Russian and Black Giant, were actually developed centuries ago north of the Black Sea in Russia. These varieties were rather noteworthy at the time because of their almost two-foot-diameter seed heads.

Interest in cultivating sunflowers began to grow here in North America during the 1880s. Progressive plantsmen and other forward-thinking individuals of the era turned to Russia for improved varieties of the very sunflowers that had originated here hundreds of years earlier. Missouri seems to have been one of the first areas in the United States to widely adopt the crop. A growers' association was established in 1926, followed shortly after by the first commercial production of sunflower oil in the country. Sunflower seed was also brought into Canada by Russian immigrants who settled on the prairies. A sunflower-breeding program was initiated by the Canadian government in 1930; the plant breeding material came from Russian Mennonite immigrants. Canada's first sunflower-seed-oil-crushing plant came online in 1946. Due to its popularity north of the border and the ideal climate of the high plains, sunflower acreage soon spread southward into the wheat country of northwestern Minnesota and North Dakota. This area has remained the epicenter of sunflower culture since World War II.

The Rise of the Hybrid

Sunflowers remained a relatively minor crop throughout the 1950s and 1960s because demand for the oil was not great and efficient harvesting machinery had not yet been developed. Acreage continued to increase slowly in the Canadian prairie provinces, and the Canadian government licensed the Russian variety Peredovik for widespread planting in 1964. The most amazing thing about sunflower culture in North America during this era was that only open-pollinated nonhybrid varieties were being planted by farmers. Hybrid corn—produced by crossing one corn variety with a mate that had been detasseled—had been the norm since the late 1930s, but this was not the case for sunflowers. This process couldn't be done with sunflowers because there was no way to turn one plant into a male and another into a female. But this all changed in the 1970s when sunflower breeders were finally able to isolate cytoplasmic male sterile lines to use as the female parent in the hybridization process. The very first sunflower hybrids were released in the mid-'70s. This corresponded with an increased public acceptance of vegetable oils as animal fats declined in popularity. European demand for sunflower oil had also begun to outstrip Russian production, and with new higher-producing sunflower hybrids and an extra-strong demand for oil in Southern Europe, American farmers began to plant more acres of this crop than ever before. The US yearly production finally exceeded five million acres for the first time in the late 1970s. Sunflower production had finally emerged from inconsequence.

Sunflowers have really come into their own since the first hybrids were released in the 1970s. Growing crops that would stand up to blustery fall winds had always been a problem in the early days of sunflower growing, and anyone who has ever grown open-pollinated Mammoth Russian sunflowers in their garden knows how these tall plants with their top-heavy seed heads like to fall over once they begin to ripen. Since hybrid sunflowers are much shorter in stature and have stronger stalks, they will stand well into late fall, which allows for excellent in-field drying and an easier harvest; potential yields have increased to more than a ton to the acre under ideal growing conditions. However, the most important advances that have been made in sunflower breeding have been in the disease- and pest-resistance departments.

Sunflowers have always been a prime target for a host of fungal and bacterial attacks, from white mold to anthracnose. Insects are also a major concern. But modern hybrid varieties have been selected and bred to withstand some of these pressures.

The heart of sunflower culture in the United States is in the northern plains. Sixty percent of the crop is grown in northwestern Minnesota and North Dakota, and all of the advances that have been made in breeding sunflowers have been made in this area as well. The USDA has a sunflower-breeding station in Fargo, North Dakota; North Dakota State University (NDSU) is also quite active in developing new and improved sunflower varieties. A small number of seed companies breed and distribute sunflower seed in the area, most of them located on the Minnesota side of the Red River. Dahlgren of Crookston, Minnesota, is one of the oldest sunflower seed companies and can be credited with many of the early advances in hybridization. Croplan Genetics of Mentor, Minnesota, a division of Land O'Lakes, is a major supplier of seed to the market. Seeds 2000 in Breckenridge, Minnesota, is a newer company that is leading the way with a lineup of certified organic sunflower seed. If you are seriously interested in raising sunflowers, you will have to source seed from this part of the country, but all of these outfits have some untreated conventional seed. Most companies are willing to take your credit card number, box up a bag of seed, and send out to us here in the Northeast by UPS.

Types of Sunflowers

Before you order seed, it might be a good idea to familiarize yourself with all the different options out there in the world of sunflowers. Why do you want to grow sunflowers and what do you plan to do with them once they are harvested? There are two basic types of sunflowers—black oil seed and confectionary. Confectionary sunflower seeds are larger (eight to twelve millimeters) and have a characteristic white stripe running across their outer seed coat. Because they are larger and contain more inner "meat," these seeds are hulled to make the sunflower seeds we eat as snack food. Confectionary sunflowers contain only 30 to 40 percent oil. For some reason, they seem to grow better out in North Dakota than here; in addition, hulling sunflower seeds is even more specialized and difficult than hulling oats or spelt. The infrastructure and knowledge for hulling confectionary sunflower seeds simply does not exist out east. As far as I know, no one has successfully managed to produce snack food sunflower seeds here in our region. This certainly doesn't mean that it cannot be done, but it will take more experimentation with varieties and hulling equipment. Black oilseed sunflowers, however, seem to be more reliable and easier to produce here in our region. Their seeds are much smaller in size (4.5 to 8 millimeters) and contain from 46 to 50 percent oil. These are the same all-black seeds that are sold for bird feeding. Black oil sunflowers contain less inner meat and are not recommended for hulling. The production of oil for human consumption and the resulting by-product of oilseed cake for livestock feeding is the primary end use for sunflowers grown here in the northeastern United States.

Until recently, the oil that was pressed from black oil sunflower seeds was linoleic oil. Linoleic sunflower oil has about 69 percent polyunsaturated fat, 20 percent monounsaturated fat, and 11 percent saturated fat. The linoleic component is a good source of omega-6 fatty acids. There are also some traditional sunflower varieties that are considered high oleic. The oil that is pressed from these seeds has at least 80 percent monounsaturated fat and is quite high in omega-9 fatty acids. Because of competition from and lost market share to the olive oil industry, the National Sunflower Association began a breeding program in 1995 to develop a mid-oleic sunflower oil that contained higher levels of omega-3 fatty acids. New sunflower varieties were released onto the market in 1996 with the trademarked name of NuSun. This oil is nutritionally superior and better for you, and hydrogenation is not required to make this oil compatible with commercial frying. It is very stable and resists the oxygenation that produces rancidity. By 2007, 85 percent of the sunflower oil market was dominated by NuSun mid-oleic oil. If you have the opportunity to buy seeds of a NuSun variety, don't be afraid of the newfangled nomenclature. It is truly superior oil seed that has been developed by traditional plant-breeding methods.

Sunflower Culture

Sunflowers have many cultural similarities to corn. They grow best when planted in well-drained loam soils, and both crops are lovers of sunshine. Corn transforms sunlight, water, and air into starch and carbohydrates; sunflowers photosynthesize these same substances in the presence of sunshine into high-quality oil and protein. Nitrogen is required for the production of protein, and one of the major differences between the two crops is that a corn plant will gobble up an inordinate amount of available nitrogen from the soil environment while sunflowers grow quite well with more moderate nitrogen levels. A soil rich in the humic substances that slowly release smaller amounts of nitrogen is best suited for sunflowers. But too much nitrogen is just as harmful to a sunflower crop as too little. I can speak from personal experience in this department. Several years ago, I fertilized a small sunflower planting with profuse amounts of dairy manure. For a while, I thought I had done the right thing—the growing crop seemed like it was making more progress than neighboring fields that had not been so heavily fertilized. But after the field flowered and began to ripen, stalks began breaking. More and more plants fell over every day, and by harvesttime over half of the sunflowers had fallen to the ground. Sunflowers are a taprooted crop with an uncanny ability to scavenge nitrogen and water from deep within the soil profile, so adequate levels of other essential elements like potassium, magnesium, and phosphorous are just as important as nitrogen to a sunflower crop. Small amounts of a balanced mineral fertilizer and a light coating of compost or aged manure should be all that is necessary. Follow the recommendations of your soil test and limit nitrogen inputs to no more than eighty units to the acre.

CHOOSING THE RIGHT VARIETY

Most of us will be planting hybrid sunflowers, and believe it or not, some of the companies out there are beginning to offer organically grown seed, Seeds 2000 and Blue River Hybrids among them. If you can't find organic seed, though, try to buy seed that is untreated. When I first began growing sunflowers in 1998, untreated seed was rare. Because sunflowers grown out in the high plains are afflicted with so many diseases, seed treatment with fungicides is standard fare. Thankfully, the organic movement has now grown enough in importance and market share that seed companies are offering untreated seed as a common practice. If you find a variety that works well for you, it might be a good idea to contact your dealer or seed company and reserve untreated seed well in advance of the coming season. These companies clean and process seed all through the winter, and a simple early reservation will ensure that you will have the seed you want the following May when it comes time to plant.

Make sure to choose a variety that fits your growing environment and microclimate. Hybrid sunflowers are usually classified as short, medium, or full season. I usually plant sunflowers that are short to medium season because of my northerly mountainous location. As with any of the crops harvested in October or November, sunflowers are subject to the weather. Cold rain, snow, and wind are of course imminent at harvesttime, and dry-down can be slow or nonexistent when the weather begins to deteriorate. When that two- or three-day window of sunny weather arrives in mid-October, you want your sunflowers to be dry and ready for combining. Not much is worse than having a high-yielding crop out there in the field that is still mushy and not yet ready to harvest when opportunity presents itself. Once again, I can speak from experience on this subject. It's better to have a moderate harvest safely in the bin than it is to have a great harvest out there being eaten by birds or knocked over by wind and snow. If you are new to growing sunflowers and want to experiment a little before diving in headfirst, I recommend doing your own little variety trial. Obtain small amounts of seed for five or six different hybrid sunflower varieties from several different companies. Plant out everything side by side in a small plot and evaluate the performance of the hybrids on your farm. By autumn, the varieties that shine in your environment will stand out. You will then be able to plant your chosen variety with confidence in the next growing season.

PLANTING SUNFLOWERS

Sunflowers usually do well after a nitrogen-contributing crop like sod or soybeans. Some professional crop advisers don't like to see sunflowers planted after soybeans because of the risk of white mold infection. If you saw any white mold (sclerotinia) in the previous year's soybeans, don't plant sunflowers in the same field, as sunflowers are very susceptible to this fungus. Work the land into a friable seedbed that is smooth and free of clods before heading to the field to plant. Sunflowers can go into the ground in mid-May just before the danger of spring frost has disappeared. Most are seeded in thirty-inch rows with a conventional corn planter. Field populations are just a bit lower than what would be normal for corn. Whereas thirty-two thousand plants to the acre might be an acceptable population for a crop of corn, twenty-five thousand would be better for sunflowers. This means spacing plants eight to ten inches apart instead of the six inches that's standard for a corn crop. John Deere plateless planters with finger-pickup seed dispensation are perfectly suited to planting sunflowers without any modification. The only adjustment necessary is to reduce the depth of planting to one and a half inches and to back off the amount of seed planted.

If sunflowers are planted too thick, they won't thrive. Heads will be small and stalks weak. Traditional plate planters need a bit more modification to plant sunflowers. First, check the tag on the bag to find out what size of sunflower seeds you have purchased. Number two is the largest sunflower seed and number five, the smallest. Once you know the seed size, you can choose the proper plate by looking at a chart on the back of the bag. Medium and large seeds travel through seed plates much more uniformly than the smallest seeds. Sunflower seed plates are quite thin and very specialized. For proper planting, they require an additional filler ring under the main plate to take up extra space. Make sure to position the filler ring under the seed plate and not above it. I made this mistake once. I wondered why there was no seed coming out of the planter, and when I stopped to inspect, I found that I had put everything together in the wrong order. No matter what kind of planter you use, it is best to take it for a dry run on a driveway or another hard-packed surface to make sure that you are laying down the right amount of seed. Count the number of seeds in seventeen and a half feet of row and multiply by a thousand to determine the number of plants to the acre. Last but not least, make sure the soil temperature is at least fifty degrees before beginning to plant sunflowers.

GERMINATION AND EARLY-SEASON WEED CONTROL

The period between planting and emergence is an ideal time to practice aggressive weed control. Constantly monitor the germination process underground by digging around for sprouted seeds. If you see a big flush of weeds coming, don't be afraid to pull a spike drag around the field on a sunny dry day to kill weed rootlets in the white hair stage. A tine weeder like a Kovar or Einbock will also do the trick, especially if the sunflower sprout is close to breaking through the ground. Post-emergence tine weeding is a little trickier. Sunflowers sprout in much the same manner as a dry bean or soybean. At emergence, you will first notice two little cotyledons; these are closely followed by the first true leaves. At the same time, the plant is beginning to sink a small taproot. It's important to wait for the appearance of the first two true leaves before charging through the field with your tine weeder. If you do plan to tine weed your sunflowers, you might want to plant a little heavier than normal because some plants will be yanked out and destroyed during the process. In the past, I have been hesitant to tine weed my sunflowers fearing that I was doing more harm than good, but experience has taught me that it's better to brutalize the little plants than it is to let the in-row weeds get a foothold. The process is much the same as it would be in corn or soybeans. Pass the weeder as many times as you feel is necessary until the plants get six inches or taller. Timely and effective early-season weed control makes all the difference later in the season by reducing in-row weeds early in the game.

GROWTH STAGES OF THE SUNFLOWER

Sunflowers have a vegetative and a reproductive stage just like any field crop. The vegetative stage begins right at emergence and continues for six to eight weeks when the terminal bud first appears at the top of the stem. Sunflowers look rather inconsequential right after emergence: Two tiny leaves make their appearance right there at soil level, and from afar it's difficult to tell the difference between a field of soybeans and one of sunflowers at this very early stage of vegetative growth. Once the roots get a foothold and the leaves begin to widen and elongate, however, it's much easier to identify the crop as a field of sunflowers. Plants gain stature rapidly through the month of June and into early July, and you might get through your sunflowers with a row-crop cultivator a couple of times before they get too tall for the process. By this time, the leaves have formed a protective canopy over the row that shades out the weeds below. If you can do a reasonably good job of keeping the weeds at bay up to now, there won't be much competition to worry about after this point. Sunflowers are uniquely characteristic in the coloring and texture of their foliage during this vegetative period. The leaves are a bit furry to the touch and light "lima bean" green in color. Most hybrid sunflowers don't grow much taller than six or seven feet. Longer-season varieties can have as many as twelve points of leaf attachment and will have a vegetative period that might be one to two weeks longer in duration than a shorter-season variety. Sunflowers are simply amazing in their ability to transform themselves from two tiny leaves hugging the earth to a veritable jungle of large green leaves on sturdy stalks in two short months.

Sometime in early to mid-July, the terminal bud will appear at the top of the stem instead of the next leaf cluster. The reproductive phase has now begun. The terminal bud elongates an inch or so above the nearest leaf and gives us the first hints that it is on its way to becoming a large beautiful flower. After the passage of a few more days, the inflorescence begins to open; the first ray flowers around the outer edge will now be visible. Flowering now begins in earnest in the disc or center of the head. Meanwhile, the sunflower head is getting larger by the day. Now is when you find out how adequate your fertility levels are for growing a crop of sunflowers. If you're a little short on nitrogen being released from the soil's humus reserves, it will definitely be reflected in the final size of the seed head. Smaller sunflower heads mean lighter yields.

Sunflower heads are actually composed of two different types of flowers. The yellow petals around the outer circumference are called the ray flowers, and the face of the head is made up of many individual disc flowers that each become a future seed. These tiny disc flowers are self-compatible for pollination and do not need the help of pollinating insects. There is a common myth that sunflowers are heliotropic—which means that the flowering heads of the plants will follow the sun throughout the day as it makes its way from the eastern horizon to the west. This may be true when the plants are quite young and still in the vegetative stage of growth, but it has been my personal experience that sunflower heads usually point to the east. I have had several sunflower fields that were situated on the east side of a north–south road. People passing by were quite disappointed because the sunflowers were all pointed in the direction opposite the highway. The peak period of brilliant sunflower inflorescence is all too short, so be sure to enjoy the sight of your sunflower field when it happens because the plants need to continue ripening to make seed. Get the camera or paintbrushes and easel out when the field is at its most beautiful state, as colors will begin to fade and petals drop off in a week to ten days.

DISEASE AND INSECT PRESSURE

Disease and insect pressures don't usually present themselves until the sunflower plants have entered the reproductive stage. Cutworms and grubs can be a problem at emergence, and there are lots of insects like grasshoppers and caterpillars that like to feed on growing sunflower foliage, but these are seldom debilitating problems. Probably the worst sunflower insect pest is the banded sunflower moth. The larvae that hatch from its eggs will migrate inside the flowering head and wreak havoc with developing seeds. We are fortunate in this part of the country because sunflowers are a relatively new crop and most of the other bothersome bugs simply haven't found us yet, but North Dakota is afflicted with all sorts of insect problems. There are little critters that eat the pith on the inside of the stalks, causing whole fields of sunflowers to fall over before they can be harvested. As far as I know, these insects have been seen only once here in Vermont. But the biggest threat to sunflowers grown anywhere is sclerotinia or white mold. Take the proper precautions against this fungal scourge by developing a long and diverse crop rotation. Sunflowers should only be grown in a field once every four or five years; if you can avoid it, don't follow soybeans or dry beans with sunflowers. Keep your sunflower field well aerated by not planting at too high a population and by keeping the weeds in check. Disease and insects are something to be aware of, but not to worry about if you are a first-time grower of sunflowers. Common sense and good agronomic practices will go a long way in this department.

MATURATION AND RIPENING

When the ray flowers that surround the head begin to wilt, the flowering stage is complete. This is the beginning of a long steady march toward maturity and ripe sunflower seeds. By this time, the inner disc flowers begin to fade in color as seed is formed. The back of the head begins to turn pale yellow in color. We are now into late August or early September, and gradually the back of the head becomes even more yellow. The bracts or ribs on the back of the seed head are the last thing to turn from green to yellow. When the yellow color turns to brown, the sunflower is considered ripe or physiologically mature. The entire reproductive process—from the first sign of a terminal bud to the physiologically mature seed head—takes close to two months. The sunflower seeds that have developed in the head are still at 35 percent moisture at this time. Dry-down will continue for at least another month. Leaves will drop and seed heads shrink a bit as the last of the yellow disc flower petals fall to the ground. As the sunflowers finish drying, the brown hues of the plant will transform into a beautiful shade that can best be described as chocolate purple. I was amazed the first time I experienced this with my beginning crop of sunflowers in the autumn of 1998. The field had an effervescence that is probably quite common out on the high plains, but not here in this part of the country. Personally, I find ripening sunflowers to be just as beautiful as that field of yellow flowers that dazzled my eyes two months earlier.

Sunflower Harvest

Sunflowers are in the same class with other mid- to late-fall-harvested crops that tend to fill you with apprehension. Like corn and soybeans, sunflowers have to dry down before the combine can enter the field. Ideal storage moisture for sunflowers is 8 percent, and it can be a long wait for physiologically mature sunflowers at 35 percent moisture to field-dry to such a low moisture level. Meanwhile the forces of nature will start to chip away at your beautiful crop of potentially high-yielding sunflowers. Large flocks of blackbirds seem to come out of nowhere to feast on the big bird feeder that you have provided for them. The rest of the wildlife in the neighborhood is also aware of that perfect balance between protein and energy available out there in your field of sunflowers. Your field has the potential to attract deer, raccoons, and even bears from miles around, and stalks keep getting pulled down one by one as you share your sunflower seeds with the local wildlife. Wind, rain, and heavy wet snow can also contribute to sunflower lodging. Since the head of a sunflower is carried at the very top of the stem, it ends up on the ground when a stalk breaks. Once a sunflower head is lying on the earth below, it's pretty much history. It certainly won't dry out anymore; it usually gets moldy within a matter of days. Corn, on the other hand, carries its ear halfway up the stalk, so chances are pretty good that a lodged corn plant might still be harvestable—the ear generally doesn't tumble all the way to the ground. With sunflowers, though, you must constantly weigh the options of earlier harvest and artificial drying against longer-term field-drying that exposes the crop to the vagaries of weather and wildlife damage. Sometimes luck will be on your side and a field of sunflowers will remain standing and untouched until Thanksgiving or afterward. In fact, the best reason I can think of for locating a field of sunflowers next to a well-traveled highway or in a residential district is that wildlife usually avoids cars and people. But depending on your particular situation, the best strategy for harvesting sunflowers is a happy medium somewhere between early and moist and later with crop losses. You will need to wait at least until the heads have shrunk down to half their size. The backs of the sunflower heads should also be dry and not mushy. Seeds will thresh from the heads quite readily when they are between 20 and 25 percent moisture, and while more drying will still be necessary, at least the crop is safe and harvested.

HARVESTING ON A HOME SCALE

If you are a garden grower of sunflowers working on a much smaller (and saner, I might add) level, the harvesting process is relatively easy and straightforward. All you need to do is go out to your sunflower plot and start clipping ripe heads off mature plants with a pair of handheld loppers or pruning shears. Bring the harvested sunflower heads into a dry and airy place like an old barn with a center drive-through alley. You can dry sunflower heads on homemade screens or hang them on an overhead wire. Suspending them on the wire is definitely more work, but it is fairly rodent-proof. Once you are satisfied the heads are nice and dry, you can thresh them out by rubbing them vigorously against hardware cloth. Cleaning is accomplished by pouring the seeds from one bucket into another in front of a box fan or in a windy doorway. If you decide to grow a small amount of sunflowers, a hand-powered harvest is a whole lot simpler than the mechanical production it will take to do larger amounts.

HARVEST WITH A COMBINE

Larger plots of sunflowers are harvested with an ordinary grain combine. Although the actual threshing process is very straightforward and easily accomplished, getting the crop into the machine can be problematic and fraught with frustration. What kind of grain head do you use to clip off a sunflower head and send it up the combine's feeder house into the main threshing cylinder? In 1998, I naively tried to harvest my first crop with a corn head. What a surprise I was in for. Corn heads actually snap ears from stalks with two specially designed metal rolls that are situated just under the gathering chains. I found out rather quickly that ripe sunflower heads are firmly attached to a very fibrous stalk; they don't just snap off like ears of corn. When I finished harvesting that first crop of sunflowers, the field was in a sorry and ragged state. Indeed, some sort of actual cutting mechanism is required to neatly remove sunflower heads from the remaining stalk. I found out many years later that there are special after-market knives you can affix to a corn head for the purpose of harvesting sunflowers.

In 1999, we used an ordinary grain platform with a steel-tooth Hume reel to harvest our sunflowers. It did a much better job than the corn head had the year before, but the results and harvesting efficiency were still far from perfect. Lots of heads fell to the ground instead of being swept back onto the platform and receiving auger, and the Hume reel teeth ended up stabbing and spearing many of the just-cut heads. A little research about this situation informed me that some sort of catch pan needs to be attached below the cutter bar to capture wayward sunflower heads that would otherwise fall to the ground below. Roger Rainville of Alburgh, Vermont, outfitted his old Case combine with the homemade plywood sunflower head. Sunflower pans are a common add-on out in North Dakota and Minnesota where the majority of the crop is grown.

Newer-style grain reels are equipped with heavy plastic teeth instead of the spring-steel versions on my machine. I have seen sunflowers harvested with very little header loss by a modern grain head at Loïc Dewavrin's Fermes Longprés in Québec. There is a type of combine attachment called a row-crop head that does an excellent job removing sunflower heads and delivering them to the inner workings of the machine. The row-crop head has points and dividers like a corn head, but it is equipped with cutting discs and rubber gathering belts instead of snapping rolls.

Combining sunflowers with a row-crop head.
PHOTO COURTESY OF JACK LAZOR

These heads were quite common in the Midwest in the 1970s for harvesting soybeans and milo planted in thirty-inch rows. I just happened to own a John Deere model 653 row-crop head that came along with an old relic of a combine I had purchased from Illinois, and we attached it to my '80s-era John Deere combine for the 2000 harvest and eliminated a lot of the field-loss problems we had experienced in the two previous years. As long as the ancient rubber gathering belts didn't break from old age and fatigue, we were able to move right along with the combine out there in the sunflower field.

The threshing adjustment chart in your owner's manual will point you in the right direction when it comes to setting your combine for sunflowers. The most important thing to remember is to keep the head in one piece as it goes through the machine. Ideally, you want to see intact heads devoid of seeds tumbling off the chaffer screen out the back of the combine. When sunflower heads break into smaller pieces because of overthreshing, little bits will end up as foreign material in the clean grain sample, and dirty sunflower seeds will hold more moisture—impeding grain flow and drying. A properly adjusted combine will put the cleanest material in the tank. Cylinder speed is a relatively slow 250 to 400 RPM. Lowering the concave about halfway will prevent heads from breaking into small pieces. Play with the concave adjustment until you get it right. If heads are traveling through only partially threshed, raise the concave a little. Cleaning fan speed should be reduced because sunflower seeds are much lighter in weight than most other grain crops. The chaffer is usually opened somewhere between half and five-eighths of an inch, and the sieve is usually closed down to three-eighths of an inch. Sunflowers are similar to other grains in that combine adjustments need to be tweaked as field and crop conditions change.

Drying Your Seeds

More often than not, moisture seems to arrive here in the fall, and wetter rather than drier harvesting of sunflowers is normal for this part of the country. Try as we might, we're just not North Dakota, and a final harvest moisture of 20 percent is not that unusual for the Northeast. But wet sunflower seeds are a ticking time bomb waiting to heat up and mold in short order. Once the crop is in the wagon, bin, or tote bag, quick and decisive action is necessary to stave off the onset of spoilage. Sunflower seeds respond quite well to aeration. Inserting a couple of screw-in aerators into a gravity box of moist sunflower seeds will stabilize the load within a couple of days. I've seen several people dry sunflower seeds by sticking one of these aerators directly into a large tote bag. A powerful fan pushing unheated autumn air through the perforated floor of a conventional grain bin will dry sunflower seeds down to the recommended 8 percent level in a few short days. Monitor the moisture closely to avoid overdrying. Sunflower seeds below 8 percent moisture will be harder to press and yield less oil. A propane-fired traditional grain dryer should be your last choice for sunflower seed drying. Dryer fires are a real concern because of the very high oil levels found in the seeds. If you decide to artificially dry sunflowers with propane, take precautions to avoid any chance of ignition, and decrease the dryer gas pressure to lower the temperature of the flame. Locate the dryer as far as possible from buildings and any other flammable materials, and position it so that the air intake fan is pointing directly into the prevailing breeze. This will ensure that sunflower dust and debris coming off the dryer is blown downwind, away from the operation, and not sucked through the air intake fan. One way or another, get those seeds drying immediately after they are harvested.

The first thing you will want to do after harvest and drying is clean your sunflower seeds. Usually one pass through the fanning mill or air screen cleaner will suffice. I usually use a homemade quarter-inch hardware cloth top screen and a small-hole bottom screen for sunflower cleaning. Numbered round-hole screens between twenty and thirty sixty-fourths will also suffice. Cleaning isn't super critical—you aren't preparing these seeds for replanting; all you really want to do is remove any excess plant material like sticks and stems. Turn the cleaning fan up just high enough to blow out empty hulled seeds. Once the seeds are clean, they are ready for the oilseed press.

Pressing Sunflower Seeds

Due to their large size and extremely high oil content, sunflower seeds are the easiest of all the oilseeds to press. There are clever tinkerers everywhere who have designed and built their own hand-powered oilseed presses for sunflowers. Carl Bielenberg of Better World Workshop in Calais, Vermont, built his own hand-pumped sunflower press twenty years ago, long before anyone around here even imagined producing their own sunflower seeds. The unit consisted of a long lever-like handle connected to a hydraulic jack that expelled sunflower seeds through a metal cage. At the time, Carl was specializing in sustainable technology for the Third World. He was (and still is) very generous with all of his information. There is also a commercially available hand-powered oilseed press available from Lehman's Homegoods in Kidron, Ohio.

Entire volumes have been written on oil extraction, and I'm not really much of an expert in this department. I have always been fortunate enough to find someone else with a press to process my sunflowers and flax. What I do know is the seed should be clean and dry, because if the sunflower seeds are too high in moisture, cloudy oil full of water droplets will be the result. The oil will look like it contains little bubbles, will be of lower quality, and will tend to go rancid quickly. Warm seeds squeeze more easily than cold seeds, and for this reason most oilseed presses are located in a heated space. Bring the seeds in from cold storage a few days in advance to warm them up to at least room temperature. If you're able to gently heat your sunflower seeds twenty or thirty more degrees up to about one hundred degrees, they will press more easily and yield higher amounts of oil. Although this process is known as cold-expeller pressing, a little extra warming will make everything work that much better. Commercial industrial oil extraction on the other hand is done under much higher heat with the aid of solvents like hexane.

There are a whole lot more oilseed presses around now than there were five years ago. Presses are available from China and from Northern Europe, and the Chinese models have been quite popular because they cost only a fraction of what you would have to pay for a German or Swedish model. Fairly large medium-scale Chinese units sell for as little as three to five thousand dollars. If you are just getting into pressing your own sunflowers and are strapped for cash, by all means buy a Chinese machine. Just remember that they aren't built to the precision tolerances found on the German Kernkraft press that sells for more than three times as much. If you are handy and have access to reasonably priced machining services, however, a Chinese press is just the ticket. You might find yourself machining components to make modifications or beefing up the design parameters. These presses have been known to have other drawbacks like a lead-based paint finish. Many farmers will start out with a Chinese model and step up to the more expensive European technology when finances permit.

I would advise finding someone with the proper equipment to press your sunflower seed into oil. Lend a hand and you will learn more than you ever wanted to know about oil extraction in the process. There is plenty of work for about one and a half people when you are pressing seeds, and sometimes it's just nice to have some company. You can always work to keep the oilseed cake by-product away from the press and bagged up. This sunflower meal is a good protein source in chicken, pig, and dairy rations. It tests somewhere around 25 percent in protein and has a good energy profile. The actual sunflower oil is caught and stored in food-grade plastic barrels, and then left to settle for a few weeks to naturally clarify. Freshly pressed sunflower oil generally has all sorts of little black particles from the hull suspended in solution. All of this will settle out over the course of several weeks, leaving you tasty and unrefined oil that is good for cooking and salads. To bottle the oil, you simply siphon or pump the clear stuff from the top of the receptacle and leave the dregs behind in the bottom. The leftover mixture of oil and hull residue can be used for treating wood or as an energy supplement in a chicken ration. There are a few individuals that prefer to send all of their freshly squeezed oil through a pressurized maple syrup filter press. This is the quick way to prepare your oil for bottling, but I'm not sure it's the best, as sunflower oil has a special aromatic quality that will disappear when it is speedily filtered under pressure. This might be a better option if you are processing large amounts of seeds to make oil that will be used for biodiesel or furnace fuel. If your sunflower oil is destined for salad dressing and cooking, though, take the time and let it settle out naturally.

Sunflowers are probably our best choice for an oilseed crop in this part of the country. Until relatively recently, the production of homegrown vegetable oil was unheard here in the Northeast. However, in the last ten years a number of enterprising and innovative individuals have begun to grow sunflowers and produce oil for a variety of uses. The yield and quality of sunflower oil in any given year is directly proportional to the amount of sunshine that is bestowed upon us. I have certainly found that no two years are the same. In a cool and cloudy season, you can expect less oil and a reduced flavor profile from a crop of sunflowers. The opposite will be true in a sunny and warm dry season. Despite wildlife pressure and a fickle climate, there are still plenty of reasons to add sunflowers to your lineup of homegrown grain crops: They can be very easily cultivated and harvested by hand on a garden scale with little or no investment in specialized machinery. Seek out others in your regions who are already growing sunflowers for advice and pointers. A little knowledge and research on your part will go a long way to ensuring your success with this new crop.

Flax

Flax (*Linum usitatissimum*) is another oilseed crop from the high plains that has possibilities here in the Northeast. Like most of the other cereals, flax dates back to ancient times. Archaeologists have found evidence of the crop in excavations of Stone Age dwellings in Switzerland. The Egyptians made fine linens from flax fiber; linen was standard for clothing throughout Europe as well. English colonists brought flax with them to the New World to be cultivated for fiber. In colonial New England, fiber flax was grown in densely planted stands that favored the development of many stems and very few seed heads. I gained some experience with flax processing when I worked on the historic farm at Old Sturbridge Village in Massachusetts in the early 1970s. Flax was cut with reaping sickles and laid into flowing streams for "retting." This process basically decomposed the outer stalk sheath, exposing the inner linen fibers. Retted flax was then dried back out and crushed with a wooden flax break to loosen the once tough outer covering. The final step in the process before the fiber could be spun into thread was to pull the fiber and crushed stalk through a series of three upright steel-tooth combs called hettles. The first hettle had large teeth spaced far apart. Once the flax bundles had been pulled through a medium and a very fine hettle, the outer stalk casing was completely removed and the fiber was ready for spinning. Fiber flax was a common crop in colonial New England, and we know that seed flax had to have been grown in small amounts during the same era because it was always needed to plant the next year's crop. The production of flax for fiber continues in Europe to this day. Special varieties that tolerate high seeding rates have been developed for this industry. These flax plants are tall, with few branches and low seed production. Seed flax is much shorter, has multiple branches, and is bred for high seed yield.

Flax is an annual broad-leafed plant with short narrow leaves. It has one main stem that branches profusely, especially when it is planted at lower populations. Average plant height is twenty-four to thirty inches. Stems terminate in a multibranched inflorescence with beautiful and very distinctive blue flowers. Flax seems to do quite well in the drier areas of the northern plains because it sends down a taproot that can grow as long as forty inches. Just about all the flax produced in the United States comes from three states: North Dakota, Minnesota, and South Dakota. As settlers moved westward, they took flax with them, and there is ample historic evidence of nineteenth-century flax cultivation from New York State into the Ohio Valley and the Upper Midwest. It is probably not the natural choice for a crop in our region, but it will grow here. Flax was successfully grown here 150 years ago, and there is no reason why it can't be done again. The length of our growing season is not a problem because the crop ripens in one hundred days.

Most of the flax planted in North America is grown for seed as opposed to fiber. Oilseeds and linen come from the same plant. If you intend to produce fiber, plant a lot more seed because you want lots of stalks instead of seedpods, but those who want to harvest seed need to back off the population to somewhere between seventy-five and a hundred plants per square foot. This is achieved by planting fifty to seventy pounds to the acre. Historically, flaxseeds have been crushed for the production of linseed oil, which is usually boiled and used in paints and as a wood preservative. Human consumption of flaxseed and cold-pressed oil is actually a relatively recent phenomenon. The lignans found in flax are proven to be anti-carcinogenic, and flaxseeds are a common ingredient found in many cereals because of their mild laxative effect. Flaxseed oil is the only major plant source of the omega-3 fatty acids commonly found in fish oils, and 55 percent of the oil contains alpha-linoleic acid, which has numerous health benefits. Flax is 40 percent oil by weight, so flax oil for human consumption can be a very lucrative enterprise even if your yield is low. The average pint of this nutritional supplement sells for more than twenty dollars at natural food outlets. The by-product of flax oil production, linseed meal (or flaxseed oil cake), is a tremendous protein source in livestock feeds; protein runs about 25 percent. Cattle fed linseed meal are well known for having shining glistening coats. Recent research has also shown that dairy rations containing flaxseed oil cake are responsible for reducing the amount of methane emitted from the bovine digestive tracts.

Planting Flax

When selecting a location to plant a crop of flax, choose a well-drained soil with reasonably good organic fertility. Most soils that will grow wheat or barley will be well suited for flax. Soil pH should be somewhere between 6.0 and 6.5 for best results. Because of its short stature and the fineness of its stems, flax does not compete well with weeds. For this reason, select a field with an empty weed seed bank if possible. Over the years, I've learned this lesson the hard way. Several times, I've established a good-looking flax seeding only to have it overtaken and totally engulfed with annual weeds. Finally, I learned that an old pasture or hay field that hadn't been plowed or heavily fertilized in years is the best place to plant flax. A plowed-down crop of red clover also works quite well. But I wouldn't advise planting flax after a row crop like soybeans or especially corn because weeds will be a major problem. Flax requires about fifty to seventy pounds of nitrogen to produce a good crop. You'll want to achieve this fertility from the breakdown of soil organic matter as opposed to soluble nitrogen from the fertilizer bag. To accomplish this, plow down some old sod.

Flax is usually the last of the cereal crops to be planted. When sod was broken on the prairies, flax was the first crop to be planted. Early planting is beneficial, but seeding need not be as early as it would be for oats and wheat in late March or early April. Recently emerged little shoots of flax can survive twenty-eight-degree frost for a few hours, and once the plant develops a second set of leaves, it can withstand temperatures as low as twenty degrees. Late April or early May is the best time to plant flax.

Good seedbed preparation is essential to establishing a great stand of flax. Fall-plowed ground will work up into a much more mellow and less lumpy soil. Flaxseeds are tiny and should be planted only a quarter to half an inch deep in firm but friable soil. Two to four passes over plowed furrows with a disc might be necessary to achieve this sort of condition. Once you've established a nice seedbed, one pass through the field with a roller or cultipacker will firm the ground enough for good soil-to-seed contact. If you have a grain drill with press wheels, pre-plant soil firming won't be necessary. Make sure that the depth control on your grain drill is functioning properly so that the flaxseeds won't be planted too deep. My twenty-year-old grain drill has an adjustable check nut on its hydraulic lift cylinder for this very purpose. I can screw in the depth-control nut to shorten the stroke of the cylinder; this in turn prevents the seed discs from dropping too far into the earth. If your drill doesn't have this feature, you can improvise by installing commercially available little metal "doughnuts" around the lift cylinder rod to shorten the stroke to achieve the desired seeding depth. Last but not least, make sure your drill is calibrated to drop the proper amount of seed to the acre. The fact that flaxseed is so tiny makes it easy to overplant. Follow the setting recommendations in your owner's manual or under the lid of the grain drill. Fifty to seventy pounds of seed to the acre is just about right for common brown flax, while golden flax doesn't have quite as much seedling vigor and can be planted a little heavier. If you are totally stumped about how to set your grain drill for flax, try planting a small amount of measured seed with a very low setting and then determine the area planted. You quickly find out if you need to increase or decrease the seeding rate.

Flax can also be sown by hand in a garden situation. All you have to do is pretend that your garden plot is a miniature version of a much larger field. Start by preparing a fine seedbed with a rototiller or other garden soil preparation tool. Broadcast between one and two pounds of flaxseed per one hundred square feet. A hand-cranked broadcaster like a Cyclone seeder will do a much more even job of distribution than the human hand and arm. Lightly cover the seed by pulling a small tree branch over the area; a final pass with a lawn roller will firm the seedbed. Hand-pushed garden seeders are even more accurate, but you must take care not to plant too deep. Pack the seedbed first with the lawn roller to promote good soil-to-seed contact for superior germination.

After the flax is planted, there is nothing to do but wait for emergence. In a week to ten days you will see the first plants beginning to appear. Recently emerged flax looks entirely different from the common cereals we are more accustomed to seeing. Flax shoots will have tiny side leaves attached to their stems, and they look a bit like miniature trees. Cultivating a crop of flax with a tine weeder is totally out of the question. Wheat or oats will stand up well to this sort of operation because planting depth is at least one inch. However, flax planted at one-quarter to one-half inch deep will not tolerate this sort of abuse. Some experiments have been done in Denmark and here in the United States with actually cultivating flax planted in wider rows, and the Danes have developed the Schmotzer steerage hoe for this very purpose. The machine is attached to the three-point hitch of a tractor and navigated through twelve- to fourteen-inch rows of flax; the seated operator steers the ranks of small cultivator shovels through the crop. Preliminary trials by UVM agronomist Heather Darby have achieved high levels of weed control in test plots of flax located in Alburgh, Vermont. Still, your best insurance against weeds in flax is to select varieties with little or no weed pressure.

Growth and Harvest of Flax

Flax is much shorter than the other more common cereals, and its early growth is much less rapid. It is a small, fine-stemmed plant with many little leaflets. By the beginning of June, it is noticeable out there in the field as a leafy three- to six-inch plant. Vegetative growth lasts between forty-five and sixty days; at the end of this period, plants have grown to their maximum height of twenty-four to thirty inches. The appearance of blue flowers marks the end of vegetative growth and the beginning of the reproductive phase. Flax is somewhat indeterminate, with a flowering period that can last between two and three weeks. There are few things in the agricultural landscape more beautiful than a field of flax in full bloom.

As flax matures, its flowers eventually turn into little round capsules called bolls. The bolls first appear as tiny green balls fixed to the ends of all the side shoots at the top of the plant. Each flax boll has

Flax in full bloom. PHOTO COURTESY OF HEATHER DARBY

the potential to produce six to ten seeds. The maturation process takes about thirty-five days, and during this time the seeds develop in the bolls. Eventually, the green color of the bolls begins to fade as the seeds ripen within. Like any other grain plant that is ripening and drying down, the green color will first fade to a creamy hue and then turn dark brown. When 90 percent of the bolls have turned brown, flax is considered physiologically mature. Stems may still appear greenish at this point. By this time, it is mid-August and we hope that the weather is sunny and dry enough for the flaxseeds to drop to a moisture level that will permit an easy harvest.

Harvested flax stores best at 8 to 10 percent moisture, but this very low level is difficult to achieve even in the drier region of North Dakota. You can speed the dry-down process by cutting and swathing the crop. If you have a garden plot of flax, mow it down with a very sharp serrated sickle and lay it out to dry further. Those with larger plantings will need to harvest the flax crop with a swather or draper head. Flax is usually mowed six to eight inches off the ground and windrowed on top of the stubble. However you cut your flax, it is extremely important to have sharp and precisely adjusted mowing equipment, because the stems are so fibrous. Several years ago, I outfitted my swather with brand-new section knives and guards before cutting my small field of flax. The effort and expense were well worth it. I cut through the flax like a hot knife through butter. I have heard numerous horror stories of people struggling to mow down

Small-scale flower and hulling facility in Ontario. PHOTO COURTESY OF JACK LAZOR

flax, and the old adage of a sharp tool being best for the job rings true once again.

Flax is not an easy crop to combine. Be patient and allow the swathed windrows sufficient time to dry down completely. Eight to 10 percent is considerably lower than the 13 to 15 percent that's customary for more common grain crops. You needn't worry about rain dampening swaths of flax and causing premature decomposition and molding. Flax straw is so tough and fibrous that it is undaunted when exposed to rainfall. The tiny seed bolls will dry back out quickly when the sunshine and breeze return. Swathed flax must be totally dry, however, to be completely threshed by the combine. Once again, I have learned this lesson the hard way. In 2008, I headed to a small one-acre field of flax with my combine and attached windrow-pickup head. The machine readily gobbled up the flax, but a close inspection of the straw coming out of the back revealed that a quarter of the bolls were unthreshed and still contained all their seeds. I'd thought the flax was dry enough, but it really wasn't. Flax threshes hard, so make sure that the internal workings of your combine are in tip-top shape. Worn cylinder rasp bars will also contribute to poor threshing. Start out by reading your manual and adjusting the combine as directed. Since the seed is so small and lightweight, the chaffer and sieve at the rear of the combine must be closed up tighter than would be necessary for a larger-seeded grain crop. Cleaning fan speed should also be reduced to prevent the seed from being blown over the chaffer and out the back of the machine. Once everything is adjusted and working properly, it will be pure joy to see flaxseed streaming into the grain tank of your combine.

After the Harvest

Flaxseed is an amazing grain. It is shiny and slippery, and if it gets wet, it has a sticky and mucilaginous quality about it. Once you've off-loaded your combine into a gravity wagon or holding bin, the next step is to clean any weed seeds and debris from the flaxseed. You will need fanning mill screens with fairly small holes and slots to accomplish this task. If your flax contains round black mustard seeds, another cleaning strategy will be in order because flax and mustard are similar in size and weight and do not separate well in an ordinary air screen cleaner. I have had pretty good luck removing mustard seeds from flax with a spiral cleaner that is more commonly used on soybeans. The round mustard seeds will readily roll down the spiral while the flatter flax seeds tumble down to a different outlet. This process is slow and tedious, but it does work. In North Dakota where flax is a very common crop, special Carter disc mills are used to clean flax. These specially designed machines have the same series of vertical discs turning on a central horizontal shaft as a regular Carter disc grain cleaner. However, the discs on a flax mill are outfitted with extra-small pockets to pick up and separate flaxseeds from grains and weed seeds of different shapes. As far as I know, there aren't any of these special machines in our part of the country. A used flax mill in North Dakota sells for well over a thousand dollars, which means you would have to be growing a lot of flax to justify buying one of these units and having it shipped halfway across the continent. Those of us growing small amounts will have to make do with our own ingenuity and the cleaning equipment that we already own.

Store your flax away from light and humidity at 8 to 10 percent moisture. Flaxseeds can be consumed directly as a topping on salads or an ingredient in breakfast cereal; some people grind them in a hand-cranked coffee grinder and add them as a healthful condiment to a variety of dishes. Be careful not to consume too many flaxseeds at one time—they have a laxative effect on the human digestive tract. If you plan to crush the seeds into oil for human consumption, a common oilseed press like a Kernkraft will do the job. Pre-warmed seed will press more easily and yield more oil than cold seed. This is especially true in the winter months, and most farmers who are operating oilseed presses like to work in a heated space for this very reason. Oilseed pressing requires patience: This is not a rapid process when done on a farm scale; it might take an hour to process one hundred pounds of flaxseed. Another important consideration is the fact that flax oil doesn't keep indefinitely. The oil is so alive with goodness and healthful benefits that it tends to oxidize and become rancid within three to four months of being pressed. You won't want to squeeze your entire flax crop into oil all at once unless you have an immediate market for it. Once pressed, flax oil definitely should be refrigerated to slow down the onset of rancidity. Store it in amber glass bottles to protect it from sunlight. The rancidity factor can be a real problem for those of us who have to depend on others to do our pressing for us; we might need to make multiple trips to the oilseed press. I have been able to solve this problem by storing my flax oil for long periods of time in the deep freeze. Flax is my favorite oil on a salad, and there is something extra special about pouring generous amounts of your own homegrown golden flax oil onto your food.

Canola

In recent years, canola has become a principal crop on the prairies of western Canada and the high plains of the north-central United States, and it has now begun to find its way eastward to our region. A number of potato farmers way up north in Aroostook County, Maine, have successfully added the crop to their rotations. Here in Vermont, several farmers who grow their own tractor fuel have begun growing canola. Roger Rainville of Alburgh, on the Canadian border, has modified his John Deere tractors to run on straight vegetable oil, and John Williamson, who farms in Shaftsbury in southwestern Vermont, grows canola to supply his small biodiesel refinery. There are numerous other individuals in New Hampshire, Maine, and New York State who have been experimenting with canola for oilseed production. Because canola prefers cool weather, it seems to grow best in northerly areas. It has become a major crop in the Saguenay region around Lac St. Jean one hundred miles north of Québec City. Canola seed is available for both spring and fall planting, but spring varieties seem to thrive best in the north, while fall-planted canola is better suited to the warmer climate found in the more southern reaches of our region. With its potential for high yields and its 40 percent oil content, canola has potential here in the Northeast. However, it is a fickle crop that is much harder to establish, grow, and harvest than any of the other cereals or more common oilseeds. If you are thinking about growing canola, remember that you are venturing into uncharted territory. You will be an experimenter and a pioneer like all of the other individuals who have recently begun to grow this modern oilseed.

Origins of Canola

Canola (*Brassica napus* L.) as we know it today has only been around since the mid-1970s. The plant is basically an edible rapeseed. Rape (*rapum*) is a member of the Brassica family along with mustard, rutabaga, cabbage, turnips, and brussels sprouts. Rape is still a major crop in England and Western Europe, and fields of brilliant yellow winter rape are evident all over England every March. Rapeseed oil was originally produced in Asia for use in industrial lubricants—the oil's ability to emulsify with water allows it to be used in steam engines and for other industrial applications. Rapeseed production began to shift to North America during World War II because the Allies needed the oil for the war effort and trade routes to the East were disrupted by the Japanese. After the war, rapeseed production dropped precipitously. By 1950, Canadian production had fallen from a high of eighty thousand acres to four hundred acres. The problem with rape was (and still is) that it's unfit for human consumption. The plant contains 45 percent erucic acid, which is toxic to livestock and humans alike. Rape also has high levels of glucosinolates, the compounds that give mustards and turnips that distinctive hot taste. Because the plant grew so well on the prairies, however, a core group of western Canadian plant breeders embarked on a program in the early 1950s to reduce the erucic acid content of rapeseed. Most of the work was done at the University of Manitoba and the Agriculture Canada Research Station at Saskatoon, Saskatchewan; the first edible rapeseed oil extract was developed by 1956. In 1958, Keith Downey and Baldur R. Stefansson began their breeding efforts at Saskatoon and Winnipeg. Several low-erucic-acid rapeseed varieties were released in the '60s, but triumph finally came in 1974 with the introduction of a variety called Tower. This new plant was called first called LEAR, which stands for "low erucic acid rapeseed." The word *lear* or *rapeseed* certainly wasn't going to sell in the marketplace, so in 1978 the Western Canadian Oilseed Crushers Association adopted and trademarked the new name of canola, which stood for "Canadian oil—low acid." By 1979, 3.4 million acres of canola were being planted in western Canada.

It is very interesting that *canola* and *controversy* both begin with the same letter. Canola oil has been touted as a new miracle substance, but there are many out there who feel it is a modern industrial creation that's not all that it is cracked up to be. After its initial introduction, canola oil became very popular in Japan, Canada, and Europe, and the FDA recognized it as safe for use in human food in 1985. Medical researchers and nutritionists first began to recognize the importance of monounsaturated fats in the 1980s. Polyunsaturated fats like corn oil had predominated until this point, but canola was considered perfect because it is 57 percent monounsaturated and only 5 percent saturated. The balance between omega-6 (24 percent) and omega-3 (10 percent) fatty acids is also considered ideal, and so canola quickly became popular as a cheaper and lighter-tasting substitute for olive oil. Canola oil does have its detractors, however. Many lovers of traditional food and artisanal eating consider it to be a man-made industrial product that has minute amounts of the same toxic substances found in mustard gas. In addition, most of the canola oil used for human consumption these days is extracted from the seed with a chemical solvent called hexane. Cold-pressed canola oil has a distinctive flavor with only a slight hint of brassica. The commercial stuff has no flavor at all and is probably better used as a biodiesel ingredient instead of salad dressing.

There are also many questions to be asked about how Drs. Downey and Stefansson achieved their original breakthrough in removing the erucic acid and glucosinolates from the rape plant. These hardworking plant breeders experimented with thousands of varieties of rapeseed from all over the world. They developed a system of "half seed" technology where they could actually split a seed into two parts and have both remain viable. By doing this, they could grow out half the seed of a particular variety; if it had low acid content, they would still have the other half to propagate and increase. Various histories of canola claim that it was developed with conventional breeding techniques, but there is very little mention of the fact that a mutation of the original rape plant was necessary to the success of this miracle cultivar. Mutations in plant breeding are imposed upon plants by means of chemical and radiation intervention. And while canola certainly wasn't a product of the modern genetic engineering that has become standard fare these days, it was genetically manipulated with some not-so-natural substances.

Last but not least is the story of canola over the last fifteen years. In 1996, Monsanto introduced its Roundup Ready canola to the marketplace in western Canada. By 1999, over half the canola grown in Canada was genetically modified. You may be familiar with the case of Saskatchewan farmer Percy Schmeiser whose conventional canola crop was contaminated by drifting pollen from neighboring Roundup Ready fields. Monsanto took Schmeiser to court for patent infringement. (As a matter of fact, Keith Downey, one of the fathers of canola, testified on Monsanto's behalf against Mr. Schmeiser.) Fifteen years later, it has become the common perception that there is no canola to found anywhere in North America that doesn't have some minute trace of Roundup Ready DNA in its genetic makeup. I'm sure there is some pure canola out there somewhere, but it is very difficult to know because genetic PCR testing is very expensive. Commercial hybrid varieties dominate the canola industry, and just about all of them are genetically modified. The industrial origins of canola and its propensity to be genetically modified all add up to questions about the crop's compatibility with organic production. Personally, I haven't grown the crop on my farm and don't think I will anytime soon. I have a lot of latent mustard seed in my soils. This mustard would become a weed in my canola and would be very hard to remove, although I have pressed some of my mustard seed cleanings into oil with good success. The GMO contamination issue is also a big red flag for me. Nevertheless, I will tell you how to grow canola here in the Northeast in case the crop still interests you.

Seeding Canola

Canola grows well on a variety of soil types, but it can be very difficult to establish. The NDSU (North Dakota State University) Canola Field Guide claims that a clay loam soil that doesn't crust is best suited for the crop. Excellent surface drainage is required, because canola doesn't like standing water or a waterlogged soil. Plant as early as you can in April, just as you would for a crop of wheat. Canola is frost-tolerant down to twenty-four degrees, so if a stand of canola is established in April when the weather is still cool and damp, the heat and drought stress that can impact it in May or early June will be minimized. For some reason, young canola plants will suffer if air temperatures climb above eighty-five degrees, although a well-established early-planted stand can stand up to this sort of stress. Soil preparation is paramount. Work the soil down to a fairly fine consistency with secondary tillage equipment like a field cultivator; a cultipacker or cultimulcher that tills and rolls the soil will give you the firmness necessary to achieve good seed-to-soil contact. Adequate moisture levels are also needed to foster complete germination. Since canola seed is so small, only five to eight pounds of seed per acre is needed to establish a good stand. Agronomists who have the means of counting seeds will tell you to plant between seventy-five and eighty-five thousand seeds to the acre. Seeding depth should be somewhere between half an inch and one inch, and a grain drill with six- or seven-inch spacing is the recommended tool for the job. Calibrate the drill and check seeding depth before beginning to plant. Attention to detail is the key ingredient to success in establishing a good stand of canola. This is a very finicky crop—you don't just "throw the seed in the ground" like you would for a crop of oats.

SOURCING YOUR SEED

Canola seed isn't readily available in our region. Begin making arrangements to procure seed well in advance of the growing season. There are plenty of seed sources in the Dakotas and in Canada. Most of the canola planted these days is hybrid and genetically modified to withstand the application of the glyphosate herbicide. These varieties are produced by seed industry giants like Monsanto and Pioneer. As an organic producer, you will have much better luck buying seed from an outfit like the Albert Lea Seedhouse in Minnesota or from another farmer who is saving seed from his own open-pollinated crop. There are two subspecies of canola to choose from—the Argentine type (*Brassica napus*) and the Polish type (*B. rapa*). Argentine varieties are generally taller, have higher oil content, and take about ninety-five days to mature. The Polish types of canola are intended for the very briefest-growing-season areas; they are shorter in stature, lower in oil content, and mature in eighty days. You will also have to decide if you want to grow spring or winter canola. Heather Darby of UVM Extension has done extensive research on winter canola on the Lake Champlain Islands, which are known for their particularly benign climate. If a stand can survive the winter months, it has the potential to yield over two tons of canola seed to the acre. Heather has found that one of the keys to ensuring the survival of a winter canola crop is early establishment, which means seeding your crop at the very end of August or during the first days of September. The canola needs to be well grown and robust when it transitions into dormancy in November or December. Most of us who live in the northern reaches of our region will be planting spring canola. There are some sources of canola out there that claim to be totally free of GMO contamination. This is the seed we want.

FERTILITY REQUIREMENTS

Canola's fertility requirements are similar to those of the small grains. Adequate mineral availability will help to grow a good crop. Nitrogen and sulfur seem to be the most essential fertility elements. Canola has an insatiable appetite for sulfur—a harvest of one ton of canola seed to the acre contains twelve pounds of sulfur. A light application of a sulfate-based fertilizer like sul-po-mag can make the difference between an ordinary and an exceptional crop. A moderate field dressing of compost or well-rotted manure will provide some of the nitrogen needed for good plant growth. We are all at the mercy of the weather when we plant crops, and canola is new and different. There are few if any mentors here in our region to give advice or tell us what to do with this man-made creation that has come to us from western Canada. But if you supply your canola with the proper organic and mineral fertility, you have a decent chance at success.

The Growth Period

Once a field of canola is established and growing well, you can stop worrying. The early-season heat and drought stress that can be such problems in the Dakotas in May and June generally create few issues in this part of the country—we are usually worried about too many rain showers making it impossible to harvest hay at this time of the year. Canola will begin to flower in late June and early July. Weather extremes with excessively high or low temperatures can be a bit of a problem at blossom stage. Excessive heat can cause blooms to "blast" or abort, and a dead flower will not produce seedpods. Low temperatures will also slow down plant development and minimize pollen shed, which has a profound effect on the ultimate yield of the crop. More so than other grain crops, canola seems to require perfect growing conditions to maximize its growth potential. I wonder if this has anything to do with the fact that it was created by humans as opposed to being selected by nature over the millennia.

WEED CONTROL

Weeds in canola are a real challenge to the organic grower. Some pre-emergence light tine weeding might be in order. Once the tender canola seedlings are up and growing, however, mechanical weed control is no longer an option because of the tender nature of the young seedlings. Weeds are even a problem to conventional growers who have the option of using herbicides on their crops. Canola is very sensitive to most herbicides and can be easily stunted or damaged when they are applied, which is the primary reason why the Roundup Ready technology was so welcomed by the canola industry. Wild mustard is also another major problem in a field of canola. It is in the same brassica family, which makes it difficult to distinguish and eliminate. Again, the conventional industry has solved this problem with Roundup. As organic farmers, we don't have this technology (and its consequences) at our disposal. Professional agronomic advice has it that we shouldn't be planting canola in a field with dormant wild mustard (we call it kale) seed in the soil. This leaves me out of the picture for sure. If you are growing canola for home biofuel use, mustard infection probably won't be too much of a problem. You can crush the mustard seeds right along with the canola; the diesel engine that you will eventually be powering with the vegetable oil will never know the difference.

DISEASES AND PESTS

Since canola is relatively new to our region, disease has not been much of a factor. The plant is susceptible, however, to fungal infection during periods of prolonged moist and cool weather. Blackleg seems to be a concern out on the high plains, but not here in the East. This is a fungus whose spores are spread by wind and rain splash as well as diseased seed. Once canola plants are infected, deep stem-girdling cankers will develop right above the soil line. Thankfully, we have not seen any of this particular infection in our region. However, we do have plenty of sclerotinia or white mold potential here in the Northeast, and it can be a major problem in canola crops. White mold can survive five to six years in a soil as little black fungal bodies called sclerotia. These sclerotia will germinate with the advent of heat and rainfall. Mushroom-like bodies called apothecia are formed in the process, and millions of white mold spores that can infect susceptible plants are released. Conventional farmers use chemical fungicides to battle white mold, but as organic farmers, we have to depend on good rotation practices to deal with this scourge. Canola should never

Oilseeds

Canola blossom. PHOTO COURTESY OF SID BOSWORTH

Canola in full blossom. PHOTO COURTESY OF SID BOSWORTH

be planted after white-mold-sensitive crops like sunflowers, soybeans, or dry beans. Balanced mineral and micronutrient fertility will also help to produce strong, rapidly growing plants with more potential to resist white mold infection.

Flea beetles are the only insect that will bother canola. They are usually only a problem in May when seedlings have just emerged and are developing their first few sets of leaves. Hot, sunny weather seems to bring on these little critters, so some cool weather and a little dampness actually seem to favor a newly seeded canola crop in the month of May. Once the plants have grown beyond the four- to six-leaf stage, flea beetle invasions are no longer a problem. Rapid and vigorous growth will outdo their onslaught.

REPRODUCTION AND RIPENING

Once the flowering period is complete, seedpods will form on the canola plant. Reproduction is now under way. Final yield will depend on how many flowers actually produce seedpods and how well they fill. Excessive heat during the flowering period may rob some yield potential by causing blasting (which we have already discussed). Canola seedpods are long and thin—they look like slightly larger versions of the pods that we see on our favorite old friend, wild mustard. Once a canola plant has begun to make seed, it's time to bring on the heat and the dry weather. This is why the crop does so well out on the semi-arid prairies. At this point in the growing season, all we can do is wait the four to six weeks for the ripening process to run its course. Mother Nature is in control; we have to be patient and let her do her job.

The Harvest

True to form, harvesting a canola crop isn't simple or easy. Ripening and combining this man-made creation isn't as straightforward as it would be for a field of oats or wheat, as canola ripens unevenly. Standard practice in the plains region is to mow down canola with a swather and then leave it in a windrow to cure for several weeks. As your field of canola begins to ripen, check the crop often to determine its stage of maturity. The bottom third of the plant sets seed and ripens earlier than the top third. When 30 or 40 percent of the seedpods on the main stem have turned from green to brown, the plant is considered to be physiologically mature. The pods at the top of the canola plant may still be green in color, but they will be filled at this point. It is now time to swath the crop. Canola is a bulky crop with lots of volume, and for this reason, the swather cutting mechanism and conveyor cross canvases must be in tip-top shape. Canola also has another problem that can complicate the harvest. The little round canola seeds inside the pods will remain green even as their moisture declines to acceptable levels. Ten to fourteen days of swath curing time is needed for the chlorophyll in the canola seeds to fade as the harvested plant transitions through a metabolic process, and seed moisture must remain at 20 percent for this to happen. If the swathed canola dries out too quickly, the ripe seeds will remain green. Commercial growers lose income when this happens: They are docked in the marketplace because of green seed. This probably isn't a major problem for us eastern novices, however, because we are most likely growing canola for biodiesel instead of the human consumption oil market. Nevertheless, the fact that we have to worry about both seed moisture and color adds extra complexity to the harvest. Swathing needs to be done at the optimum time to achieve a maximize yield, and if you wait too long, seed shattering is almost guaranteed. The experts will tell you that if you are growing canola, expect to find it as a weed in your next year's crop.

COMBINING CANOLA

There is also the option of direct-combining canola. We can get lots of information about swathing and other harvest strategies from the experts in North Dakota, but when it comes right down to it, we are limited by our regional climate and the machinery that is available locally. Swathers are rare in these parts. Roger Rainville from Alburgh, Vermont, is the only person I know who grows canola in this area. Several years ago, he borrowed my eighteen-foot swather to harvest his first crop. The whole swathing process gave him fits, and he swore he would never do it that way again: Seed shattering turned into a major problem. Roger has since taken to direct-combining his canola crops with excellent success. His yields always exceed a ton to the acre, and there is very little seed wasted from shattering. None of us is an expert with canola, but we do know how to run a combine to direct-cut a grain crop. So, unless you are some sort of self-styled canola expert, forget about all the swathing and curing business. Wait until half or more of the seedpods have turned from green to brown, then take the combine to the field and direct-cut the crop.

To combine canola, take the necessary precautions you would with any small-seeded crop. Check the metal on the sides and bottom of the machine for small holes and worn spots. Canola seeds are tiny, round, and slippery; they will escape from your combine wherever they can find a way out, so cover any little pinholes in the exterior sheet metal with duct tape to prevent leakage. Threshing cylinder speed should be lowered to half to two-thirds the recommended settings for regular cereal crops. Drop the concave down as you would for soybeans. You can avoid overthreshing by providing just enough force and abrasion to crack open the canola pods. If you get too aggressive, broken pods and stems will end up in the clean grain sample. Cleaning fan speed will also need to be cut back to prevent seeds from being blown out the back over the chaffer screen. Cut the canola crop as high off the ground as you can and still get all the grain—there is no need to put any more material through the combine than you must to get the job done. Ten percent is the best moisture for harvesting canola. The crop will thresh nicely, and there will be only minimal seed loss. Canola dries quite rapidly in the field once the leaves drop from the plants; a prolonged period of hot dry weather can dry the crop fast enough to drop the moisture below the 10 percent level. This is when the potential for shattering and seed loss increases. Canola is such a new crop to us here in the Northeast that there are very few experienced mentors we can call on for help and advice, but there is no substitute for simply going to the field and trying our best to harvest something brand new. Common sense and good observational skills will go a long way in helping us figure out what to do.

After the Harvest

Post-harvest treatment of canola isn't really any different than it would be for any other cereal grain. The crop must be dry before it goes into the bin for long-term storage. Canola will keep well at 8 to 10 percent moisture, which is the exact same level you would like to see for flax or sunflower seeds. Some sort of drying will be needed if the moisture of the harvested crop is higher than 10 percent. Forced-air bin drying works well for canola, but there are a few caveats. The seeds are so small that they can lodge in the perforated flooring and plug the aeration. Special grain bin flooring with extra-small holes might be needed in a situation like this; you can also improvise by spreading finely woven burlap on the existing grain bin floor before filling the silo with canola. A little supplemental heat will speed the canola-drying process inside a forced-air grain bin. If you choose to batch-dry your canola in a freestanding grain dryer, take every precaution possible to prevent a fire. Seed with a 40 percent oil level will readily combust if something goes wrong in the drying process. Exercise extreme caution and care with your harvested canola. A well-preserved crop will be there for you when you want to unload the silo and bring seeds to the oilseed press.

Oilseeds are up and coming here in the Northeast. Sunflowers and flax are old-fashioned crops that have potential for making high-quality locally produced food-grade oil. Sunflowers lend themselves best to small-scale low-tech production on a garden scale because the whole process can be done by hand. Flax poses a bit more of a problem because it is so small-seeded and so prone to weed infestation. Canola, the new kid on the block, offers us the best total oil yield per acre, but it is not without its problems. If you are still bound and determined to grow this finicky crop that has been a bit overmanipulated by humankind, beware of the perils and pitfalls. You might want to start out with a small experimental acreage to get some experience under your belt.

CHAPTER SEVENTEEN
The Minor Grains

Every plant species must reproduce in order to continue existing in the natural world, and just about all of the more common farm-based plants make seeds that we can harvest for grain. Up until this point, we have concentrated on the most popular farm crops. But what about those other purchased farm seeds that we take for granted? The clover, timothy, and alfalfa seeds that we put in the grass seed box of the grain drill are usually produced in distant, drier regions, but they will grow in the Northeast. The millet seed that you might plant on your dairy farm in early July for a late-summer forage boost was probably produced on the arid plains of North Dakota. Once planted, this warm-weather annual grass will thrive here in the Northeast because of our more abundant rainfall. So if millet grows like blazes here in our warm, humid summer climate, why not harvest it as a seed crop? The same holds true for buckwheat—acres and acres of this crop are planted for green manure and cover crops throughout the region. Buckwheat seed is quite expensive and sometimes in short supply. Here is another opportunity to produce the seed we need closer to home. These forage and cover crops grow and produce seed in much the same manner as the major grain crops that we have already discussed. There are, however, some differences in the growth habits of these plants; seed harvest can also present its own unique set of challenges. In the hope of being as self-sufficient as possible on the farm, let's explore the possibilities of producing seed from some of these less common crops.

Buckwheat

Buckwheat (*Fagopyrum sagittatum*) isn't a true cereal; it's a member of the Polygonaceae family, which includes a number of unpopular plants like sorrel, dock, bindweed, and smartweed. With its green-and-red tuberous stem, buckwheat looks more like a begonia or rhubarb plant than any of the common cereal grains. It is an agronomic crop that produces small hard-shelled pyramid-shaped seeds on three- to four-foot-tall plants. But since buckwheat is sown and harvested like all the other cereals, it is considered a grain nonetheless. The crop has a myriad of advantages. It is a vigorous grower that requires little fertility; buckwheat will go from seed to grain in a mere ten weeks. For this reason, it is a great choice for summer seeding in late June or July. Buckwheat has always been an afterthought crop for me, but there always seems to be a field that slips by cereal-grain-planting season because it may be infertile, too wet, rocky, or otherwise disadvantaged. Cereals planted in late May or early June don't usually thrive, so if you find yourself in this sort of situation, you might want to consider planting buckwheat on that particular piece instead. In most areas of our region, July fourth is the cutoff date for seeding buckwheat with the hope of making grain. So basically, you have almost the entire month of June to procrastinate yet still have a little time to plant something that will give you a food crop from the earth.

Benefits abound when we choose to plant buckwheat. It doubles as a cover crop because it suppresses weeds and builds soil simultaneously, so if you are trying to bring a run-out piece of ground back to productivity, buckwheat is an excellent choice. A moderate crop of an edible grain that will feed both humans and livestock can be produced while we improve soil structure and fertility. This unusual plant with its tuberous stems, prolific white flowers, and ivy-like triangular leaves has the uncanny ability to sequester long-forgotten phosphorous reserves deep within the soil profile. So if you are plagued with quack grass rhizomes that choke your crops, buckwheat is the knight in shining armor: It outgrows and eliminates this bothersome scourge.

The Basics of Buckwheat

Buckwheat isn't a member of the ancient class of grains like wheat or barley—it was unknown to the Greeks, Romans, and Egyptians. The first evidence of this unusual plant being used as a food seems to appear in China about AD 1000. Buckwheat soon became popular in Japan and Korea as well—the soba noodles consumed by these cultures are made from buckwheat. The grain made its way to Europe in the Middle Ages; by the seventeenth century it had been brought to North America by early settlers.

The word *buckwheat* comes from the German word *buchweisen*, which means "beech wheat." The tetrahedron-like shape of the seed is very similar to a beechnut, while its food constituents are more similar to wheat. Indeed, buckwheat found a ready place in the diet and agriculture of late-nineteenth-century America. The peak of production in the United States occurred in 1866 when twenty-two million bushels were produced. By the early twentieth century, two-thirds of this crop was grown in New York and Pennsylvania. The buckwheat pancake with its heavy taste and dark appearance was a unique American preparation during this period. The galette, a French Canadian buckwheat crêpe, is still quite popular in Québec cuisine. Buckwheat is lower in protein than wheat, but much higher in fiber. Its white endosperm is easily milled into fine flour, which usually contains brown flecks from the outer hull. Buckwheat is also relished by poultry and is particularly good for egg production. According to Census of Agriculture records, rather large acreages were being grown for chicken feed in northern New England between 1880 and 1920. Buckwheat has declined in popularity since then, but it continues to be used in the preparation of ethnic foods like Asian noodles and kasha, which is part of the Jewish tradition.

Preparation for Planting

Buckwheat will grow on just about any soil type, but it prefers a well-drained loam like most other crops. Fair yields of ten to twenty bushels per acre can be achieved on soils too poor for other crops that have higher fertility requirements. Although buckwheat has been called the "poor man's crop," it does respond well when grown on better ground under favorable growing conditions. Professional growers in New York State regularly harvest crops that exceed a ton to the acre, so buckwheat can produce a passable harvest under marginal conditions. At the same time, the more effort and inputs we can muster, the better our results will be with buckwheat. This certainly holds true when it comes to soil preparation.

Buckwheat seeds will sprout in a hastily prepared, not very well worked seedbed. If you spend the extra time discing and breaking up clods and sod, your buckwheat will do that much better in a mellow seedbed. You'll have to make your decisions about buckwheat planting sometime in early June. Hopefully, by this time all of your other crops will be in the ground; if hay is part of your farming regime, it should be well under way. It's a lot more difficult to prepare land for a crop at this time of the year because all the plant life you're tilling under is in full-speed-dead-ahead growth mode. There are also numerous other distractions that might prevent you from taking the plows and disc back out to what is supposed to be a field of buckwheat. You also might be wondering why you are waiting until late June to plant this crop instead of seeding it at the same time as barley, oats, or wheat. The reason is that buckwheat is so succulent and tuberous, it will not survive even a light frost. Another reason to put off seeding until late June or early July is to protect the future buckwheat flowers from heat blasting. Buckwheat flowers will abort and never set seed if they are exposed to hot sun and temperatures of ninety degrees. Later planting postpones flowering until early August, when excessive temperatures and scorching sunshine have diminished somewhat. Planting around the Fourth of July is also advisable because buckwheat seeds will germinate and emerge in three to four days when the soil temperature is seventy degrees. Your final results will be directly proportionate to the amount of effort you make, so if you can find time to prepare a nice seedbed and add a little compost or farm manure, you will be rewarded with higher yields.

Planting Buckwheat

Buckwheat seed can be broadcasted or drilled. I prefer using my grain drill because I know that I am evenly distributing the seed on the land in a uniform and orderly manner, but some of the New York growers I know who plant hundreds of acres of buckwheat as a cash crop would rather broadcast seed because it is a whole lot quicker. These individuals have the experience and the skill to evenly distribute seed and cover it perfectly with the proper equipment. Buckwheat for a cover crop is usually planted at a rate of one hundred pounds to the acre. You'd need to halve that rate to fifty to sixty pounds to the acre if you intend to harvest a grain crop. A common mistake is planting too much seed. Buckwheat doesn't tiller and put out more shoots like the common cereals; each seed produces one single stem. The crop is self-regulating because the number of lateral side branches on each plant will vary in relation to the thickness of the stand, so thinner stands will have more fruit on more side branches. Plant buckwheat seeds half an inch to one inch deep. When you are finished planting, put your equipment away. Stand back and relax. There is no other crop that blows out of the earth like buckwheat.

Growth Cycle

Those of us unfamiliar with buckwheat will be amazed by its rapidity of growth. Truly, this crop thickens up fast! Usually the ground between recently sprouted plants is no longer visible a week after emergence. Buckwheat differs from the more grass-like cereals in that it sends down a central root that grows deep into the soil for moisture and nutrients, whereas cereals grow with an extensive network of lateral roots. A rather minimal and much less extensive lateral root system also develops from buckwheat's central taproot. According to a North Dakota State University production manual, buckwheat produces a root system that is less than 3 percent of its total weight; most cereals produce a root mass that is between 6 and 14 percent of total plant weight.

You can almost see buckwheat grow before your eyes. Its leaves are triangular in shape and measure from two to four inches in length. Three to four weeks after the initial date of seeding, buckwheat will already be close to eighteen inches tall and begin to flower. Plant growth is indeterminate, so from this point onward it will just keep growing and blooming. Warm and rainy July weather promotes this rocket-ship-style growth. Buckwheat flowers are self-sterile and require cross-pollination from insects to set fruit. A field of blooming buckwheat is a beekeeper's prize because the pollen from this crop is excellent bee food. Buckwheat honey is dark in color and has the same strong and unmistakable fragrance that emanates from a field in full bloom on a sunny, humid August afternoon. Ideally, daytime temperatures in the sixties and seventies promote the best flowering conditions. Seed set usually begins by week six or seven; little green triangular fruit bodies will begin to appear behind the flowers at the very top of the plant. Meanwhile in true indeterminate fashion, new flowers are forming on the bottom branches. Buckwheat ripens quickly and unevenly, and the unusually shaped green fruits called akenes begin filling by the eighth week of growth. As the month of August moves into September, more and more of these akenes turn brown, harden up, and ripen. There are very few other starch-producing plants that can produce edible grain in two months.

Considerations for Harvest

As the month of September progresses, buckwheat will fade from full flower. The green akenes or buckwheat fruits will begin to turn brown from the top down to the bottom of the four-foot-tall plants. The thick buckwheat "forest" will begin to take on a brownish hue as more and more ripe grain begins to appear. But harvesting a slowly ripening crop of buckwheat presents its challenges. There is ripe grain ready to shatter at the top while the plant is just beginning to flower at the bottom. Unlike cereal grains, whose straw dries up and turns brown during ripening, buckwheat remains moist and juicy during this period—the tuberous stems and branches remain filled with water. This situation necessitates some sort of pre-cutting action to dry out and cure the succulent straw before the crop can be threshed. Careful hand-cutting with a sickle will have to do in a garden situation, but swathing is the harvest method of choice for most commercial producers of buckwheat. Most growers wait until three-quarters of the akenes have turned brown before cutting down the crop with the swather. The thick tuberous stems of the buckwheat are lopped off by the swather cutter bar and transported by cross canvases to a fluffy windrow.

Buckwheat should be cut as high off the ground as possible; a windrow that sits on top of tough eighteen-inch buckwheat stubble will dry quickly if it remains suspended in the air. Early-morning windrowing (when the dew is still on) is recommended to prevent some of the very ripe grain from shattering. The remaining quarter of the buckwheat plant that is just finishing flowering will continue to ripen by pulling nutrients from the straw during the curing process. Hopefully, September rains and thunderstorms will be minimal and allow for a rapid dry-down process—a driving thunderstorm with heavy rain and wind can drive a windrow of swathed buckwheat down into the stubble closer to the ground where drying will take longer. I generally have to leave buckwheat in the field for two weeks to get the crop dry enough to combine. Normal rainfall doesn't seem to bother swathed buckwheat because it is so wet and juicy to start with, and grain quality isn't lost by an extended time in the swath, either. Most production manuals put swath curing at a week to ten days, and while this might be the case in North Dakota or western New York, it's not on my farm where we always seem to get some rain through September. Monitor your swathed buckwheat closely—when the straw is no longer ringing wet and the brown akenes crunch between your teeth, it's time to combine the crop.

Using Your Combine for Buckwheat Harvest

Buckwheat must be carefully handled to avoid losing seed through shattering. If you are doing a small amount with a hand flail or threshing machine, use extra care to handle the ripe plants because the seeds shake off easily. Combine windrow-pickup attachments are designed to treat swathed grain with care. An endless belt with specially designed spring teeth gently lifts the windrow of dry swathed buckwheat and delivers it to the feeder house of the combine. It is important to always operate the combine down the windrow toward the seed heads in the swath. Traveling in this direction will prevent shattering and make for a smoother flow of the crop into the threshing chamber. Some seed loss is inevitable when harvesting buckwheat, so be prepared for some volunteer buckwheat in your crop next season. Even professional growers have this problem. For this reason, I like to follow buckwheat with a row crop that can be cultivated, as volunteer buckwheat in a solid-seeded cereal crop like wheat or barley is a recipe for disaster.

Buckwheat is difficult to remove in the grain-cleaning process. A little buckwheat in an animal feed crop like barley is tolerable, but not in a grain like wheat that is destined for human consumption. Combine cylinder speed should be set at between six and eight hundred RPM, and you only want to use enough threshing force to remove the grains of buckwheat from the akenes. If you find that the threshed grain is being damaged from cracking, slow the cylinder speed down and lower the concave grate. Combine buckwheat slowly and deliberately. Driving too fast can overload the machine and cause poor and incomplete threshing. Efficient buckwheat combining requires a fairly strong air blast; the cleaning fan should be run just fast enough to avoid blowing clean grain over the chaffer and out the back of the machine. Adjust the chaffer teeth openings so grain will drop to the sieve below before it has passed two-thirds of the way across the top screen. The sieve, or lower screen in the combine boot, should be closed down as much as possible so as not to overload the return elevator with excess grain. These are just a few ideas to get you started adjusting your combine for buckwheat harvest; careful observation coupled with a little trial and error will teach you how to do it yourself.

Not everyone has ready access to specialized grain-harvesting machines like swathers and combines with windrow-pickup heads. In most cases, it's hard enough to find a plain old combine with a normal grain platform. For many first-time growers of buckwheat, direct-combining will be the only option. If you choose this method, wait until three-quarters of the seeds are brown and the inner buckwheat groats are firm. It is also advisable to wait until a frost or two has killed the crop and taken away some of the moisture in the stalks and stems. However, frost does make the buckwheat more brittle and less able to hold on to its seeds, so some shattering and seed loss is a possibility when you're direct-combining buckwheat. This scenario is still better than leaving a nice-looking crop in the field because you don't have special equipment to harvest it the "right" way.

Cleaning Your Crop

Good final harvest moisture for buckwheat is 13 percent, and you'll want to dry the crop with unheated forced air in a bin with a perforated drying floor or in loosely woven burlap bags. Standard cleaning machinery like a fanning mill will remove 90 percent of the unwanted foreign material from a sample of buckwheat. I've had a bit of trouble separating out little bits of broken stems from my harvested buckwheat. I usually have to pass the crop over the gravity table after the air screen to finish the job. I have also discovered that companies like A. T. Ferrell (which makes Clipper grain cleaners) offer special cleaning screens with triangular holes that are specifically designed for buckwheat. I have tried several times to speak with professional buckwheat millers in the Finger Lakes of New York to gather more hints and information about buckwheat cleaning, and I remember standing on the street once outside Brickett Mills in Penn Yan, New York, asking an employee in the cleaning division what sorts of screens he used for cleaning buckwheat. The fellow came right out and told me that it was a secret and he certainly wasn't going to tell me anything. Don't let this sort of thing be an obstacle. You can always send a sample of your dirty buckwheat to the folks at A. T. Ferrell in Bluffton, Indiana. They will test-clean your sample on special handheld sieves and tell you what screens you will need for that specific application.

Uses for Buckwheat

After your buckwheat crop is dry, stable, and in storage, you can begin to think about what to do with it next. The simplest thing is to sell it as cover crop seed to other farmers and gardeners. Buckwheat seed is usually in short supply and retails for over a dollar per pound. If you can wholesale it as seed for seventy to eighty cents per pound, a twenty- to thirty-bushel-to-the-acre crop of close to fifteen hundred pounds can be a rather lucrative undertaking. Maybe all that trouble was worth it after all. The other option is to mill your buckwheat into flour or some other product for human consumption. A rather large proportion of buckwheat is its fibrous hull, however, and serious commercial buckwheat millers have special hulling machinery to remove this outer hull from the inner groat. Because of its unusual shape, buckwheat can only be hulled on specially designed machinery used exclusively for this purpose. Some steaming and redrying might be necessary. Once the groats and hulls have been separated, ethnic products like kasha can be produced. Hulled buckwheat can also be ground into special extra-white flour. For the rest of us who lack fancy milling equipment and are flying by the seat of our pants, we will have to grind our buckwheat with its hull on. Then the mixture of hulls and finely ground endosperm can be sifted and cleaned to produce peasant-style flour that is full of little brown bits of hull and plenty of extra buckwheat flavor.

In a short three months or less, you have been able to produce a viable crop in a spot that may have been intended for something else earlier in the growing season. You may have lost your chance to grow your first choice due to a variety of circumstances—from wet soil to lack of time and opportunity—but buckwheat is a great crop choice because it gives you a second chance to grow something else that is unique and of fair value.

Grass Seed

Cool-season forage grasses and cereal grains are both in the same family, and hundreds of thousands of pounds of grass seed are planted in our farm fields each season. Most of the perennial rye grass, orchard grass, and brome planted for hay by northeastern farmers is produced in the Willamette Valley of Oregon, a part of the country that has a favorable climate for forage seed production and is known as the grass seed capital of the world. Indeed, mild wet winters coupled with hot dry summers are ideal for this type of production. Grass seed production is monoculture farming that is very dependent on chemical nitrogen, herbicides, and fungicides. Our wet and humid summer climate in the Northeast is a whole lot better suited to producing these cool-season grasses for hay and silage; the commercial production of specific grass seeds here in our region is out of the question. We would be hard-pressed to establish a pure-enough stand of perennial rye or orchard grass to even think of harvesting seed. This sort of monoculture works a lot better in Oregon with the help of chemicals and a predictably arid summer climate. Plus, we simply don't have a dry period long enough to cure a crop of ripe grass for seed production. All is not lost, however. There are a few possibilities for the production of seed from some of the basic grasses like timothy and millet. This particular agricultural activity, however, requires creativity, hard work, and imagination, but if you're hell-bent on saving some seed from the basic elementary forage grasses, it can be done. The final product might not be Oregon Number One certified grass seed, but it will be organic and all our own. Please be advised that this crazy activity is only for the hardest-core of farm seed savers out there.

Timothy

I've had the best luck saving timothy seed. Timothy is the prince of all the cool-season forage grasses. It holds off flowering and heading much longer into the growing season than orchard grass, which is usually overripe by the end of May. Timothy is softer and much more palatable, and is a natural mate to clover and alfalfa in a hay field because the seed head doesn't emerge from the boot of the plant until well into June. The crop is much less apt to get coarse and woody, which helps to preserve hay quality, as cows prefer hay that is less fibrous and softer in texture. Several years ago, I combined a field of winter spelt. There were very few annual weeds in the crop, but there sure were a lot of ripe timothy heads waving around in the spelt. When we ran the spelt through our screen cleaner, I was very pleasantly surprised to find copious amounts of pure timothy seed pouring out the number three shoot. I had to buy a special screen with tiny holes from Clipper to further process the timothy seed. I ended up with close to five hundred pounds of it from the ten-acre spelt field. This felt like true wealth in so many ways, as organic timothy seed sells for close to two dollars per pound. It sure meant a lot to me the following spring to plant hay crops using my own timothy seed. Fanning mills and air screen cleaners are valuable assets on a farm that saves and plants its own seeds. It pays to develop proficiency with grain cleaning, and the only way to accomplish this is through experimentation and trial and error. Grass seed cleaning requires less air and a greater assortment of specialty screens.

Millet

Millet is a midsummer annual grass that was once used as a source of supplemental forage on dairy farms when pastures began to dry up and run short in August. Millet is very similar to buckwheat in that it grows quickly and vigorously when seeded in the heat of high summer; if you plant millet in late June or early July, it will be ready to cut for hay in August when the cool-season pasture grasses have slipped into dormancy. Millet grows tall and rank, providing many tons of dry matter to the acre. Thirty years ago, farmers would mow a little every day and bring it to the cows in the stable during milking. With the advent of rotational grazing, some organic dairy farmers have taken to strip-grazing a new small area of millet each day. Brent Beidler from Randolph, Vermont has had great success with millet on his organic dairy farm, and several years ago, he decided to let some of his crop go to seed. While millet seed production is common in Minnesota and the Upper Midwest, Brent decided to try swathing and direct-combining the crop here. It didn't take him long to settle on the direct-combining approach. He's been so successful with this endeavor that he has actually been able to sell his own home-produced millet seed to other farmers.

Legume Seed Production

Legumes, particularly forage plants, are an organic farmer's best friend. These very special plants have the ability to remove nitrogen from the atmosphere with the help of a unique relationship between nodules on their roots and specific strains of rhizobium bacteria that populate the root zone. Various varieties of clover, alfalfa, vetch, and peas can either provide first-class high-protein hay and silage for dairy cows or tremendous green manure crops for plow-down and organic soil building. Crops like corn and wheat that have a tremendous appetite for nitrogen really thrive when planted after a lush legume has been incorporated into the soil. Nitrate nitrogen that has been added to the earth in this green manure process is much more stable and long lasting than mineral nitrogen like urea (or even so-called organic Chilean nitrate) that comes from a fertilizer bag. Cost is also a concern when it comes to nitrogen fertilization. Chemical nitrogen fertilizers have doubled and tripled in price in recent years because they are synthesized from natural gas. Chilean nitrate, which is mined in the Atacama Desert of northern Chile, travels many miles by boat to get here and is even more expensive than its chemical cousins. The National Organic Program has withdrawn its approval of sodium nitrate; its use was no longer allowed after the 2012 growing season. The point is, we all need more legumes on our farms. These plants not only supply us with free nitrogen but also send roots deep into the soil to break up compaction and loosen the earth. Plants like alfalfa and sweet clover have the ability to send taproots twenty feet or more down through the soil profile, as these deep roots bring up water and minerals from below. Legumes have all kinds of benefits and are ideal mates to cereals and hay crops, which are both grasses.

If you are practicing a diverse crop rotation that includes grain and hay, legumes have a very important part to play. Just about all of us have been consumers of legume seeds at one time or another. I think back to the countless times and number of bags of clover and alfalfa seed that I have inoculated and poured into the grass seed box of my grain drill. Ten years ago, organic clover and alfalfa seed was a rare commodity, as the conventional stuff was untreated and quite inexpensive; in many cases, clover seed could be purchased for less than a dollar per pound. Older varieties of alfalfa like Vernal didn't sell for much more, and yellow blossom sweet clover was even cheaper. A quick look at the seed tags on these bags told me that most of this seed was produced on the vast prairies of Saskatchewan and Manitoba. As the organic movement developed and matured, certified organic legume seed also became commercially available. I switched over to planting organic clover, sweet clover, and alfalfa underneath my cereal grains, but the pricing structure for forage seed in general is rather volatile. If seed growers in western Canada have a poor production year because of extended wet weather, prices can triple and quadruple. This has definitely been the case over the past five years. Thanks to supply and demand, conventional legume seed has doubled in price, and organic legume seed has quadrupled. The run-of-the-mill organic Vernal alfalfa seed that I used to buy for a dollar a pound now sells for close to three dollars, and fancier modern varieties can cost as much as five dollars per pound. To get a good catch, many farmers will plant somewhere between fifteen and twenty pounds of alfalfa or clover seed to the acre, as forage seeding can now cost a hundred dollars or more per acre.

This situation sent me on a search for more reasonably priced clover seed. I was delighted to find a source of Demeter certified biodynamic clover seed from the Hack Farm in Kincardine, Ontario. I found Martin Hack with the help of my close friends Wilhelm and Alex Brand of North Hatley, Québec. Ten years earlier, Alex had done his youth *stagière* (apprenticeship) at the Hack Farm on the shores of Lake Huron; Martin was a grower of flax and clover seed in addition to his herd of beef cows on his thousand-acre farm. The seed was plump and beautiful and about a dollar a pound cheaper than the commercial organic clover I had been buying from my local supplier. This was also an inspiration to me—if I really wanted, I could grow my own clover seed. But this certainly wasn't anything new, as Addison County, Vermont, farmers Ben Gleason and Ken Van Hazinga had been doing it for years. So while alfalfa seed production is much better suited to the arid West, there is no reason why we can't produce our own clover seed here. At this point, I see farm-based legume seed production as an avenue to local self-reliance instead of a commercial venture. We might not be able to meet the standards of the certified seed industry, but we can grow our own high-quality clover seed right here where we live.

Red Clover

Red clover (*Trifolium pratense*) is probably the most popular of all the legumes. This wonderful plant is a relatively short-lived perennial. Many in the farming world consider red clover to be a biennial because most stands die out after two years. This was certainly the case thirty-five years ago when I first started crossing the border into Québec's

Eastern Township farming country. Most farmers at that time practiced a cereal grain/timothy and red clover hay rotation. After the clover in a new hay field began to die out in year two or three, the land would be plowed in the fall and reseeded to oats/timothy and clover the next spring. The grain would be harvested the first year and the straw baled. A good catch of the underseeded hay crop would provide a red-clover-dominated bumper crop of hay the following year, and it wouldn't take too many more years for the clover portion of the hay crop to fade while the timothy grass component increased and filled in the gaps.

Medium red clover has been pretty well supplanted by alfalfa on most dairy farms over the past few decades, however. After switching to alfalfa myself, I have begun planting a lot more red clover for hay in the past few years, as I got tired of trying to do everything "right" when planting alfalfa and still losing crops to winterkill and other maladies. If an alfalfa stand is only going to survive two or three years in my mountain location, I might as well plant clover for half the price and half the aggravation. There is also something magical about red clover. It makes fine legume hay if cut early enough. It has a chemical composition and a plant structure that promote rotting and breakdown—clover lying in a mowed windrow can turn black in an instant if gets soaked by a passing shower. Couple this plant's ability to produce top growth that is thigh-high with an amazingly extensive root system, and you have the recipe for a whole lot of soil conditioning and improvement. We should be planting as much red clover as we can into our annual cereal grains, which gives us all the more reason to harvest some of this beautiful pink- and red-blossomed legume for seed. Having a ready supply of farm-produced red clover seed will save you a ton of money and make it easier to keep filling the grass seed box on the grain drill.

If you wish to grow some clover seed, you will first need to start out with a really nice thick stand of this legume. Plant it the year before with a companion crop of oats, wheat, or barley, and don't forget to inoculate your seed with the proper strain of rhizobium nitrogen-fixing bacteria. I plant some timothy at the same time, as I like mating a grass and a legume—they were meant to be together. It is standard practice in the clover seed industry to take a pre-bud early first cutting of hay, and get the hay off the field by the first of June if possible. This early cut promotes more blossoming and greater growth for the second cut, which will be the seed crop. Clover flowers are self-sterile and require cross-pollination by insects to set seed. Bumblebees are the primary pollinators, and there are more of them in July and early August at the height of the summer season. The nectar in clover flowers is difficult for regular honeybees to reach, however, so you'll have to be patient if you want to harvest red clover for seed. Be ready for a field of red blossoms by the end of July, and your dairy farmer neighbors will be wondering why you let this incredible field of hay go to full blossom. The crop will get coarser and tougher as it matures. Clover stems get large and woody, which increases whole-plant fiber levels and decreases hay quality. Red blossoms will turn brown after pollination and begin to develop seed. Tiny pods that each contain one seed will develop in the flowers. It's time to think about harvest when three-quarters of the blossoms have turned dark brown. Swathing is preferable to direct-combining because the whole plant needs to dry considerably before going through the combine. It will take a week to ten days of hot August weather to dry the windrow of clover and sufficiently ripen the seed.

Combining Clover

August is the best month to combine clover. I've heard it said by a few older experienced grain men that combining red clover seed is the worst harvesting job on the farm. It is certainly a dirty and dusty business. After a week or two in the field, that once lush large windrow of mowed clover shrinks and dries into a tinder-dry shadow of its former self.

As in all swathed crops, you'll want to drive the combine and windrow-pickup attachment toward the seed heads. Clover seed threshes hard, so drive slowly and don't overload the combine. The concave needs to be closed up tight with a gap of only three thirty-seconds to three-sixteenths of an inch between it and the cylinder. Operate the cylinder between nine hundred and fourteen hundred RPM as you would for wheat. The cleaning fan speed also has to be reduced for clover. And get ready for dust! The bone-dry clover entering the combine gets pretty well mashed in the threshing process. You might even want to make sure that you are driving the combine into the prevailing wind to keep the dust and chaff blowing away from the back of the machine. Combine fires are a common occurrence during clover seed harvest, as all those fine particles in the air can ignite easily when exposed to engine exhaust or the surface of a hot bearing on the side of the machine. Farmer Martin Hack carries several giant squirt guns full of water in the cab of his combine to extinguish flare-ups during clover harvest. This whole process is a giant labor of love, and the clover seed that you do harvest is hard-won because yields usually range between two and five hundred pounds per acre. This might not seem like much when compared with a couple of tons of cereal grain per acre, but remember how valuable clover seed is at three bucks a pound.

Storage and Seed Saving

Twelve percent is considered a safe level of moisture for the long-term storage of clover seed. A bone-dry windrow of swathed clover will very likely yield seed of acceptable moisture. Clover seed is easily cleaned in a common fanning mill or air screen cleaner. The top scalping screen is usually a one fifteenth. Depending on the plumpness of the seed, you can choose between 6 × 22 or a 6 × 24 slotted screen for the bottom. Large industrial seed establishments have lots more specialized and very expensive cleaning equipment for processing clover seed. We will have to be satisfied with simple and elementary cleaning machinery for our farm-produced end product. Clover seed at 12 percent or lower in moisture can be safely stored in grain bags in a cool, dry location out of direct sunlight.

You also might be wondering about saving seed from alfalfa and other clovers. Alfalfa flowers are even more difficult for honeybees to access than clover blossoms. Special cutter bees are employed for this purpose on the prairies of western Canada. Low-growing white clovers are difficult to harvest because they are so short, but yellow blossom sweet clover can certainly be swathed and combined just like red clover. Sweet clover is a tremendous soil builder that can grow as tall as six or seven feet if left unchecked, but swathing a crop this thick and tall can be difficult for an old machine like mine that has seen better days. It is more important than ever to take a first cutting of sweet clover to prevent it from getting too rank to harvest later in the season. Another consideration is the problem of seed loss during the harvesting process. You may have red clover and sweet clover volunteering in your fields for years to come. Still, this doesn't seem like too big a problem to me. Last but not least, don't expect to harvest multiple crops of clover from a field over a period of years. You usually only get one opportunity to get a seed crop from clover in that first production year when the stand is strong and has not yet been invaded by too many grass species.

Canadian Field Pea

The Canadian field pea (*Pisum sativum* L.) is another leguminous plant that can provide tremendous advantages to organic grain and hay growers. I discovered this crop quite by accident over twenty-five years ago, when my Québec seed dealer asked me in French if I had ever considered planting *céréales mélangés*. I asked him just what mixed cereals were, and he said that many farmers plant a mixture of 60 percent wheat, 20 percent oats, and 20 percent field peas for livestock feed. The three different grains all have the same maturity ratings and all go through the grain drill together. He told me that planting three very different grains together would provide a distinct advantage because some part of the mix was bound to thrive even if growing conditions turned out less than favorable for the other two. Mixed grain became a regular part of my spring grain lineup after this introduction, and my love affair with field peas continues to this day. The seed dealer was correct when he said that the crop never turns out the same twice. Some growing seasons produced a higher percentage of peas in the final mix, while other years saw higher proportions of wheat and oats in the harvested crop. But adding field peas to a cereal grain seeding is a simple way to boost the protein level of a grain ration. Peas test between 21 and 25 percent protein, and it wasn't unusual to harvest a mixed-grain crop with protein levels of 18 percent or better. This was almost like having a complete and balanced dairy ration coming right out the spout of the combine. With protein levels of 18 to 20 percent, there was no need to be adding expensive (and much-harder-to-grow and -process) roasted soybeans to a dairy ration. The mixed grain could be ground up with a little corn and barley and fed directly to dairy cows.

I also had good success mixing one part of a shorter-season variety of peas with three parts barley. Field peas were a popular choice for new forage seedings, too. In the early 1990s, several seed companies began to offer fifty–fifty blends of peas and triticale as companion crops for seeding down. These forage blends were given names like Tripper and Peakal and planted about one hundred pounds to the acre with a grain drill. Clover, timothy, and alfalfa were seeded simultaneously through the grass seed box on the drill. The whole mix was harvested as a high-quality and very voluminous silage crop in early July. Field peas are very well adapted to our climate, and are a natural choice in an organic system because they are high in protein and easy to grow.

Growth Habits of Peas

The field pea is a cool-season annual legume. It has been around for at least nine thousand years, originating in the southwest corner of Asia in the area that is now Turkey, Syria, Iran, and Iraq. This very versatile crop has an amino acid profile that gives it superior status as a protein source in both human and animal nutrition. Field peas are high in both lysine and tryptophan, which are both severely lacking in most of the cereal grains. The trypsin inhibitor that necessitates the heating and roasting of soybeans is much reduced in peas, which offer complete and very digestible protein without having to be cooked, roasted, or extruded like soybeans. Fiber levels are low in peas, making them very easy to digest by both man and beast. In fact, the green and yellow split peas from which we make hearty soups come from the same plant that we grow with our cereals to round out dairy cow rations. The center for pea production in North America stretches from the Palouse of eastern Washington through Montana and the Canadian prairie provinces, ending in the Dakotas and Minnesota.

Peas need to be planted early in moist firm soil. Adequate water and good seed-to-soil contact will foster better germination. Once a plant is up and growing, however, water is not a problem. As a matter of fact, peas are very efficient users of soil moisture, which is why they are so popular in more arid regions. Their low water requirements make them ideal field mates to wheat and barley, which both require more moisture. Pea roots can grow to a depth of three to four feet, and are also quite resistant to other early-season stresses. When a pea germinates, its two growing cotyledons remain under the soil surface. If a plant gets badly damaged by heavy spring frost or some mechanical mishap, the root zone will send up another growing shoot. Young pea plants can tolerate freezing temperatures as low as twenty degrees.

Pulse crop plant breeders out in pea country have worked long and hard developing pea plants with characteristics that are beneficial to a wide variety of growers. Peas are also available with both determinate and indeterminate growing habits. Determinate varieties are best adapted for mixed-grain production because they flower for a set period of time and usually ripen in eighty to ninety days, while indeterminate peas have vines as tall as six feet and flower over a much longer period of time. These varieties are much better grown alone, because ripening can be prolonged to as much as one hundred days in cool wet weather. Plant breeders have also improved the standability of pea vines by developing "leafless" varieties with smaller leaves and increased tendrils to hold the crop up. Peas will flower forty to fifty days after being planted, and are very sensitive to heat stress at time of flowering. Pod set and grain yield can be severely reduced by a spell of hot, humid weather with temperatures near ninety. This gives us all the more reason to plant early in April. Peas are also self-pollinating. Each flower produces a pod that contains four to nine seeds. It is a joy to see pea plants in full white or purple flower scattered throughout a field of ripening barley, wheat, or oats—the pea plants wrap their tendrils around neighboring cereal plants for support. Pea and cereal production in a mixed-grain system is a match made in heaven.

The Value of Inoculation

There a few management factors that will help you increase your success when growing field peas. Inoculation with the proper rhizobium bacteria is essential, but most northeastern farmers who are planting peas don't bother with this step. Peas have a greater ability to fix atmospheric nitrogen than any other legume crop. According to the North Dakota State University growing manual, peas can provide up to 80 percent of their own nitrogen needs if they are properly inoculated. This means finding and applying the specific *Rhizobium leguminosarum* strain of bacteria to the seed. Inoculants for other crops like clover, alfalfa, soybeans, and dry beans will not work on peas. For some reason, seed suppliers in our part of the country don't stock large packages of lentil/pea inoculant, and it is not very cost-effective to buy small garden-sized packages of it, so most people planting field peas skip inoculation. Every spring, I get on the telephone and order my complete year's supply of inoculants in large packages from Agassiz Seed in West Fargo, North Dakota. I use a combination of liquid and dry peat-based inoculant on my peas. The peas are first dampened with the liquid and then covered with the dry stuff—I like to see my yellow or beige pea seed pretty well blackened. Peas have an extremely fragile seed coat and are very easily damaged if care is not taken during drilling, so don't overdampen the seed when you inoculate it.

Make sure that there is adequate space for the pea seed through the seed delivery passages of the grain drill. The little fluted gears that meter seed at the top of each grain tube can easily tear off or crack the seed coat of peas if there isn't enough room for the seed to pass through. Drop the adjustable little seed cups at the top of each delivery tube on the drill down a notch or two if you're planting peas. You will also have to figure out how much seed to plant. If I am planting some sort of mixed grain, I usually limit the amount of peas in the mixture to no more than 25 percent. Pea seed comes in a wide variety of sizes. There can be anywhere from sixteen hundred large peas to five thousand tiny ones in a pound of seed. Having some knowledge of seed size will help you figure out just how much seed to plant per acre. If for some reason you decide to plant straight peas, the experts recommend a final population of three hundred thousand plants to the acre or seven to eight plants per square foot. Depending on seed size, this can represent planting anywhere from 125 pounds to more than 200 pounds per acre.

Sourcing (or Saving) Seed

Most of us, me included, are consumers of commercial pea seed, and buy seed that comes from

Canada or the high plains that we mix with other cereals for both grain and forage crops. Pea seed is considerably more expensive than seed for the more common cereal grains, as it's higher in protein and produced thousands of miles away. I like having a supply of it around in the springtime. In the past, I've purchased large totes of the stuff from my friend Blaine Schmaltz in North Dakota. If I have enough pea seed on hand, I will add it to every cereal grain mix that I plant to feed to my cows. I constantly ask myself if there is any way that I could produce my own pea seed, and I have made a few noble attempts over the years with moderate success. If I harvest a bountiful crop of mixed grain, I can set some aside and run it through my cleaner to separate out the peas. Peas will roll off the top scalping screen of the fanning mill along with chaff, stems, and other foreign material. Separating a round seed from other cereals isn't a straightforward and quick process, however. Once you have collected enough material off the scalping screen, you need to run it through the cleaner a second time with larger round-hole screens to clean the peas. Depending on the percentage of peas in the mixed grain, you might have to clean several tons of grain to end up with five hundred pounds of peas. You also might begin to ask yourself, *Is this really worth it? Didn't I just grow this crop of wheat, oats, and peas to increase the protein of my livestock ration? And here I am removing the protein fraction from the mix. Maybe it would be better to simply buy my pea seed.* The intensity of our desire for total self-sufficiency will have to guide each of us as we make a decision about cleaning peas out of a crop of mixed grain. This is certainly the easiest way to accomplish this task.

Planting Straight-Seeded Peas

You can also plant straight field peas just like they do in North Dakota. This seems like the simplest method for producing pea seed, but there are a few caveats. For one thing, pea vines don't support themselves very well. I usually plant this crop with a cereal so the pea plants have something to climb on and attach their tendrils to. Jacques Beauchesne, the owner of Semican in Plessisville, Québec, is the only person I know here in the East who grows straight peas for seed. He plants his peas as early in April as soon he can get on the land, and sets his grain drill to plant one hundred pounds to the acre. The field is planted in one direction, then Jacques goes over it a second time at ninety degrees to the first pass. This seeding rate of two hundred pounds to the acre crowds out weeds and produces a very thick stand of peas that is somewhat self-supporting. Once the peas are up and growing, it's time to stand back and hope for reasonably dry, calm weather. It doesn't take too many severe thunderstorms to flatten a nice field of peas. In most years, Jacques ends up combining his pea crop right down at soil level with a soybean flex head. Keep this in mind if you plan to try a direct seeding of peas—always roll your field after seeding and pick every stone. If you have to "shave" the crop off the ground with a flexible cutter bar, you don't want to be picking up rocks with combine. The biggest obstacle to direct-planting peas here in our region is that we get a whole lot more summer precipitation than the Dakotas. If you are adventurous and don't mind watching your crop get flattened by wind and rain, give this method a try. You might want to experiment with a small acreage before taking the plunge into bi-time field pea production.

Harvesting Field Peas

If you do get the opportunity to harvest a crop of direct-seeded peas, take extreme care with the combine to avoid damaging these tender little fruits. Attention to detail starts right at the cutter bar. Special vine lifters can be bolted to the guards to gently lift prostrate pea plants for cutting; a hume reel with spring teeth will also aid in the pickup of a downed crop. Start combining when most of the pods have turned from yellow to a darker shade of brown. Slow down the reel speed to minimize shattering, as peas can be "skinned" or otherwise damaged if threshed too aggressively. A cylinder speed between 350 and 600 RPM will thresh peas without seed coat cracking or splitting. Lower the concave to five-eighths of an inch in the front and three-eighths of an inch in the rear, and don't be afraid to turn the cleaning fan up: Peas are heavy and require a high flow of air for good cleaning action. The chaffer is usually opened up to five-eighths of an inch and the sieve to three-eighths. Peas seem to combine well between at 14 and 20 percent moisture. Field losses due to shattering and splitting will increase if the crop is too dry.

Low moisture is more of a problem out on the arid high plains, but here in our higher-rainfall part of the country, we are usually more concerned about getting the crop down to 20 percent moisture. A simple rule of thumb for harvest readiness with peas is to actually use your thumbnail on the surface of a pea. If the pea is a bit rubbery, but you are unable to leave a thumbnail mark in its outer skin, its moisture is low enough to combine without doing any damage to the crop. Once peas are harvested, avoid moving them around by auger if possible, as sharp auger flighting can cut and bruise seed coats. Split and cracked peas are fine for livestock feed, but not for replanting. Peas will store well at a bit higher moisture than the common cereal grains. The NDSU Pea Crop Manual claims that peas at 18 percent moisture will keep up to twenty weeks at or below sixty-eight degrees. If the temperature climbs up to seventy-seven degrees, the safe storage window for an 18 percent moisture crop is closer to a month. If you use plenty of forced air in a bin full of field peas, the moisture will drop to between 12 and 14 percent, which is perfectly acceptable for long-term storage. Straight pea seed production is certainly a possibility here where we live, but it still remains uncharted territory to many of us who have been growing organic grains here for the last three-plus decades. I tried a small plot of straight peas once, only to have the crop ruined by river flooding. My very last bit of advice in the field pea department is to plant your peas in a high and dry place where drying winds are strong and prevalent and high water is not a consideration.

Vetch

The very last of the legume crops for our consideration belongs to the vetch family. Vetch is a very popular cover crop choice for many organic vegetable and corn grain producers. A lush crop of hairy or common vetch will add as much as two hundred pounds of nitrogen to the acre in a single season. Vetch is a native of the Mediterranean region of Southern Europe. This viny little plant is a native in many of our hay fields, and stands out because of its beautiful purple flowers and stems loaded with many tiny pinnate leaflets. There are two basic types of vetch that thrive in the Northeast. Hairy vetch (*Vicia villosa* Roth), which is winter-hardy, is the most common variety. It is seeded into rye and other winter grain cover crops to provide nitrogen when plowed down the following spring. Common vetch (*V. sativa* L.) is an annual usually seeded by itself in the spring as a nitrogen-fixing cover crop. Several of my Canadian grain grower friends have taken to planting common vetch early in the spring. They let the crop grow until late July or early August, when it bears seed in tiny pods that look like miniature peas. The crop is quite green and difficult to harvest. The strategy is to intentionally take only a part of the seed. The remainder of the crop ends up being lost on the ground. After harvest, the land is lightly tilled and the wasted seed grows up into a second crop of vetch, which will winterkill late in the fall. Enough seed is harvested to replace what was used in the initial planting of the vetch. This double crop method of growing common vetch will provide more than enough residual soil nitrogen to comfortably grow a crop of corn without any additional inputs the following year.

But using vetch (especially hairy vetch) in an organic crop system is a bit of a double-edged sword. Many wheat growers don't want anything to do with it because hairy vetch readily sheds its seed and has the potential to turn into an unwanted weed in a grain crop. Vetch seeds look like little round black peppercorns and can be extremely difficult to remove from grain with a standard fanning mill. I speak from experience in this department—I've had to use spiral cleaners and indent sizers to remove vetch seed from wheat. However, it is possible to achieve this separation. This should give you plenty of home-produced pure vetch seed, which sells for well over a dollar a pound in the organic marketplace.

Legumes and grasses for cover cropping and forage production are a way of life on most organic farms, where soil quality and good animal nutrition are important. There is certainly nothing wrong with buying good high-quality organic seed to meet your needs. However, if you are fanatical and like trying new things, a foray into homegrown seed production can be fun, challenging, and rewarding.

CHAPTER EIGHTEEN
Preparing Livestock Rations with Farm-Raised Grains

For most of us, feed grains have been something that is purchased at a farm supply store or other retail outlet, but as the backyard raising of livestock has increased, the choices of specific ground livestock rations have proliferated. Fifty-pound paper bags of starter, grower, and finisher rations are available for just about every class of livestock from cows, pigs, and chickens right down to emus, llamas, and ostriches, and most of this commercial grain is available as certified organic. All you have to do is pull a handy little string and presto—there is grain ready to feed to your animals. Most of us give hardly a thought as to just what is in these commercial offerings. You might see an occasional oat kernel or steam-rolled kernel of flaked corn, but that is the only evidence that what's in the bag actually came from a field. Pelletized grain is even more mysterious. Finely ground grains are pressed and extruded into petite cylindrical shapes with the help of heat, pressure, and a clay binding material. When you pour this pelletized product into your pig trough or chicken feeder, you have absolutely no idea what you are feeding aside from a manufactured mixture of grain and other additives. I remember buying bags of Blue Seal Coarse 16 for our first family cow back in 1975. The grain was a mixture of oats, flaked corn, and pellets covered with a generous coating of molasses. The stuff seemed more like candy than it did feed grain for a dairy cow. Cows would simply inhale it. My impression at the time was that the commercial feed mills knew a whole lot more about grain rations than I did, and it would be impossible for me to grow and prepare anything even close to this molasses-covered mystery grain. But store-bought grain is certainly not a new phenomenon, and it is one of the main reasons why we don't have a tradition of feeding farm-raised grains to our stock. Railcar shipments of commercial dairy rations from the Midwest began arriving here in the late nineteenth and early twentieth centuries. The feeding of locally produced oats and barley was forgone for manufactured "complete feeds" over one hundred years ago.

Although I grew my first crops of oats, wheat, and barley in the late 1970s, it took me a number of years to learn about livestock rations and build my confidence. I thought back a few years earlier to my very brief experience with dairying in southern Wisconsin, where ear corn and oats were taken to the mill and came back to the farm ground and mixed with protein concentrates. I took a look through my collection of old agricultural textbooks and found several different editions of *Feeds and Feeding: A Handbook for the Student and Stockman*. This book was first published in 1898 by W. A. Henry, who was dean of agriculture and director of the Agricultural

Experiment Station at the University of Wisconsin–Madison. It was rewritten and updated several decades later by F. B. Morrison, a professor of animal husbandry at UW Madison. *Feeds and Feeding* was just what a lover of the agricultural past like me needed for direction and information in regard to grinding and formulating my own livestock rations from the grains I had raised on my farm. Yes, there was life after store-bought sweet dairy grain. This book was and still is the bible for livestock feeding. It is almost eight hundred pages in length and contains countless chapters on the raising and feeding of every type of farm animal. The appendix at the end is loaded with tables that describe the components of every type of forage and grain known to man, and there are numerous recipes for all sorts of grain rations to suit every type of stock at any age or stage of development. If you need to put a grain mix together for baby piglets or for a laying flock, the recipe is there. I remembered my Wisconsin farmer mentor John Ace telling me how common rolled oats were the best feed choice for recently weaned piglets. Sure enough, Morrison concurred. *Feeds and Feeding* is a good read and an even better reference book for anyone who is raising livestock and wants to look beyond the grain store. This book also comes in an abridged edition. Although it is out of print, millions of copies and numerous editions of *Feeds and Feeding* were published over a seventy-year period. Keep your eyes peeled at used-book stores and at Amazon for a copy. I prefer the later editions from the 1940s and '50s.

My Foray into Formulating Grain Rations

My first foray into formulating grain rations for my farm animals took place in a little barn/garage that we attached to our house in 1979. Armed with recipes from *Feeds and Feeding*, homegrown barley and oats, and a supply of concentrates from nearby Québec feed mills, I went to work with a measuring scoop and a sap bucket. Two scoops of ground barley mixed with one scoop of purchased dairy mix or soybean meal was the basis of my cow grain. The concentrate already contained the necessary minerals, and we mixed a separate pail for each of our six milk cows. Poultry grain required a different special concentrate and the addition of some purchased Canadian ground corn. We were lucky to have four different Shurgain mills less than thirty miles north of the border near our farm. Shurgain was the brand name for Canada Packers, a large agricultural product supplier based in Ontario, and every week we drove our 1966 hand-me-down Plymouth station wagon "to the mill" across the border to buy the extra grains and concentrates we needed to mix with our farm-grown fare. While at the Viens et Frères mill in Ayers Cliff, we met many local Eastern Townships farmers who were bringing their own barley, wheat, and oats in for grinding and mixing with concentrates right on site, exactly as we had seen in Wisconsin. This sort of service certainly didn't exist on our side of the border—all of the commercially available grain in northern Vermont was manufactured at distant mills and transported to retail outlets. So we decided to cross the border with our grain and avail ourselves of this wonderful service. We loaded up the pickup truck with sacks of oats and barley and were on our way. Sure, there were two customs stations and an international border en route, but this was over thirty years ago and the process was relatively simple. (This sort of importation/exportation would never be allowed today.) We enjoyed these trips for a short while until we were able to source some of these ingredients closer to home, when Poulin Grain in nearby Newport began grinding and mixing feed grains in 1983. Jeff Poulin very kindly let us into his bulk bins to shovel by-products like corn gluten and distiller's grains into sacks. We no longer had access to specifically formulated premixes for various types of livestock like chickens and pigs, but we made do by buying in some extra minerals.

The Importance of Proper Grinding Equipment

Throughout this whole adventure of learning how to put a livestock ration together, I tried a number of different grain grinders, many of which have been described in previous chapters. The antique steel buhr or plate mill that I bought for twenty-five dollars in 1981 was my first mode of feed grinding, employed very early in my farming career. This machine was a belt-driven affair that consisted of two vertical grinding plates encased in a cast-iron housing. An external hand wheel could be turned out or in to regulate the distance between the two grinding plates, which in turn governed the coarseness or fineness of the grist. This was the ideal machine to grind harder-textured grain like barley into a medium-coarse meal. Of course, there was no safety shield or cyclone on the machine, so feed grinding was a brutally dusty task. The only respiratory protection I ever used was an occasional diaper wrapped around my head, and I'm sure that this might have something to do with the asthma that I developed ten years later. Take this little bit of advice from me—protect yourself from airborne grain dust whenever possible. We now buy dust masks by the case, but we sure didn't back then. The old buhr mill did the job quite well, but the ground grains, minerals, and concentrates still had to be mixed together. By this time we had graduated from the scoop and sap bucket to a scoop shovel and a pile of grain on the floor. Grinding was hard on the lungs and mixing was hard on the back; nevertheless, this seat-of-the-pants method worked for us thirty years ago and didn't require a lot of extra investment.

During these early days of my farming adventure, I was constantly longing for better grinding equipment than I had. I picked up a couple of marginal belt-driven hammermills that didn't work very well. Hammermills are self-enclosed units with four sets of thin-profile pieces of steel that revolve around a central shaft.

As seen in the illustration on page 392, these "hammers" literally pound feed grains through a curved screen. The texture of the ground feed is controlled by the size of the holes in the screen. Large five-eighth-inch holes are for coarse grinding of materials like ear corn. Screens with smaller three-eighth- and half-inch openings are for fine grinding of cereal grains for pigs and chickens. I wanted a decent hammermill so badly, but never found a reasonably priced one in my travels. Finally, I stumbled onto a machine called a grinder-mixer. I kept seeing these things advertised in farm papers and auction flyers. These self-contained units consisted of a hammermill and a mix tank mounted together on a set of wheels powered by a PTO shaft from a tractor. It seemed like the most popular brand out there was made by New Holland and sold used for somewhere in the five-thousand-dollar range. In the early fall of 1985, my eyes lit up when I found one being sold at a nearby auction. I had high hopes of buying this machine for a reasonable price because I figured that no one around my neighborhood would be interested in it. I went to the sale and was disappointed when the machine sold for forty-five hundred dollars to a farmer who lived on the shore of Lake Champlain and intended to use it to grind ear corn. I struck up a conversation with the fellow and learned that he had an older model he was replacing with his recently purchased prize. I asked him if he wanted to sell his grinder-mixer and he thought he would. Thus began two months of dickering and negotiating. I drove down to Bridport, Vermont, the day after Thanksgiving and purchased an International Harvester Model 1150 grinder-mixer for eighteen hundred dollars. I had finally entered the modern age. My sixty-five-horsepower farm tractor had just enough power to run the thing. I could dump in corn, barley, minerals, and concentrates and end up with a finished ration all mixed in the tank. This was living!

My IH 1150 turned out to be a honey of a piece of equipment. It was ten years old when I bought it

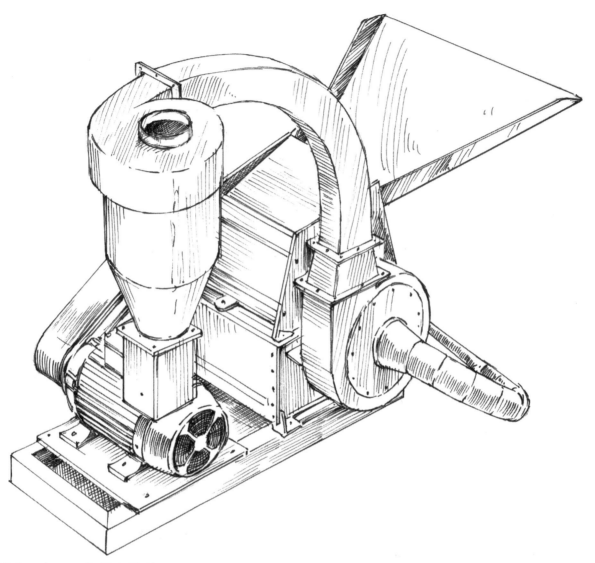

Stationary hammermill with electric drive.

in 1985, and it still serves me well today over twenty-five years later. I ran into the former owner five years after buying it and learned that the New Holland machine he had picked up at the farm sale was only half the machine that he sold to me. This seems to be the case with most used farm machinery—when you buy something like this, it is either a gem or a lemon. I was lucky this time; my feed grinder has an intake auger and a ground feed delivery auger. These older International models were custom-made by the Arts Way Co. from Iowa, and they are some of the best machines on the market. By 1985, we were milking and feeding close to twenty-five cows, and it was nice to know how to grind and mix feed rations the old-fashioned way with antique technology. However, it was even nicer to graduate to a piece of equipment that could crank out two tons of completely mixed cow grain in a couple of hours without the backbreaking labor of yesteryear.

Grinder-Mixer Troubleshooting

I've learned a few hard lessons over the years about my grinder-mixer. The learning curve in the beginning was especially steep. Undercover storage space was in short supply at the time; the machine had to remain outdoors, exposed to all sorts of weather. I tried covering the thing with a large tarp, but the relentless winds on this hilltop blew it off and shredded it, allowing rainwater to get down inside the hammermill. Rainfall in the late fall and early winter was particularly problematic because it would get into the inner workings of the grinder and freeze. Gravity would take all of this moisture to the lowest point, an auger beneath the grinding screen. This auger needed to turn freely to transport ground grains horizontally into the mixing tank. Never underestimate the strength of ice—it's as strong as steel. I'll never forget the first time I experienced this problem on a cold December day not long after I first got the new machine. I engaged the PTO lever to start the thing up, and something sounded really out of the ordinary. I could hear a drive chain slipping over the immovable sprocket that was responsible for driving the delivery auger. It doesn't take long to ruin a sprocket or a roller chain in this sort of situation. I turned the PTO off instantly and got off the tractor to investigate. Upon taking off the grinder cover and removing the large curved screen, I found the auger trough in the bottom to be totally full of ice. It took me over three hours with a slender steel bar and a propane torch to chip and melt it out. This was hard lesson number one. I'm not proud to tell you this, but I let this happen more than once in those early days when my storage shed was the "Lord's garage." I made a few other mistakes, too, like feeding grain into the hammermill too quickly and stalling the tractor. There were also times when I wasn't paying attention to the level of grain inside the mixing tank—which could be easily monitored by looking through little glass inspection windows on its sides. All of a sudden there would come a waterfall of ground grain out the top cover of the tank and falling over the edge onto the ground below. I have to say that this older piece of equipment must have been built quite well to stand up to the abuse it got from a novice like me. If you are lucky enough to get a grinder-mixer, always keep it in a shed when it isn't in use.

Putting Together a Ration

Animal nutrition consultants are standard fare in high-production agriculture these days. These individuals fine-tune livestock rations to help farmers squeeze extra milk, eggs, and meat from their animals. In the world of commodity milk production, nutritionists are integral to the success of the operation, putting together rations based on the laboratory analysis of farm-grown grains and forages. By-product feeds like whole cotton seed, canola oil cake, and brewer's and distiller's grains are delivered by the truckload to large farms to be mixed with corn, cereal, and forages into a TMR (total mixed ration). Hominy and corn gluten, which are by-products of the corn oil and cornstarch industries, regularly find their way into hog feeds; other additives like antibiotics, growth regulators, and insecticides are commonplace in swine and poultry feeds. This high-tech route isn't for us. Aside from wheat middlings from organic flour processors, certified organic grain by-products are just about nonexistent. As organic farmers, we need a simple, straightforward method to take our farm-raised grains and feed them to our livestock. So your basic ingredients will be the corn, oats, wheat, barley, peas, soybeans, peas, sunflowers, and flax that you coax from your fields. You will have to figure out the best grinding texture to suit the needs of your stock and how to proportion different grains in relation to one another. This needn't be the complicated and daunting task the feed industry wants us to think it is. To begin with, you'll have to use the grains that you have on hand. Very few of us will have adequate supplies of all the grains just mentioned, so it's important to take an inventory of what you have available and learn some basic concepts about feed grains and how to use them.

Achieving the Proper Protein Levels

Farm animals need both energy and protein. Carbohydrates supply the starch and calories necessary for motive power and bodily functions; these are the "go" feeds that provide energy. Protein is the "grow" feed responsible for building stature, muscle, flesh, and bone. Corn and oats are the most common energy grains that we can easily grow and use as the basis for just about any standard animal feed. Barley and wheat fall into a middle category; with protein levels of 13 to 16 percent respectively, these two cereals provide moderate amounts of protein coupled with high levels of carbohydrates. Smaller amounts of high-protein legumes like soybeans and peas or oilseeds like flax and sunflowers can be added to a grain mix to raise protein levels from the low teens to the mid- to high teens. Adding one part of roasted soybeans at 42 percent protein to three or four parts of 9 percent corn and 13 percent barley will round a ration to a comfortable 16 to 18 percent protein level. Rule number one in feeding your farm animals with homegrown grains is to use what you have on hand and mix it with other grains to reach the protein level you desire in a finished grain. Protein levels can also creep downward as farm animals make their way through life. Starter rations for young piglets and meat birds are always 18 to 20 percent protein when the little critters are just starting out; once chickens and pigs are a third of the way through their growth cycles, the amount of protein in their grain can be reduced gradually. Finisher grains for pigs and meat birds are somewhere around 14 percent in protein. You simply adjust the amount of high protein in relation to what you want for a final level. This can be done mathematically using proportions. The first thing you need to know is the relative protein levels of all the grains that you have at your disposal. Some of this grain can be homegrown, and some can be brought in from the outside. All of this information is available in Morrison's *Feeds and Feeding*. But for simplicity, let's make a short list right here and now.

PROTEIN LEVELS FOR THE COMMON GRAINS

Grain	Protein
Corn	8–9%
Barley	13–14%
Oats	11%
Wheat	13–16%
Mixed Cereal Grains	13–16%
Roasted Soybeans	40–44%
Field Peas	25%
Flaxseed Meal	25%
Sunflower Seed Meal	25–28%

Rule number two is this: To determine the final protein of a grain mix, multiply each individual grain's percentage of the ration by its percentage of protein. Add together these figures to obtain the final protein level of the mix. What would happen if we mixed one part of roasted soy with two parts barley and two parts corn? First, figure the volume of each individual grain in relation to the finished mix. In this particular case, the roasted soy would make up 20 percent of the total; the corn and barley would each make up 40 percent. So let's do the numbers for our theoretical grain ration.

corn @ 9% protein × 40% of the mix (.09 × .40) = .036
barley @ 13% protein × 40% of the mix (.13 × .40) = .052
roasted soy @ 42% protein × 20% of the mix (.42 × .20) = .084

Add .036, .052, and .084 together to come up with .172, which is a protein level of 17.2 percent.

This little bit of math will help you decide how much of what to add together to achieve the protein level you want in a finished ration. This formula will work on any scale; it doesn't matter if you are mixing together individual scoops or buckets of each component or working with tons of ingredients going through a grinder-mixer. A proportion is a proportion no matter what you do. Use this formula to tweak protein levels upward or downward—all you have to do is play with the numbers on paper. For example, if 17.2 percent protein is a little low for you, increase the amount of roasted soy in the previous ration to 25 percent and reduce the barley or corn to 35 percent. Multiply everything out again and you will have your answer.

Achieving high-enough protein levels in grain rations can be a stumbling block for many of us. The common cereals are a lot easier to grow than the higher-protein legumes and oilseeds, so you might have adequate supplies of corn or barley but have to buy some higher-protein materials like roasted beans or oilseed cake. The protein concentrates that we bought in from the Canadian feed mill in our beginning days of farming simplified this process immensely. These commercial mixes came complete with the proper added vitamins and minerals for each class of livestock. But since we organic producers don't have this conventional option, we'll have to be a little more creative. The first thing you can do to raise protein levels in the least painful and easiest manner is to replace as much of the soy as possible with field peas. I have already praised the Canadian field pea in the last chapter. Peas grow well in the same field with cereals like oats and barley, and do not need to be roasted, like soybeans. The addition of peas to a field of barley can raise the final protein level of the harvested mixed grain from 13 to 18 percent, and 18 percent is pretty close to an acceptable level for most standard grain rations. For this reason, as I said before, I have become a great fan of growing mixed cereal grains, and I always seed peas with my barley. Remember, though, it's not the end of the world if you have to purchase some of the protein components of your ration. Thankfully, roasted soy usually doesn't amount to any more than a quarter of any grain mix.

Additional Ingredients

You will, however, need to add small amounts of other necessary ingredients to most farm-produced livestock rations. Salt and kelp are two rather important additions needed by every domesticated farm animal out there. For some reason, I omitted salt from the first cow grain that I produced all those years ago, and I couldn't figure out why my cows were listless and not milking well when they were being fed excellent hay. Stew Gibson from UVM Extension came to visit the farm one day, and I asked him what the problem might be. He asked me if I was using salt in my homemade grain mix, and when I told him no, he recommended using a fifty-pound bag of common livestock salt for each two-ton batch of cow grain that came from the grinder-mixer. I took his advice and realized an immediate five- to eight-pound-per-day, per-cow rise in milk production as a result. We now use Redmond Natural Trace Mineral Salt from Utah in every batch of cow grain. This product has become commercially available and widely distributed in our region, and just about every farm supply store now carries it. We also add a fifty-pound bag of kelp to each two-ton batch of cow feed. Kelp is an amazing source of minerals because it comes from the ocean—which contains every element known to humans. There are many brands of kelp out there for organic livestock farmers to choose from. Kelp comes from Maine, Nova Scotia, Iceland, and Norway. Prices range from thirty-five to forty dollars for fifty pounds of kelp from the North American Atlantic coast to fifty dollars a bag for geothermally dried material from Iceland. I'm rather partial to the Icelandic stuff because it is dried at a much lower heat.

The last minor addition you'll need to make to most farm-produced grain mixes is calcium. The feed industry uses standard high-calcium ground agricultural limestone as a calcium source. Standard practice is to add about twenty-five pounds of calcium carbonate to every ton of finished grain. I used regular lime for years, but never felt like my animals could get the calcium they needed from an inert ground rock. Finally, I found out about a material called aragonite that was recommended as a calcium source for laying hens. Aragonite is ground fossilized coral from a beach in the Bahamas. Because this material once came from living bodies, it turns out to be a very available and wonderful calcium source for just about every class of farm livestock. Aragonite is added to grain rations at the same rate as kelp and salt; a fifty-pound bag will work for a two-ton batch of feed. Prices for aragonite are under twenty dollars for a fifty-pound bag, and while limestone might cost only a third as much, its calcium availability to farm animals is much less.

The very last detail to attend to in formulating a grain ration is the addition of some sort of commercial mineral pre-mix. Thirty years ago, a good organic livestock mineral was not yet in existence. Things changed in the early to mid-1990s as the number of organic dairy farms increased across the country, and pioneers in alternative animal nutrition like Jerry Brunetti and Gary Zimmer appeared on the scene and began to educate us about mineral balances in soil, crops, and cows. Companies like Midwestern Bio Ag in Wisconsin and Agri-Dynamics in Pennsylvania began to offer high-quality livestock minerals. These outfits also emphasized balancing soil minerals, promoting biology, and the development of good soil structure. The purpose of this new and alternative system was to produce forage crops that were rich in available minerals. It sure would be nice to get minerals from the feed as opposed to a bag. Theoretically, minerals in cow grain might not be necessary if they were in the forage consumed by the dairy herd, but most of us have a long way to go in meeting the mineral needs of a high-producing dairy herd through hay alone. These companies and others offer all sorts of livestock minerals in many different forms and combinations. You can buy cow minerals with a calcium-to-phosphorous ratio of 2:1 or 1:2. Poultry and swine pre-mixes are also available to add to starter and grower rations. There are numerous places to obtain these feed additives, and some local farm supply dealers can order them in for you. If you are farming on any sort of scale, it might be wise to buy your minerals and other additives by the pallet and have them shipped by common carrier. These products aren't cheap; a fifty-pound bag will take care of one ton of grain and can cost between thirty-five and fifty dollars. This may seem rather expensive, but I think it is money well spent. You might be able to omit adding a pre-mix to a dairy ration if the cows are eating nutrient-dense forages. Don't try this with chickens and pigs, however, because these monogastric animals are much more sensitive to nutrient shortages than ruminants. Chickens need methionine and a host of other vitamins and special minerals. Ask other people in the business of feeding organic livestock what they use and recommend; a little research and inquiry will help you figure out what might work best for your particular situation. Above all else, make sure to obtain some high-quality minerals. You will be rewarded with healthy and trouble-free livestock for your efforts.

Grain Mixes for Dairy Cows

Cows were meant to eat forages, not grain, and grass-fed cows most likely have higher CLA (conjugated linoleic acid) levels in their milk. Grain makes rumen acidic and causes digestive upsets like acidosis. Nevertheless, grain feeding is standard practice in both the organic and conventional dairy sectors, and small to moderate amounts of feed grains can provide extra energy to the microbes in the rumen of a dairy cow. The dietary needs of lactating dairy cows change as they transition from the winter feeding of stored forages to spring and early-summer pasture in the grazing season. Cows on lush early pasture will turn up their noses at grain, which is a good reminder that animals are getting adequate levels of protein from the premium pasture that they are consuming. If you find yourself in this sort of situation, you can omit expensive purchased protein like ground roasted soybeans from the dairy ration, but corn, barley, and oats coupled with late-May premium pasture will supply a dairy cow with needed levels of energy. Closely monitor your cows' feeding habits. If the hay or pasture is primo, extra protein is not needed in the grain. Many organic dairy farmers with excellent pastures purchase 12 to 14 percent grain when the cows first begin grazing. A mixture of corn and barley will give you the right amount of energy and a grain with these lower protein levels. Once you raise the protein of a dairy ration over 16 percent, be prepared to add nine-hundred-dollar-per-ton roasted soybeans to the mix. Medium- to poor-quality hay demands higher levels of protein in the accompanying feed grain, so you'd be wise to remember that field peas are a cheaper and easier-to-use source of supplemental protein in a dairy ration.

Grinding and Mixing Feed

Now that we have covered the basics of developing a recipe for cow grain, let's do the actual work of grinding and mixing the feed. If you have a very small dairy herd or a family cow, you can hand-grind the various grain components with a large Diamant grinder and mix things together scoop by scoop in a pail for each individual cow to be fed. Your cows will really appreciate it if you grind fresh grain before each milking. If you are machine-grinding with an old buhr mill, a small hammermill, or a roller mill, you will have to grind the various grains separately, weigh things out, and mix them together on a floor with a large lightweight shovel. This is the low-investment "hard way" of doing things. I did it this way for years, and it always worked well as long as I had the energy. Since those early days of ration making, we have ground hundreds of batches of cow grain here on our farm with the grinder-mixer that I purchased in 1986. I recommend this method for ease and simplicity and will try to give you a quick lesson in the operation of the standard grinder-mixer. There are a number of perils and pitfalls that you might be able to avoid by hearing all about the mistakes I have made over the years.

First, determine how much of each type of grain you will be using in your final mix of cow feed. Refer back to the proportional formula to determine how much protein and carbohydrate you will need to create an acceptable grain mix. First and foremost, try to use what you have on hand. Corn and barley are the best cereals, but don't be afraid of wheat and oats if you have them on hand, and make sure to have enough roasted soy or peas on hand to achieve the protein level you want in the finished grain. Determine how many pounds of each component you will need to make your special mix. The tank on my old grinder-mixer holds 105 bushels, which is slightly more than two tons. If you are doing a four-thousand-pound batch that is 20 percent roasted soy, 40 percent corn, and 40 percent barley, you will need eight hundred pounds of soybeans, sixteen hundred pounds of barley, and sixteen hundred pounds of corn. All of this grain will have to travel up the intake auger, through the hammermill, and into the mixing tank. How can you do this easily without actually physically weighing all of the ingredients? We have several ways to make this happen. First, remember that this isn't a totally precise operation; flying by the seat of your pants is quite all right. The easiest way to determine the weight of each grain going into the mix is to have a grinder-mixer equipped with a digital electronic scale attachment. Simply zero the thing out before you start grinding and watch it like a hawk. When you get to sixteen hundred pounds of corn, switch to the next component and keep watching the scale. If you are grinding roasted soy next, add the eight hundred pounds to the total and grind until the scale says twenty-four hundred pounds. If you are in the market for a used grinder-mixer, try to buy one with a scale. My machine has a bracket for an electronic scale, but no scale. I use the eyeball method when grinding a batch of grain. The side of the tank is outfitted with three elliptical glass viewing windows and number decals that go from 1 near the bottom to 9 near the top. Stenciled onto the side of the mix tank is a weight chart that tells you the poundage for each type of standard grain at each number level. For example, barley at number 5 might weigh eighteen hundred pounds. The window viewing system will suffice, but it isn't entirely accurate because the ground grain inside the tank isn't dead level as it is being mixed. If precision is the name of your game, you might have to load the grinder with five-gallon pails of grain and keep count. (A five-gallon pail of corn weighs just over thirty pounds.) This definitely slows the process of grinding. If you're not sure of everything, use the pail method the first few times you grind to get a feel for things. After you've dumped fifty-three pails of corn in the intake auger, look at the window scale on the side of the tank. This will give you an idea of how accurate the weight chart is. Proper proportioning of ingredients is essential to making a dairy ration with the protein and energy levels you want.

When operating a grinder-mixer to produce your own grain mix, there are a few secrets that will make the whole process foolproof. First, select the right grinding screen for the job. I generally like to use a screen with three-eighth- to half-inch holes, which will give you a medium grind that is not too coarse or too fine. If you are adding ear corn to your mix, a screen with five-eighth- to three-quarter-inch holes is perfectly fine, and a little bit of cob in the finished product won't be a problem. Next, consider the order in which you add ingredients to the batch. Paying attention to this detail will permit the machine to do a more thorough and complete job of mixing. I like to start with my 40 percent of corn. I look at the batch as if it's a sandwich. The amount of protein is smaller and is akin to the meat in the middle, while the barley and the corn are the two pieces of bread on the top and the bottom. The feed additives can be added with the soy in the middle of the operation, which ensures more even and uniform mixing.

The other important detail is to not overfill the mixing tank. Mixing is accomplished by means of a shielded central vertical auger that grabs the ground grain at the bottom of the tank, elevates it to the top, and flings it outward against the sides. Extra room at the top is essential for this mixing to take place. We have talked about adding kelp, aragonite, and Redmond Salt to grain ration; I also like to add a fine clay product called Desert Dynamin to the mix. Dynamin is so fine in texture and flowable that it acts like an anti-caking agent in a dairy mix. It also contains trace amounts of selenium, which is absolutely essential for good livestock nutrition. It's available from Agri-Dynamics in Martins Creek, Pennsylvania. Once you have all the ingredients ground and in the mixing tank, take the hammermill out of gear and allow the batch of grain to sit there and mix for half an hour. If you've done everything properly, you'll have two tons of great cow grain to unload into your bin. The grinder-mixer has a feed delivery auger for this purpose.

Rations for Poultry

There are many different schools of thought when it comes to feeding chickens. Commercial layer and broiler feeds are fine-tuned for maximum egg and meat production. The ingredient list attached to a bag of modern-day chicken grain is rather long because there are so many added special components like riboflavin, deactivated sterols, and animal by-products. At first glance, it seems just about impossible to produce anything this complicated on the farm. However, once we delve into the world of small-scale poultry husbandry, we find several different approaches to feeding broiler and layer flocks. First, there is the whole-grain crowd who only want to feed their chickens whole corn, wheat, and other unground grains like flax and soy. The logic behind this approach is that chickens are able to preprocess whole grains with the help of the stones and grit that they have in their crops. Ground layer and broiler mash is seen as wasteful because every little bit that gets spilled by the birds onto the henhouse floor is wasted; whole grains are much better because poultry will peck away at whole-grain "scratch feed" scattered about. Another group of poultry lovers out there don't want to feed soybeans to their birds. Soy is seen as a producer of estrogen-like compounds and as much too strong a protein source. And then there are the rest of us who feed ground broiler and layer mash to our birds. We might supplement the diet of our poultry with addition of some extra cracked wheat and corn broadcasted out in the chicken yard. I'm definitely not going to tell you which feeding method will work best for your flock, but I will try to enlighten you on the basic concepts of feeding poultry and help you figure out how to use as much home-produced grain as possible in the process.

How I Stumbled upon a Good Ration

I entered the world of poultry ration making quite by accident through the back door. About fifteen years ago, an organic farmer named Eric in the Intervale of Burlington, Vermont, asked me if I could provide him with grain for his meat bird business. Eric had a special recipe for me that he had gotten from Joel Salatin, the well-known poultry aficionado from the Shenandoah Valley of Virginia. I had plenty of corn, oats, and soybeans at the time, so I figured I would give it a try, and all I had to do was follow directions. Here is the recipe on a one-ton basis:

  975 pounds ground shelled corn
  650 pounds ground roasted soybeans
  200 pounds ground oats
  50 pounds aragonite
  50 pounds crab or fish meal
  50 pounds Poultry Nutri-Balancer from Fertrell
  25 pounds of kelp

I had tried many times before this to make my own chicken grain without much luck, but this particular ration was an instant success. The learning came afterward. I had never used aragonite or crab meal before, and these additions, along with the specialized mineral from Fertrell Company of Bainbridge, Pennsylvania, made all the difference. Eric loved the chicken grain that I formulated for him, and I was happy to have a workable recipe that I could use to supply other people with a high-quality poultry ration. I ground this particular mix through a hammermill screen with rather small holes to achieve a fairly fine consistency. This mixture was especially good for starter for young broiler chicks. The protein level was close to 20 percent. I very quickly found that all poultry thrives on the addition of some animal protein. The whole crab meal was just the ticket, and it made so much sense to me—the shells served as a source of extra protein and calcium. I thought back to all the times I had seen my chickens devour meat scraps from the compost bucket. If you find that your chickens are eating their eggs, you should increase the protein of the ration by using some sort of animal protein. Chickens are carnivores in addition to being grain eaters. I once bought a whole tractor-trailer-load of crab meal from Nova Scotia, and it doubled as a chicken feed additive and a starter fertilizer for my field corn.

The aragonite was another pleasant surprise. For years, we had been using ground limestone as a calcium source in our animal feeds, but aragonite was much softer in texture and much more available to the metabolism of poultry and other farm animals. I did learn that the amount of aragonite could be adjusted upward or downward depending on what sorts of birds were being fed. Broilers could get away with twenty-five pounds per ton, while laying hens needed the full fifty pounds per ton to produce eggs with strong shells. I ended up producing a dual-purpose ration that could be fed to meat birds and layers alike. I left the aragonite at fifty pounds per ton, which worked quite well for both broilers and layers.

When I first obtained this recipe, I wondered why oats were included, as I had never been a big fan of ground whole oats in any sort of animal feed. They always seemed too fibrous and unpalatable to me before this, but after grinding a few batches, I came to realize that the oats were essential because they provided fiber and loft to the ration. All the other ingredients were rather dense and heavy when mixed together, but the oats lightened things up quite a bit. Over the years, however, I have found barley to be almost as good as oats in farm-ground chicken grain.

Corn and roasted soy are definitely the two main ingredients in this particular ration. You will find that you might need to increase or decrease the overall protein of your mix, depending on what kinds of birds you are feeding and their stage of development. Young chicks in their first month of life need lots of extra protein. Six hundred pounds of roasted soy per ton will give you 20 percent protein in the finished grain. You will have to add at least seven hundred pounds of soy to a turkey starter feed, because young turkey poults need even more protein than meat birds, but if you increase the soy component of the ration, the corn will have to be cut back so all the numbers still add up to two thousand pounds. Once your flock of young meat birds or laying hens is a month old, you can begin to drop their protein by cutting back the soybeans. This is always a delight because roasted soy is so much more expensive than corn. Grinding your own poultry feed needn't be an exact science; this recipe is meant to be tweaked and adjusted to meet your needs as well. Use the grains you have in a close-enough ratio to achieve the desired protein level necessary for your situation. You can replace some of the corn with barley if need be, and sunflower oilseed cake and peas can be substituted for the soybeans. Just remember to use the formula described earlier to ensure that your birds are getting an adequate amount of protein.

Last but not least is the mineral component. I used the Fertrell Poultry Nutri-Balancer for years because it was the only product that I knew about. After a while, the company could no longer sell their product as an additive in certified organic grain rations because of a change in the National Organic Program rules that no longer allowed synthetic sources of methionine. I began to look around for another poultry mineral and found an organically approved one from Lancaster Agricultural Products in Pennsylvania. I finally discovered that Agri-Dynamics offered a product called Poultry-Lytes, which I found to be of very good quality. It doesn't matter where you source your chicken minerals, as long as you use something to supplement the grain diet of your birds. Dairy products such as skim milk and whey are also good sources of complete protein and minerals like calcium and phosphorous. Remember to provide some sort of grit like oyster shells to your birds to aid in the pre-digestive process. The most important thing to do when grinding and feeding your own chicken grain is to have fun while you're doing it. Step out and be your own experimenter. You might just mix together the next miracle ration without even knowing it.

Feeding Farm-Grown Grains to Pigs

Preparing rations for pigs offers us the most opportunities to use our own farm-produced grains. Swine convert grain dry matter to meat better than any other class of livestock, and their ratio of usable meat to total body weight is also the highest. Take the extra care to get recently weaned piglets off to a good start by providing them with proper nutrition. Rolled oats soaked in milk are a perfect follow-up once milk from the mother sow is no longer being consumed. You'll want to grind starter pig feed to a very fine consistency and mix into a "slop" with milk or skim milk if possible. Corn and barley combined with a non-soy source of protein like sunflower meal will work just fine. Once baby piglets are well started and growing, you can relax your vigilance and begin to use a wide variety of feeds to foster rapid growth. Like most other farm animals being raised for meat, younger pigs require higher levels of protein—shoot for 16 to 18 percent protein levels in hog starter feeds.

As the piglets transition into their rapid growth phase, your options increase. Protein levels can drop down to 14 percent after a very short time. Corn, barley, and roasted soybeans work quite well to make a good hog ration; three to four hundred pounds of extra soy protein per ton of feed is all that is necessary. Oilseed meals from flax, sunflowers, and canola can also be used. This makes pig feed a whole lot more economical to produce than chicken grain. The other great thing about pigs is that they will consume lots of other non-grain farm products like waste milk, garden vegetables, and small amounts of legume forage and pasture. I like to grind my pig grain fairly fine and mix it with a liquid like water, whey, or milk. The hog slop is then poured into troughs at feeding time. Commercial confinement hog operations feed dry grain, but the beauty of the slop system of hog feeding lies in the fact that feed grains can be made more digestible through a pre-soaking process. Pigs are monogastric just like humans and will benefit from softened and soaked grains. If you don't have access to grinding equipment, whole grains can be slow-cooked and fed to pigs as an alternative.

I started off grinding pig grain with absolutely no idea what I was doing. All I did was process a mixture of corn, barley, and a protein source into a fine-ground feed. At the time, my principal customer wanted me to use sunflower cake instead of roasted soy. I didn't add minerals and didn't appear to have any problems. As time passed, however, I began to feel guilty that I was shortchanging the pigs in the nutrition department, especially if the animals had no access to milk rinsings and other dairy products. So I began buying a hog mineral pre-mix and adding it my rations. I have had the best luck with minerals from Agri-Dynamics, but there are plenty of other high-quality hog minerals out there in the marketplace. Now I didn't have to worry anymore that I might be shortchanging my animals.

A ton of good solid 16 percent hog ration can be made with the following finely ground ingredients:

900 pounds ground shelled corn
800 pounds ground barley
200 pounds ground roasted soy
50 pounds kelp
50 pounds swine mineral

As your pigs begin to mature, begin cutting back on the soy and increasing the corn and barley. If you have access to milk sources, you may be able to eliminate supplemental protein inputs entirely. Keep the kelp and mineral at the same level. Lactating and gestating sows require more protein and a special mineral additive. Farrowing pigs is an entirely different sort of operation. Mineral balance and calcium-to-phosphorous ratios are extremely important, so you'd be wise to buy the special blend of minerals for sows and keep grain protein levels sufficiently high.

There are other creative ways of feeding farm-grown grains to pigs. If you are a small-time operator with garden-sized fields, you can raise ear corn for your pigs. Choose a variety of corn with low-enough heat-unit requirements that it will mature in your area. Plant the corn in thirty-inch rows as you would if you were planting it for grain or silage, but raise the corn with grain production in mind. When the corn begins to dent in late August or early September, you can begin handpicking and husking ears to feed to your pigs. Feed the whole cob by throwing ears right in the pig yard. You'll be surprised at how well your pigs will eat the partially ripe kernels right off the cob. Who needs a corn sheller when you have hogs? Feed your pigs as much ear corn as they will eat through the month of September as the grain gets riper and riper. I remember pulling a gravity box dribbling just-picked ear corn through John Ace's pig pasture in Wisconsin back in September 1974. Standing corn can also be "hogged down" right in place. This old-time practice works quite well if your pigs are trained to stay inside a low-hung electric fence. Use temporary portable fencing to cordon off a new row of standing corn for your pigs each day, and pigs will pull down cornstalks and eat the ears right in place. This is the swine version of rotational grazing—instead of bringing the feed to your animals, bring the animals directly to the feed.

If you are not growing every last little thing that your pigs need to eat, don't worry about it. You can use small amounts of any sorts of grain and other by-products to mix with purchased inputs. Maybe you have some barley or corn, but no roasted soybeans. You might be able to use cheese whey instead of extra purchased beans. The important thing is to be creative and open-minded. Some folks get food scraps from restaurants and institutions. Don't be afraid to experiment: Do whatever you can to replace purchased inputs with things you can produce yourself on your own farm.

Some Last Thoughts on Feeding Your Own Grain

There are numerous ways to feed your own grain to your own stock. Infrastructure and scale may vary from farm to farm, but there is always a way to make it work. Small amounts of grain can be hand-ground with a Diamant or similar hand grinder. Slightly larger and more mechanized operations can grind up grains by means of buhr mill, hammermills, or roller mills. Cooking and soaking are also good options. There also might be the possibility that someone could custom-grind your grain for you. I believe this service will most likely become more commonplace in the coming years as various regions work on food security issues.

The beauty of feeding farm-grown grains to livestock is that grain quality need not be top-notch. You certainly don't want to feed moldy or mycotoxin-infected grains to farm animals, but grain of medium quality that has low test weight or has experienced an early frost at harvesttime is perfectly fine. Grains that don't make the grade for human consumption make great livestock feed. This is the main reason why many individuals with flour mills keep a few pigs or beef cows. You need animals to eat your mistakes. As with just about every other facet in the world of grains, I recommend that you find someone who has more experience and knows more than you do about growing and feeding grains. Everyone needs a mentor to help them develop their farming craft.

CHAPTER NINETEEN
Where Do We Go from Here?

There is no doubt that grains can be grown profitably in our region, but for many reasons they haven't been the crop of choice for the agricultural sector here in the Northeast. Farmers here have found more profits in traditional crops like hay, vegetables, and fruits. Despite this situation, the farming subculture that does plant and harvest grain crops is alive and well and growing in size every year. So what is it that fascinates people so much about the culture of grain? Is it the beauty of amber waves of golden grain moving to and fro in a summer breeze—or is it just novelty? The self-reliance factor has been a big draw for me—I want to know where all the food I eat comes from. Growing and processing my own wheat and oats into flour and breakfast cereal contributes to the wholeness of my life. I can't seem to shake this addiction even though for health reasons we eat a lot less processed grain products than we used to, and I have to be content with rye bread these days because wheat seems to have a bit of a detrimental effect on me. But this isn't going to stop me from planting lots of wheat. It must be romance that keeps me going.

In this last chapter, I would like to give some broader perspective to the subject that we have been discussing for the last several hundred pages. Can we develop a viable regional food system that includes local grain? What sort of infrastructure will have to be put in place to accomplish this dream? How does grain growing fit into the context of an increasingly erratic climate that makes agricultural activities even more challenging? Can we grow grain crops and improve soil fertility at the same time? There are also issues of seed sovereignty and corporate control to discuss. I hope that I have provided you with enough basic information to help you get started with grains, and I have certainly rambled on long enough. But I would like to wrap things up with some practical conversation about the future. There is a whole lot more to this than just planting and harvesting seeds.

The Human Consumption Factor

Each of the particular grains that we have discussed requires varying degrees of processing before it makes its way to your table. Wheat and rye need minimal post-harvest treatment to enter the human food chain: Cleaning and drying are all that is necessary to prepare these two grains for milling into flour. Some of the other common grains like oats, barley, and spelt need to have hulls or husks removed before they can be milled, flaked, or otherwise processed into breakfast cereals. Here is where things get complicated and expensive. The lion's share of this sort of grain processing takes place behind closed doors in large industrial settings in distant facilities owned by large corporations. The nearest oat-hulling and -flaking plant that I know of is in Iowa, for instance, and it certainly isn't open for public inspection. I find it rather ironic that oats are the most naturally adapted grain crop to our region, yet the most difficult to turn into an edible product. I speak from experience in this matter. We have spent well over forty thousand dollars on oat-processing equipment and so far haven't turned out any salable oat groats or flakes. We would probably need to spend three times that amount to seriously get into the oat-processing business. You can re-read the hulling and flaking section of my oat chapter to familiarize yourself with all the details, but suffice it to say that my collection of antique oat-processing equipment might work better in a museum than in my granary. My personal example is pretty representative of the difficulties that might be incurred in setting up small- to medium-scale grain-processing facilities on local farms. We can't afford the best machinery, and the knowledge base for this sort of activity lies in faraway places like Minneapolis, Minnesota.

This predicament is exactly the same for barley and spelt. A Satake barley huller imported from Japan is simply not an affordable item for any of us just starting out in the grain business. We have tried putting spelt through our old Roskamp Champion hulling machine with varying degrees of success, as spelt is quite soft and very easily damaged in the hulling process. Our old machine needs to have the RPMs of its hulling component slowed down to prevent the spelt berries from being chipped and broken. We would need a special and very costly retrofit to turn this piece of equipment into a variable-speed huller. After spending close to thirty-five thousand already, there is no money left in the coffers for this kind of improvement. Then, of course, there is the Franz Horn spelt huller from Germany for fifty thousand dollars. After considering all of these expensive options, I decided that we would have to be satisfied with the less-than-perfect spelt berries that we can produce with our old Roskamp Champion. We simply grind the broken-up spelt directly into flour after the hulling process. I would like to be able to offer whole spelt berries to our customers, but I'll have to wait until we have the wherewithal to refine the process. More often than not, this seems like the best approach, so do what you can afford and make the best out of what you have.

Processing infrastructure is a complicated subject. The fact that the machinery needed for turning grain into human food is very specialized and expensive almost precludes this sort of activity on a farm scale. This has been the primary reason why farmers have remained producers of a commodity that is sold to an industrialized processing sector. The investment required to build a modern white-flour mill or to install a state-of-the-art hulling and flaking line is beyond the means of most small- to medium-sized farm operations, but *value adding* is the catchphrase of agricultural policy makers these days. There are a lot more on-farm processing grants available from private foundations and state departments of agriculture than there used to be. Unfortunately, much of the grant money that is available is for market and feasibility studies instead of bricks-and-mortar projects on farms, and most USDA Rural Development grant programs are tailored to study a market, not to erect and outfit an on-farm grain mill. The few grants that are awarded to individuals for specific projects are very narrow in scope—you might receive money to hire labor or implement product deliveries, but not to buy specialized grain-processing machinery. We've been rather fortunate here in Vermont, however. The Vermont Housing and Conservation Board has awarded a number of small grants for processing infrastructure to farmers with well-prepared business plans. The competition is tough, but the process has been well worth it for many of us with dreams of turning raw grains into food for our neighbors. There is a lot to be said for doing a business plan with the help of an experienced adviser. You can sharpen your pencil and take a very realistic look at your dreams. In the end, all the frills and fluff are pared away, leaving you with a pretty realistic assessment of the costs and benefits of a potential on-farm processing project. I would say if you have this sort of opportunity where you live, jump on it. You might not get the help you need in the first round of applications, but the experience of taking a hard look at your financials will stand you well for future participation in these sorts of programs.

There are numerous other methods for financing an on-farm grain processing project. Angel investors do exist. These folks are looking to capitalize projects that enhance local food systems where they live. Unlike your typical venture capitalist who is looking for a 20 or 30 percent return on his money, these individuals are interested in social capital and willing to accept a much lower rate of return. There certainly aren't too many banks or other conventional financial institutions that will take a chance lending money to a starry-eyed grain farmer who wants to process his grain for local consumption. Five to 7 percent interest is considered "slow money." Return payments in these situations can be deferred for a number of years until a grain-processing operation is on solid ground and making money. I hope to see more of this kind of social investing in years to come. Informal community-supported agriculture models are an even better bet for cash-strapped grain growers looking to set up infrastructure. You can offer buy-in shares to a larger group of consumers, who each contribute to a community-supported investment in a small flour mill or hulling facility. Each localvore investor might put up a thousand dollars and receive eleven hundred dollars' worth of grain products in return over the period of a couple of years. I like this model best because it spreads the risk out over a larger group of people. You aren't beholden to one generous individual—who just might want to exert some benign control over you and your operation. Like it or not, a serious grain-processing facility requires serious money for building and start-up.

And then there is the pay-as-you-go, seat-of-the-pants method of putting infrastructure in place on your farm. Most of us are familiar with this approach when it comes to buying a combine or grain drill, but we can generally buy farm machinery pretty reasonably if we don't go for the brand-new stuff. It always amazes me when I see a ten-year-old combine advertised in a farm paper for a tenth to a quarter of what it originally sold for. Agricultural "iron" seems to depreciate rather quickly over time, and the same holds true with grain-processing equipment. There are all kinds of machines out there that are considered obsolete by the present-day industrial milling establishment, but would be just right for a farm-scale operation. These pieces of equipment were in place in grain-processing facilities thirty to fifty years ago when the general scale of operations was much smaller, and economies were more regional and localized. After paying top dollar for my Roskamp Oat Huller, I found out that there were three or four of these machines free for the taking on the third floor of an old oat mill on the Welland Canal between Niagara Falls, Ontario, and Buffalo, New York. This old cement factory building was due to be demolished, and everything had to go in a hurry. The only problem is figuring out how to take advantage of this opportunity when I lived a day's drive away and didn't have access to a crane and rigging to extract a heavy machine from three floors up in a derelict old building. Once in a while, luck and good fortune will fall upon us and we might just end up with the equipment we need for a very reasonable price. You certainly have to have your ear to the rail if you're looking for inexpensive grain-processing machinery. A little trip to North Dakota or Kansas City where grain and grain processing are a way of life might just be in order, so keep those lines of communication open and get to know people in the business. Sometimes I think that life is just one big research project. You'll be surprised at what you find out there once you begin looking. You might not find the exact piece of equipment you are looking for right away, but you will garner lots of valuable information that will help you to plan and design your dream facility.

There are a number of dealers out there who specialize in finding, refurbishing, and reselling old grain equipment. As agriculture scaled up in the Midwest and the plains in the 1960s and '70s, lots of small, independent seed-cleaning and grain-processing facilities fell by the wayside. There are abandoned establishments stretching from western New York all the way out to the Dakotas. Scrap iron and salvage operations will go in and clean out these once prosperous local mills. These are the people you need to get to know. Two individuals come to mind right away—Charles Stodden Sr. and Charles Stodden Jr. This father-and-son team travels all over the Midwest taking apart old grain facilities. They have formed a business called Commodity Traders International based in the small hamlet of Trilla, in east-central Illinois. I first learned of the two Charlies fifteen years ago from Tom Stearns of High Mowing Seeds. I was looking for some seed corn sizing and grading equipment at the time, and these two fellows had everything I needed and more for very reasonable prices. I ended up buying a whole trailer-load of old equipment from them. I got some indent sizers for corn seed and the elevator system that is presently in place in my granary as well as a few other rare items that I still haven't found a use for. I have to say that my wife Anne was pretty miffed when the trailer from Illinois showed up on a snowy December afternoon. In the years that have followed since my first purchase from Commodity Traders, however, I have been able to source many specialized pieces of grain equipment from these two characters. Unfortunately, prices for all of this old stuff have risen considerably because there are more of us who want to process grain out east. The Stoddens have turned their once informal business into a thriving concern, and they have a rather significant online presence complete with several yearly auctions. Bidding by computer and telephone line has become quite popular, and prices have risen accordingly as well. A decade and a half ago, this business sold items for a little more than the scrap-iron price, but this is no longer the case. Still, Commodity Traders' offerings are reasonably priced compared with the cost of new. We should be very thankful that someone is running around the Midwest saving all this equipment. I encourage you to check out their extensive website and contact Commodity Traders if you are in need of a specific item to process grains on your farm.

The subject of infrastructure dominates many conversations these days, especially among agricultural policy makers. There is a common underlying assumption that if there were more specialized equipment and processing facilities around, more grain would be grown in our region. But I have my doubts. When it comes to this sort of infrastructure, we'd all like to think that if we build it, they will come, but the situation is akin to the old metaphor of the chicken and the egg. Which one came first? Grain-processing facilities of any kind are extremely specialized and very expensive to construct, and the fact that state and federal government coffers are running low these days is a pretty good indicator that there won't be too much money for grain infrastructure coming from these sources. Larger industrial development projects that create numerous jobs and an expanded tax base seem to get the most help and attention. If you go to an economic development authority in your state with the idea of building a new flour mill, you probably will not be taken very seriously. The powers that be will want to know where the raw materials will be coming from and who will benefit economically. I think that the critical mass of farmers raising grains will have to increase considerably before we can expect government agencies to jump in and provide funds to help grow what is now a still fledgling industry.

Other Infrastructure Considerations

Processing farm-grown grains into finished products for human consumption may be out of reach for most of us, but the infrastructure needed to actually grow the crops on our farms is not. There is plenty of reasonably priced and serviceable farm equipment available out there, and as I've said many times before, you don't have to go out and buy brand-new stuff. Grain drills are plentiful and cheap, and a good combine can be purchased for five thousand dollars or less. Once you've decided that grain growing is something you want to do, start looking for machinery. Publications like *Fastline*, *Country Folks*, and *Lancaster Farming* are full of used grain machinery being sold by individuals and dealers alike. You might have to drive to Pennsylvania, New York State, or Québec to get what you want, but the machinery is available and it is usually affordable. There are also a number of individuals and dealers right in our region who specialize in bringing in loads of machinery from the Midwest that is too small for the heartland, but just right for our scale here in the East. The one outfit here in Vermont that comes to mind is Rene J. Fournier Farm Equipment in Swanton. These folks have two yearly consignment auctions—one in April and one in August. People come from hundreds of miles away to buy items like row-crop cultivators, grain drills, combines, and even old threshing machines. The quarterly newsletter of the Northern Grain Growers Association is another good place to look for equipment. Once you decide to begin looking for the machinery you need to grow and harvest grain, you will enter a world filled with individuals who are passionate about this type of agriculture. Networking is a wonderful thing. The person you call in search of a particular piece of equipment might not have what you are looking for, but he will tell you about several others whom you can call, and sooner or later, after lots of research and good conversation, you will find what you need. There are also numerous farm machinery locator websites on the Internet. The beauty of equipping yourself for grain farming is that most of the machinery available can be used to produce any of a number of different crops. The farm equipment you need for crop production is much less specialized than what you'd require to turn raw commodities into finished goods for human consumption.

The movers and shakers in the agricultural policy world will continue their advocacy for grain infrastructure, but I think that we have to start the process ourselves on our own farms—the seat-of-the-pants approach is the only thing I know that really works. Food systems proponents can make a real contribution to the on-farm grain movement, however, by supporting and funding a post-harvest grain-cleaning and -stabilization service that has the ability to travel from farm to farm. I've seen this sort of thing on a much larger scale in North Dakota. Itinerant grain cleaners trailer their equipment all over the state to clean combine-run grain for seed and for entry into the human consumption market. I picture this same type of service on a much smaller scale here in our part of the country. A trailer with several seed cleaners and a small dryer can be pulled from farm to farm with a pickup truck. This would allow smaller entry-level operations to actually produce salable grain crops for local consumption. Perhaps a small Meadows flour mill could be added to the mix. Organizations like the Northern Grain Growers Association and the Extension Service are poised to help interested individuals learn the skills required to successfully produce grains, so look for meetings, conferences, and on-farm workshops offered in your region. You can learn so much by visiting the people who are already doing it. Networking opportunities abound at these events. The subculture of grain lovers and producers is small, friendly, and tightly knit, so join in and reap the benefits. Small scale is where we need to begin to build the infrastructure for growing grain here in our region. A top-down industrial-scale flour mill or hulling and flaking facility would run counter to the informal local and regional agricultural economy that many of us envision.

Weather, Climate, and the Future

Destructive weather events like floods, droughts, and wet summers seem to be more prevalent these days than they were thirty-five years ago when I planted my first grain crops. In the 1970s and early '80s, we simply planted seeds and harvested crops with seemingly little trouble. Our yields may not have been bin busting, but we always got a crop. I didn't even know what mycotoxins in grains were at the time. But things are different today. All it takes is a prolonged spell of wet weather at pollination time to infect a field of wheat with fusarium and render it unsalable for human consumption. We've discussed DON levels in the wheat chapter; anything over one part per million cannot be made into flour. When you send your wheat away for testing and find that it has a DON level of two or three, it is a major heartbreak. All that work and expectation and all you have is a crop for livestock feed. A prolonged wet spell at combining time also has the potential to reduce the quality of any grain crop. I'm not sure if it's my imagination, but it seems like the weather has gotten increasingly finicky over the past several decades. It's either too dry or too wet, with nothing in between. This real or perceived shift in climate has taken place in the context of ever more people wanting to grow grain crops. I'm glad that I started when I did because it is a whole lot more difficult these days to reliably raise that perfect crop of grain. Nevertheless, this situation isn't going to force us to give up just yet. We have discussed lots of different strategies—like early harvest and bin drying—to avoid the potential damaging effects of prolonged wet weather. As a matter of fact, the onset of difficult weather across the entire country may be just what we need to encourage and boost the production of locally grown grains in our region. As energy prices skyrocket and grain growing is affected by poor growing conditions in the Corn Belt and on the high plains, production here at home becomes more viable. As I write this in the summer of 2012, the central part of the United States is afflicted with its worst drought in over twenty years. Farmers are plowing down failed crops of corn from Ohio all the way out to Nebraska. In this part of the country, we've had scant but adequate rainfall and our corn crops look fantastic. Between the high cost of transportation and the potential for drastically reduced yields in the Grain Belt, localized production of grains looks better than ever this season.

My own personal impression of the natural world in which we farm these days is that a massive decarbonization of the earth's crust has taken place and continues to do so at an alarming rate. Carbon dioxide levels in the atmosphere have been on the increase since the advent of the Industrial Revolution in England late in the eighteenth century, and most scientists concur that the burning of fossil fuels like coal and oil has been the largest major contributor to the accumulation of greenhouse gases in our atmosphere. The readily accepted view is that most of this carbon dioxide is released from the smokestacks of factories and power plants, the chimneys of our homes, and the tailpipes of our cars. Everyone, including climate scientists, fails to consider how much carbon and humus we have burned up out of our soils as a result of the effects of high-input industrialized agriculture. Soil organic matter levels on this continent once averaged between 5 and 10 percent; now the national average for soil organic matter in the United States has declined to the 1 percent level because we have cropped our farmland way too hard with too much added synthetic nitrogen. It takes twenty parts of soil carbon to assimilate one part of added nitrate. (You can review this concept in the chapter on soil fertility.) We have figured out how to boost yields far beyond what was ever deemed possible a generation ago. This looks good on paper and in farm magazines, but the true cost of all of this so-called productivity is lost in the counting of bins and bushels. As soil organic matter levels have declined, the ground's ability to hold excess water has also declined. This depletion has resulted in poor soil structure and much-diminished soil biology. Crops are thirstier in times of low rainfall, and soil erosion is on the increase as applied crop nutrients leach into streams and waterways. We've all heard plenty about the dead zone in the Gulf of Mexico that is the result of soil erosion and nutrient runoff from cropland in the Mississippi River basin. No-till agricultural methods are becoming increasingly popular in conventional corn, soybean, and cereal agriculture, and there is no doubt that improvements in soil quality have resulted from reducing and eliminating tillage, but at the same time the use of toxic chemical herbicides like Roundup has skyrocketed. My impression of the situation in my brief third of a century of farming is that things are getting worse out there. There is less respect for the earth than ever, and natural systems are ignored and forgotten. Petroleum is the number one agricultural input, in the form of diesel fuel and agricultural chemicals. I have felt this way for a long time. It seems like it's time for a change before we lose everything and starve to death as a civilization.

Doom and gloom, however, don't make for effective change, and it's no fun to be such a pessimist. Fortunately, organic agriculture offers us some hope in an otherwise desperate situation. Grains and forages need to be grown together again in a sustainable rotation, and cattle need to get out of feedlots and back onto farms and grass. Fundamental changes need to happen on every level of society, and it all starts with what we eat and how it is produced, distributed, and consumed. Concentrated animal feeding operations (CAFOs) depend on the centralized agricultural model presently in place here in North America. From there we go on to large centralized meat-slaughtering and -packing plants and chain-store distribution. It has taken half a century or more for a very few corporate players to come to dominate the food system. This mode of production and consumption is now firmly entrenched and very difficult to change. However, the local food movement offers us hope of breaking this cycle, which is so hard on Mother Earth. The handful of large multinational corporations that are in control of everything are beginning to lose market share as food consumers are voting for local sources with their food dollars. Farmer's markets, CSAs, food cooperatives, and other forms of direct marketing are all on the rise, and this bodes well for localized grain production and consumption here in the Northeast. There is no doubt that declining petroleum reserves and higher fuel prices will give us increased advantages in local grain production. Commodity prices nearly doubled in 2011 and will most likely double again if the Midwest loses its corn and soybean crop to drought. Things will have to change all over the country. More and better-quality forages will have to replace purchased grains on our dairy farms in the region. Farm-produced feed grains will become a very valuable commodity. Economics coupled with the challenges of a climate that wants to humble us will be the drivers of change.

The organic approach has a lot to offer us in this time of need. Crop rotation can be our salvation. We can no longer ignore the Golden Rule when it comes to caring for Mother Earth, and we have to give back as much as or more than we take. It's so simple, but we seem to have forgotten the lessons of our grandparents. Always follow a heavy-feeding crop like corn or wheat with a lighter feeder like oats or barley. A truly sustainable farmer will plant hay and use legumes whenever possible, and alfalfa and clover sod plow-downs eliminate the need for purchased nitrogen fertilizers. Sufficiently mineralize your soil to provide the proper environment for a diverse biology and an efficient plant metabolism, as crop plants with excellent photosynthetic function and root systems resist stress and produce fruit with superior nutritional qualities. This is what is becoming known as the nutrient-dense form of agriculture. Trace minerals like copper, zinc, manganese, and boron will need to be applied in small amounts along with adequate phosphorous and balanced levels of calcium, magnesium, and potassium. I needn't totally review these broader concepts of balanced soil fertility. However, I do want to say that the earth in which we grow our crops is the foundation, and it must be treated as something sacred. I've learned this lesson over and over again in my farming tenure. The more you give, the more you get back. For some of us (myself included), this will translate into doing a better job with crops on less acres. I tend to go for the extensive approach, but often I have planted more than I could properly fertilize or care for. I've proved it to myself so many times. I can harvest just as much grain corn from thirty-five well-tended acres as I can from fifty sloppily cared for acres. This past season I spent the extra time foliar feeding and weeding my spring wheat crop, and I was rewarded handily with fifty bushels (three thousand pounds) of beautiful golden red wheat to the acre. I cannot emphasize enough the importance of taking good care of your soil. Be a first-class organic farmer. Test your soil and then give it everything it needs in the proper balance. There are plenty of consultants and other knowledgeable people out there to help you. This might mean farming fewer acres because of costs, but you'll be a happier farmer with better-looking fields to present to the public if you take the time to pay attention to these important details. You'll produce higher-quality crops with myriad advantages for the livestock and people that consume them. I hope that you will take organic farming into your heart and make it work for you.

Corporations and Seed Sovereignty

Concentration is the name of the game in just about every sector of the agricultural economy. Power over most of the farming world has slipped into the hands of just a few major corporate players, and agricultural seed is no exception. About twenty years ago, I began to notice the names of large chemical and pharmaceutical companies in small print in corn seed catalogs. Names like Novartis and Sandoz were associated with established companies like Northrup King and Jacques. The 1990s were the decade of acquisitions—Monsanto bought the ever-famous Dekalb Corn Seed Company. I thought the world of Pioneer Hi-Bred at the time because they remained independent for a long while after every other company had been gobbled up by a major chemical firm, but eventually Pioneer bit the dust as well and became a subsidiary of DuPont. After this came the onslaught of genetically modified technology in corn, soybeans, cotton, and canola, and laws were put in place to protect the intellectual property of corporate giants. And while seed saving has been in existence since civilizations first started cultivating crops with sticks thousands of years ago, our highly concentrated corporate seed sector doesn't want us saving seeds on our farms; they want us to buy their products every crop season. Public plant-breeding programs at land grant universities have taken a backseat to the plant variety development activities of these major corporate players. Lots of seed varieties, especially the genetically modified ones, are protected by patents and cannot legally be replanted on the farms that choose to grow them. Plant Variety Protection laws have been put in place to prevent us from saving and selling seeds without paying royalties to a corporate entity. Fortunately, we can still save and use non-GMO seed on our own farms. We'd better not share that seed with anyone else, however. It's against the law.

In light of this situation, we are between a rock and a hard place as organic farmers in the world of seed. We need our own varieties of grain that are bred for organic conditions, and the National Organic Program requires us to plant organic seed whenever possible. There has been some small progress in the world of organic seed production, and there are a handful of organic plant breeders scattered across the country at various universities and foundations. These people need our support, along with a fledgling production sector that is struggling. An organization called the Organic Seed Growers and Trade Association (OSGATA) was organized several years ago to advance the interests of this very small group of companies and individuals. I think we need to emphasize the importance of being a "seedsman" in today's world of modern agriculture. We all need to know how to save and select seed for our next year's crop. Get to know as much as you can about the various varieties available for the particular grains that you grow. Here in Vermont, we are quite fortunate to have Dr. Heather Darby's yearly variety trial plots to study and inspect each season, but find the varieties that work best on your farm. Once you've done this, find out the status of a particular variety. Is it a public variety or is it PVP? You'll be surprised at how many popular varieties of cereal grains are still in the public domain, and these are the ones you will want to plant. Participatory plant-breeding programs are beginning to happen in a few areas out there. University plant breeders are partnering with farmers to try out new germplasm on individual farms.

Corn and soybeans are much more in the realm of corporate control. Hybrids predominate in the world of corn. Lots of work needs to be done to improve available open-pollinated varieties to give us more seed choices with corn. My friend Frank Kutka, a plant breeder from Dickinson, North Dakota, tells me that open-pollinated corn breeding came to a screeching halt in the 1930s when hybrids became widespread. According to Frank, we have eighty years of catch-up work to do to improve OP corn and make it competitive with hybrids. The will and the need are there. Soybean varieties come and go from the scene very rapidly, but most of today's beans are genetically modified to tolerate being sprayed by the herbicide Roundup. This leaves organic farmers with very few choices for soybean varieties. In my own particular situation, I have found some good non-GMO soybean varieties north of me in Québec. We need organic seed more than ever. The point is, we need to make the choice to support an organic seed sector. Many conventional varieties of farm seeds are bred for large inputs, but we need plants that can scavenge nutrients as opposed to being spoon-fed.

The Long Look Back and a Peek into the Future

I began my farming adventure with only a dream, although I had some knowledge from reading old agricultural texts and from working at a living historical farm. Looking back thirty-five years, I have to say that it has been fun and exciting. I didn't really know much when I started. I needed experience, and I got it by actually going out there and trying to plant crops. My first crop of wheat in 1977 wasn't very impressive by my standards of today, but at least it was all mine and we were able to make flour for our own bread. Over the years, I met many mentors and tried lots of different approaches. Some things worked; others didn't. The lessons have piled up over the years, and they keep coming every day. Throughout my entire adult life I have been driven by the need for independence from a system that I considered dominating and controlling. Maybe I'm simply deceiving myself into some sort of false quest for true freedom that doesn't even exist. I still need diesel fuel and electricity to do what I do in agriculture. In the 1970s, I wanted to be totally self-sufficient. I wanted to grow my own corn and soybeans and save all my own seeds in this rather cool and short-season part of northern New England. But I have to say that most of my dreams have come true during the past three decades. Seed production for just about every type of grain happens on my farm these days. Over the years, I've constantly pushed for perfection in my operation, but of course there have been many failed crops and flops. I've been driven, however, by a passion that keeps me looking forward and hoping for improvement.

I have been overjoyed to share my lessons with you. I hope that I can contribute to your passion as so many of the great people I have met did for me over the years. Some of these folks have now departed this life, and I am ever so grateful for what they gave to me. There are many other individuals out there in the world of organic agriculture with whom I share lots of inspiration. We have so much to be thankful for these days. More people than ever want to grow grain crops, and they want to do it using organic practices. I hope the tide is turning away from earth exploitation and toward a deeper organic agriculture. We are still the minority in the grand scheme of things, but our movement is rising fast. I say follow your dreams and passion and you will succeed.

References

"Abrasion Debranner VTA-5," Satake, www.satake.com.au/pdf/VTA5_Debranner_Dehuller.pdf.

"A Cross Section of a Typical Oliver Evans Mill," T. R. Hazen, 1999, www.angelfire.com/journal/millrestoration/section.html.

"A Dehuller Attachment for the Corona Grain Mill," Bilagaana, 1999, www.bilagaana.com/dehuller/Sunflower%20Dehuller.html.

Albers, Jan, *Hands on the Land: A History of the Vermont Landscape*, Cambridge, MA: MIT Press, 2002.

"A List of Mill Parts Suppliers for Flour Mill Machinery," T. R. Hazen, 1999, www.angelfire.com/journal/pondlilymill/suppliers.html.

"The Art of the Millstones, How They Work," T. R. Hazen, 2001, www.angelfire.com/journal/millrestoration/millstones.html.

"Austrian Flour Mills," Osttiroler Getreidemmuhlen, 2012, www.getreidemuehlen.com/en/grain-mills/combi-mills/combi-mills.php.

Bain, Angela Goebel, Lynne Manring, and Barbara Mathews, "Native Peoples in New England," Memorial Hall Museum Online, 2008, www.americancenturies.mass.edu/classroom/curriculum_6th/lesson2/bkgdessay.html.

"Barley," Canadian Food Inspection Agency, last modified December 1, 2012, www.inspection.gc.ca/english/plaveg/variet/barorge.shtml.

Berglund, Duane, Kent McKay, and Janet Knodel, "Canola Production—A686," North Dakota State University, last modified October 2, 2012, www.ag.ndsu.edu/publications/landing-pages/crops/canola-production-a-686.

Berglund, Duane, and Richard K. Zollinger, "Flax Production in North Dakota A-1038," North Dakota State University, last modified November 2, 2012, www.ag.ndsu.edu/publications/landing-pages/crops/flax-production-in-north-dakota-a-1038.

Christmas, E. P., "Plant Populations and Seeding Rates for Soybeans," Purdue University, last modified February 1993, www.extension.purdue.edu/extmedia/AY/AY-217.html.

Cognition: The Voice of Canadian Organic Growers, "Canadian Organic Growers Organic Field Crop Handbook: 3.3 Buckwheat," Ecological Agriculture Projects, McGill University, 1992, http://eap.mcgill.ca/MagRack/COG/COGHandbook/COGHandbook_3_3.htm.

"Corn Growth Stage Development," University of Illinois, http://weedsoft.unl.edu/documents/growthstagesmodule/corn/corn.htm.

Cyclopedia of American Agriculture, edited by Liberty Hyde Bailey, volume 2, New York: Macmillian, 1907.

Darby, Heather, "2010 Heirloom Wheat Performance Trials," University of Vermont Extension, http://northerngraingrowers.org/wp-content/uploads/2010-Heirloom-Wheat-Variety-Trial.pdf.

———, "2010 Oat Variety Trial Report," University of Vermont Extension, March 2011, www.uvm.edu/extension/cropsoil/wp-content/uploads/2010-Oat-Report.pdf.

"Diseases of Barley," West Virginia University Extension Service, 2012, www.wvu.edu/~exten/infores/pubs/pest/pcerti10.pdf.

"Example of Photographic Documentation of a Mill," T. R. Hazen, 2002, www.angelfire.com/folk/molinologist/photo7.html.

"Falling Number," North American Export Grain Association, www.wheatflourbook.org/doc.aspx?Id=121.

"Field Pea Production," North Dakota State University, last modified March 2009, www.ag.ndsu.edu/pubs/plantsci/rowcrops/a1166.pdf.

Flory, Paul B., "Millstones and Their Varied Usage," last modified August 23, 2011, www.angelfire.com/journal/pondlilymill/flory2.html.

Gibson, Lance, and Garren Benson, "Origin, History and Uses of Oat and Wheat," Iowa State University, Department of Agronomy, last modified January 2002, www.agron.iastate.edu/courses/agron212/readings/oat_wheat_history.htm.

"Grain Drying and Storage," North Dakota State University, last modified September 4, 2012, www.ag.ndsu.edu/graindrying.

Gupta, Mahesh, Nisreen Abu-Ghannam, and Eimear Galaghar, "Barley for Brewing: Characteristic Changes During Malting, Brewing and Application of Its Byproducts," *Comprehensive Reviews in Food Science and Food Safety*, Institute of Food Technologies, volume 9, 2010, http://onlinelibrary.wiley.com/doi/10.1111/j.1541-4337.2010.00112.x/pdf.

Harlan, Harry Vaughn, *Barley: Culture, Uses, and Varieties*, Washington, DC: US Department of Agriculture, 1925, UNT Digital Library, 2013, http://digital.library.unt.edu/ark:/67531/metadc6191/m1/1/.

Hirst, Kris, "Barley," About.com, 2013, http://archaeology.about.com/od/domestications/g/barley.htm.

"History of the Canola Plant," Canola Council of Canada, 2007, www.canolainfo.org/canola/index.php?page=5.

"How the Roller Mills Changed the Milling Industry," T. R. Hazen, www.angelfire.com/journal/millrestoration/roller.html.

Hunt, Thomas F., *The Cereals in America*, New York: Orange Judd, 1904.

"Information for Buckwheat Growers," Cornell University College of Agriculture and Life Sciences New York State Agriculture Experiment Station, last modified 2009, http://calshort-lamp.cit.cornell.edu/bjorkman/buck/main.php.

"International Harvester's First Generation of Self-Propelled Combines," Toy Tractor Show, last modified April 27, 2000, www.toytractorshow.com/ih_conventional_combine_history.htm.

Janzen, Kristi Bahrenburg, "Farmer Finds His Niche with Spelt," Agricultural Marketing Resource Center, 2013, www.agmrc.org/media/cms/SpeltStory_0E4AE35188557.pdf.

Kandel, Hans, editor, "Soybean Production: Field Guide for North Dakota and Northwestern Minnesota," North Dakota State University, June 2010, www.ag.ndsu.edu/pubs/plantsci/rowcrops/a1172.pdf.

Lardy, Gary, "Feeding Corn to Beef Cattle," North Dakota State University Extension Service, December 2002, www.ag.ndsu.edu/publications/gsearch?cx=018281009562415871852%3Ajpqauwrs0sa&cof=FORID%3A10&ie=UTF-8&q=feeding+Corn+to+Beef+catle&sa=+&siteurl=www.ag.ndsu.edu%2Fpublications&ref=&ss=7632j2373320j30.

Lyon, Drew J., and Robert N. Klein, "Rye Control in Winter Wheat," University of Nebraska, last modified May 2007, www.wintercereals.us/Documents/Growing%20WW/Production%20Articles/Weeds/Rye%20Control%20in%20Winter%20Wheat.pdf.

Martens, Klaas, and Mary-Howell Martens, "The Basics of Effective Tillage Techniques," Rodale Institute, last modified January 27, 2005, http://newfarm.rodaleinstitute.org/features/2005/0105/earlyweeds/index1.shtml.

———, "Blind Cultivation," Rodale Institute, last modified February 10, 2005, http://newfarm.rodaleinstitute.org/features/2005/0205/earlyweeds/index2.shtml.

———, "In-Row Cultivation," Rodale Institute, last modified March 17, 2005, http://newfarm.rodaleinstitute.org/features/2005/0305/earlyweeds/index3.shtml.

References

"Model 15-D Impact Huller," Forsbergs, Inc., http://forsbergs.com/products/15-d-ih.html.

Neate, Stephen, and Marcia McMullen, "Barley Disease Handbook," North Dakota State University, 2005, www.ag.ndsu.nodak.edu/aginfo/barleypath/barleydiseases/index.htm.

"Oat," Canadian Food Inspection Agency, last modified January 01, 2013, www.inspection.gc.ca/english/plaveg/variet/oatavoe.shtml.

"Oat Production in North Dakota," North Dakota State University, last modified November 5, 2012, www.ag.ndsu.edu/publications/gsearch?cx=018281009562415871852%3Ajpqauwrs0sa&cof=FORID%3A10&ie=UTF-8&q=Oat+Production+In+North+dakota&sa=+&siteurl=www.ag.ndsu.edu%2Fpublications&ref=&ss=8928j3116544j35.

"The Old Red Mill," Jericho Historical Society, www.jerichohistoricalsociety.org/Omill.htm.

"Organic Seed Processing: Threshing, Cleaning and Storage," University of Vermont Extension, last modified March 15, 2010, www.extension.org/pages/18350/organic-seed-processing:-threshing-cleaning-and-storage.

"Pulses and Healthy Living," Saskatchewan Pulse Growers, 2013, www.saskpulse.com/media/pdfs/ppm-field-pea.pdf.

Putnam, D. H., E. S. Oplinger, D. R. Hicks, B. R. Durgan, D. M. Noetzel, R. A. Meronuck, J. D. Doll, and E. E. Schulte, *Alternative Field Crops Manual*, "Sunflower," Purdue University, last modified January 2013, www.hort.purdue.edu/newcrop/afcm/sunflower.html.

Schmitz, Karl, "Spelt—A Grain for the Future," Schapfenmühle, November 2004, www.greekmills.com/pdfs/Schmitz-Dinkel-Vortrag-EN1.pdf.

Schneiter, A. A., J. F. Miller, and D. R. Berglund, "Stages of Sunflower Development," North Dakota State University, February 1998, www.ag.ndsu.edu/pubs/plantsci/rowcrops/a1145.pdf.

Sherman, Michael, Gene Sessions, and P. Jeffrey Potash, *Freedom and Unity: A History of Vermont*, Barre: Vermont Historical Society, 2004.

Shurtleff, William, and Akiko Aoyagi, "History of Soybeans and Soyfoods in Sweden," in *History of Soybeans and Soyfoods: 1100 BC to the 1980s*, Lafayette, CA: Soy Info Center, 2007, www.soyinfocenter.com/HSS/europe5.php.

"Small Scale Maize Milling," New Zealand Digital Library, www.nzdl.org/gsdlmod?e=d-00000-00---off-0fnl2.2--00-0----0-10-0---0---0direct-10---4-------0-1l--11-en-50---20-about---00-0-1-00-0--4----0-0-11-10-0utfZz-8-00&cl=CL3.69&d=HASH0c49ff2.

Stallknecht, G. F., K. M. Gilbertson, and J. E. Ranney, "Alternative Wheat Cereals as Food Grains: Einkorn, Emmer, Spelt, Kamut and Triticale," in *Progress in New Crops*, Purdue University, www.hort.purdue.edu/newcrop/proceedings1996/V3-156.html.

Steinhoff, Dan, "A. W. Gray and Sons," *Farm Collector Magazine Online*, January–February 1979, http://steamtraction.farmcollector.com/Farm-life/A-W-GRAY-and-SONS.aspx.

Wells, Frederic Palmer, *History of Newbury, Vermont*, St. Johnsbury, VT: Caledonian Company, 1902.

Willis, Harold L., *How to Grow Super Soybeans*, Kansas City: Acres USA Press, 1989.

Wilson, Harold Fisher, *The Hill Country of Northern New England: Its Social and Economic History 1790–1930*, New York: Columbia University Press, 1936.

Index

Note: Page numbers appearing in italic type refer to pages in the color insert (e.g. CI:2)

A. T. Ferrell Company, 103, 108–9, 377–78
abrasion husking, barley, 241
Ace, John, xiv, 120, 143, 390, CI:2
acidosis, 163, 218
actinomycetes, 23
aeration fans, 99–100. *See also* aerators, screw-in grain; bin aeration of grain
aerators, screw-in grain, 86, 92, 101, 204, 288, 321, 345, 357, CI:*12*
agronomy texts, reference use of, 21–22, 24
air aspiration, 101–2. *See also* aspirators
air screen cleaners, 102–9, 170, 205, 357, 379, 382
    Clipper grain cleaner, 6, 103, 108–9, 379
    fan adjustments, 108
    Forano, 103–9
    screen sizes for, 107–8
air seeders, 53
Albrecht, William A., 24
Albrecht Papers, The, 24
alfalfa, 27, 53, 373, 379–82. *See also* hay
algae, 23
algicide, barley as, 228
Allen, Eric, 13–14
ammonium, 27
amylase enzyme, levels of, 57, 60
anaerobic soil, 202
Angier, J. Francis, 7–8, 11, 284
animal consumption, grains for. *See* livestock feed
anthracnose, 314, 331, 334, 336, 350
anti-fungal seed treatments, 37–38, 124, 130, 192, 351
aphids, soybean, 315
aragonite, 396, 399–400
aspirators, 111, 264

bacteria
    seed inoculation with (*See Rhizobium*, seeds inoculated with)
    soil, 23
bacterial blight, 314, 336
bags, grain storage in, 77, 78
Bailey, Liberty Hyde, 22, 327
banded sunflower moth, 354
barley, xv, 10, 60, 114, 219–43, CI:*24*
    agronomic considerations for, 224–33
    as algicide, 228
    cleaning, 111, 233–34
    debearding, 111, 233
    diseases, 226, 229–30
    dry-down, 231, 232
    drying, 94, 233–34
    fines, 235, 236, 269
    harvesting, 231–33
    history of, 4–6, 8, 11–12, 14–16, 219–20
    hulling, 220, 404, CI:*25*
    for human consumption, 220, 223, 233, 240–43
    interplanted with other crops, 225, 227, 231–32, 381
    as livestock feed, 168, 220–21, 223–24, 226–27, 233–37, 383, 394–95, 397, 400, 401
    lodging (*See* lodged grain)
    malting (*See* barley, malting)
    pearled, 220
    photosynthesis, efficiency of, 113–14
    planting (*See* barley, planting)
    problems of, 228
    processing (*See* barley processing)
    protein levels, 220, 224, 226, 227, 234, 394
    starch, 234–35
    storage, 80, 233
    straw, 226, 228
    types of, 221–24, CI:*24*
    varieties, 219, 223, 225–26, 228, 233
    weeding, 229
    winter, 225
    yield, 128, 223, 226, 228, 230
Barley Disease Handbook, 230

barley, malting, 221–24, 229, 232–34, 237–40
    industrial, 238–39
    kilning, 237–38
    micro-malting, 239–40
    microbrew industry, 239
    Warminster Malting, 238
barley midge, 194–95
barley, planting, 30, 224–27
    crop rotation, 230
    quality of seed, 225
    seed sources, 35–36
    selecting varieties, 225–26, 228
    soil fertility, 226, 228–31
barley processing
    abrasion husking, 241
    hulling, 241–43
    for human consumption, 240–43
    for livestock, 234–37
    malting (*See* barley, malting)
    roller-mill method, 235–36
    sprouting and hand-grinding, 236–37
    tempering, 236
base saturation, 24, 26–27, 180
bean blight, 332
bean threshers, 339
bean weevil, 336
beans. *See* dry beans
beggar-ticks, 142
Beidler, Brent, 379
Bergstrasser, Bob, 341
Bidwell Bean Thresher Company, 341
Bielenberg, Carl, 357
bin aeration of grain, 15, 60, 87–95, 99–101
    barley, 232, 234
    canola, 371
    corn, 162, 170
    homemade grain dryer, 92–94
    oats, 259
    soybeans, 321
    sunflowers, 357
    wheat, 204

binding, 57–58
biodiesel, oilseeds used for, 358, 365, 366, 370
birds, 160, 355
bird's-foot trefoil seed, 7, 8
blackleg, 368
Blake, Newton, 103–4
blind harrowing, 51–53
blood meal, 182
blotches, 190, 229–30
Bobs Equipment Company, 341–42, 344
Bonsall, Will, 223
boron, 27, 180, 195, 229, 410
Boucher, Denis, 322
Bowen, Samuel, 299
broadcasters, seed, 226–27, 254, 280, 361
    mechanics of seeding with, 38
    side-dressing, used for, 149
    underseeding grass seed with, 53
broadcasting seed, 38
    barley, 226–27, 229
    buckwheat, 375
    flax, 361
    oats, 254
    rye, 279, 280
brown rust, 193–94
brown spot, 314
Bt *(Bacillus thuringiensis)*, 119
buckwheat, xvi, 373–78, CI:32
    cleaning, 377–78
    as cover crop, 32, 375
    in crop rotation, 377
    dry-down, 376
    growth cycle, 375–76
    harvesting, 376–77
    history of, 5, 6, 8, 374
    for human consumption, 378
    as livestock feed, 374
    planting, 30, 374–75
    protein levels, 374
    uses for, 378
    yield, 374–75
buhr mills (plate mills), 209–10, 237, 261–62, 397. *See also* hand mills
bulgur, 217
bulk bags, grain storage in, 77, 78
bushel, weight of, 82. *See also* test weights
    corn, 82, 110, 170
    oats, 248, 254
    spelt, 286
Butterworks Farm, xv–xvi, 12, 15
    combine, use of, 70–73
    corn cultivation, 145–46
    dry beans, growing, 342–44
    granary, 100–101
    threshing machines, use of, 63

C-shank cultivator, 148–49
calcium, 23, 24, 27, 142, 329
    for barley crop, 229
    in livestock ration, 395–96, 399–400
    for wheat crop, 180
Callan, Mark, 331, 332
Canada
    corn as cash crop, 123–25
    grain production in, 9–11
    as seed source, 35–36, 196–98, 251–52
Canada Malt, 224
Canadian field peas. *See* field peas, Canadian
Canadian heat units (CHUs), 120, 123, 130, 152, 153, 303–4
canola, 80, 347, 364–72, 401
    breeding, 365–66
    in crop rotation, 368–70
    diseases, 368–70
    drying, 371
    genetically modified, 366, 368
    growth cycle, 368–70
    harvesting, 370–71
    history of, 365–66
    open-pollinated, 367
    planting, 366–68, 370
    seed sources/saving, 367
    varieties, 367
carbon, 23, 27, 32, 128, 180–82, 228, 408
Carter disc cleaner, 112, 205, 265, 266, 364
castor bean pomace, 182
cation exchange capacity (CEC), 24–25, 180
cattle. *See* cows and cattle, feed for
Cereals in America, The, 22, 277
certified seed, 36–37, 185, 196, 332, 380
certified seed laboratory, 173
chemical fertilizers. *See* fertilizers, commercial/chemical
chicken feed. *See* poultry feed
chicken manure, 149, 181–82, 228
Chilean nitrate, 149, 182–83, 228, 379
chisel plows, 34
chlorosis, 253
clay, 24–25, 29, 366

barley crop, soil for, 224–25
bird's foot trefoil, soil for, 7
canola, soil for, 366
claybottom moldboard plow, 31
corn, soil for, 130, 131
culti-mulcher, use of, 34
grain binder, use of, 58
grain drill, use of, 42
wheat, soil for, 4, 185
winter grains *vs.* spring grains in, 30
cleaning grain, 101–12, 403, CI:*13–16*. *See also* air screen cleaners; Carter disc cleaner; gravity table separator; *subhead* "cleaning" *under specific crops*
    aspirator, 111, 264
    augur, 112
    basics of, 101–2
    debearder, 106–7, 111, 233
    destoner, 111, 323
    elevator legs, 112
    from home garden harvest, 101–2
    precleaning, 83–84, 101
climate issues, 408–11
Clipper grain cleaner. *See* air screen cleaners
clover, 179, 373, 379–81. *See also* hay
    cover crop, 53
    in crop rotation, 129, 360, 381
    interplanted with other crops, 53, 381, CI:*4, 12*
    seed growing, saving, and storage, 6, 380–82
    wheat harvest flavored by, 83
CMC (Custom Marketing Corporation) grain drying system, 91
cobalt, 195
Codema, LLC, 263–64
Coe, Stephen and Tunis H., 327
coleoptile (grain plant), 45
Collaborative Oat Research Enterprise (CORE), 272
combine, harvesting with, 7, 10, 15, 65–75
    barley, 233
    buckwheat, 376–77
    Butterworks Farm, use at, 70–73
    canola, 370–71
    clover, 381–82
    corn, 8, 124, 160–64, CI:*18–19*
    dry beans, 341–44, CI:*30*
    filling of grain storage containers, 78–79

fires, 382
flax, 363
Hume reel, 66
millet, 379
oats, 258–61, CI:26
operation, 66–70
rye, 281–83
selection of combine, 73–75
soybeans, 299, 318–21, 323
spelt, 287–88
sunflowers, 355–57
troubleshooting, 69
wheat, CI:10
Coming Home to Eat: The Pleasures and Politics of Local Foods, 16
commercial fertilizers. *See* fertilizers, commercial/chemical
Commodity Traders International, 109, 406
common bunt, 190, 193
common take-all, 193
community supported agriculture, 405
companion crops. *See* cover crops; hay; *subhead* "interplanted with other crops" *under specific crops*
compost, 23, 32, CI:4
    for barley, 224, 230
    for buckwheat, 375
    for canola, 367
    for corn, 129
    for dry beans, 329, 330
    for oats, 253
    for rye, 278
    for sunflowers, 351
    for wheat, 180–82, 184, 202
    for winter grains, 275
concentrated animal feed operations (CAFOs), 21
Conway, Chuck and Carla, 211
Cooking with Whole Grains, 8
copper, 195, 229, 409
corn, xv, xvi, 113–74, 272, 347, CI:17–20. *See also* dent corn; flint corn; open-pollinated (OP) varieties
    biology of, 114–16
    black layer, 152–54
    blister stage, 150
    breeding, 116–17, 173–74
    Butterworks Farm granary, 100–101
    cleaning, 108, 170, 172–73
    colder climates, growing in, 113–14
    in crop rotation (*See* crop rotation)
    cultivation (*See* corn, cultivation of)

density and test weight, 151–52, 170
dent stage, 151–52
diversity *vs.* uniformity, 118, 127
dry-down, 153–54
drying, 15, 91–92, 94–96, 162, 164–67, CI:17, 19–20
ear corn, harvest and storage of, 154–59, CI:17
fines, 162, 164
flour, 114–15
as full-season crop, 152–53
genetically modified (*See* genetically modified organisms (GMOs))
germination and emergence, 136–37
grinding, 171
growth cycle, 114, 139–53
harvesting (*See* corn, harvesting)
high-moisture, 9, 11, 15, 162–63
history of, 1–6, 8, 11–15, 113–16
human consumption, processing for, 169–71
hybrids, 117–18, 122–25, 153, 411
interplanted with other crops, CI:4
late vegetative and reproductive stages, 149–53
as livestock feed, 162–63, 168–69, 394–95, 397, 399–402
lodging (*See* lodged grain)
maturity ratings, 129–30
milk line and milk stage, 150–52, CI:18
photosynthesis, efficiency of, 113
planting (*See* corn, planting)
pollination, 150, 152
protein levels, 394
seed saving, 171–74
shelled, 8, 11, 15, 92, 94–95, 99–100, 123, 124, 159–64, 170
shelling, 154–55, 172
side dressing, 149
silage (ensilage), 116, 152, 162, 286
specific density, 82
starch, 234
storage, 80, 99–100, 154, 170 (*See also* corncribs)
varieties (*See* corn varieties)
yield, 118, 128
corn borer, European, 119, 128
corn, cultivation of, 131, 137–49
    at Butterworks Farm, 145–46
    equipment for, 138–48
    field preparation, 129
    lay-by, 148–49

post-emergence, 139–40
pre-emergence, 137–39
row width for, 131
Corn Culture, 172
corn, harvesting, 122–23, 153–64, 170, CI:17
    combine, use of, 8, 124, 160–64, CI:18–19
    corn picker, use and maintenance of, 155–58
    corn sheller, use of, 154–55
    ear shank size and, 153
    handpicking, 154
    picker-sheller, use of, 161–62
    small scale, 154–55
    timing of, 153–54
corn heat units. *See* Canadian heat units (CHUs)
corn picker, 123, 155–58
corn picker-sheller, 161–62
corn planters, 132–36
    dry beans, planting, 332–33
    fertilizer hopper, 135–36
    planting seed with, 134–35
    plate, 132
    plateless, 133
    soybeans, planting, 304, 307, 310–11
    sunflowers, planting, 352
    use of, 134
corn, planting, 30, 31, 128–36
    corn planter, using (*See* corn planters)
    fertilizer hopper, using, 135–36
    field preparation, 129
    methods, 131
    no-till, 281
    row spacing, 131, 139–41, 145, 156
    seed sources, 35, 116
    selecting varieties, 129–30, 170
    timing of, 130–31
    treated seed, 124, 130
corn rootworm, 119, 128
corn sheller, 154–55, 170, 172
corn varieties
    dent corn (*See* dent corn)
    flint corn (*See* flint corn)
    Mandan Bride, 170
    selecting, 129–30, 170
    Vermont, suited for, 119–28
corncribs, 122–23, 154, 169, CI:20
    drawbacks of, 159–60
    for larger harvests, 158–59

for Northeast, 159
cornmeal, 170–71
cottonseed meal, 182
Couture, Jean, 239
cover crops, 32, 379. *See also* hay
    after soybeans, 252
    buckwheat, 32, 375
    clover, 53
    dry beans, 330
    before dry beans, 330
    oats, 32, 250, 306
    rye, 279, 280, 281, 283
cows and cattle, feed for, 168–69, 218, 379. *See also* forage crops
    barley, 221, 224, 234–35
    DON (deoxynivalenol) in, 56, 188
    flaxseed oil cake (linseed meal), 360
    livestock ration ingredients, 396–98
    millet, 379
    oats, 257–58, 260
    peas, 385
    rye, 283
    soybeans, 299, 300
    triticale, 294
    wheat, 188
crab meal, 182
crop diversity, 21
crop residue, 32
crop rotation, 14, 23, 27, 124, 272, 380, 409–10
    barley, 230
    buckwheat, 377
    canola, 368–70
    clover, 129, 360, 381
    corn, 124, 128–29, 252, 280, 285, 286, 306, 330, 360
    disease prevention and, 230
    dry beans, 330, 336, 370
    flax, 360
    hay, 128–29, 285, 360, 381
    oats, 129, 252
    rye, 278, 280
    soybeans, 124, 252, 280, 299, 306, 352, 360, 370
    spelt, 285, 286
    sunflowers, 352, 370
crop selection, 30–31
cross breeding, heirloom wheat, 56
cross-pollination, 277–78, 281, 375, 381
culti-mulcher, 34, 367
cultipacker, 306, 312, 360, 367
cultivation, 32. *See also* corn, cultivation of; weed control

dry beans, 333–38
flax, 361
rye, 280–81
soybeans, 311–13
in wet weather, 139
cultivators. *See* field cultivator; row-crop cultivators
cutworms, 354
Cyclone hand-cranked broadcaster, 38, 226, 254, 280, 361
Cyclopedia of Agriculture, 22, 327
Cyclopedia of Horticulture, 22

dairy farms, grain production and, 4–6, 8, 10, 13, 15–16. *See also* cows and cattle, feed for
Danish tine cultivator, 148, 313
Darby, Heather, xiv, 17–18, 127, 201, 203, 331, 411
    canola research, 367
    durum wheat trials, 189
    falling number machine, funding of, 203
    flax test plots, 361
    loose smut, wheat trial fields affected by, 193
    nitrogen fertilizer study, 183
    oat trials, 251
    wheat/nitrogen research, 202
    winter wheat trials, 198, 199
debearder, 106–7, 111, 233
dehulling. *See* hulling
density, grain. *See* grain density
dent corn, 6, 114–16, 171, CI:20
    Early Riser, 126–28, 174, CI:17, 20
    Mathesen, 125–26
    Vermont, varieties suited for, 120, 122, 125–27
    Wapsie, 126
deoxynivalenol (DON). *See* DON (deoxynivalenol)
destoner, 111, 323
Dewavrin, Loïc, 53, 347–48, 356
disc, 33–34, 38–39, CI:5, 6
    for buckwheat, 374
    for corn, 129
    for flax, 360
    for oats, 252, 254
    for rye, 278, 280
    for soybeans, 306
disc cleaner. *See* Carter disc cleaner
diseases, 7, 118, 190. *See also* fungal diseases

barley, 226, 229–30
canola, 368–70
dry bean, 331–32, 335–36
rye, 278, 282–84
seed test for, 173
seedborne, 37–38
soybean, 314–15
sunflower, 349–50, 354
wheat, 187–95, 202
dolomite, 27
DON (deoxynivalenol), 56, 191–92, 229
    testing for, 56, 192, 229
    unsafe levels of, 56, 188, 192
Dong, Allen, 289
dormant period
    rye, 279–80
    spelt, 287
dough stage, 57–59, 65
    barley, 231, 232
    corn, 151
    oats, 257–59
    rye, 281
    wheat, 192, CI:21
Downey, Keith, 365–66
downy mildew, 314
drag harrow, 33
drainage, 24–25, 27, 29, 54
    for barley crop, 226
    for canola crop, 366–67
    for dry bean crop, 328–29
    for spelt crop, 286
    weed pressure and, 142
    for winter or spring crop, 31
dry beans, xvi, 11, 94, 327–46, CI:30
    benefits of growing, 328
    Butterworks Farm, growing at, 342–44
    cleaning, 108–11, 339, 345
    as cover crop, 330
    in crop rotation, 330, 336, 370
    cultivation of, 333–38
    determinate and indeterminate, 328–29
    diseases, 331–32, 335–36
    drying, 345
    growth cycle, 333–38
    harvesting (*See* dry beans, harvesting)
    history of, 6, 327–28
    for human consumption, 328
    insects, 335–36
    lodging, 310
    nitrogen-fixing by (*See* nitrogen-fixing)
    photosynthesis, efficiency of, 113

Index

"pick out" process, 345–46
planting (*See* dry beans, planting)
as protein crop, 328
seed saving, 331
staining, issues with, 345
underdevelopment, issues with, 345
varieties, 328–29, 331–33, 338, 343, 345, CI:*30*
dry beans, harvesting, 338–44, CI:*30*
bean threshers, 339
combine, 341–44, CI:*30*
hand-threshing, 338–39, 344
mechanical bean pulling, 339–41
dry beans, planting, 30, 31, 329–33
row spacing, 332
seed sources, 331–32
soil fertility, 329–30
soil preparation, 332–33
dry-down, 58
barley, 231, 232
buckwheat, 376
corn, 153–54
flax, 362–63
oats, 258, 259
soybeans, 306, 317
sunflowers, 349, 351, 354, 355
drying grain, 85–97, 403, CI:*12*.
See also aerators, screw-in grain;
bin aeration of grain; drying silos;
propane-powered grain dryer; *subhead*
"drying" *under specific crops*
basic method for smallest crops, 85–86
other techniques, 95–97
pressure-cure, 90–92
small amounts, homemade dryer for, 92–94
stationary dryers, 94–95
drying silos, 60, 86–90
adding supplemental heat, 89–90
construction, 87
removing grain from, 89
use of, 87–89
Duncan, Hank, 171, 211–13
Dunn, Brian, 93–94
durum wheat, 188–89
dust mulching, 137–38
Dyck, Elizabeth, 178

ear corn, 123
ear savers, combine, 164
early growth period, grains, 45–46
corn, 139
oats, 256–57

rye, 278
soybeans, 312–13
Earthway seeder. *See* hand-pushed garden seeder
East Tyrolean grain mills, 213–14
Edberg, Roger J., 289
Einbock tine weeder. *See* tine weeder
einkorn, 176
elevator legs, 112
emmer, 176–78
endosperm, wheat, 205
Enhancing Farmers' Capacity to Produce High quality Bread Wheat, 18
ensilage. *See* silage
enzymatic activity, wheat kernel, 201, 203–4
ergot fungus, 278, 282–84
Eric and Andy's Homemade Oats, 13–14, 263
erosion, 32, 128, 252
European corn borer, 119, 128
European settlers, grain production by, 1–9

Fair, Roy, 125
falling number analysis
barley, 232
wheat, 203–4
fanning mills, use of, 5, 102–3, CI:*15, 16*
for barley, 234
for buckwheat, 377
for clover seed, 382
for corn, 170, 172
for dry beans, 345
for field peas, 385
for grass seed, 379
for oilseeds, 357, 364
for rye, 284
for wheat, 205
fans
aeration, 99–100 (*See also* aerators, screw-in grain; bin aeration of grain)
air screen cleaner, 108
greenhouse, 154
farro (emmer), 176–78
feather meal, 149, 182
Fedco Seeds, 125
feed grinder, 123, 237
Feeds and Feeding: A Handbook for the Student and Stockman, 389–90, 394
fertilizer hoppers, 135–36
fertilizers. *See also* sul-po-mag; *specific soil amendments*

for barley, 228
for canola, 367
for corn, 135–36, 149, 399
for dry beans, 329
for sunflowers, 351
for wheat, 180–82, 184, 195, 198
fertilizers, commercial/chemical, 118, 149, 178, 379
for corn, 118, 135, 149, 161
history of, 7, 9, 22–23
field cultivator, 32–34, 50, 129, 143, 224, 280, 306, 367, CI:*5, 7*
field-drying. *See* dry-down
field peas, Canadian, xvi, 379, 382–86
cleaning, 108, 385
determinate and indeterminate varieties, 384
growth cycle, 383–84
harvesting, 385–86
inoculation, 384
interplanted with other crops, 227, 294, 382–83
as livestock feed, 382–83, 385, 394–95, 400
planting, 227, 294, 382–85
protein levels, 227, 383, 394
seed sources/saving, 384–85
straight-seeded, 385
Fife, David, 200
fines
barley, 235, 236, 269
corn, 162, 164
oats, 260, 269, 271
fish meal, 182
flag leaf (grain plant), 55
flail, ix, xiv, 3, 60, 232, 287, 318, 338, 376
flax, xvi, 347, 359–64, 371–72, CI:*31*
cleaning, 363–64
dry-down, 362–63
growth cycle, 361–62
harvesting, 362–64, CI:*31*
history of, 359
for human consumption, 360, 364
livestock feed, flaxseed oil cake as, 360, 394, 401
planting, 360–61
pressing oil from seeds, 364
protein levels, 394
storage, 80, 364
flea beetles, 370
flint corn, xv, 2, 3, 6, 8, 11, 12, 114–15, 153, 171, CI:*20*

Garland, 122, 128, 170
Gaspé Yellow, 120–21
Longfellow, 170
Rhode Island Whitecap, 119–20
Roy's Calais, 125, 128, 170
Vermont, varieties suited for, 120–22, 125
flour corn, 114–15, 153, 170, CI:20
flour milling. *See* wheat processing and flour production
flour sifters, 215–17
foliar feeding, 46, 55, 122, 195, 230, 410
forage crops, 396–97. *See also* cover crops; hay
barley, seeded into, 225, 229
field peas, 383
grasses, 378–79
legumes, 379
oats, 247, 257
rye, 278–79, 281
soybeans, 299
triticale, 294
Forano 150 air screen cleaner, 103–8, CI:16
Ford, Henry, 299
Forsbergs Inc., 109–11, 241, 264, 323, CI:15
Franklin, Benjamin, 299
French, W. B., 285
frost, effect of, 35, 402
barley, 224, 225
buckwheat, 375, 377
canola, 367
corn, 114, 122, 130, 137, 151–53, 162, 173
field peas, 383
flax, 360
rye, 279, 280
soybeans, 301, 303–5, 317, 319, 323, 328
sunflowers, 352
wheat, 184, 185, 200
frost seeding, 38, 45, 285
fungal diseases, 37–38, 56, 118, 230, 314–15, 331–32, 350. *See also* anthracnose; anti-fungal seed treatments; fusarium head blight
blackleg, 368
ergot, 278, 282–84
seed test for, 173
fungi, soil, 23
fungicides, seed treated with. *See* anti-fungal seed treatments

fusarium head blight, 56, 187–92, 201, 229, 248. *See also* DON (deoxynivalenol)

garden hoe, 141, 306–7
garden seeder. *See* hand-pushed garden seeder
Garland, George, 128
Gaudreau, Michel, 249, 262–63, 265, 271–72, 289, 347
genetic markers, 273
genetically modified organisms (GMOs), 21, 410
canola, 366, 368
corn, 21, 118–19, 127–28, 170
identity preserved (IP) nonmodified seeds, 325
PCR test to detect in seed, 173, 366
soybean seed inoculants, 311
soybeans, 21, 119, 300, 324–25, 411
genotype, 173
germ, preservation of, 101
germination of seeds
barley, 224, 225, 229
canola, 367
corn, 136
diseases, effects of, 37–38 (*See also* anti-fungal seed treatments)
dry beans, 330–31, 333, 334
field peas, 383
oats, 246, 253, 254
potential for, 37–38
rye, 279
soybeans, 306, 307, 311, 312
spelt, 287
sunflowers, 352–53
temperature, optimal, 45
tests for, 37, 185
triticale, 295
wheat, 185
Gilles, Rick, 263–67
glumes
barley, 220
corn, 155, 170
einkorn and emmer, 176
oats, 246, 247
rye, 276
goat feed, 260
Goldstein, Walter, 126, 172
grading, seed, 101–2, 110, 112, 172–73, 205, CI:14
Graham, Rodney, 90–91
grain aerator, screw-in. *See* aerators, screw-in grain

grain augur, 84–85
grain binder (reaper), 7, 12, 58, 258
grain cleaning. *See* cleaning grain
grain density, 80, 151, 159, 202. *See also* test weights
grain sorted by, 110, 170, 192, 205, 291, 345
measuring, 82
grain drills, 39–45, 407, CI:7
barley, planting, 226–27
buckwheat, planting, 375
canola, planting, 367
caring for, 43–44
dry beans, planting, 330, 332–33
field peas, planting, 383
flax, planting, 360–61
oats, planting, 254, 256
peas, planting, 384
rye, planting, 278, 279
soybeans, planting, 304, 307–9, 312, 313
spelt, planting, 286–87
triticale, planting, 294
types of, 39–43
grain dryer. *See* propane-powered grain dryer
grain drying. *See* drying grain
grain filling, 56–57
grain header, 61, 66–67
grain moisture. *See* moisture, grain
grain swathers. *See* swathing grain
grains, research on, 17–18, CI:32. *See also* Darby, Heather
grasses, 27, 378–79. *See also* hay; timothy
gravity boxes, 79–80, 321, CI:11, 13
gravity table separator, 109–11, 170, 205, 282–84, 323, 345, CI:14–15
green manure, 379. *See also* cover crops
greenhouse, drying grains in, 154, 170, 204
grinder-mixer, 123, 159–60, 169, 391–93, 397–98, CI:32
grits, 171
growing degree days (GDDs), 130. *See also* Canadian heat units (CHUs)
growth cycle, grains, 55–57, 256–58. *See also specific grains*
early growth period (*See* early growth period, grains)
grain filling, 56–57
pollination (*See* pollination)
reproductive period (*See* reproductive period, grain)

Index

vegetative growth period (*See* vegetative growth period, grain)
grubs, 336, 354
guano, 22
gypsum, 195, 329

Hack, Martin, 380, 382
hair, as nitrogen source, 182
hairy vetch, 205, 386–87
hammermill, 13, 391–93, 397–99, 402
    for barley, 235
    for corn, 123, 159, 169
    for oats, 260–61
    for spelt, 292
    for wheat bran, 216
hand mills, 207–9, 262, 271, 289–90
hand-pushed garden seeder, 30, 39, 131, 226–27, 307, 332, 361
hand-seeding, 38, 226–27, 254. *See also* broadcasting seed
hand-threshing, 232–33, 287, 318, 338–39, CI:*23*. *See also* flail
hard dough harvest, 59
hard red spring and winter wheat, 30, 185–87, 197–99, 202, CI:*10, 22*
hard white spring wheat, 185–86
harvesting grains, xv, 57–75. *See also* combine; subhead "harvesting" *under specific crops*
    bin-drying of grain (*See* bin aeration of grain)
    binding and reaping, 57–58
    dry-down (*See* dry-down)
    hard dough, 59
    lodged grain, 57, 153, 156, 157, 164, 231, 259
    moisture percentage of grain, acceptable, 60, 66, 85, 91
    readiness for harvest, determining, 57, 60, 65–66, 85, 91
    row crops, 30
    stooking grain, 58–59
    swathing grain (*See* swathing grain)
    threshing (*See* threshing)
    transporting harvested grain, 78–80
    weather and, 57, 59–60, 65–66
    winter grains, 276, 281–83, 287–88
Hatzenbichler weeder, 51
hay, 14, 27, 180, 378–79, 381, 382, CI:*12*
    barley interplanted with, 225, 227, 229, 231–32
    corn interplanted with, CI:*4*

in crop rotation (*See* crop rotation)
oats interplanted with, 251–54, 256, 257
oats used as, 247, 257–58
planting, 40
rye interplanted with, 280
rye used as, 281
triticale interplanted with, 294
wheat interplanted with, 200
hemp nettle, 142
Henry, W. A., 389–90
herbicides, 46–47, 119, 127–28, 304, 368. *See also* Roundup Ready plants
Herd Seed Broadcaster, 53, 254
Hessian fly, 2, 4, 194–95
high-moisture grain
    barley, 236
    corn, 9, 11, 15, 162–63
High Mowing Seeds, 125
history
    of grain growing and consumption in Northeast (*See* Northeast, history of grain growing and consumption in)
    of specific grains (*See* subhead "history of" *under specific grains*)
Holmberg, Sven A., 300
hominy grits, 171
honey, buckwheat, 375–76
horse-drawn cultivators, 141–42
horse feed, 260
Huff, Clarence, xii, 47, 71, 103, 251
hulless grains
    barley, 221, 223, 241
    oats, 249–50, 271
hulling, 403–4
    barley, 220, 241–43, 404, CI:*25*
    oats, 262–68, 403–4
    spelt, 287, 289–92, 404, CI:*26–27*
    sunflowers, 350
hulling equipment, 241, 264, 404, CI:*25–27*. *See also* Roskamp Champion Oat Huller
    antique equipment, use of, 291–92
    European, 292, 404, CI:*27*
    hand-cranked, homemade, 262, 289–90
    Stutzman's mill, 290–91
human consumption, grain for, 403–6. *See also* subhead "for human consumption" *under specific crops*
Hume reel, combine, 66
humus, 23, 27, 54, 128, 351, 408

Hunt, Thomas F., 22, 277
hydrogen, 24

Indian corn, 2–3
industrial agriculture, 21, 408–11
infrastructure, acquiring, 404–7
insect pests, 2, 4
    canola, 370
    corn, 119, 128
    dry beans, 335–36
    soybeans, 315
    sunflowers, 350, 354
interplanting
    with hay (*See* hay)
    with specific crops (*See* subhead "interplanted with other crops" *under specific crops*)

Johnny's Selected Seeds, 122, 125, 300
Johnston, Rob, 125
jointing (grain plant), 55
Jones, D. F., 117
Jones, Steve, 198

kelp, 395, 399, 401
Kneading Conference, 18–19
Kovar, Pete, 50
Kovar tine weeder. *See* tine weeder
Kucyk, Victor, 126
Kutka, Frank, 126–27, 172, 173, 411

ladybugs, 315
LaFrancois, Bob, 50–51, 52
lamb's-quarters, 142, 187, 257, 275, 319, 336–37
lay-by cultivation, corn, 148–49
lead contamination, 235–36
leaf rust, 190, 193–94, 202, 204, 229, 282
legumes, 27, 379–87. *See also* alfalfa; clover; dry beans; field peas; nitrogen-fixing; soybeans
    in crop rotation (*See* crop rotation)
    vetch, 386–87
Lehman's Catalog, 209, 237, 261, 271, 357
Leinoff, Andy, 13–14, 263
Lely tine weeder, 48–50
Liberty herbicide, 119, 128
Liebig, Justus von, 22–23
lime, 25, 27, 142, 275, 329
linoleic oil, 350
linseed meal, 360

livestock bedding material, 7, 120, 228, 247, 284, 339
livestock feed, 101. *See also subhead* "as livestock feed" *under specific crops*
    combining grains for, 168–69, 218
    DON (deoxynivalenol) in, 188
    flaxseed oil cake (linseed meal), 360, 394, 401
    grain mixtures, 221, 382–83, 385, 394
    preparing (*See* livestock rations, preparing)
    protein in, 168
    soybeans (*See* soybeans)
    sunflower cake as (*See* sunflower cake)
livestock rations, preparing, 389–402
    for cows, 396–98
    determining ingredients for, 393–96
    grinding equipment for, 391–93, 397–98, CI:32
    non-grain ingredients, 395–96
    for pigs, 400–402
    for poultry, 398–400
    protein levels, 394–95, 397, 400
loam, 24–25, 278, 328, 351, 366, 374
local food movement, 16–17
lodged grain, 57
    barley, 224, 226, 228, 230–33, CI:24
    corn, 153, 156, 157, 164
    dry beans, 310
    einkorn and emmer, 176, 178
    oats, 9, 252, 253, 258, 259, 273
    rye, 276, 282
    sunflowers, 355
    wheat, 194, 198
L'Oiselle, Marc, 200
loose smut, 37–38, 190, 192–93

MacDonald, Keith, 123–24
magnesium, 23, 24, 27
    for barley, 229
    for dry beans, 329
    for oilseeds, 351, 367
    for wheat, 180, 195
maize (Indian corn), 2–3
Mallory, Ellen, 18
Maltex Co., 8
malting barley. *See* barley, malting
manganese, 195, 229, 410
manure, 14, 23, 24
    for barley, 224
    for buckwheat, 375
    for canola, 367
    for corn, 129
    for oats, 253
    for rye, 278
    for sunflowers, 351
    for wheat, 180–82, 202
manure spreader, 181
Martens, Klaas, 50, 126–27, 289
masa harina, 171
maturity
    canola, 367, 370
    corn seed maturing ratings, 129–30
    dry beans' days to maturity, 328
    flax, 361–62
    oats' days to maturity, 251, 258
    rye, 281–82
    soybean maturity phase and zones, 301–3, 305, 306, 310, 317
    sunflowers, 354
McFarlane Flexible Tine Tooth Harrow, 137–38, 333
McIlvain, Dean, 292
Meadows Mills, 171, 210–13, CI:22
    flour sifter, 216–17
    history of, 210–11
    maintenance of, 212
    newer mills, problems of, 212–13
mesocotyl, 136
metal drums, grain storage in, 77–78
Mexican bean beetles, 336
Meyer-Morton grain dryer, 94–95
microclimates, 15, 130, 152, 302, 303, 351
micronization, 142, 195
Midwest, corn grown in, 114–16
millet, 18, 113, 373, 379
minerals, livestock ration, 396, 400
minerals, soil, 23–25, 27, 54, 409–10
    for barley crop, 229
    for canola crop, 367, 370
    for dry bean crop, 329–30
    for oat crop, 253
    plant tissue tests, 230
    for rye crop, 278
    for sunflower crop, 351
    for wheat crop, 180, 184
Mitchell-Fetch, Jennifer, 273
moisture, grain, 57, 78, 80–82, 96. *See also* drying
    absorbed by stored grain, 78, 99–101
    barley, 80, 231–34, 236
    canola, 80, 370, 371
    clover seed, 382
    corn, 9, 11, 15, 80, 150–54, 158–59, 162–63, 167, 170
    determining, 81–82, 85, CI:12
    dry beans, 344–45
    flax, 80, 362
    at harvest, 57, 60, 65–66, 85, 91
    oats, 248, 259
    peas, 386
    rye, 282
    soybeans, 317–21, 323
    spelt, 288
    sunflowers, 80, 357, 358
    weeds, imparted by, 47
    wheat, 80, 204
moisture, soil, 23–25
    for barley, 224–25
    for canola crop, 367
    for dry beans, 329
    for field pea crop, 383
    for soybeans, 303, 306
    for spelt, 286
moisture tester, grain, 66, 81–82, 154, 288
mold, 402. *See also* white mold
    barley, 231
    canola, 368, 370
    oats, 248, 254
    rye, 279
    soybeans, 315, 321
    sunflowers, 350, 352, 354, 355, 357
    wheat, 187–92
moldboard plow, 31, 129, 252, 280, CI:5
molds, 47, 65, 80–81, 99. *See also* fusarium head blight; snow mold
    barley, 231
    corn, 115, 153, 159, 162
    wheat, 184, 187
molybdenum, 195
Monahan, Susan, 183, 202
Monsanto Company, 119, 410
Morrison, F. B., 390
mustard plant, 46–49, 54, 187, 223, 257, 275, 366, 368
mustard seed
    in harvested grain, 47–48, 54, 101, 106–7, 233, 364, 366, 368
    in soil, 54, 366
musty smell of grain, 81, 94, 99, 233, 254, 281
mychorrhizal fungi, 23, 142

Index

mycotoxins, 56, 162, 205. *See also* DON (deoxynivalenol)

Nabhan, Gary Paul, 16
National Farmers Organization (NFO), 10
National Organic Program, 379–80, 410
Native Americans, 1–2
natto soybeans, 305
Natural Resources Conservation Service, 29
nematodes, 23
net blotch, 229–30
Neves, Tony, 109, 331, 344
New Idea Farm Machinery Company
    corn picker, 123, 155–56
    picker-sheller, 161–62
Nickerson, Ginger, 16
nitrate nitrogen (NO_3), 22–23, 27, 32, 54. *See also* Chilean nitrate
    for corn, 128
    from legume cover crops, 379
    weed growth and, 54, 142
    for wheat, 178, 180
nitrite, 27
nitrogen, 9, 22, 24, 32, 228. *See also* nitrate nitrogen (NO_3); nitrogen-fixing
    adequate, plant color as indicator of, 253
    for barley, 228, 230
    for canola, 367
    for corn, 149
    for oats, 253
    Pre Sidedress Nitrogen (PSN) test, 149
    from rye cover crop, 280
    sources of, 27–28, 149, 181–82, 280, 386
    for spelt, 285–86
    for sunflowers, 351, 352, 353
    supplemental, 55
    synthetic, 178, 408
    timing of applications, 183
    for wheat, 178–83, 202
nitrogen-fixing, 27. *See also* soybeans, nitrogen-fixing by
    clover, 27, 381
    dry beans, 27, 329, 330, 332
    field peas, 227
    vetch, 386
no-till seeding, 184, 278, 281, 409

Northeast, history of grain growing and consumption in, 1–18
    Canada, 9–11
    early America, 2–9
    heirloom wheat research, 17–18
    local food movement, 16–17
    Vermont, organic grain production in, 11–15
Northeast Organic Farming Association, 12
Northern Grain Growers Association, xiv, 17, 172, 196, 200, 201, 407

oats, 60, 114, 245–73, CI:25–26
    biology of, 246–50
    bran, 248–49
    breeding, 272–73
    bushel weights, variations in, 248
    cleaning, 108, 111, 264, 266–68
    as cover crop, 32, 250, 306
    in crop rotation, 129, 252
    dry-down, 258, 259
    drying, 259
    fat content, 249, 268, 273
    fines, 260, 269, 271
    groats, 247–49, 251, 268, 271
    growth cycle, 246–47, 256–58, CI:25
    harvesting, 258–61, CI:26
    as hay crop, 247, 257–58
    history of, 4–9, 11, 13–15, 245–46
    hulless, 249–50, 271
    hulls, 247–48
    for human consumption, 248–49, 262–72, 403–4
    interplanted with other crops, 251–54, 256–57, 381–83
    kernel size (plumpness), 248, 252, 273
    kernels, description of, 247–48
    as livestock feed, 168, 248, 250–51, 257–62, 266, 382–83, 394, 397, 399–400
    lodging (*See* lodged grain)
    photosynthesis, efficiency of, 113–14
    planting, 30
    processing (*See* oats, processing)
    protein levels, 248, 269, 394
    quality of crop, 248, 251
    specific density, 82
    straw, 251, 273
    test weight, 251
    varieties, 250–52, 273
    winter, 250

    yield, 249, 251, 262–63
oats, processing, 260–72, 403–4
    aspiration, 264
    flaking, 269–72
    flour, 272
    hulling, 262–68, 403–4
    for human consumption, 262–72
    for livestock, 260–62
    for oatmeal, 268–72
    plate mill method, 261–62
    roller-mill method, 260–61, 263, 269–70
    "shrink," concept of, 262–63
    toasting oats, 268–69, 271
oats, planting, 252–56
    field preparation, 252–53
    methods, 254–56
    quality of seed, 253–54
    seed sources, 35–36, 250–51
    selecting varieties, 250–52
    soil fertility, 253
Oechsner, Troy, 178
oilseed cake, 182, 360, 394, 400, 401. *See also* sunflower cake
oilseed press, 347–48, 357–59, 364
oilseeds, 80, 96, 113, 347–72. *See also* canola; flax; sunflowers
Oliver Manufacturing Company, 110–11
omega-3 fatty acids, 350, 360
open-pollinated (OP) farmer networks, 172
open-pollinated (OP) varieties
    canola, 367
    corn, 118, 125–28, 153, 170–72, 411, CI:20
    sunflowers, 349
Organic Crop Improvement Association, 12
Organic Equipment Technology, 50–52
Organic Growers Research and Information Network (OGRIN), 178
organic matter, soil, 23, 32, 128, 180, 202, 253, 408–9
Organic Seed Growers and Trade Association, 411
Orton, Vrest and Mildred Allen, 8
oxidation, 32
oxygen, 202

packing (rolling seedbed), 254, 256, 286, 306, 312, 360
paddy table separator, 266–69
pasta, wheat for, 188–89

pastry flour, wheat for. *See* soft red and white winter wheat
PCR test, 173, 366
peanut meal, 149, 182
peas. *See* field peas, Canadian
Peaslee, Bert, 94, 282
pencil-point weeder, 140–41
pesticides, 119, 315
pests, 160, 334, 355. *See also* insect pests; raccoons
Petrini, Carlo, 16
petroleum usage, 118, 409
phenotype, 173
phosphate, 25, 329
phosphorous, 23, 24, 27
    for barley, 229, 230
    for corn, 135–36
    for dry beans, 329
    for oats, 253
    for sunflowers, 351
    for wheat, 180
photoperiod sensitivity, soybeans, 301, 306
photosynthesis
    barley plants, 230, 231
    corn, 113–14
    dry beans, 330
    sunflowers, 351
    wheat, 194, 202
phytophthora root rot, 314–15
pig feed, 394
    barley, 221
    corn, 168–69
    oats, 250, 260
    rye, 283
    soybeans, 299, 300
    triticale, 294
    wheat, 218
pigweed, 336–37
Pioneer Hi-Bred Seed Company, 117, 123–25, 410
Planet Jr. seeder. *See* hand-pushed garden seeder
plant tissue tests, 230
Plant Variety Protection laws, 410
planting flax, 360–61
planting grain, 30–31, 38–45, CI:6. *See also* broadcasting seed; no-till seeding; seeding equipment; seeding rates; subhead "planting" *under specific crops*
    frost seeding, 38, 45, 285
    hand-seeding, 38, 226–27, 254
    in home or research plot, 39

mechanics of, 38–39
    row crops, 30
    spring grains, 30–31, 45
    winter grains, 30–31
plate mills (buhr mills). *See* buhr mills (plate mills)
plowing, 129, CI:5–6. *See also* moldboard plow; tillage
    fall, 31, 252, CI:6
    spring, 31–32, 252, CI:5
pollination, 56
    buckwheat, 375
    clover, 381
    corn, 150, 152
    dry beans, 329
    peas, 384
    rye, 277–78, 280–81
    triticale, 294
    wheat, 201
potassium, 23, 24, 27
    for dry beans, 329
    for oats, 253
    for oilseeds, 351, 367
    for wheat, 195
potatoes, 11
Poulin, Jeff, 390
poultry feed, 394
    barley, 221, 223
    buckwheat, 374
    corn, 168–69
    livestock ration ingredients, 398–400
    oats, 250, 260
    rye, 283
    soybeans, 299, 300
    triticale, 294
    wheat, 218
powdery mildew, 190, 230
power harrows, 34
Pre Sidedress Nitrogen (PSN) test, 149
precleaning grain, 83–84, 101
pressure-cure drying, 90–92
Pro Gro fertilizer, 182
Prograin, 196
propane-powered grain dryer, 95–96
    for barley, 234
    for canola, 371
    for corn, 159, 160, 162, 164–67
    monitoring, 166–67
    for oats, 259
    shutting down and unloading, 167
    for sunflower seeds, 357

PVP (Plant Variety Protection), 36
pyrethrum spray, 315

Qua, Patty, 331
quack grass, 142, 145

Rabher weeder, 51
raccoons, 120–21, 128, 160, 170, 355
radicle (grain plant), 45
Ragosa, Eli, 176
ragweed, 47, 83, 142, 205
Rainville, Roger, 356, 365, 371
reaper (grain binder). *See* grain binder (reaper)
reaping, 57–58, CI:2, 3
reciprocating screens. *See* screens, reciprocating
red clover, 179, 380–81. *See also* clover
red dog. *See* glumes
redroot pigweed, 142, 319
Rene J. Fournier Farm Equipment, 407
reproductive period, grain, 56. *See also* pollination
    canola, 370
    corn, 114, 149–53
    dry beans, 329, 335, 337–38
    flax, 361
    grain, 56
    oats, 258
    rye, 280–81
    soybeans, 301, 317
    sunflowers, 353–54
Rhizobium, seeds inoculated with
    clover, 381
    R. japonicum, for soybeans, 311–12
    R. leguminosarum, for peas, 227, 384
    R. phaseoli, for dry beans, 332
rhizoctonia stem rot, 314
riding cultivators, horse-drawn, 141–42
roasters, soybean, 322–23, 325, CI:29
Robillard, Guy, 71–72, 236
rock phosphate, 25, 329
rock powders, 25, 329
rodents, 77, 160
roller-crimpers, 281
roller mills, 397
    for barley, 234–37
    lead contamination from, 235–36
    for oats, 260, 263, 269–70
    for wheat, 214–15
rolling seedbed (packing). *See* packing (rolling seedbed)
root fungus, 7

Index

roots, grain plant, 45–46
Roskamp Champion Oat Huller, 264–66, 291–92, 404, 405, CI:25
rotary hoe, 138–39, 312–13, 334, CI:9
rotary screen cleaners, 83–84, 162, 164, 234, 259–60, CI:13
rototillers, 32, 34
Roundup Ready plants
    canola, 366, 368
    chicken manure products, gene in, 181
    corn, 119, 128
    soybeans, 119, 300, 324–25, 411
row-crop cultivators, 46, 141–49, 313–14, 335, CI:8, 9
    C-shank, 148–49
    horse-drawn walking and riding, 141–42
    side dressing applicator units, 149
    six-row, 145
    three-point hitch, 146–48
    tractor-mounted, 142–43
row crops. *See also* row-crop cultivators; *specific crops*
    planting, 30–32
    seed sources for, 35
    weed control, 32, CI:9
Rural Science, 22
rust, 190, 193–94, 202, 204, 226, 229–30, 273
rye, 32, 276–84, CI:26–27
    cleaning, 111, 112, 284
    as cover crop, 279–81, 283
    in crop rotation, 128, 278, 280
    dormant period, 279–80
    ergot infection, 278, 282–83
    flour, processing, 284
    as forage crop, 278–79, 281
    growth cycle, 278–81
    harvesting, 281–83
    as hay crop, 281
    history of, 2–5, 8, 11, 14, 15, 276–77
    for human consumption, 284
    interplanted with other crops, 280
    as livestock feed, 283–84
    lodging, 276, 282
    no-till planting, as bed for, 281
    planting, 30, 35–36, 278–79, 283
    seed sources/saving, 35–36, 283
    specific density, 82
    straw, 281, 284
    uses for, 283–84
    varieties, 277–78
    yield, 3, 278

Sachs, Paul, 182
salt, 395
sandy soils, 25, 29
Saunders, Charles E., 200
scab. *See* fusarium head blight
scald, 230
scalping, 102, 103, 106–8
Scatterseed Project, 223
Schmaltz, Blaine, 177–78, 224, 226, 295, 385
Schmeiser, Percy, 366
Schmotzer hoe, 46, 361
sclerotinia stem rot. *See* white mold
screens, reciprocating, 101. *See also* air screen cleaners
    hand, 102, 109
    homemade, 102
    size variations, 107–8
screw-in grain aerator. *See* aerators, screw-in grain
seaweed/sea mineral concentrates and sprays, 195
seed broadcasters. *See* broadcasters
seed germination. *See* germination of seeds
seed grading. *See* grading, seed
seed inoculation, 142, 380
    clover, 381
    dry beans, 332
    peas, 227, 384
    soybeans, 311–12
seed planting. *See* planting grain
Seed Savers Exchange, 172
seed saving, 36–37, 101, 107, 410. *See also* grading, seed
    buckwheat, 378
    canola, 367
    clover, 380–82
    corn, 171–74
    dry beans, 331
    loose smut, seeds infected with, 193
    millet, 379
    peas, 384–85
    rye, 283
    spelt, 287
    testing, 173
    timothy, 378–79
    vetch, 386
seed sources, 35–38. *See also* seed saving
    barley, 35–36
    buckwheat, 378
    Canadian varieties, 35–36, 196–98, 251–52
    certified seed (*See* certified seed)
    clover, 380
    corn, 35, 116
    dry beans, 331–32
    farm-produced seed, 36–37, 185
    field peas, 384–85
    germination potential, 37–38, 185
    Midwest, row crop seeds from, 35
    millet, 379
    oats, 35–36, 250–51
    oilseeds, 350, 367
    patented seeds, 410–11
    quality of seed, importance of, 185, 253–54
    rye, 283
    shipping costs, 251
    soybeans, 35, 325, 411
    spelt, 286
    triticale, 295
    wheat, 193, 196–97
seed sovereignty, 21
seed spinner, 254
seedbed, preparing, 30, 34
    for barley, 224
    for buckwheat, 374–75
    for canola, 367
    for corn, 129
    for dry beans, 330, 332–33
    for flax, 360
    for oats, 252–53
    for rye, 278
    for soybeans, 306
    for spelt, 286
    for sunflowers, 352
    for wheat, 184
seeding equipment. *See also* broadcasters, seed; grain drills, hand-pushed garden seeder
    air seeder, 53
    grass seed attachment for, 40
seeding rates, 37–41
    barley, 225
    buckwheat, 375
    corn, 134, 161
    dry beans, 333
    field peas, 385
    flax, 360, 361
    oats, 256
    oilseeds, 359, 361, 367
    rye, 279
    soybeans, 43, 308–10

spelt, 286, 287
triticale, 295
wheat, 199
selenium, 195
Semican Inc., 188, 196, 198
separators. *See also* gravity table separator
paddy table, 266–68
sheep feed, 260
Sherman, Dick, 11
shipping costs, seed, 251
Shull, George H., 117
sickle, 258, 376
sifters, flour, 215–17
silage, 378, 379, 383
corn (*See* corn)
oats, 257
silos, 15, 99–100, 161. *See also* drying silos
aeration of grain in, 99–100
airtight, 9
condensation in, 99–100
temperature changes, effect of, 99
silts, 25, 29
six-row barley, 221–24, CI:24
Skiold flour mills, 214
Skrdla, Willis H., 120
Smith, Margaret, 126, 127, 173–74
smoothing harrow, 33
snow mold, 184, 187, 279
sod, 27, 409. *See also* cover crops; hay
barley, planted prior to, 230
beans, planted prior to, 332, 336
buckwheat, planted prior to, 374
corn, planted prior to, 128, 129, 135
fall or spring plowing of, 31–32, 129
flax, planted prior to, 360
oats, planted prior to, 252, 253
rye, planted prior to, 278
secondary tillage, 33
sunflowers, planted prior to, 352
sodium, 24
soft red and white spring wheat, 188
soft red and white winter wheat, 12, 14, 30, 187–88, 197, 199, 218, CI:22
soil. *See also* loam
anaerobic, 202
cation exchange capacity (CEC), 24–25, 180
compaction, 24, 54, 142
drainage (*See* drainage)
erosion, 32, 128, 252
organic matter in (*See* organic matter, soil)

types, 23–25, 29–30 (*See also* clay)
soil aggregation (structure), 23, 25, 128, 180
soil amendments, 23, 25, 54
soil biology, 23–24, 202, 396, 409. *See also* organic matter, soil; soil microbes
for oat crop, 252, 253
weeds and, 54, 142
for wheat crop, 180
soil fertility, 21–28, 408–10
for barley, 226, 228–31
for canola, 367–68, 370
chemical fertilizers and, 22–23
disease prevention and, 195, 230
for dry beans, 329–30
in holistic system, 27–28
importance of, 23–24
managing organically, 24–25
mineral fertility, 24–25
for oats, 253
for rye, 278
for soybeans, 306
for spelt, 285–86
tests, 25–27
weeds as indicators of, 54, 142
for wheat, 173–83, 202
for winter grains, 275
soil inoculants, 142, 306
soil microbes, 27, 32, 54, 180, 181, 306, 330
soil microbiology. *See* soil biology
soil moisture. *See* moisture, soil
soil tests, 25–27, 180
sorghum, 113
Sorrells, Mark, 197
southern corn leaf blight, 118
soybean cyst nematode, 314–15
soybean oil
as food product, 299
industrial uses of, 299–300
meal as livestock feed, 149, 182, 299
soybean varieties, 299–301, 305
foliage and plant structure, 304
fruit characteristics, 304–5
for human consumption, 305
selecting, 303–5, 309, 310
soybeans, xvi, 272, 297–325, 411, CI:28–29
cleaning, 111
in crop rotation (*See* crop rotation)
cultivation of, 312–13
dry-down, 306, 317
drying, 321, 323

genetically modified (*See* genetically modified organisms (GMOs))
growth cycle, 312–14, 316–17
harvesting (*See* soybeans, harvesting)
hilum color, 305
history of, 10, 11, 13–16, 297–300
for human consumption, 297, 299, 305, 319–20, 323–24
as livestock feed, 149, 168, 182, 297, 299, 300, 304–5, 320, 322, 394–95, 397, 399–401
maturity zones, 301–3, 305, 306, 310
natto, 305
nitrogen-fixing by, 297–99, 306, 314, 316, 317
no-till planting, 281
oil content, 297
photoperiod sensitivity, soybeans, 301, 306
planting (*See* soybeans, planting)
protein levels, 297, 300, 304–5, 322, 394
roasting, 322–23, 325, CI:29
varieties (*See* soybean varieties)
yield, 299, 303–6, 309, 311
soybeans, harvesting, 299, 318–21, CI:29
combine, 318–21, 323
by hand, 318
timing of, 319–20
soybeans, planting, 31, 305–12
methods, 304, 306–11
row spacing, 299, 304, 307, 311, 313, 315, CI:29
seed inoculation, 311–12
seed sources, 35, 325, 411
selecting varieties, 303–5, 309, 310
soil preparation, 306
timing of, 305–6
speckled leaf blotch, 229
spelt, 284–93, CI:26–27
cleaning, 111
in crop rotation, 285, 286
fat content, 293
growth cycle, 287
harvesting, 287–88
history of, 285
hulling, 287, 289–92, 404, CI:26
for human consumption, 289, 292–93
as livestock feed, 288–89
planting, 30, 285–87

Index

protein levels, 289, 293
varieties, 286
Spencer, Earl, 137, 138
spike-tooth drag, 33–34, 38–39, 129, 137–38, 224, 278, 333, 352
  barley, planting, 224
  corn, planting, 129, 137–38
  dry beans, planting, 333
  McFarlane Flexible Tine Tooth Harrow, 137–38, 333
  oats, planting, 254
  rye, planting, 278
  sunflowers, planting, 352
spike-tooth threshing, 339
spiral cleaners, 364, 387
spot blotch, 226, 229
spring grains, 30–31, 45, 294, 367. *See also* wheat
spring-tooth harrow, 33–34, 129
sprouted grains
  barley, 236–37
  wheat, 217
Stanley, Christian and Andrea, 240
Stearns, Tom, 125
steerage hoe, 46
Stefanson, Baldur R., 365–66
stem elongation (grain plant), 55
stem rust, 202, 226, 229–30, 273, 282
stinking smut (common bunt), 190, 193
Stir-Ator, 90
Stodden, Charles, Sr. and Charles, Jr., 109, 406
stone flour mills, 207, CI:23
stooking grain, 58–59, CI:3, 27
storage, 77–97, 99–101. *See also* silos; storage container options
  barley, 80, 233
  bins, inside *vs.* outside, 100–101
  on cement floors, 99
  cleaning grain for (*See* cleaning grain)
  clover seed, 382
  corn (*See* corn)
  drying for (*See* drying grain)
  flax, 80, 364
  grain augur, 84–85
  grain density, measuring, 82
  grain moisture and, 78, 99–101
  precleaning grain, 83–84
  rodents, problem of, 77
  sunflowers, 80, 355
storage container options, 77–80
  automatic filling, 78–79
  bulk bags, 77, 78

gravity bags, 79–80
trash cans/metal drums, 77–78
Stutzman, Monroe, 290–91
sugarcane, 113
sul-po-mag, 181, 195, 329, 367
sulfur, 27, 180, 181, 195, 329, 367
sunflower cake, 348, 350, 358, 394, 400, 401
sunflowers, xvi, 347–59, 371–72, CI:31
  black oil seed, 350–51
  cleaning seeds, 357
  confectionary, 350
  in crop rotation, 352, 370
  diseases, 349–50, 354
  dry-down, 349, 351, 354, 355
  drying heads and seeds, 96, 355, 357
  growth cycle, 352–54
  harvesting, 349, 355–57
  hulling, 350
  for human consumption, 350
  hybrids, 349–51
  insect pests, 349–50, 354
  as livestock feed (*See* sunflower cake)
  lodged grain, 355
  open-pollinated, 349
  planting, 30, 31, 350–53
  pressing oil from seeds, 347–48, 357–59
  protein levels, 394
  Russia, developed as oil seed crop in, 348–49
  seed sources, 350
  standability, 348–49, 351
  storage, 80, 355
  types of, 350–51
  varieties, 349–51
  yield, 348–49
swathing grain, 7, 59–60
  buckwheat, 376
  canola, 370–71
  clover, 382
  flax, 362–63
  millet, 379
  wheat, CI:10

tempering grains, 236, 269, 271
test weights, 82, 110
  barley, 231
  corn, 124, 151–52, 159, 162, 170
  oats, 248, 251, 254, 266, 273
  scale, CI:11
  triticale, 295
  wheat, 194, 201–2

three-point hitch cone-style broadcaster, 149
three-point hitch cultivator, 146–48, CI:9
threshing, xv, 60–64. *See also* combine; flail; harvesting grains
  barley, 232–33
  dry beans, 338–42
  flax, 363
  soybeans, 318–20
  spelt, 287–88
  sunflowers, 355–57
threshing, hand. *See* hand-threshing
threshing machines, CI:2, 287, 318
  Butterworks Farm, use at, 63
  filling of grain storage containers from, 78
  history of, 61–63
  old machinery, working with, 63–64
  small units for home use, 64
tillage, 29–34, CI:6. *See also* plowing; seedbed, preparing
  cover crops, benefit of, 32
  crop selection, 30–31
  dangers to soil from, 32
  final fitting, 33–34
  for grain crops following other crops, 34
  primary, 32
  of rye cover crop, 280
  secondary, 33, 184, 252, 278, 367
  seedbed preparation, 30, 34
  soil preparation, 31–32
  soil types and, 29–30
  for soybean crop, 306
  wet soil, avoiding, 29
  wheat, 184
  for winter grains, 275
tiller (grain plant), 46, 55
tilth. *See* soil aggregation (structure)
timothy, 53, 129, 373, 378–79, 381. *See also* hay
tine weeder, 48–54, 140, 229, 257, 312–13, 332–35, 352–53, CI:8–9
Tisquantum (Squanto), 1, 2
trace elements, 195, 229, 330, 409–10. *See also specific elements*
tractor-mounted cultivators, 142–43
tram lines, field, 55
transportation of harvested grain, 78–80
trash cans, grain storage in, 77–78
trefoil seed. *See* bird's-foot trefoil seed
triticale, 8, 293–95

breeding, 293–94
history of, 13, 14, 293
interplanted with other crops, 294, 383
as livestock feed, 294–95
planting, 30, 294–95
protein levels, 294, 295
yield, 294, 295
Tull, Jethro, 39
two-row barley, 221–24

USDA Farm Service Administration (FSA), 29–30

Vachon, Bruno, 239–40
vegetative growth period, grain, 55–56
corn, 114, 129, 139, 149–52
dry beans, 329, 335
flax, 361–62
oats, 257–58
rye, 280
soybeans, 301, 303, 306, 313–14, 316–17
sunflowers, 353–54
velvet leaf, 142
Vermont
corn varieties for, 119–28
grains, research on, 17–18 (See also Darby, Heather)
local food movement, 16
organic grains in, 11–15
Vermont Housing and Conservation Board Farm Viability Program, 267–68, 404
Vermont Organic Farmers, 12
Vermont's Local Food Landscape: An Inventory and Assessment of Recent Local Food Initiatives, 16
vernalization, 279–80, 287
Verschelden, Rémi, 239–40
vetch, 379, 386–87

walking cultivators, horse-drawn, 141
Wallace, Henry, 117
Warden, Robert, xiv
Warminster Malting, 238
water, potable, 22–23
weather, effect of, 37, 45, 56, 59–60, 85, 408–11. *See also* frost, effect of; moisture, grain; wet weather
on barley, 226, 228
on buckwheat, 375, 376
on canola, 367, 368
on corn, 150, 152–53
on dry beans, 330–31
on harvest, 57, 59–60, 65–66
on oats, 256
on rye, 278–80
on soybeans, 314, 317–18
on sunflowers, 351–52, 355
on wheat, 175, 176, 189, 201
weed control, 30, 32, 33, 46–48. *See also* cultivation; mustard plant; tine weeder; weed seeds
barley, 229
canola, 368
cover crops, 53
cultivators, use of, 32, 46
dry beans, 330, 333–37
flax, 360, 361
herbicides, 46–47
oat mulch for, 32
oats, 246, 254, 256, 257
post-emergence, 53–54
pre-emergence, 51–53
rye, 277
seed planting rate and, 39
soil condition and, 54, 142
soybeans, 304, 309, 311–14, 319, 323, CI:9
spring grains, 30–31, 202
sunflowers, 352–53
wheat, CI:9
wide-row planting, 46
winter grains, 30–31, 187, 275
weed seeds. *See also* mustard seed; weed seeds in harvested grains
in oat seed, 253
in rye seed, 278
weed pressure resulting from, 54, 360
weed seeds in harvested grains, 47, 80, 83
in barley, 233–34
in dry beans, 345
in early-harvested rye, 281
in flaxseed, 365
in livestock ration, 101, 234
in oats, 259
removing from grain, 83–84, 101, 102, 106–8
in wheat, 205
weedy crops, harvesting, 47, 59, 80, 204–5. *See also* weed seeds in harvested grains
wet weather, 8–9, 37

barley crop and, 224, 225, 228, 229, 231, 232
buckwheat and, 376
corn crop and, 114, 115, 149, 150, 152, 162, 164
cultivation in, 139
dry beans and, 331, 334
harvesting in, 59–60, 162, 164, 259
oats and, 259
rye crop and, 278
soybeans and, 306, 314, 317
spelt crop and, 286, 288
sunflowers and, 351–52, 355
wheat crop and, 56, 91–92, 188, 192–94, 198, 201–4
wheat, xv, 60, 80, 114, 175–218, 394, CI:21–23
agronomic considerations for, 173–83
allergies to, 177
bran, 205, 216, 218
breeding, 197–99, CI:23
cleaning, 111, 112, 204–5
diseases, 187–95, 202
drying, 92
dwarf varieties, 55
endosperm, 205
enzymatic activity, wheat kernel, 201, 203–4
flour production (*See* wheat processing and flour production)
grinding, 171
hard, 30, 185–87, 197–99, 202, CI:10, CI:22
harvesting (*See* wheat, harvesting)
history of, 3–5, 7–8, 11, 12, 14–17, 175–78
for human consumption, 217 (*See also* wheat processing and flour production)
interplanted with other crops, 200, 381–83
as livestock feed, 168, 188, 217–18, 382–83, 394
for pastry flour, 187, 188
photosynthesis, efficiency of, 113–14
planting (*See* wheat, planting)
processing (*See* wheat processing and flour production)
protein levels, 185, 187, 189, 201–2, 217–18
research grants, 17–18

Index

soft, 12, 14, 30, 187–88, 197, 199, 218, CI:22
spring, 30–31, 45, 175, 183–86, 194, 198, 202, CI:22
types of (*See* wheat varieties)
varieties (*See* wheat varieties)
 selecting varieties, 197–98
winter, 12, 14, 30, 175, 183–88, 197, 199, 218, CI:21–22
yield, 128, 183, 187
wheat, harvesting, 60, 91–92, 201–4, CI:*10*, 23
 falling number, achieving high, 203–4
 high quality grain, components of, 201–2
 pre-harvest sprouting, avoiding, 203
 protein content, 201–2
 test weight, 202
wheat midge, 4
wheat, planting, CI:2, 183–85, 7–8
 seed quality, importance of, 185
 seed sources, 35–36, 196–97
 soil fertility, 173–83
 spring wheat, 183–84
wheat processing and flour production, 7–8, 12, 16–17, 204–17, CI:22–23
 bran, 205
 buhr mills, 209–10
 cleaning grain, 204–5
 dust, protection from, 209–10
 electric mills, 210
 endosperm, 205
 fundamentals of flour milling, 205–7
 hand mills, 207–9
 home-scale, 207–9, CI:22
 horizontal flour mills, European, 213–14
 industrial, 214–15
 Meadows Mill, 210–13
 roller mills, 214–15
 sifters, flour, 215–17
 stone flour mills, 207, CI:23
wheat seed, 7
wheat varieties, 194, 196–200
 durum wheat, 188–89
 dwarf, 198
 hard red and white spring wheat, 185–86, 198, 202, CI:*10*
 hard red winter wheat, 186–87, 198–99
 heirloom, 56, 199–200, CI:*21*
 selecting, 197–98
 soft red and white spring wheat, 188
 soft red winter wheat, 188
 soft white winter wheat, 187–88, 199
 sourcing seed for, 35–36, 196–97
wheel hoes, 141
wheel tracks, field, 55
white mold, 315, 350, 352, 368–70
Williams, Matt, 16–17, 193, 267, 271
Williamson, John, 365
windrower, 340–44, CI:31–32
winnowing, 3, 61, 101–3, 339. *See also* cleaning
winnowing mills, 5, 102–3
winter grains, 30–31, 128, 275–95. *See also* rye; spelt; triticale; wheat
 barley, 225
 canola, 367
 oats, 250
 yields, 276
winterkill, 30, 32, 187, 225, 287, 295, 381, 386
wireworms, 336
woodchucks, 334
Wright, George, 250

yellow blossom sweet clover, 382
yellow stripe rust, 194
yield, crop. *See specific crop*

Zimmerman, Ken, 291
zinc, 195, 229, 410
Zwinger, Steve, 177–78

About the Author

Jack Lazor is co-owner of Butterworks Farm in Westfield, Vermont, with his wife, Anne, and cofounder of the Northern Grain Growers Association. Jack has been growing organic grains in the mountains of Vermont's Northeast Kingdom since 1975 and is considered a leader in the movement for growing grains in cold climates. Lazor grows grains both for human consumption and for feed for their herd of Jersey cows, including corn, oats, barley, soybeans, legumes, alfalfa, and oilseeds such as flax and sunflower. Butterworks Farm also produces organic Jersey-milk yogurt, buttermilk, sweet Jersey cream, cheddar cheese, and grain products. He is the recipient of many agricultural awards.